LINEAR OPTIMAL
CONTROL SYSTEMS

Linear Optimal Control Systems

HUIBERT KWAKERNAAK
Twente University of Technology
Enschede, The Netherlands

RAPHAEL SIVAN
Technion, Israel Institute of Technology
Haifa, Israel

WILEY-INTERSCIENCE, *a Division of John Wiley & Sons, Inc.*
New York Chichester Brisbane Toronto Singapore

Copyright © 1972, by John Wiley & Sons, Inc.

All rights reserved. Published simultaneously in Canada.

Reproduction or translation of any part of this work beyond that permitted by Sections 107 or 108 of the 1976 United States Copyright Act without the permission of the copyright owner is unlawful. Requests for permission or further information should be addressed to the Permissions Department, John Wiley & Sons, Inc.

Library of Congress Cataloging in Publication Data:

Kwakernaak, Huibert.
Linear optimal control systems.

Bibliography: p.
1. Control theory. 2. Automatic control.
I. Sivan, Raphael, joint author. II. Title

QA402.3.K89   629.8′32   72-3576
ISBN 0-471-51110-2

Printed in the United States of America

20 19 18 17 16 15

To Hélène, Annemarie, and Martin

H. K.

In memory of my parents Yehuda and Tova and to my wife Ilana

R. S.

PREFACE

During the last few years modern linear control theory has advanced rapidly and is now being recognized as a powerful and eminently practical tool for the solution of linear feedback control problems. The main characteristics of modern linear control theory are the state space description of systems, optimization in terms of quadratic performance criteria, and incorporation of Kalman–Bucy optimal state reconstruction theory. The significant advantage of modern linear control theory over the classical theory is its applicability to control problems involving multiinput multioutput systems and time-varying situations; the classical theory is essentially restricted to single-input single-output time-invariant situations.

The use of the term "modern" control theory could suggest a disregard for "classical," or "conventional," control theory, namely, the theory that consists of design methods based upon suitably shaping the transmission and loop gain functions, employing pole-zero techniques. However, we do not share such a disregard; on the contrary, we believe that the classical approach is well-established and proven by practice, and distinguishes itself by a collection of sensible and useful goals and problem formulations.

This book attempts to reconcile modern linear control theory with classical control theory. One of the major concerns of this text is to present design methods, employing modern techniques, for obtaining control systems that stand up to the requirements that have been so well developed in the classical expositions of control theory. Therefore, among other things, an entire chapter is devoted to a description of the analysis of control systems, mostly following the classical lines of thought. In the later chapters of the book, in which modern synthesis methods are developed, the chapter on analysis is recurrently referred to. Furthermore, special attention is paid to subjects that are standard in classical control theory but are frequently overlooked in modern treatments, such as nonzero set point control systems, tracking systems, and control systems that have to cope with constant disturbances. Also, heavy emphasis is placed upon the stochastic nature of control problems because the stochastic aspects are so essential.

We believe that modern and classical control theory can very well be taught simultaneously, since they cover different aspects of the same problems. There is no inherent reason for teaching the classical theory first in undergraduate courses and to defer the modern theory, particularly the stochastic part of it, to graduate courses. In fact, we believe that a modern course should be a blend of classical, modern, and stochastic control theory. This is the approach followed in this book.

The book has been organized as follows. About half of the material, containing most of the analysis and design methods, as well as a large number of examples, is presented in unmarked sections. The finer points, such as conditions for existence, detailed results concerning convergence to steady-state solutions, and asymptotic properties, are dealt with in sections whose titles have been marked with an asterisk. *The unmarked sections have been so written that they form a textbook for a two-semester first course on control theory at the senior or first-year graduate level.* The marked sections consist of supplementary material of a more advanced nature. The control engineer who is interested in applying the material will find most design methods in the unmarked sections but may have to refer to the remaining sections for more detailed information on difficult points.

The following background is assumed. The reader should have had a first course on linear systems or linear circuits and should possess some introductory knowledge of stochastic processes. It is also recommended that the reader have some experience in digital computer programming and that he have access to a computer. We do not believe that it is necessary for the reader to have followed a course on classical control theory before studying the material of this book.

A chapter-by-chapter description of the book follows.

In Chapter 1, "Elements of Linear System Theory," the description of linear systems in terms of their state is the starting point, while transfer matrix and frequency response concepts are derived from the state description. Topics important for the steady-state analysis of linear optimal systems are carefully discussed. They are: controllability, stabilizability, reconstructibility, detectability, and duality. The last two sections of this chapter are devoted to a description of vector stochastic processes, with special emphasis on the representation of stochastic processes as the outputs of linear differential systems driven by white noise. In later chapters this material is extensively employed.

Chapter 2, "Analysis of Control Systems," gives a general description of control problems. Furthermore, it includes a step-by-step analysis of the various aspects of control system performance. Single-input single-output and multivariable control systems are discussed in a unified framework by the use of the concepts of mean square tracking error and mean square input.

Chapter 3, "Optimal Linear State Feedback Control Systems," not only presents the usual exposition of the linear optimal regulator problem but also gives a rather complete survey of the steady-state properties of the Riccati equation and the optimal regulator. It deals with the numerical solution of Riccati equations and treats stochastic optimal regulators, optimal tracking systems, and regulators with constant disturbances and nonzero set points. As a special feature, the asymptotic properties of steady-state control laws and the maximally achievable accuracy of regulators and tracking systems are discussed.

Chapter 4, "Optimal Linear Reconstruction of the State," derives the Kalman–Bucy filter starting with observer theory. Various special cases, such as singular observer problems and problems with colored observation noise, are also treated. The various steady-state and asymptotic properties of optimal observers are reviewed.

In Chapter 5, "Optimal Linear Output Feedback Control Systems," the state feedback controllers of Chapter 3 are connected to the observers of Chapter 4. A heuristic and relatively simple proof of the separation principle is presented based on the innovations concept, which is discussed in Chapter 4. Guidelines are given for the design of various types of output feedback control systems, and a review of the design of reduced-order controllers is included.

In Chapter 6, "Linear Optimal Control Theory for Discrete-Time Systems," the entire theory of Chapters 1 through 5 is repeated in condensed form for linear discrete-time control systems. Special attention is given to state deadbeat and output deadbeat control systems, and to questions concerning the synchronization of the measurements and the control actuation.

Throughout the book important concepts are introduced in definitions, and the main results summarized in the form of theorems. Almost every section concludes with one or more examples, many of which are numerical. These examples serve to clarify the material of the text and, by their physical significance, to emphasize the practical applicability of the results. Most examples are continuations of earlier examples so that a specific problem is developed over several sections or even chapters. Whenever numerical values are used, care has been taken to designate the proper dimensions of the various quantities. To this end, the SI system of units has been employed, which is now being internationally accepted (see, e.g., Barrow, 1966; IEEE Standards Committee, 1970). A complete review of the SI system can be found in the *Recommendations* of the International Organization for Standardization (various dates).

The book contains about 50 problems. They can be divided into two categories: elementary exercises, directly illustrating the material of the text; and supplementary results, extending the material of the text. A few of the problems require the use of a digital computer. The problems marked with an asterisk are not considered to belong to the *textbook* material. Suitable term projects could consist of writing and testing the computer subroutines listed in Section 5.8.

Many references are quoted throughout the book, but no attempt has been made to reach any degree of completeness or to do justice to history. The fact that a particular publication is mentioned simply means that it has been used by us as source material or that related material can be found in it. The references are indicated by the author's name, the year of publication, and a letter indicating which publication is intended (e.g., Miller, 1971b).

HUIBERT KWAKERNAAK
RAPHAEL SIVAN

Enschede, The Netherlands
Haifa, Israel
January 1972

ACKNOWLEDGMENTS

The first author wishes to express his thanks to the Department of Applied Physics at the Delft University of Technology, where he worked until April, 1970, and to the Department of Applied Mathematics at the Twente University of Technology for invaluable support during the writing of this book in terms of time granted and facilities made available. The second author extends his thanks to the Technion, the Israel Institute of Technology, for supporting the writing of the book. Time on the preparation of the manuscript was spent by the second author while he was a National Research Council Senior Research Associate at the NASA Langley Research Center, Hampton, Virginia, during the academic year 1970–1971. Without the assistance of these institutions, and their help in financing various trips to Israel, the Netherlands, and the United States, it would not have been possible to complete this book.

Several typists spent their efforts on the various versions of the manuscript. Special mention should be made of the extremely diligent and competent work of Miss Marja Genemans of Delft and Mrs. Dini Rengelink of Twente. The line drawings were made by Mr. M. G. Langen of Delft, who is commended for his accurate and careful work.

Final thanks are due to one of the first author's former students, Mr. J. H. van Schuppen, for his comments on the text and for programming and working examples, and to Mr. R. C. W. Strijbos of Twente and Prof. J. van de Vegte, Toronto, for their comments on early versions of the manuscript. The final manuscript was read by Prof. L. Hasdorff of the Virginia Polytechnic Institute and Dr. Paul Alper of Twente; their constructive criticism and remarks are greatly appreciated. The second author is grateful to his graduate students, in particular to Victor Shenkar, for helping to correct early versions of the manuscript.

H. K.
R. S.

CONTENTS

Notation and Symbols   21

Chapter 1   **Elements of Linear System Theory**   1

     1.1   *Introduction*, 1

     1.2   *State Description of Linear Systems*, 1
         1.2.1   State Description of Nonlinear and Linear Differential Systems, 1
         1.2.2   Linearization, 2
         1.2.3   Examples, 3
         1.2.4   State Transformations, 10

     1.3   *Solution of the State Differential Equation of Linear Systems*, 11
         1.3.1   The Transition Matrix and the Impulse Response Matrix, 11
         1.3.2   The Transition Matrix of a Time-Invariant System, 13
         1.3.3   Diagonalization, 15
         1.3.4*   The Jordan Form, 19

     1.4   *Stability*, 24
         1.4.1   Definitions of Stability, 24
         1.4.2   Stability of Time-Invariant Linear Systems, 27
         1.4.3*   Stable and Unstable Subspaces for Time-Invariant Linear Systems, 29
         1.4.4*   Investigation of the Stability of Nonlinear Systems through Linearization, 31

     1.5   *Transform Analysis of Time-Invariant Systems*, 33
         1.5.1   Solution of the State Differential Equation through Laplace Transformation, 33

* See the Preface for the significance of the marked sections.

    1.5.2 Frequency Response, 37
    1.5.3 Zeroes of Transfer Matrices, 39
    1.5.4 Interconnections of Linear Systems, 43
    1.5.5* Root Loci, 51

1.6* *Controllability*, **53**
    1.6.1* Definition of Controllability, 53
    1.6.2* Controllability of Linear Time-Invariant Systems, 55
    1.6.3* The Controllable Subspace, 57
    1.6.4* Stabilizability, 62
    1.6.5* Controllability of Time-Varying Linear Systems, 64

1.7* *Reconstructibility*, **65**
    1.7.1* Definition of Reconstructibility, 65
    1.7.2* Reconstructibility of Linear Time-Invariant Systems, 67
    1.7.3* The Unreconstructible Subspace, 70
    1.7.4* Detectability, 76
    1.7.5* Reconstructibility of Time-Varying Linear Systems, 78

1.8* *Duality of Linear Systems*, **79**

1.9* *Phase-Variable Canonical Forms*, **82**

1.10 *Vector Stochastic Processes*, **85**
    1.10.1 Definitions, 85
    1.10.2 Power Spectral Density Matrices, 90
    1.10.3 The Response of Linear Systems to Stochastic Inputs, 91
    1.10.4 Quadratic Expressions, 94

1.11 *The Response of Linear Differential Systems to White Noise*, **97**
    1.11.1 White Noise, 97
    1.11.2 Linear Differential Systems Driven by White Noise, 100
    1.11.3 The Steady-State Variance Matrix for the Time-Invariant Case, 103
    1.11.4 Modeling of Stochastic Processes, 106
    1.11.5 Quadratic Integral Expressions, 108

1.12 *Problems*, **113**

Chapter 2 Analysis of Linear Control Systems   **119**

   2.1   *Introduction,* 119

   2.2   *The Formulation of Control Problems,* 121
       2.2.1   Introduction, 121
       2.2.2   The Formulation of Tracking and Regulator Problems, 121
       2.2.3   The Formulation of Terminal Control Problems, 127

   2.3   *Closed-Loop Controllers; The Basic Design Objective,* 128

   2.4   *The Stability of Control Systems,* 136

   2.5   *The Steady-State Analysis of the Tracking Properties,* 140
       2.5.1   The Steady-State Mean Square Tracking Error and Input, 140
       2.5.2   The Single-Input Single-Output Case, 144
       2.5.3   The Multiinput Multioutput Case, 155

   2.6   *The Transient Analysis of the Tracking Properties,* 165

   2.7   *The Effects of Disturbances in the Single-Input Single-Output Case,* 167

   2.8   *The Effects of Observation Noise in the Single-Input Single-Output Case,* 174

   2.9   *The Effect of Plant Parameter Uncertainty in the Single-Input Single-Output Case,* 178

   2.10*   *The Open-Loop Steady-State Equivalent Control Scheme,* 183

   2.11   *Conclusions,* 188

   2.12   *Problems,* 189

Chapter 3 Optimal Linear State Feedback Control Systems   **193**

   3.1   *Introduction,* 193

   3.2   *Stability Improvement of Linear Systems by State Feedback,* 193
       3.2.1   Linear State Feedback Control, 193

3.2.2* Conditions for Pole Assignment and Stabilization, 198

3.3 The Deterministic Linear Optimal Regulator Problem, 201
3.3.1 Introduction, 201
3.3.2 Solution of the Regulator Problem, 207
3.3.3 Derivation of the Riccati Equation, 216

3.4 Steady-State Solution of the Deterministic Linear Optimal Regulator Problem, 220
3.4.1 Introduction and Summary of Main Results, 220
3.4.2* Steady-State Properties of Optimal Regulators, 230
3.4.3* Steady-State Properties of the Time-Invariant Optimal Regulator, 237
3.4.4* Solution of the Time-Invariant Regulator Problem by Diagonalization, 243

3.5 Numerical Solution of the Riccati Equation, 248
3.5.1 Direct Integration, 248
3.5.2 The Kalman–Englar Method, 249
3.5.3* Solution by Diagonalization, 250
3.5.4* Solution by the Newton–Raphson Method, 251

3.6 Stochastic Linear Optimal Regulator and Tracking Problems, 253
3.6.1 Regulator Problems with Disturbances—The Stochastic Regulator Problem, 253
3.6.2 Stochastic Tracking Problems, 257
3.6.3 Solution of the Stochastic Linear Optimal Regulator Problem, 259

3.7 Regulators and Tracking Systems with Nonzero Set Points and Constant Disturbances, 270
3.7.1 Nonzero Set Points, 270
3.7.2* Constant Disturbances, 277

3.8* Asymptotic Properties of Time-Invariant Optimal Control Laws, 281
3.8.1* Asymptotic Behavior of the Optimal Closed-Loop Poles, 281

		3.8.2*	Asymptotic Properties of the Single-Input Single-Output Nonzero Set Point Regulator, 297

            3.8.3*   The Maximally Achievable Accuracy of Regulators and Tracking Systems, 306

    3.9* *Sensitivity of Linear State Feedback Control Systems,* **312**

    3.10 *Conclusions,* **318**

    3.11 *Problems,* **319**

Chapter 4   **Optimal Linear Reconstruction of the State**   **328**

    4.1 *Introduction,* **328**

    4.2 *Observers,* **329**
        4.2.1   Full-Order Observers, 329
        4.2.2* Conditions for Pole Assignment and Stabilization of Observers, 334
        4.2.3* Reduced-Order Observers, 335

    4.3 *The Optimal Observer,* **339**
        4.3.1   A Stochastic Approach to the Observer Problem, 339
        4.3.2   The Nonsingular Optimal Observer Problem with Uncorrelated State Excitation and Observation Noises, 341
        4.3.3* The Nonsingular Optimal Observer Problem with Correlated State Excitation and Observation Noises, 351
        4.3.4* The Time-Invariant Singular Optimal Observer Problem, 352
        4.3.5* The Colored Noise Observation Problem, 356
        4.3.6* Innovations, 361

    4.4* *The Duality of the Optimal Observer and the Optimal Regulator; Steady-State Properties of the Optimal Observer,* **364**
        4.4.1* Introduction, 364
        4.4.2* The Duality of the Optimal Regulator and the Optimal Observer Problem, 364

4.4.3* Steady-State Properties of the Optimal Observer, 365
4.4.4* Asymptotic Properties of Time-Invariant Steady-State Optimal Observers, 368

4.5 *Conclusions*, 373

4.6 *Problems*, 373

Chapter 5 Optimal Linear Output Feedback Control Systems   377

5.1 *Introduction*, 377

5.2 *The Regulation of Linear Systems with Incomplete Measurements*, 378
    5.2.1 The Structure of Output Feedback Control Systems, 378
    5.2.2* Conditions for Pole Assignment and Stabilization of Output Feedback Control Systems, 388

5.3 *Optimal Linear Regulators with Incomplete and Noisy Measurements*, 389
    5.3.1 Problem Formulation and Solution, 389
    5.3.2 Evaluation of the Performance of Optimal Output Feedback Regulators, 391
    5.3.3* Proof of the Separation Principle, 400

5.4 *Linear Optimal Tracking Systems with Incomplete and Noisy Measurements*, 402

5.5 *Regulators and Tracking Systems with Nonzero Set Points and Constant Disturbances*, 409
    5.5.1 Nonzero Set Points, 409
    5.5.2* Constant Disturbances, 414

5.6* *Sensitivity of Time-Invariant Optimal Linear Output Feedback Control Systems*, 419

5.7* *Linear Optimal Output Feedback Controllers of Reduced Dimensions*, 427
    5.7.1* Introduction, 427
    5.7.2* Controllers of Reduced Dimensions, 428
    5.7.3* Numerical Determination of Optimal Controllers of Reduced Dimensions, 432

5.8   *Conclusions*, 436

5.9   *Problems*, 438

Chapter 6*   Linear Optimal Control Theory for Discrete-Time Systems   442

6.1   *Introduction*, 442

6.2   *Theory of Linear Discrete-Time Systems*, 442
    6.2.1   Introduction, 442
    6.2.2   State Description of Linear Discrete-Time Systems, 443
    6.2.3   Interconnections of Discrete-Time and Continuous-Time Systems, 443
    6.2.4   Solution of State Difference Equations, 452
    6.2.5   Stability, 454
    6.2.6   Transform Analysis of Linear Discrete-Time Systems, 455
    6.2.7   Controllability, 459
    6.2.8   Reconstructibility, 462
    6.2.9   Duality, 465
    6.2.10   Phase-Variable Canonical Forms, 466
    6.2.11   Discrete-Time Vector Stochastic Processes, 467
    6.2.12   Linear Discrete-Time Systems Driven by White Noise, 470

6.3   *Analysis of Linear Discrete-Time Control Systems*, 475
    6.3.1   Introduction, 475
    6.3.2   Discrete-Time Linear Control Systems, 475
    6.3.3   The Steady-State and the Transient Analysis of the Tracking Properties, 478
    6.3.4   Further Aspects of Linear Discrete-Time Control System Performance, 487

6.4   *Optimal Linear Discrete-Time State Feedback Control Systems*, 488
    6.4.1   Introduction, 488
    6.4.2   Stability Improvement by State Feedback, 488
    6.4.3   The Linear Discrete-Time Optimal Regulator Problem, 490

xx Contents

      6.4.4   Steady-State Solution of the Discrete-Time Regulator Problem, 495
      6.4.5   The Stochastic Discrete-Time Linear Optimal Regulator, 502
      6.4.6   Linear Discrete-Time Regulators with Nonzero Set Points and Constant Disturbances, 504
      6.4.7   Asymptotic Properties of Time-Invariant Optimal Control Laws, 509
      6.4.8   Sensitivity, 520

6.5 *Optimal Linear Reconstruction of the State of Linear Discrete-Time Systems,* **522**
      6.5.1   Introduction, 522
      6.5.2   The Formulation of Linear Discrete-Time Reconstruction Problems, 522
      6.5.3   Discrete-Time Observers, 525
      6.5.4   Optimal Discrete-Time Linear Observers, 528
      6.5.5   Innovations, 533
      6.5.6   Duality of the Optimal Observer and Regulator Problems; Steady-State Properties of the Optimal Observer, 533

6.6 *Optimal Linear Discrete-Time Output Feedback Systems,* **536**
      6.6.1   Introduction, 536
      6.6.2   The Regulation of Systems with Incomplete Measurements, 536
      6.6.3   Optimal Linear Discrete-Time Regulators with Incomplete and Noisy Measurements, 539
      6.6.4   Nonzero Set Points and Constant Disturbances, 543

6.7 *Conclusions,* **546**

6.8 *Problems,* **547**

References   553

Index   563

NOTATION AND SYMBOLS

Chapters are subdivided into sections, which are numbered 1.1, 1.2, 1.3, and so on. Sections may be divided into subsections, which are numbered 1.1.1, 1.1.2, and so on. Theorems, examples, figures, and similar features are numbered consecutively within each chapter, prefixed by the chapter number. The section number is usually given in parentheses if reference is made to an item in another section.

Vectors are denoted by lowercase letters (such as x and u), matrices by uppercase letters (such as A and B) and scalars by lower case Greek letters (such as α and β). It has not been possible to adhere to these rules completely consistently; notable exceptions are t for time, i and j for integers, and so on. The components of vectors are denoted by lowercase Greek letters which correspond as closely as possible to the Latin letter that denotes the vector; thus the n-dimensional vector x has as components the scalars $\xi_1, \xi_2, \cdots, \xi_n$, the m-dimensional vector y has as components the scalars $\eta_1, \eta_2, \cdots, \eta_m$, and so on. Boldface capitals indicate the Laplace or z-transform of the corresponding lowercase time functions [$\mathbf{X}(s)$ for the Laplace transform of $x(t)$, $\mathbf{Y}(z)$ for the z-transform of $y(i)$, etc.].

Operations

x^T	transpose of the vector x
col $(\xi_1, \xi_2, \cdots, \xi_n)$	column vector with components $\xi_1, \xi_2, \cdots, \xi_n$
$(\eta_1, \eta_2, \cdots, \eta_n)$	row vector with components $\eta_1, \eta_2, \cdots, \eta_n$
$\begin{pmatrix} x_1 \\ x_2 \end{pmatrix}$, col (x_1, x_2)	partitioning of a column vector into subvectors x_1 and x_2
$\|x\|$	norm of a vector x
dim (x)	dimension of the vector x
A^T	transpose of the matrix A
A^{-1}	inverse of the square matrix A
tr (A)	trace of the square matrix A
det (A)	determinant of the square matrix A

diag $(\lambda_1, \lambda_2, \cdots, \lambda_n)$	diagonal matrix with diagonal entries $\lambda_1, \lambda_2, \cdots, \lambda_n$
$(e_1, e_2, \cdots, e_n)$	partitioning of a matrix into its columns $e_1, e_2, \cdots, e_n$
$\begin{pmatrix} f_1 \\ f_2 \\ \cdot \\ \cdot \\ \cdot \\ f_n \end{pmatrix}$	partitioning of a matrix into its rows $f_1, f_2, \cdots, f_n$
$(T_1, T_2, \cdots, T_m)$	partitioning of a matrix into column blocks $T_1, T_2, \cdots, T_n$
$\begin{pmatrix} U_1 \\ U_2 \\ \cdot \\ \cdot \\ \cdot \\ U_m \end{pmatrix}$	partitioning of a matrix into row blocks $U_1, U_2, \cdots, U_m$
$\begin{pmatrix} A & B \\ C & D \end{pmatrix}$	partitioning of a matrix into blocks A, B, C, and D
diag $(J_1, J_2, \cdots, J_m)$	block diagonal matrix with diagonal blocks $J_1, J_2, \cdots, J_m$
$M > 0$, $M \geq 0$	the real symmetric or Hermitian matrix M is positive-definite or nonnegative-definite, respectively
$M > N$, $M \geq N$	the real symmetric or Hermitian matrix $M - N$ is positive-definite or nonnegative-definite, respectively
$\dot{x}(t)$ or $\dfrac{dx(t)}{dt}$	time derivative of the time-varying vector $x(t)$
$\mathscr{L}\{x(t)\}$	Laplace transform of $x(t)$
Re (α)	real part of the complex number α
Im (α)	imaginary part of the complex number α
min (α, β)	the smallest of the numbers α and β
$\min_{\alpha}$	the minimum with respect to α
$\max_{\alpha}$	the maximum with respect to α

Commonly used symbols

0	zero; zero vector; zero matrix
$A(t)$, $A(i)$, A	plant matrix of a finite-dimensional linear differential system

Notation and Symbols

$B(t)$, $B(i)$, B	input matrix of a finite-dimensional linear differential system (B becomes b in the single-input case)
$C(t)$, $C(i)$, C	output matrix of a finite-dimensional linear differential system; output matrix for the observed variable (C becomes c in the single-output case)
$C_e(t)$, $C_e(i)$, $C_{e\infty}$	mean square tracking or regulating error
$C_u(t)$, $C_u(i)$, $C_{u\infty}$	mean square input
$D(t)$, $D(i)$, D	output matrix for the controlled variable (D becomes d in the single-output case)
e	base of the natural logarithm
$e(t)$ or $e(i)$	tracking or regulating error; reconstruction error
e_i	i-th characteristic vector
E	expectation operator
$E(i)$	gain matrix of the direct link of a plant (Ch. 6 only)
f	frequency
$F(t)$, $F(i)$, F, $\bar{F}$	regulator gain matrix (F becomes f in the single-input case)
$G(s)$, $G(z)$	controller transfer matrix (from y to $-u$)
$H(s)$, $H(z)$	plant transfer matrix (from u to y)
i	integer
I	unit matrix
j	$\sqrt{-1}$; integer
$J(s)$, $J(z)$	return difference matrix or function
$K(t)$, $K(i)$, K, $\bar{K}$	observer gain matrix (K becomes k in the single-output case)
$K(s)$	plant transfer matrix (from u to z)
$H_c(s)$, $H_c(z)$	closed-loop transfer matrix
n	dimension of the state x
$N(s)$, $N(z)$	transfer matrix or function from r to u in a control system
P	controllability matrix
$P(t)$, $P(i)$, $\bar{P}$	solution of the regulator Riccati equation
$P(s)$, $P(z)$	controller transfer matrix (from r to u)
P_1	terminal state weighting matrix
Q	reconstructibility matrix
$Q(t)$, $Q(i)$, $\bar{Q}$	variance matrix; solution of the observer Riccati equation
Q_0	initial variance matrix
$Q'(t)$, $Q'(i)$	second-order moment matrix
$r(t)$, $r(i)$	reference variable
$R_1(t)$, $R_1(i)$, R_1	weighting matrix of the state
$R_2(t)$, $R_2(i)$, R_2	weighting matrix of the input

Notation and Symbols

$R_3(t)$, $R_3(i)$, R_3	weighting matrix of the tracking of regulating error
$R_v(t_1, t_2)$, $R_v(t_1 - t_2)$, $R_v(i,j)$, $R_v(i-j)$	
	covariance function of the stochastic process v
s	variable of the Laplace transform
$S(s)$, $S(z)$	sensitivity matrix or function
t	time
$T(s)$ or $T(z)$	transmission
$u(t)$, $u(i)$	input variable
$v(t)$, $v(i)$	stochastic process
$v_m(t)$, $v_m(i)$	observation noise, measurement noise
v_0	constant disturbance
$v_0(t)$, $v_0(i)$	equivalent disturbance at the controlled variable
$v_p(t)$, $v_p(i)$	disturbance variable
$V(t)$, $V(i)$	intensity of a white noise process
$w(t)$, $w(i)$	white noise process
$W_e(t)$, $W_e(i)$, W_e	weighting matrix of the tracking or regulating error
$W_u(t)$, $W_u(i)$, W_u	weighting matrix of the input
$x(t)$, $x(i)$	state variable
$\hat{x}(t)$, $\hat{x}(i)$	reconstructed state variable
x_0	initial state
$y(t)$, $y(i)$	output variable; observed variable
z	z-transform variable
$z(t)$, $z(i)$	controlled variable
Z	compound matrix of system and adjoint differential equations
$\delta(t)$	delta function
Δ	sampling interval
$\zeta(t)$, $\zeta(i)$	scalar controlled variable
$\eta(t)$, $\eta(i)$	scalar output variable; scalar observed variable
θ	time difference; time constant; normalized angular frequency
λ_i	i-th characteristic value
$\mu(t)$, $\mu(i)$	scalar input variable
$\nu(t)$, $\nu(i)$	scalar stochastic process
ν_i	i-th zero
$\xi(t)$, $\xi(i)$	scalar state variable
π_i	i-th pole
ρ	weighting coefficient of the integrated or mean square input
$\Sigma_v(\omega)$, $\Sigma_v(\theta)$	spectral density matrix of the stochastic process v
$\phi(s)$, $\phi(z)$	characteristic polynomial
$\phi_c(s)$, $\phi_c(z)$	closed-loop characteristic polynomial

$\Phi(t, t_0)$, $\Phi(i, i_0)$	transition matrix
$\psi(s)$, $\psi(z)$	numerator polynomial
ω	angular frequency

SI units

A	ampere
Hz	hertz
kg	kilogram
kmol	kilomole
m	meter
N	newton
rad	radian
s	second
V	volt
Ω	ohm

LINEAR OPTIMAL
CONTROL SYSTEMS

1 ELEMENTS OF LINEAR SYSTEM THEORY

1.1 INTRODUCTION

This book deals with the analysis and design of linear control systems. A prerequisite for studying linear control systems is a knowledge of linear system theory. We therefore devote this first chapter to a review of the most important ingredients of linear system theory. The introduction of control problems is postponed until Chapter 2.

The main purpose of this chapter is to establish a conceptual framework, introduce notational conventions, and give a survey of the basic facts of linear system theory. The starting point is the state space description of linear systems. We then proceed to discussions of the solution of linear state differential equations, the stability of linear systems, and the transform analysis of such systems. The topics next dealt with are of a more advanced nature; they concern controllability, reconstructibility, duality, and phase-variable canonical forms of linear systems. The chapter concludes with a discussion of vector stochastic processes and the response of linear systems to white noise. These topics play an important role in the development of the theory.

Since the reader of this chapter is assumed to have had an introduction to linear system theory, the proofs of several well-known theorems are omitted. References to relevant textbooks are provided, however. Some topics are treated in sections marked with an asterisk, notably controllability, reconstructibility, duality and phase-variable canonical forms. The asterisk indicates that these notions are of a more advanced nature, and needed only in the sections similarly marked in the remainder of the book.

1.2 STATE DESCRIPTION OF LINEAR SYSTEMS

1.2.1 State Description of Nonlinear and Linear Differential Systems

Many systems can be described by a set of simultaneous differential equations of the form

$$\dot{x}(t) = f[x(t), u(t), t]. \qquad \text{1-1}$$

2   Elements of Linear System Theory

Here t is the time variable, $x(t)$ is a real n-dimensional time-varying column vector which denotes the *state* of the system, and $u(t)$ is a real k-dimensional column vector which indicates the *input variable* or *control* variable. The function f is real and vector-valued. For many systems the choice of the state follows naturally from the physical structure, and **1-1**, which will be called the *state differential equation*, usually follows directly from the elementary physical laws that govern the system.

Let $y(t)$ be a real l-dimensional system variable that can be observed or through which the system influences its environment. Such a variable we call an *output variable* of the system. It can often be expressed as

$$y(t) = g[x(t), u(t), t]. \qquad \textbf{1-2}$$

This equation we call the *output equation* of the system.

We call a system that is described by **1-1** and **1-2** a *finite-dimensional differential system* or, for short, a *differential system*. Equations **1-1** and **1-2** together are called the *system equations*. If the vector-valued function g contains u explicitly, we say that the system has a *direct link*.

In this book we are mainly concerned with the case where f and g are linear functions. We then speak of a (*finite-dimensional*) *linear differential system*. Its state differential equation has the form

$$\dot{x}(t) = A(t)x(t) + B(t)u(t), \qquad \textbf{1-3}$$

where $A(t)$ and $B(t)$ are time-varying matrices of appropriate dimensions. We call the dimension n of x the *dimension* of the system. The output equation for such a system takes the form

$$y(t) = C(t)x(t) + D(t)u(t). \qquad \textbf{1-4}$$

If the matrices A, B, C, and D are constant, the system is *time-invariant*.

1.2.2   Linearization

It is the purpose of this section to show that if $u_0(t)$ is a given input to a system described by the state differential equation **1-1**, and $x_0(t)$ is a known solution of the state differential equation, we can find approximations to neighboring solutions, for small deviations in the initial state and in the input, from a linear state differential equation. Suppose that $x_0(t)$ satisfies

$$\dot{x}_0(t) = f[x_0(t), u_0(t), t], \qquad t_0 \leq t \leq t_1. \qquad \textbf{1-5}$$

We refer to u_0 as a *nominal input* and to x_0 as a *nominal trajectory*. Often we can assume that the system is operated close to nominal conditions, which means that u and x deviate only slightly from u_0 and x_0. Let us therefore write

$$\begin{aligned} u(t) &= u_0(t) + \tilde{u}(t), \qquad t_0 \leq t \leq t_1, \\ x(t_0) &= x_0(t_0) + \tilde{x}(t_0), \end{aligned} \qquad \textbf{1-6}$$

1.2 State Description of Linear Systems

where $\tilde{u}(t)$ and $\tilde{x}(t_0)$ are small perturbations. Correspondingly, let us introduce $\tilde{x}(t)$ by

$$x(t) = x_0(t) + \tilde{x}(t), \qquad t_0 \le t \le t_1. \qquad \text{1-7}$$

Let us now substitute x and u into the state differential equation and make a Taylor expansion. It follows that

$$\dot{x}_0(t) + \dot{\tilde{x}}(t) = f[x_0(t), u_0(t), t] + J_x[x_0(t), u_0(t), t]\tilde{x}(t)$$
$$+ J_u[x_0(t), u_0(t), t]\tilde{u}(t) + h(t), \qquad t_0 \le t \le t_1. \qquad \text{1-8}$$

Here J_x and J_u are the Jacobian matrices of f with respect to x and u, respectively, that is, J_x is a matrix the (i,j)-th element of which is

$$(J_x)_{i,j} = \frac{\partial f_i}{\partial \xi_j}, \qquad \text{1-9}$$

where f_i is the i-th component of f and ξ_j the j-th component of x. J_u is similarly defined. The term $h(t)$ is an expression that is supposed to be "small" with respect to $\tilde{x}$ and $\tilde{u}$. Neglecting h, we see that $\tilde{x}$ and $\tilde{u}$ approximately satisfy the *linear* equation

$$\dot{\tilde{x}}(t) = A(t)\tilde{x}(t) + B(t)\tilde{u}(t), \qquad t_0 \le t \le t_1, \qquad \text{1-10}$$

where $A(t) = J_x[x_0(t), u_0(t), t]$ and $B(t) = J_u[x_0(t), u_0(t), t]$. We call **1-10** the *linearized state differential equation*. The initial condition of **1-10** is $\tilde{x}(t_0)$.

The linearization procedure outlined here is very common practice in the solution of control problems. Often it is more convenient to linearize the system differential equations before arranging them in the form of state differential equations. This leads to the same results, of course (see the examples of Section 1.2.3).

It can be inferred from texts on differential equations (see, e.g., Roseau, 1966) that the approximation to $x(t)$ obtained in this manner can be made arbitrarily accurate, provided the function f possesses partial derivatives with respect to the components of x and u near the nominal values x_0, u_0, the interval $[t_0, t_1]$ is finite, and the initial deviation $\tilde{x}(t_0)$ and the deviation of the input $\tilde{u}$ are chosen sufficiently small.

In Section 1.4.4 we present further justification of the extensive use of linearization in control engineering.

1.2.3 Examples

In this section several examples are given which serve to show how physical equations are converted into state differential equations and how linearization is performed. We discuss these examples at some length because later they are extensively used to illustrate the theory that is given.

4 Elements of Linear System Theory

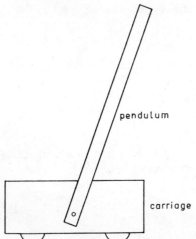

Fig. 1.1. An inverted pendulum positioning system.

Example 1.1. *Inverted pendulum positioning system.*

Consider the inverted pendulum of Figure 1.1 (see also, for this example, Cannon, 1967; Elgerd, 1967). The pivot of the pendulum is mounted on a carriage which can move in a horizontal direction. The carriage is driven by a small motor that at time t exerts a force $\mu(t)$ on the carriage. This force is the input variable to the system.

Figure 1.2 indicates the forces and the displacements. The displacement of the pivot at time t is $s(t)$, while the angular rotation at time t of the pendulum is $\phi(t)$. The mass of the pendulum is m, the distance from the pivot to the center of gravity L, and the moment of inertia with respect to the center of gravity J. The carriage has mass M. The forces exerted on the pendulum are

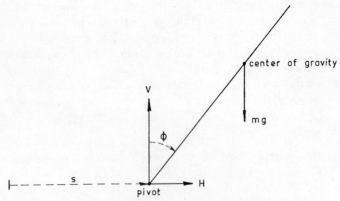

Fig. 1.2. Inverted pendulum: forces and displacements.

1.2 State Description of Linear Systems

the force mg in the center of gravity, a horizontal reaction force $H(t)$, and a vertical reaction force $V(t)$ in the pivot. Here g is the gravitational acceleration. The following equations hold for the system:

$$m \frac{d^2}{dt^2}[s(t) + L \sin \phi(t)] = H(t), \qquad \text{1-11}$$

$$m \frac{d^2}{dt^2}[L \cos \phi(t)] = V(t) - mg, \qquad \text{1-12}$$

$$J \frac{d^2 \phi(t)}{dt^2} = LV(t) \sin \phi(t) - LH(t) \cos \phi(t), \qquad \text{1-13}$$

$$M \frac{d^2 s(t)}{dt^2} = \mu(t) - H(t) - F \frac{ds(t)}{dt}. \qquad \text{1-14}$$

Friction is accounted for only in the motion of the carriage and not at the pivot; in **1-14**, F represents the friction coefficient. Performing the differentiations indicated in **1-11** and **1-12**, we obtain

$$m\ddot{s}(t) + mL\ddot{\phi}(t) \cos \phi(t) - mL\dot{\phi}^2(t) \sin \phi(t) = H(t), \qquad \text{1-15}$$

$$-mL\ddot{\phi}(t) \sin \phi(t) - mL\dot{\phi}^2(t) \cos \phi(t) = V(t) - mg, \qquad \text{1-16}$$

$$J\ddot{\phi}(t) = LV(t) \sin \phi(t) - LH(t) \cos \phi(t), \qquad \text{1-17}$$

$$M\ddot{s}(t) = \mu(t) - H(t) - F\dot{s}(t). \qquad \text{1-18}$$

To simplify the equations we assume that m is small with respect to M and therefore neglect the horizontal reaction force $H(t)$ on the motion of the carriage. This allows us to replace **1-18** with

$$M\ddot{s}(t) = \mu(t) - F\dot{s}(t). \qquad \text{1-19}$$

Elimination of $H(t)$ and $V(t)$ from **1-15**, **1-16**, and **1-17** yields

$$(J + mL^2)\ddot{\phi}(t) - mgL \sin \phi(t) + mL\ddot{s}(t) \cos \phi(t) = 0. \qquad \text{1-20}$$

Division of this equation by $J + mL^2$ yields

$$\ddot{\phi}(t) - \frac{g}{L'} \sin \phi(t) + \frac{1}{L'} \ddot{s}(t) \cos \phi(t) = 0, \qquad \text{1-21}$$

where

$$L' = \frac{J + mL^2}{mL}. \qquad \text{1-22}$$

This quantity has the significance of "effective pendulum length" since a mathematical pendulum of length L' would also yield **1-21**.

Let us choose as the nominal solution $\mu(t) \equiv 0$, $s(t) \equiv 0$, $\phi(t) \equiv 0$. Linearization can easily be performed by using Taylor series expansions for $\sin \phi(t)$ and $\cos \phi(t)$ in **1-21** and retaining only the first term of the series. This yields the linearized version of **1-21**:

$$\ddot{\phi}(t) - \frac{g}{L'} \phi(t) + \frac{1}{L'} \ddot{s}(t) = 0. \qquad \text{1-23}$$

We choose the components of the state $x(t)$ as

$$\begin{aligned}
\xi_1(t) &= s(t), \\
\xi_2(t) &= \dot{s}(t), \\
\xi_3(t) &= s(t) + L'\phi(t), \\
\xi_4(t) &= \dot{s}(t) + L'\dot{\phi}(t).
\end{aligned} \qquad \text{1-24}$$

The third component of the state represents a linearized approximation to the displacement of a point of the pendulum at a distance L' from the pivot. We refer to $\xi_3(t)$ as the displacement of the pendulum. With these definitions we find from **1-19** and **1-23** the linearized state differential equation

$$\begin{aligned}
\dot{\xi}_1(t) &= \xi_2(t), \\
\dot{\xi}_2(t) &= \frac{1}{M} \mu(t) - \frac{F}{M} \xi_2(t), \\
\dot{\xi}_3(t) &= \xi_4(t), \\
\dot{\xi}_4(t) &= g\phi(t) = \frac{g}{L'} [\xi_3(t) - \xi_1(t)].
\end{aligned} \qquad \text{1-25}$$

In vector notation we write

$$\dot{x}(t) = \begin{pmatrix} 0 & 1 & 0 & 0 \\ 0 & -\dfrac{F}{M} & 0 & 0 \\ 0 & 0 & 0 & 1 \\ -\dfrac{g}{L'} & 0 & \dfrac{g}{L'} & 0 \end{pmatrix} x(t) + \begin{pmatrix} 0 \\ \dfrac{1}{M} \\ 0 \\ 0 \end{pmatrix} \mu(t), \qquad \text{1-26}$$

where $x(t) = \text{col } [\xi_1(t), \xi_2(t), \xi_3(t), \xi_4(t)]$.

Later the following numerical values are used:

$$\frac{F}{M} = 1 \text{ s}^{-1},$$

$$\frac{1}{M} = 1 \text{ kg}^{-1},$$

$$\frac{g}{L'} = 11.65 \text{ s}^{-2},$$

$$L' = 0.842 \text{ m}.$$

1-27

Example 1.2. *A stirred tank.*

As a further example we treat a system that is to some extent typical of process control systems. Consider the stirred tank of Fig. 1.3. The tank is fed

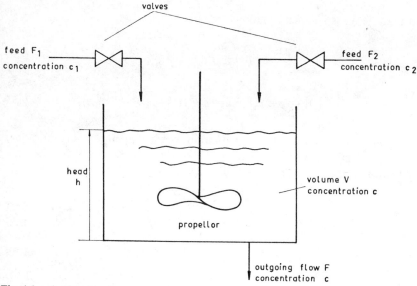

Fig. 1.3. A stirred tank.

with two incoming flows with time-varying flow rates $F_1(t)$ and $F_2(t)$. Both feeds contain dissolved material with constant concentrations c_1 and c_2, respectively. The outgoing flow has a flow rate $F(t)$. It is assumed that the tank is stirred well so that the concentration of the outgoing flow equals the concentration $c(t)$ in the tank.

The mass balance equations are

$$\frac{dV(t)}{dt} = F_1(t) + F_2(t) - F(t), \qquad \text{1-28}$$

$$\frac{d}{dt}[c(t)V(t)] = c_1 F_1(t) + c_2 F_2(t) - c(t)F(t), \qquad \text{1-29}$$

where $V(t)$ is the volume of the fluid in the tank. The outgoing flow rate $F(t)$ depends upon the head $h(t)$ as follows

$$F(t) = k\sqrt{h(t)}, \qquad \text{1-30}$$

where k is an experimental constant. If the tank has constant cross-sectional area S, we can write

$$F(t) = k\sqrt{\frac{V(t)}{S}}, \qquad \text{1-31}$$

so that the mass balance equations are

$$\frac{dV(t)}{dt} = F_1(t) + F_2(t) - k\sqrt{\frac{V(t)}{S}}, \qquad \text{1-32}$$

$$\frac{d}{dt}[c(t)V(t)] = c_1 F_1(t) + c_2 F_2(t) - c(t)k\sqrt{\frac{V(t)}{S}}. \qquad \text{1-33}$$

Let us first consider a steady-state situation where all quantities are constant, say F_{10}, F_{20}, and F_0 for the flow rates, V_0 for the volume, and c_0 for the concentration in the tank. Then the following relations hold:

$$0 = F_{10} + F_{20} - F_0, \qquad \text{1-34}$$

$$0 = c_1 F_{10} + c_2 F_{20} - c_0 F_0, \qquad \text{1-35}$$

$$F_0 = k\sqrt{\frac{V_0}{S}}. \qquad \text{1-36}$$

For given F_{10} and F_{20}, these equations can be solved for F_0, V_0, and c_0. Let us now assume that only small deviations from steady-state conditions occur. We write

$$\begin{aligned} F_1(t) &= F_{10} + \mu_1(t), \\ F_2(t) &= F_{20} + \mu_2(t), \\ V(t) &= V_0 + \xi_1(t), \\ c(t) &= c_0 + \xi_2(t), \end{aligned} \qquad \text{1-37}$$

1.2 State Description of Linear Systems

where we consider μ_1 and μ_2 input variables and ξ_1 and ξ_2 state variables. By assuming that these four quantities are small, linearization of **1-32** and **1-33** gives

$$\dot{\xi}_1(t) = \mu_1(t) + \mu_2(t) - \frac{k}{2V_0}\sqrt{\frac{V_0}{S}}\,\xi_1(t), \qquad \text{1-38}$$

$$\dot{\xi}_2(t)V_0 + c_0\dot{\xi}_1(t) = c_1\mu_1(t) + c_2\mu_2(t) - c_0\frac{k}{2V_0}\sqrt{\frac{V_0}{S}}\,\xi_1(t) - k\sqrt{\frac{V_0}{S}}\,\xi_2(t). \qquad \text{1-39}$$

Substitution of **1-36** into these equations yields

$$\dot{\xi}_1(t) = \mu_1(t) + \mu_2(t) - \frac{1}{2}\frac{F_0}{V_0}\,\xi_1(t), \qquad \text{1-40}$$

$$\dot{\xi}_2(t)V_0 + c_0\dot{\xi}_1(t) = c_1\mu_1(t) + c_2\mu_2(t) - \frac{1}{2}c_0\frac{F_0}{V_0}\,\xi_1(t) - F_0\xi_2(t). \qquad \text{1-41}$$

We define

$$\frac{V_0}{F_0} = \theta, \qquad \text{1-42}$$

and refer to θ as the *holdup time* of the tank. Elimination of $\dot{\xi}_1$ from **1-41** results in the linearized state differential equation

$$\dot{x}(t) = \begin{pmatrix} -\dfrac{1}{2\theta} & 0 \\ 0 & -\dfrac{1}{\theta} \end{pmatrix} x(t) + \begin{pmatrix} 1 & 1 \\ \dfrac{c_1 - c_0}{V_0} & \dfrac{c_2 - c_0}{V_0} \end{pmatrix} u(t), \qquad \text{1-43}$$

where $x(t) = \text{col}\,[\xi_1(t), \xi_2(t)]$ and $u(t) = \text{col}\,[\mu_1(t), \mu_2(t)]$. If we moreover define the output variables

$$\eta_1(t) = F(t) - F_0 \simeq \frac{1}{2}\frac{F_0}{V_0}\,\xi_1(t) = \frac{1}{2\theta}\,\xi_1(t),$$

$$\eta_2(t) = c(t) - c_0 = \xi_2(t), \qquad \text{1-44}$$

we can complement **1-43** with the linearized output equation

$$y(t) = \begin{pmatrix} \dfrac{1}{2\theta} & 0 \\ 0 & 1 \end{pmatrix} x(t), \qquad \text{1-45}$$

10 Elements of Linear System Theory

where $y(t) = \text{col} \, [\eta_1(t), \eta_2(t)]$. We use the following numerical values:

$$F_{10} = 0.015 \text{ m}^3/\text{s},$$
$$F_{20} = 0.005 \text{ m}^3/\text{s},$$
$$F_0 = 0.02 \text{ m}^3/\text{s},$$
$$c_1 = 1 \text{ kmol/m}^3,$$
$$c_2 = 2 \text{ kmol/m}^3,$$
$$c_0 = 1.25 \text{ kmol/m}^3,$$
$$V_0 = 1 \text{ m}^3,$$
$$\theta = 50 \text{ s}.$$

1-46

This results in the linearized system equations

$$\dot{x}(t) = \begin{pmatrix} -0.01 & 0 \\ 0 & -0.02 \end{pmatrix} x(t) + \begin{pmatrix} 1 & 1 \\ -0.25 & 0.75 \end{pmatrix} u(t),$$

$$y(t) = \begin{pmatrix} 0.01 & 0 \\ 0 & 1 \end{pmatrix} x(t).$$

1-47

1.2.4 State Transformations

As we shall see, it is sometimes useful to employ a transformed representation of the state. In this section we briefly review linear state transformations for time-invariant linear differential systems. Consider the linear time-invariant system

$$\dot{x}(t) = Ax(t) + Bu(t),$$
$$y(t) = Cx(t).$$

1-48

Let us define a transformed state variable

$$x'(t) = Tx(t),$$

1-49

where T is a constant, nonsingular transformation matrix. Substitution of $x(t) = T^{-1}x'(t)$ into **1-48** yields

$$T^{-1}\dot{x}'(t) = AT^{-1}x'(t) + Bu(t),$$
$$y(t) = CT^{-1}x'(t),$$

1-50

or

$$\dot{x}'(t) = TAT^{-1}x'(t) + TBu(t),$$
$$y(t) = CT^{-1}x'(t).$$

1-51

These are the state differential equation and the output equation of the system in terms of the new state $x'(t)$. It is clear that the transformed representation is completely equivalent to the original system, since we can always reconstruct the behavior of the system in terms of the original state by the relation $x(t) = T^{-1}x'(t)$. This derivation shows that the choice of the state is to some extent arbitrary and therefore can be adapted to suit various purposes. Many properties of linear, time-invariant systems remain unchanged under a state transformation (Problems 1.3, 1.6, 1.7).

1.3 SOLUTION OF THE STATE DIFFERENTIAL EQUATION OF LINEAR SYSTEMS

1.3.1 The Transition Matrix and the Impulse Response Matrix

In this section we discuss the solution of the linear state differential equation

$$\dot{x}(t) = A(t)x(t) + B(t)u(t). \qquad \textbf{1-52}$$

We first have the following result (Zadeh and Desoer, 1963; Desoer, 1970).

Theorem 1.1. *Consider the homogeneous equation*

$$\dot{x}(t) = A(t)x(t). \qquad \textbf{1-53}$$

Then if $A(t)$ is continuous for all t, **1-53** *always has a solution which can be expressed as*

$$x(t) = \Phi(t, t_0)x(t_0), \quad \text{for all } t. \qquad \textbf{1-54}$$

The **transition matrix** $\Phi(t, t_0)$ *is the solution of the matrix differential equation*

$$\frac{d}{dt}\Phi(t, t_0) = A(t)\Phi(t, t_0), \quad \text{for all } t,$$

$$\Phi(t_0, t_0) = I, \qquad \textbf{1-55}$$

where I is the unit matrix.

For a general time-varying system, the transition matrix rarely can be obtained in terms of standard functions, so that one must resort to numerical integration techniques. For time-invariant systems of low dimensions or of a simple structure, the transition matrix can be computed by any of the methods discussed in Sections 1.3.2, 1.3.3, and 1.5.1. For complicated time-invariant problems, one must employ numerical methods such as described in Section 1.3.2.

The transition matrix can be shown to possess the following properties (Zadeh and Desoer, 1963).

Theorem 1.2. *The transition matrix $\Phi(t, t_0)$ of a linear differential system has the following properties:*

(a) $\Phi(t_2, t_1)\Phi(t_1, t_0) = \Phi(t_2, t_0)$ *for all* t_0, t_1, t_2; 1-56
(b) $\Phi(t, t_0)$ *is nonsingular for all* t, t_0; 1-57
(c) $\Phi^{-1}(t, t_0) = \Phi(t_0, t)$ *for all* t, t_0; 1-58
(d) $\dfrac{d}{dt}\Phi^T(t_0, t) = -A^T(t)\Phi^T(t_0, t)$ *for all* t, t_0, 1-59

where the superscript T denotes the transpose.

Property (d) shows that the system $\dot{x}(t) = -A^T(t)x(t)$ has the transition matrix $\Phi^T(t_0, t)$. This can be proved by differentiating the identity $\Phi(t, t_0)\Phi(t_0, t) = I$.

Once the transition matrix has been found, it is easy to obtain solutions to the state differential equation **1-52**.

Theorem 1.3. *Consider the linear state differential equation*

$$\dot{x}(t) = A(t)x(t) + B(t)u(t). \quad \text{1-60}$$

Then if $A(t)$ is continuous and $B(t)$ and $u(t)$ are piecewise continuous for all t, the solution of **1-60** *is*

$$x(t) = \Phi(t, t_0)x(t_0) + \int_{t_0}^{t} \Phi(t, \tau)B(\tau)u(\tau)\, d\tau \quad \text{1-61}$$

for all t.

This result is easily verified by direct substitution into the state differential equation (Zadeh and Desoer, 1963).

Consider now a system with the state differential equation **1-60** and the output equation

$$y(t) = C(t)x(t). \quad \text{1-62}$$

For the output variable we write

$$y(t) = C(t)\Phi(t, t_0)x(t_0) + C(t)\int_{t_0}^{t} \Phi(t, \tau)B(\tau)u(\tau)\, d\tau. \quad \text{1-63}$$

If the system is initially in the zero state, that is, $x(t_0) = 0$, the response of the output variable is given by

$$y(t) = \int_{t_0}^{t} K(t, \tau)u(\tau)\, d\tau, \qquad t \geq t_0, \quad \text{1-64}$$

where

$$K(t, \tau) = C(t)\Phi(t, \tau)B(\tau), \qquad t \geq \tau. \quad \text{1-65}$$

The matrix $K(t, \tau)$ is called the *impulse response matrix* of the system because the (i,j)-th element of this matrix is the response at time t of the i-th component of the output variable to an impulse applied at the j-th component of the input at time $\tau > t_0$ while all other components of the input are zero and the initial state is zero. The *step response matrix* $S(t, \tau)$ is defined as

$$S(t, \tau) = \int_\tau^t K(t, \tau') \, d\tau', \qquad t \geq \tau. \qquad \textbf{1-66}$$

The (i,j)-th element of the step response matrix is the response at time t of the i-th component of the output when the j-th component of the input is a step function applied at time $\tau > t_0$ while all other components of the input are zero and the initial state is the zero state.

1.3.2 The Transition Matrix of a Time-Invariant System

For a time-invariant system, the transition matrix can be given in an explicit form (Zadeh and Desoer, 1963; Desoer, 1970; Polak and Wong, 1970).

Theorem 1.4. *The time-invariant system*

$$\dot{x}(t) = Ax(t) \qquad \textbf{1-67}$$

has the transition matrix

$$\Phi(t, t_0) = e^{A(t-t_0)}, \qquad \textbf{1-68}$$

where the exponential of a square matrix M is defined as

$$e^M = I + M + \frac{1}{2!} M^2 + \frac{1}{3!} M^3 + \cdots. \qquad \textbf{1-69}$$

This series converges for all M.

For small dimensions or simple structures of the matrix A, this result can be used to write down the transition matrix explicitly in terms of elementary functions (see Example 1.3). For high dimensions of the matrix A, Theorem 1.4 is quite useful for the computation of the transition matrix by a digital computer since the repeated multiplications and additions are easily programmed and performed. Such programs must include a stopping rule to truncate the infinite series after a finite number of terms. A usual stopping rule is to truncate when the addition of a new term changes each of the elements of the partial sum by less than a specified fraction. Numerical difficulties may occur when M is too large; this means that $t - t_0$ in **1-68** cannot be chosen too large (see Kalman, 1966; Kalman and Englar, 1966). Having a program for computing matrix exponentials is essential for anyone who wishes to simulate linear time-invariant systems. There are numerous references on the computation of matrix exponentials and simulating linear

14   Elements of Linear System Theory

systems; some of these are: Everling (1967), Liou (1966a,b, 1967, 1968), Whitney (1966a–c), Bickart (1968), Fath (1968), Plant (1969), Wallach (1969), Levis (1969), Rohrer (1970), Mastascusa and Simes (1970), and Krouse and Ward (1970). Melsa (1970) gives listings of FORTRAN computer programs.

By using **1-68** the time-invariant version of **1-63** becomes

$$y(t) = Ce^{A(t-t_0)}x(t_0) + C\int_{t_0}^{t} e^{A(t-\tau)}Bu(\tau)\,d\tau. \qquad \textbf{1-70}$$

Comparing **1-64** and **1-70** we see that the impulse response matrix of a time-invariant linear differential system depends on $t - \tau$ only and can be expressed as

$$K(t-\tau) = Ce^{A(t-\tau)}B, \qquad t \geq \tau. \qquad \textbf{1-71}$$

Example 1.3. *Stirred tank.*

The homogeneous part of the linearized state differential equation of the stirred tank of Example 1.2 is given by

$$\dot{x}(t) = \begin{pmatrix} -\dfrac{1}{2\theta} & 0 \\ 0 & -\dfrac{1}{\theta} \end{pmatrix} x(t). \qquad \textbf{1-72}$$

It is easily found that its transition matrix is given by

$$\Phi(t, t_0) = e^{A(t-t_0)}, \qquad \textbf{1-73}$$

where

$$e^{At} = \begin{pmatrix} 1 - \dfrac{t}{2\theta} + \dfrac{1}{2!}\left(\dfrac{t}{2\theta}\right)^2 - \dfrac{1}{3!}\left(\dfrac{t}{2\theta}\right)^3 + \cdots & 0 \\ 0 & 1 - \dfrac{t}{\theta} + \dfrac{1}{2!}\left(\dfrac{t}{\theta}\right)^2 - \dfrac{1}{3!}\left(\dfrac{t}{\theta}\right)^3 + \cdots \end{pmatrix}$$

$$= \begin{pmatrix} e^{-t/2\theta} & 0 \\ 0 & e^{-t/\theta} \end{pmatrix}. \qquad \textbf{1-74}$$

The impulse response matrix of the system is

$$K(t-\tau) = \begin{pmatrix} \dfrac{1}{2\theta}e^{-(t-\tau)/2\theta} & \dfrac{1}{2\theta}e^{-(t-\tau)/2\theta} \\ \dfrac{c_1 - c_0}{V_0}e^{-(t-\tau)/\theta} & \dfrac{c_2 - c_0}{V_0}e^{-(t-\tau)/\theta} \end{pmatrix}. \qquad \textbf{1-75}$$

1.3 Solution of State Equation

We find for the step response matrix of the stirred tank:

$$S(t-\tau) = \begin{pmatrix} 1 - e^{-(t-\tau)/2\theta} & 1 - e^{-(t-\tau)/2\theta} \\ \dfrac{c_1 - c_0}{F_0}(1 - e^{-(t-\tau)/\theta}) & \dfrac{c_2 - c_0}{F_0}(1 - e^{-(t-\tau)/\theta}) \end{pmatrix}. \quad \text{1-76}$$

In Fig. 1.4 the step responses are sketched for the numerical data of Example 1.2.

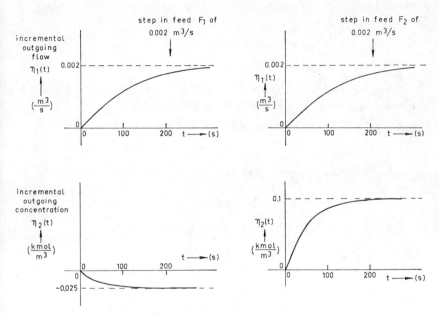

Fig. 1.4. Response of the stirred tank to a step of 0.002 m³/s in the feed F_1 (left column) and to a step of 0.002 m³/s in the feed F_2 (right column).

1.3.3 Diagonalization

An explicit form of the transition matrix of a time-invariant system can be obtained by diagonalization of the matrix A. The following result is available (Noble, 1969).

Theorem 1.5. *Suppose that the constant $n \times n$ matrix A has n distinct characteristic values $\lambda_1, \lambda_2, \cdots, \lambda_n$. Let the corresponding characteristic vectors be $e_1, e_2, \cdots, e_n$. Define the $n \times n$ matrices*

$$T = (e_1, e_2, \cdots, e_n), \quad \text{1-77a}$$
$$\Lambda = \text{diag}(\lambda_1, \lambda_2, \cdots, \lambda_n). \quad \text{1-77b}$$

Then T is nonsingular and A can be represented as

$$A = T\Lambda T^{-1}. \qquad \text{1-78}$$

Here the notation **1-77a** implies that the vectors $e_1, e_2, \cdots, e_n$ are the columns of the matrix T, and **1-77b** means that Λ is a diagonal matrix with $\lambda_1, \lambda_2, \cdots, \lambda_n$ as diagonal elements. It is said that T *diagonalizes* A.

The following fact is easily verified.

Theorem 1.6. *Consider the matrix A that satisfies the assumptions of Theorem 1.5. Then*

(a) $e^{At} = T e^{\Lambda t} T^{-1}$,   **1-79**

(b) $e^{\Lambda t} = \text{diag}\,(e^{\lambda_1 t}, e^{\lambda_2 t}, \cdots, e^{\lambda_n t})$.   **1-80**

This result makes it simple to compute $\exp(At)$ once A is diagonalized. It is instructive to present the same result in a different form.

Theorem 1.7. *Consider the time-invariant system*

$$\dot{x}(t) = A x(t), \qquad \text{1-81}$$

where A satisfies the assumptions of Theorem 1.5. Write the matrix T^{-1} in the form

$$T^{-1} = \begin{pmatrix} f_1 \\ f_2 \\ \vdots \\ f_n \end{pmatrix}, \qquad \text{1-82}$$

that is, the row vectors $f_1, f_2, \cdots, f_n$ are the rows of T^{-1}. Then the solution of **1-81** *can be written as*

$$x(t) = \sum_{i=1}^{n} e^{\lambda_i t} e_i f_i x(0). \qquad \text{1-83}$$

This is easily shown by expanding $x(t) = T \exp(\Lambda t) T^{-1} x(0)$ in terms of e_i, f_i, and $\exp(\lambda_i t)$, $i = 1, 2, \cdots, n$. We write **1-83** in the form

$$x(t) = \sum_{i=1}^{n} \mu_i e^{\lambda_i t} e_i, \qquad \text{1-84}$$

where the μ_i are the scalars $f_i x(0)$, $i = 1, 2, \cdots, n$. This clearly shows that the response of the system **1-81** is a composition of motions along the characteristic vectors of the matrix A. We call such a motion a *mode* of the system. A particular mode is *excited* by choosing the initial state to have a component along the corresponding characteristic vector.

1.3 Solution of State Equation

It is clear that the characteristic values $\lambda_i, i = 1, 2, \cdots, n$, to a considerable extent determine the dynamic behavior of the system. We often refer to these numbers as the *poles of the system*.

Even if the matrix A has multiple characteristic values, it can be diagonalized provided that the number of linearly independent characteristic vectors for each characteristic value equals the multiplicity of the characteristic value. The more complicated case, where the matrix A cannot be diagonalized, is discussed in Section 1.3.4.

Example 1.4. *Inverted pendulum.*

The homogeneous part of the state differential equation of the inverted pendulum balancing system of Example 1.1 is

$$\dot{x}(t) = \begin{pmatrix} 0 & 1 & 0 & 0 \\ 0 & -\dfrac{F}{M} & 0 & 0 \\ 0 & 0 & 0 & 1 \\ -\dfrac{g}{L'} & 0 & \dfrac{g}{L'} & 0 \end{pmatrix} x(t). \qquad \text{1-85}$$

The characteristic values and characteristic vectors of the matrix A can be found to be

$$\lambda_1 = 0, \quad \lambda_2 = -\frac{F}{M}, \quad \lambda_3 = \sqrt{\frac{g}{L'}}, \quad \lambda_4 = -\sqrt{\frac{g}{L'}},$$

$$e_1 = \begin{pmatrix} 1 \\ 0 \\ 1 \\ 0 \end{pmatrix}, \quad e_2 = \begin{pmatrix} 1 \\ -\dfrac{F}{M} \\ \alpha \\ -\alpha\dfrac{F}{M} \end{pmatrix}, \quad e_3 = \begin{pmatrix} 0 \\ 0 \\ 1 \\ \sqrt{\dfrac{g}{L'}} \end{pmatrix}, \quad e_4 = \begin{pmatrix} 0 \\ 0 \\ 1 \\ -\sqrt{\dfrac{g}{L'}} \end{pmatrix},$$

$$\text{1-86}$$

where

$$\alpha = \frac{\dfrac{g}{L'}}{\dfrac{g}{L'} - \dfrac{F^2}{M^2}}, \qquad \text{1-87}$$

18   Elements of Linear System Theory

and where we assume that the denominator of α differs from zero. The matrix T and its inverse are

$$T = \begin{pmatrix} 1 & 1 & 0 & 0 \\ 0 & -\dfrac{F}{M} & 0 & 0 \\ 1 & \alpha & 1 & 1 \\ 0 & -\alpha\dfrac{F}{M} & \sqrt{\dfrac{g}{L'}} & -\sqrt{\dfrac{g}{L'}} \end{pmatrix},$$

$$T^{-1} = \begin{pmatrix} 1 & \dfrac{M}{F} & 0 & 0 \\ 0 & -\dfrac{M}{F} & 0 & 0 \\ -\dfrac{1}{2} & -\dfrac{\frac{1}{2}}{\dfrac{F}{M}+\sqrt{\dfrac{g}{L'}}} & \dfrac{1}{2} & \dfrac{1}{2}\sqrt{\dfrac{L'}{g}} \\ -\dfrac{1}{2} & \dfrac{\frac{1}{2}}{-\dfrac{F}{M}+\sqrt{\dfrac{g}{L'}}} & \dfrac{1}{2} & -\dfrac{1}{2}\sqrt{\dfrac{L'}{g}} \end{pmatrix}. \qquad \text{1-88}$$

The modes of the system are

$$\begin{pmatrix} 1 \\ 0 \\ 1 \\ 0 \end{pmatrix}, \quad \begin{pmatrix} 1 \\ -\dfrac{F}{M} \\ \alpha \\ -\alpha\dfrac{F}{M} \end{pmatrix} e^{-(F/M)t}, \quad \begin{pmatrix} 0 \\ 0 \\ 1 \\ \sqrt{\dfrac{g}{L'}} \end{pmatrix} e^{t\sqrt{g/L'}}, \quad \begin{pmatrix} 0 \\ 0 \\ 1 \\ -\sqrt{\dfrac{g}{L'}} \end{pmatrix} e^{-t\sqrt{g/L'}}.$$

$$\text{1-89}$$

The first mode represents the indifference of the system with respect to horizontal translations, while the third mode exhibits the unstable character of the inverted pendulum.

1.3.4* The Jordan Form

In the preceding section we saw that the representation of the transition matrix can be facilitated by diagonalizing the matrix A. This diagonalization is not possible if the $n \times n$ matrix A does not have n linearly independent characteristic vectors. In this case, however, it is possible to bring A into the so-called Jordan normal form which is almost diagonal and from which the transition matrix can easily be obtained.

We first recall a few facts from linear algebra. If M is a matrix, the *null space* of M is defined as

$$\mathcal{N}(M) = \{x \colon x \in \mathscr{C}^n, Mx = 0\}, \qquad \text{1-90}$$

where $\mathscr{C}^n$ is the n-dimensional complex vector space. Furthermore, if $\mathcal{M}_1$ and $\mathcal{M}_2$ are two linear subspaces of an n-dimensional space, a linear subspace $\mathcal{M}_3$ is said to be the *direct sum* of $\mathcal{M}_1$ and $\mathcal{M}_2$, written as

$$\mathcal{M}_3 = \mathcal{M}_1 \oplus \mathcal{M}_2, \qquad \text{1-91}$$

if any vector $x_3 \in \mathcal{M}_3$ can be written in one and only one way as $x_3 = x_1 + x_2$, where $x_1 \in \mathcal{M}_1$ and $x_2 \in \mathcal{M}_2$.

We have the following result (Zadeh and Desoer, 1963).

Theorem 1.8. *Suppose that the* $n \times n$ *matrix A has k distinct characteristic values λ_i, $i = 1, 2, \cdots, k$. Let the multiplicity of each characteristic value λ_i in the characteristic polynomial of A be given by m_i. Define*

$$M_i = (A - \lambda_i I)^{m_i}, \qquad \text{1-92}$$

and let

$$\mathcal{N}_i = \mathcal{N}(M_i). \qquad \text{1-93}$$

Then

(a) *The dimension of the linear subspace $\mathcal{N}_i$ is m_i, $i = 1, 2, \cdots, k$;*
(b) *The whole n-dimensional complex space $\mathscr{C}^n$ is the direct sum of the null spaces $\mathcal{N}_i$, $i = 1, 2, \cdots, k$, that is,*

$$\mathscr{C}^n = \mathcal{N}_1 \oplus \mathcal{N}_2 \oplus \cdots \oplus \mathcal{N}_k. \qquad \text{1-94}$$

When the matrix A has n distinct characteristic values, the null spaces $\mathcal{N}_i$ reduce to one-dimensional subspaces each of which is spanned by a characteristic vector of A.

We have the following fact (Noble, 1969).

Theorem 1.9. *Consider the matrix A with the same notation as in Theorem 1.8. Then it is always possible to find a nonsingular transformation matrix T*

---

* See the Preface for the significance of the sections marked with an asterisk.

which can be partitioned as
$$T = (T_1, T_2, \cdots, T_k), \quad \text{1-95}$$
such that
$$A = TJT^{-1}, \quad \text{1-96}$$
where
$$J = \text{diag}(J_1, J_2, \cdots, J_k). \quad \text{1-97}$$

The block J_i has dimensions $m_i \times m_i$, $i = 1, 2, \cdots, k$, and the partitioning of T matches that of J. The columns of T_i form a specially chosen basis for the null space $\mathcal{N}_i$, $i = 1, 2, \cdots, k$. The blocks J_i can be subpartitioned as
$$J_i = \text{diag}(J_{i1}, J_{i2}, \cdots, J_{il_i}), \quad \text{1-98}$$
where each subblock J_{ij} is of the form
$$J_{ij} = \begin{pmatrix} \lambda_i & 1 & 0 & \cdots & & \\ 0 & \lambda_i & 1 & 0 & \cdots & \\ \cdots & \cdots & \cdots & \cdots & \cdots & \cdots \\ 0 & \cdots & \cdots & 0 & \lambda_i & 1 \\ 0 & \cdots & \cdots & \cdots & 0 & \lambda_i \end{pmatrix}. \quad \text{1-99}$$

J is called the **Jordan normal form** of A.

Expression **1-96** suggests the following practical method of computing the transformation matrix T (Noble, 1969). From **1-96** it follows
$$AT = TJ. \quad \text{1-100}$$

Let us denote the columns of T as $q_1, q_2, \cdots, q_n$. Then from the form of J, it follows with **1-100** that
$$Aq_i = \lambda q_i + \gamma_i q_{i-1}, \quad \text{1-101}$$
where γ_i is either 0 or 1, depending on J, and where λ is a characteristic value of A. Let us subpartition the block T_i of T corresponding to the subpartitioning **1-98** of J_i as $T_{i1}, T_{i2}, \cdots, T_{il_i}$. Then the number γ_i is zero whenever the corresponding column q_i of T is the first column of a subblock. Since if $\gamma_i = 0$ the vector q_i is a characteristic vector of A, we see that we can find the first columns of each subblock as the characteristic vectors of A. The remaining columns of each subblock then follow from **1-101** with $\gamma_i = 1$. Those remaining columns are known as *generalized* characteristic vectors of the matrix A. We stop this process when **1-101** fails to have a solution. Example 1.5 at the end of this section illustrates the procedure.

Once the matrix A has been brought into Jordan normal form, the exponential of A is easily found.

Theorem 1.10. *Consider the matrix A with the same notation as in Theorems 1.8 and 1.9. Then*

(a) $e^{At} = Te^{Jt}T^{-1}$,   1-102

(b) $e^{Jt} = \text{diag}\,(e^{J_1 t}, e^{J_2 t}, \cdots, e^{J_k t})$,   1-103

(c) $e^{J_i t} = \text{diag}\,(e^{J_{i1} t}, e^{J_{i2} t}, \cdots, e^{J_{il_i} t})$,   1-104

(d) $e^{J_{ij} t} = e^{\lambda_i t} \begin{pmatrix} 1 & t & \dfrac{t^2}{2!} & \cdots & \dfrac{t^{n_{ij}-1}}{(n_{ij}-1)!} \\ 0 & 1 & t & \cdots & \dfrac{t^{n_{ij}-2}}{(n_{ij}-2)!} \\ \cdots\cdots\cdots\cdots\cdots\cdots\cdots \\ 0 & & & \cdots & 1 \end{pmatrix}$,   1-105

where n_{ij} is the dimension of J_{ij}.

It is seen from this theorem that the response of the system

$$\dot{x}(t) = Ax(t) \qquad \text{1-106}$$

may contain besides purely exponential terms of the form $\exp(\lambda_i t)$ also terms of the form $t \exp(\lambda_i t)$, $t^2 \exp(\lambda_i t)$, and so on.

Completely in analogy with Section 1.3.3, we have the following fact (Zadeh and Desoer, 1963).

Theorem 1.11. *Consider the time-invariant linear system*

$$\dot{x}(t) = Ax(t). \qquad \text{1-107}$$

Express the initial state $x(0)$ as

$$x(0) = \sum_{i=1}^{k} v_i \quad \text{with } v_i \in \mathcal{N}_i, \quad i = 1, 2, \cdots, k. \qquad \text{1-108}$$

Write

$$T^{-1} = \begin{pmatrix} U_1 \\ U_2 \\ \vdots \\ U_k \end{pmatrix}, \qquad \text{1-109}$$

where the partitioning corresponds to that of T in Theorem 1.9. Then the response of the system can be expressed as

$$x(t) = \sum_{i=1}^{k} T_i \exp(J_i t) U_i v_i. \qquad \text{1-110}$$

From this theorem we see that if the initial state is within one of the null spaces $\mathcal{N}_i$, the nature of the response of the system to this initial state is completely determined by the corresponding characteristic value. In analogy with the simple case of Section 1.3.3, we call the response of the system to any initial state within one of the null spaces a *mode* of the system.

Example 1.5. *Inverted pendulum.*

Consider the inverted pendulum of Example 1.1, but suppose that we neglect the friction of the carriage so that $F = 0$. The homogeneous part of the linearized state differential equation is now given by $\dot{x}(t) = Ax(t)$, where

$$A = \begin{pmatrix} 0 & 1 & 0 & 0 \\ 0 & 0 & 0 & 0 \\ 0 & 0 & 0 & 1 \\ -\dfrac{g}{L'} & 0 & \dfrac{g}{L'} & 0 \end{pmatrix}. \qquad \text{1-111}$$

The characteristic values of A can be found to be

$$\lambda_1 = 0, \quad \lambda_2 = 0, \quad \lambda_3 = \sqrt{\dfrac{g}{L'}}, \quad \lambda_3 = -\sqrt{\dfrac{g}{L'}}. \qquad \text{1-112}$$

It is easily found that corresponding to the double characteristic value 0 there is only one characteristic vector, given by

$$\begin{pmatrix} 1 \\ 0 \\ 1 \\ 0 \end{pmatrix}. \qquad \text{1-113}$$

To λ_3 and λ_4 correspond the characteristic vectors

$$\begin{pmatrix} 0 \\ 0 \\ 1 \\ \sqrt{\dfrac{g}{L'}} \end{pmatrix} \quad \text{and} \quad \begin{pmatrix} 0 \\ 0 \\ 1 \\ -\sqrt{\dfrac{g}{L'}} \end{pmatrix}. \qquad \text{1-114}$$

1.3 Solution of State Equation

Since the characteristic values λ_3 and λ_4 are single, the corresponding null spaces have dimension one and are spanned by the corresponding characteristic vectors. Since zero is a double characteristic value, the corresponding null space is two-dimensional. The fact that there do not exist two linearly independent characteristic vectors gives rise to one subblock in the Jordan form of size 2×2. Let the characteristic vector **1-113** be the first column q_1 of the transformation matrix T. Then the second column q_2 must follow from

$$Aq_2 = 0 \cdot q_2 + q_1. \qquad \text{1-115}$$

It is easily found that the general solution to this equation is

$$q_2 = \begin{pmatrix} 0 \\ 1 \\ 0 \\ 1 \end{pmatrix} + \beta \begin{pmatrix} 1 \\ 0 \\ 1 \\ 0 \end{pmatrix}, \qquad \text{1-116}$$

where β is an arbitrary constant. We take $\beta = 0$. Since q_3 and q_4 have to be the characteristic vectors given by **1-114**, we find for the transformation matrix T,

$$T = \begin{pmatrix} 1 & 0 & 0 & 0 \\ 0 & 1 & 0 & 0 \\ 1 & 0 & 1 & 1 \\ 0 & 1 & \sqrt{\frac{g}{L'}} & -\sqrt{\frac{g}{L'}} \end{pmatrix}. \qquad \text{1-117}$$

The corresponding Jordan normal form of A is

$$J = \begin{pmatrix} 0 & 1 & 0 & 0 \\ 0 & 0 & 0 & 0 \\ 0 & 0 & \sqrt{\frac{g}{L'}} & 0 \\ 0 & 0 & 0 & -\sqrt{\frac{g}{L'}} \end{pmatrix}. \qquad \text{1-118}$$

The exponential of A can now easily be found from **1-102**, **1-117**, and **1-118**.

1.4 STABILITY

1.4.1 Definitions of Stability

In this section we are interested in the overall time behavior of differential systems. Consider the general nonlinear state differential equation

$$\dot{x}(t) = f[x(t), u(t), t]. \quad \text{1-119}$$

An important property of the system is whether or not the solutions of the state differential equation tend to grow indefinitely as $t \to \infty$. In order to simplify this question, we assume that we are dealing with an autonomous system, that is, a system without an input u or, equivalently, a system where u is a fixed time function. Thus we reduce our attention to the system

$$\dot{x}(t) = f[x(t), t]. \quad \text{1-120}$$

Just as in Section 1.2.2 on linearization, we introduce a *nominal solution* $x_0(t)$ which satisfies the state differential equation:

$$\dot{x}_0(t) = f[x_0(t), t]. \quad \text{1-121}$$

A case of special interest occurs when $x_0(t)$ is a constant vector x_e; in this case we say that x_e is an *equilibrium state* of the system.

We now discuss the stability of *solutions* of state differential equations. First we have the following definition (for the whole sequence of definitions that follows, see also Kalman and Bertram, 1960; Zadeh and Desoer, 1963; Brockett, 1970).

Definition 1.1. *Consider the state differential equation*

$$\dot{x}(t) = f[x(t), t] \quad \text{1-122}$$

with the nominal solution $x_0(t)$. Then the nominal solution is **stable in the sense of Lyapunov** *if for any t_0 and any $\varepsilon > 0$ there exists a $\delta(\varepsilon, t_0) > 0$ (depending upon ε and possibly upon t_0) such that $\|x(t_0) - x_0(t_0)\| \leq \delta$ implies $\|x(t) - x_0(t)\| < \varepsilon$ for all $t \geq t_0$.*

Here $\|x\|$ denotes the norm of a vector x; the Euclidean norm

$$\|x\| = \sqrt{\sum_{i=1}^{n} \xi_i^2}, \quad \text{1-123}$$

where the ξ_i, $i = 1, 2, \cdots, n$, are the components of x, can be used. Other norms are also possible.

Stability in the sense of Lyapunov guarantees that the state can be prevented from departing too far from the nominal solution by choosing the initial state close enough to the nominal solution. Stability in the sense of

Lyapunov is a rather weak form of stability. We therefore extend our concept of stability.

Definition 1.2. *The nominal solution $x_0(t)$ of the state differential equation*

$$\dot{x}(t) = f[x(t), t] \qquad \text{1-124}$$

*is **asymptotically stable** if*

(a) *It is stable in the sense of Lyapunov;*
(b) *For all t_0 there exists a $\rho(t_0) > 0$ (possibly depending upon t_0) such that $\|x(t_0) - x_0(t_0)\| < \rho$ implies*

$$\|x(t) - x_0(t)\| \to 0 \text{ as } t \to \infty.$$

Thus asymptotic stability implies, in addition to stability in the sense of Lyapunov, that the solution always approaches the nominal solution, provided the initial deviation is within the region defined by

$$\|x(t_0) - x_0(t_0)\| < \rho.$$

Asymptotic stability does not always give information for large initial deviations from the nominal solution. The following definition refers to the case of arbitrary initial deviations.

Definition 1.3. *The nominal solution $x_0(t)$ of the state differential equation*

$$\dot{x}(t) = f[x(t), t] \qquad \text{1-125}$$

*is **asymptotically stable in the large** if*

(a) *It is stable in the sense of Lyapunov;*
(b) *For any $x(t_0)$ and any t_0*

$$\|x(t) - x_0(t)\| \to 0 \qquad \text{1-126}$$

as $t \to \infty$.

A solution that is asymptotically stable in the large has therefore the property that all other solutions eventually approach it.

So far we have discussed only the stability of *solutions*. For nonlinear systems this is necessary because of the complex phenomena that may occur. In the case of linear systems, however, the situation is simpler, and we find it convenient to speak of the *stability of systems* rather than that of solutions. To make this point clear, let $x_0(t)$ be any nominal solution of the linear differential system

$$\dot{x}(t) = A(t)x(t), \qquad \text{1-127}$$

and denote by $x(t)$ any other solution of **1-127**. Since both $x_0(t)$ and $x(t)$ are solutions of the linear state differential equation **1-127** $x(t) - x_0(t)$ is also a solution, that is,

$$\frac{d}{dt}[x(t) - x_0(t)] = A(t)[x(t) - x_0(t)]. \qquad \textbf{1-128}$$

This shows that in order to study the stability of the nominal solution $x_0(t)$, we may as well study the stability of the zero solution, that is, the solution $x(t) \equiv 0$. If the zero solution is stable in any sense (of Lyapunov, asymptotically or asymptotically in the large), any other solution will also be stable in that sense. We therefore introduce the following terminology.

Definition 1.4. *The linear differential system*

$$\dot{x}(t) = A(t)x(t) \qquad \textbf{1-129}$$

is stable in a certain sense (of Lyapunov, asymptotically or asymptotically in the large), if the zero solution $x_0(t) \equiv 0$ is stable in that sense.

In addition to the fact that all nominal solutions of a linear differential system exhibit the same stability properties, for linear systems there is no need to make a distinction between asymptotic stability and asymptotic stability in the large as stated in the following theorem.

Theorem 1.12. *The linear differential system*

$$\dot{x}(t) = A(t)x(t) \qquad \textbf{1-130}$$

is asymptotically stable if and only if it is asymptotically stable in the large.

This theorem follows from the fact that for linear systems solutions may be scaled up or down without changing their behavior.

We conclude this section by introducing another form of stability, which we define only for linear systems (Brockett, 1970).

Definition 1.5. *The linear time-varying differential system*

$$\dot{x}(t) = A(t)x(t) \qquad \textbf{1-131}$$

*is **exponentially stable** if there exist positive constants α and β such that*

$$\|x(t)\| \leq \alpha e^{-\beta(t-t_0)} \|x(t_0)\|, \qquad t \geq t_0, \qquad \textbf{1-132}$$

for every initial state $x(t_0)$.

A system that is exponentially stable has the property that the state converges exponentially to the zero state irrespective of the initial state.

We clarify the concepts introduced in this section by some examples.

Example 1.6. *Inverted pendulum.*

The equilibrium position $s(t) \equiv 0$, $\phi(t) \equiv 0$, $\mu(t) \equiv 0$ of the inverted pendulum of Example 1.1 (Section 1.2.3) obviously is not stable in any sense.

Example 1.7. *Suspended pendulum.*

Consider the pendulum discussed in Example 1.1 (Section 1.2.3). Suppose that $\mu(t) \equiv 0$. From physical considerations it is clear that the solution $s(t) \equiv 0$, $\phi(t) \equiv \pi$ (corresponding to a suspended pendulum) is stable in the sense of Lyapunov; by choosing sufficiently small initial offsets and velocities, the motions of the system can be made to remain arbitrarily small. The system is not asymptotically stable, however, since no friction is assumed for the pendulum; once it is in motion, it remains in motion. Moreover, if the carriage has an initial displacement, it will not return to the zero position without an external force.

Example 1.8. *Stirred tank.*

Consider the stirred tank of Example 1.2 (Section 1.2.3). For $u(t) \equiv 0$ the linearized system is described by

$$\dot{x}(t) = \begin{pmatrix} -\dfrac{1}{2\theta} & 0 \\ 0 & -\dfrac{1}{\theta} \end{pmatrix} x(t), \qquad \text{1-133}$$

which has the solution

$$\begin{aligned} \xi_1(t) &= e^{-t/2\theta} \xi_1(0), & t \geq 0, \\ \xi_2(t) &= e^{-t/\theta} \xi_2(0), & t \geq 0. \end{aligned} \qquad \text{1-134}$$

Obviously $\xi_1(t)$ and $\xi_2(t)$ always approach the value zero as t increases since $\theta > 0$. As a result, the linearized system is asymptotically stable. Moreover, since the convergence to the equilibrium state is exponential, the system is exponentially stable.

In Section 1.4.4 it is seen that if a linearized system is asymptotically stable then the equilibrium state about which the linearization is performed is asymptotically stable but not necessarily asymptotically stable in the large. Physical considerations, however, lead us to expect that in the present case the system is also asymptotically stable in the large.

1.4.2 Stability of Time-Invariant Linear Systems

In this section we establish under what conditions time-invariant linear systems possess any of the forms of stability we have discussed. Consider the system

$$\dot{x}(t) = Ax(t), \qquad \text{1-135}$$

where A is a constant $n \times n$ matrix. In Section 1.3.3 we have seen that if A has n distinct characteristic values $\lambda_1, \lambda_2, \cdots, \lambda_n$ and corresponding characteristic vectors $e_1, e_2, \cdots, e_n$, the response of the system to any initial state can be represented as

$$x(t) = \sum_{i=1}^{n} \mu_i e^{\lambda_i t} e_i, \qquad \text{1-136}$$

where the scalars μ_i, $i = 1, 2, \cdots, n$ follow from the initial state $x(0)$. For systems with nondiagonizable A, this expression contains additional terms of the form $t^k \exp(\lambda_i t)$ (Section 1.3.4). Clearly, the stability of the system in both cases is determined by the characteristic values λ_i. We have the following result.

Theorem 1.13. *The time-invariant linear system*

$$\dot{x}(t) = Ax(t) \qquad \text{1-137}$$

is stable in the sense of Lyapunov if and only if

(a) *all of the characteristic values of A have nonpositive real parts, and*
(b) *to any characteristic value on the imaginary axis with multiplicity m there correspond exactly m characteristic vectors of the matrix A.*

Condition (b) is necessary to prevent terms that grow as t^k (see Section 1.3.4). This condition is always satisfied if A has no multiple characteristic values on the imaginary axis. For asymptotic stability we need slightly stronger conditions.

Theorem 1.14. *The time-invariant system*

$$\dot{x}(t) = Ax(t) \qquad \text{1-138}$$

is asymptotically stable if and only if all of the characteristic values of A have strictly negative real parts.

This result is also easily recognized to be valid. We furthermore see that if a time-invariant linear system is asymptotically stable the convergence of the state to the zero state is exponential. This results in the following theorem.

Theorem 1.15. *The time-invariant system*

$$\dot{x}(t) = Ax(t) \qquad \text{1-139}$$

is exponentially stable if and only if it is asymptotically stable.

Since it is really the matrix A that determines whether a time-invariant system is asymptotically stable, it is convenient to use the following terminology.

Definition 1.6. *The $n \times n$ constant matrix A is asymptotically stable if all its characteristic values have strictly negative real parts.*

The characteristic values of A are the roots of the characteristic polynomial $\det(\lambda I - A)$. Through the well-known Routh–Hurwitz criterion (see, e.g.,

Schwarz and Friedland, 1965) the stability of A can be tested directly from the coefficients of the characteristic polynomial without explicitly evaluating the roots. With systems that are not asymptotically stable, we find it convenient to refer to those characteristic values of A that have strictly negative real parts as the *stable poles* of the system, and to the remaining ones as the *unstable poles*.

We conclude this section with a simple example. An additional example is given in Section 1.5.1.

Example 1.9. *Stirred tank.*

The matrix A of the linearized state differential equation of the stirred tank of Example 1.2 has the characteristic values $-(1/2\theta)$ and $-(1/\theta)$. As we concluded before (Example 1.8), the linearized system is asymptotically stable since $\theta > 0$.

1.4.3* Stable and Unstable Subspaces for Time-Invariant Linear Systems

In this section we show how the state space of a linear time-invariant differential system can be decomposed into two subspaces, such that the response of the system from an initial state in the first subspace always converges to the zero state while the response from a nonzero initial state in the other subspace never converges.

Let us consider the time-invariant system

$$\dot{x}(t) = Ax(t) \qquad \textbf{1-140}$$

and assume that the matrix A has distinct characteristic values (the more general case is discussed later in this section). Then we know from Section 1.3.3 that the response of this system can be written as

$$x(t) = \sum_{i=1}^{n} \mu_i e^{\lambda_i t} e_i, \qquad \textbf{1-141}$$

where $\lambda_1, \lambda_2, \cdots, \lambda_n$ are the characteristic values of A, and $e_1, \cdots, e_n$ are the corresponding characteristic vectors. The numbers $\mu_1, \mu_2, \cdots, \mu_n$ are the coefficients that express how the initial state $x(0)$ is decomposed along the vectors $e_1, e_2, \cdots, e_n$.

Let us now suppose that the system is not asymptotically stable, which means that some of the characteristic values λ_i have nonnegative real parts. Then it is clear that the state will converge to the zero state only if the initial state has components only along those characteristic vectors that correspond to stable poles.

If the initial state has components only along the characteristic vectors that correspond to unstable poles, the response of the state will be composed of nondecreasing exponentials. This leads to the following decomposition of the state space.

30 Elements of Linear System Theory

Definition 1.7. *Consider the n-dimensional system $\dot{x}(t) = Ax(t)$ with A a constant matrix. Suppose that A has n distinct characteristic values. Then we define the **stable subspace** for this system as the real linear subspace spanned by those characteristic vectors of A that correspond to characteristic values with strictly negative real parts. The **unstable subspace** for this system is the real subspace spanned by those characteristic vectors of A that correspond to characteristic values with nonnegative real parts.*

We now extend this concept to more general time-invariant systems. In Section 1.3.4 we saw that the response of the system can be written as

$$x(t) = \sum_{i=1}^{k} T_i \exp(J_i t) U_i v_i, \qquad \text{1-142}$$

where the v_i are in the null spaces $\mathcal{N}_i$, $i = 1, 2, \cdots, k$. The behavior of the factor $\exp(J_i t)$ is determined by the characteristic value λ_i; only if λ_i has a strictly negative real part does the corresponding component of the state approach the zero state. This leads us in analogy with the simple case of Definition 1.7 to the following decomposition:

Definition 1.8. *Consider the n-dimensional linear time-invariant system $\dot{x}(t) = Ax(t)$. Then we define the **stable subspace** for this system as the real subspace of the direct sum of those null spaces $\mathcal{N}_i$ that correspond to characteristic values of A with strictly negative real parts. Similarly, we define the **unstable subspace** of A as the real subspace of the direct sum of those null spaces $\mathcal{N}_i$ that correspond to characteristic values of A with nonnegative real parts.*

As a result of this definition the whole real n-dimensional space $\mathcal{R}^n$ is the direct sum of the stable and the unstable subspace.

Example 1.10. *Inverted pendulum.*

In Example 1.4 (Section 1.3.3), we saw that the matrix A of the linearized state differential equation of the inverted pendulum has the characteristic values and vectors:

$$\lambda_1 = 0, \quad \lambda_2 = -\frac{F}{M}, \quad \lambda_3 = \sqrt{\frac{g}{L'}}, \quad \lambda_4 = -\sqrt{\frac{g}{L'}}, \qquad \text{1-143}$$

$$e_1 = \begin{pmatrix} 1 \\ 0 \\ 1 \\ 0 \end{pmatrix}, \quad e_2 = \begin{pmatrix} 1 \\ -\dfrac{F}{M} \\ \alpha \\ -\alpha\dfrac{F}{M} \end{pmatrix}, \quad e_3 = \begin{pmatrix} 0 \\ 0 \\ 1 \\ \sqrt{\dfrac{g}{L'}} \end{pmatrix}, \quad e_4 = \begin{pmatrix} 0 \\ 0 \\ 1 \\ -\sqrt{\dfrac{g}{L'}} \end{pmatrix}.$$

$$\text{1-144}$$

Apparently, the stable subspace of this system is spanned by the vectors e_2 and e_4, while the unstable subspace is spanned by e_1 and e_3.

Example 1.11. *Inverted pendulum without friction.*

In Example 1.5 (Section 1.3.4), we discussed the Jordan normal form of the A matrix of the inverted pendulum with negligible friction. There we found a double characteristic value 0 and the single characteristic values $\sqrt{(g/L')}$ and $-\sqrt{(g/L')}$. The null space corresponding to the characteristic value 0 is spanned by the first two columns of the transformation matrix T, that is, by

$$\begin{pmatrix} 1 \\ 0 \\ 1 \\ 0 \end{pmatrix} \quad \text{and} \quad \begin{pmatrix} 0 \\ 1 \\ 0 \\ 1 \end{pmatrix}. \qquad \text{1-145}$$

These two column vectors, together with the characteristic vector corresponding to $\sqrt{(g/L')}$, that is,

$$\begin{pmatrix} 0 \\ 0 \\ 1 \\ \sqrt{\frac{g}{L'}} \end{pmatrix}, \qquad \text{1-146}$$

span the unstable subspace of the system. The stable subspace is spanned by the remaining characteristic vector

$$\begin{pmatrix} 0 \\ 0 \\ 1 \\ -\sqrt{\frac{g}{L'}} \end{pmatrix}. \qquad \text{1-147}$$

1.4.4* Investigation of the Stability of Nonlinear Systems through Linearization

Most of the material of this book is concerned with the design of linear control systems. One major goal in the design of such systems is stability. In later chapters very powerful techniques for finding stable linear feedback control systems are developed. As we have seen, however, actual systems are never linear, and the linear models used are obtained by linearization.

This means that we design systems whose linearized models possess good properties. The question now is: What remains of these properties when the actual nonlinear system is implemented? Here the following result is helpful.

Theorem 1.16. *Consider the time-invariant system with state differential equation*

$$\dot{x}(t) = f[x(t)]. \qquad \textbf{1-148}$$

Suppose that the system has an equilibrium state x_e and that the function f possesses partial derivatives with respect to the components of x at x_e. Suppose that the linearized state differential equation about x_e is

$$\dot{\tilde{x}}(t) = A\tilde{x}(t), \qquad \textbf{1-149}$$

where the constant matrix A is the Jacobian of f at x_e. Then if A is asymptotically stable, the solution $x(t) = x_e$ is an asymptotically stable solution of **1-148**.

For a proof we refer the reader to Roseau (1966). Note that of course we cannot conclude anything about stability in the large from the linearized state differential equation.

This theorem leads to a reassuring conclusion. Suppose that we are confronted with an initially unstable system, and that we use linearized equations to find a controller that makes the linearized system stable. Then it can be shown from the theorem that the actual nonlinear system with this controller will at least be asymptotically stable for small deviations from the equilibrium state.

Note, however, that the theorem is reassuring only when the system contains "smooth" nonlinearities. If discontinuous elements occur (dead zones, stiction) this theory is of no help.

We conclude by noting that if some of the characteristic values of A have zero real parts while all the other characteristic values have strictly negative real parts no conclusions about the stability of x_e can be drawn from the linearized analysis. If A has some characteristic values with positive real parts, however, x_e is not stable in any sense (Roseau, 1966).

An example of the application of this theorem is given in Chapter 2 (Example 2.6, Section 2.4).

1.5 TRANSFORM ANALYSIS OF TIME-INVARIANT SYSTEMS

1.5.1 Solution of the State Differential Equation through Laplace Transformation

Often it is helpful to analyze time-invariant linear systems through Laplace transformation. We define the Laplace transform of a time-varying vector $z(t)$ as follows

$$\mathbf{Z}(s) = \mathscr{L}[z(t)] = \int_0^\infty e^{-st} z(t) \, dt, \qquad \text{1-150}$$

where s is a complex variable. A boldface capital indicates the Laplace transform of the corresponding lowercase time function. The Laplace transform is defined for those values of s for which **1-150** converges. We see that the Laplace transform of a time-varying vector $z(t)$ is simply a vector whose components are the Laplace transforms of the components of $z(t)$.

Let us first consider the homogeneous state differential equation

$$\dot{x}(t) = Ax(t), \qquad \text{1-151}$$

where A is a constant matrix. Laplace transformation yields

$$s\mathbf{X}(s) - x(0) = A\mathbf{X}(s), \qquad \text{1-152}$$

since all the usual rules of Laplace transformations for scalar expressions carry over to the vector case (Polak and Wong, 1970). Solution for $\mathbf{X}(s)$ yields

$$\mathbf{X}(s) = (sI - A)^{-1} x(0). \qquad \text{1-153}$$

This is the equivalent of the time domain expression

$$x(t) = e^{At} x(0). \qquad \text{1-154}$$

We conclude the following.

Theorem 1.17. *Let A be a constant $n \times n$ matrix. Then $(sI - A)^{-1} = \mathscr{L}[e^{At}]$, or, equivalently, $e^{At} = \mathscr{L}^{-1}[(sI - A)^{-1}]$.*

The Laplace transform of a time-varying matrix is obtained by transforming each of its elements. Theorem 1.17 is particularly convenient for obtaining the explicit form of the transition matrix as long as n is not too large, irrespective of whether or not A is diagonalizable.

The matrix function $(sI - A)^{-1}$ is called the *resolvent* of A. The following result is useful (Zadeh and Desoer, 1963; Bass and Gura, 1965).

Theorem 1.18. *Consider the constant $n \times n$ matrix A with characteristic polynomial*

$$\det(sI - A) = s^n + \alpha_{n-1}s^{n-1} + \cdots + \alpha_1 s + \alpha_0. \qquad \text{1-155}$$

Then the resolvent of A can be written as

$$(sI - A)^{-1} = \frac{1}{\det(sI - A)} \sum_{i=1}^{n} s^{i-1} R_i, \qquad \text{1-156}$$

where the matrices R_i are given by

$$R_i = \sum_{j=i}^{n} \alpha_j A^{j-i}, \qquad i = 1, 2, \cdots, n, \qquad \text{1-157}$$

with $\alpha_n = 1$. The coefficients α_i and the matrices R_i, $i = 1, 2, \cdots, n$ can be obtained through the following algorithm. Set

$$\alpha_n = 1, \qquad R_n = I. \qquad \text{1-158}$$

Then

$$\alpha_{n-k} = -\frac{1}{k} \operatorname{tr}(AR_{n-k+1}), \qquad \text{1-159}$$

$$R_{n-k} = \alpha_{n-k} I + AR_{n-k+1}, \qquad \text{1-160}$$

for $k = 1, 2, \cdots, n$. For $k = n$ we have

$$R_0 = 0. \qquad \text{1-161}$$

Here we have employed the notation

$$\operatorname{tr}(M) = \sum_{i=1}^{n} M_{ii}, \qquad \text{1-162}$$

if M is an $n \times n$ matrix with diagonal elements M_{ii}, $i = 1, 2, \cdots, n$. We refer to the algorithm of the theorem as *Leverrier's algorithm* (Bass and Gura, 1965). It is also known as *Souriau's method* or *Faddeeva's method* (Zadeh and Desoer, 1963). The fact that $R_0 = 0$ can be used as a numerical check. The algorithm is very convenient for a digital computer. It must be pointed out, however, that the algorithm is relatively sensitive to round-off errors (Forsythe and Strauss, 1955), and double precision is usually employed in the computations. Melsa (1970) gives a listing of a FORTRAN computer program.

Let us now consider the inhomogeneous equation

$$\dot{x}(t) = Ax(t) + Bu(t), \qquad \text{1-163}$$

where A and B are constant. Laplace transformation yields

$$sX(s) - x(0) = AX(s) + BU(s), \qquad \text{1-164}$$

which can be solved for $\mathbf{X}(s)$. We find
$$\mathbf{X}(s) = (sI - A)^{-1}x(0) + (sI - A)^{-1}B\mathbf{U}(s). \qquad \textbf{1-165}$$
Let the output equation of the system be given by
$$y(t) = Cx(t), \qquad \textbf{1-166}$$
where C is constant. Laplace transformation and substitution of **1-165** yields
$$\mathbf{Y}(s) = C\mathbf{X}(s) = C(sI - A)^{-1}x(0) + C(sI - A)^{-1}B\mathbf{U}(s), \qquad \textbf{1-167}$$
which is the equivalent in the Laplace transform domain of the time domain expression **1-70** with $t_0 = 0$:
$$y(t) = Ce^{At}x(0) + C\int_0^t e^{A(t-\tau)}Bu(\tau)\,d\tau. \qquad \textbf{1-168}$$
For $x(0) = 0$ the expression **1-167** reduces to
$$\mathbf{Y}(s) = H(s)\mathbf{U}(s), \qquad \textbf{1-169}$$
where
$$H(s) = C(sI - A)^{-1}B. \qquad \textbf{1-170}$$

The matrix $H(s)$ is called the *transfer matrix* of the system. If $H(s)$ and $\mathbf{U}(s)$ are known, the zero initial state response of the system can be found by inverse Laplace transformation of **1-169**.

By Theorem 1.17 it follows immediately from **1-170** that the transfer matrix $H(s)$ is the Laplace transform of the matrix function $K(t) = C \exp(At)B$, $t \geq 0$. It is seen from **1-168** that $K(t - \tau)$, $t \geq \tau$, is precisely the impulse response matrix of the system.

From Theorem 1.18 we note that the transfer matrix can be written in the form
$$H(s) = \frac{1}{\det(sI - A)}P(s), \qquad \textbf{1-171}$$
where $P(s)$ is a matrix whose elements are polynomials in s. The elements of the transfer matrix $H(s)$ are therefore rational functions of s. The common denominator of the elements of $H(s)$ is $\det(sI - A)$, *unless* cancellation occurs of factors of the form $s - \lambda_i$, where λ_i is a characteristic value of A, in *all* the elements of $H(s)$.

We call the roots of the common denominator of $H(s)$ the *poles of the transfer matrix* $H(s)$. If no cancellation occurs, the poles of the transfer matrix are precisely the poles of the system, that is, the characteristic values of A.

If the input $u(t)$ and the output variable $y(t)$ are both one-dimensional, the transfer matrix reduces to a scalar *transfer function*. For multiinput multioutput systems, each element $H_{ij}(s)$ of the transfer matrix $H(s)$ is the transfer function from the j-th component of the input to the i-th component of the output.

36   Elements of Linear System Theory

Example 1.12.   *A nondiagonizable system.*
Consider the system

$$\dot{x}(t) = \begin{pmatrix} 0 & 1 \\ 0 & 0 \end{pmatrix} x(t). \qquad 1\text{-}172$$

It is easily verified that this system has a double characteristic value 0 but only a single characteristic vector, so that it is not diagonizable. We compute its transition matrix by Laplace transformation. The resolvent of the system can be found to be

$$(sI - A)^{-1} = \frac{1}{s^2} \begin{pmatrix} s & 1 \\ 0 & s \end{pmatrix} = \begin{pmatrix} \frac{1}{s} & \frac{1}{s^2} \\ 0 & \frac{1}{s} \end{pmatrix}. \qquad 1\text{-}173$$

Inverse Laplace transformation yields

$$e^{At} = \begin{pmatrix} 1 & t \\ 0 & 1 \end{pmatrix}. \qquad 1\text{-}174$$

Note that this system is not stable in the sense of Lyapunov.

Example 1.13.   *Stirred tank.*
The stirred tank of Example 1.2 is described by the linearized state differential equation

$$\dot{x}(t) = \begin{pmatrix} -\frac{1}{2\theta} & 0 \\ 0 & -\frac{1}{\theta} \end{pmatrix} x(t) + \begin{pmatrix} 1 & 1 \\ \dfrac{c_1 - c_0}{V_0} & \dfrac{c_2 - c_0}{V_0} \end{pmatrix} u(t) \qquad 1\text{-}175$$

and the output equation

$$y(t) = \begin{pmatrix} \frac{1}{2\theta} & 0 \\ 0 & 1 \end{pmatrix} x(t). \qquad 1\text{-}176$$

The resolvent of the matrix A is

$$(sI - A)^{-1} = \begin{pmatrix} \dfrac{1}{s + \dfrac{1}{2\theta}} & 0 \\ 0 & \dfrac{1}{s + \dfrac{1}{\theta}} \end{pmatrix}. \qquad 1\text{-}177$$

The system has the transfer matrix

$$H(s) = \begin{pmatrix} \dfrac{\dfrac{1}{2\theta}}{s + \dfrac{1}{2\theta}} & \dfrac{\dfrac{1}{2\theta}}{s + \dfrac{1}{2\theta}} \\[2ex] \dfrac{c_1 - c_0}{V_0} \dfrac{1}{s + \dfrac{1}{\theta}} & \dfrac{c_2 - c_0}{V_0} \dfrac{1}{s + \dfrac{1}{\theta}} \end{pmatrix}. \qquad \textbf{1-178}$$

The impulse response matrix **1-75** of the system follows immediately by inverse Laplace transformation of **1-178**.

1.5.2 Frequency Response

In this section we study the frequency response of time-invariant systems, that is, the response to an input of the form

$$u(t) = u_m e^{j\omega t}, \qquad t \geq 0, \qquad \textbf{1-179}$$

where u_m is a constant vector. We express the solution of the state differential equation

$$\dot{x}(t) = Ax(t) + Bu(t) \qquad \textbf{1-180}$$

in terms of the solution of the homogeneous equation plus a particular solution. Let us first try to find a particular solution of the form

$$x_p(t) = x_m e^{j\omega t}, \qquad \textbf{1-181}$$

where x_m is a constant vector to be determined. It is easily found that this particular solution is given by

$$x_p(t) = (j\omega I - A)^{-1} B u_m e^{j\omega t}, \qquad t \geq 0. \qquad \textbf{1-182}$$

The general solution of the homogeneous equation $\dot{x}(t) = Ax(t)$ can be written as

$$x_h(t) = e^{At} a, \qquad \textbf{1-183}$$

where a is an arbitrary constant vector. The general solution of the inhomogeneous equation **1-180** is therefore

$$x(t) = e^{At} a + (j\omega I - A)^{-1} B u_m e^{j\omega t}, \qquad t \geq 0. \qquad \textbf{1-184}$$

The constant vector a can be determined from the initial conditions. If the system **1-180** is *asymptotically stable*, the first term of the solution will eventually vanish as t increases, and the second term represents the *steady-state response* of the state to the input **1-179**. The corresponding steady-state

response of the output
$$y(t) = Cx(t) \qquad \text{1-185}$$
is given by
$$\begin{aligned} y(t) &= C(j\omega I - A)^{-1} B u_m e^{j\omega t} \\ &= H(j\omega) u_m e^{j\omega t}. \end{aligned} \qquad \text{1-186}$$

We note that in this expression the transfer matrix $H(s)$ appears with s replaced by $j\omega$. We call $H(j\omega)$ the *frequency response matrix* of the system.

Once we have obtained the response to complex periodic inputs of the type **1-179**, the steady-state response to real, sinusoidal inputs is easily found. Suppose that the k-th component $\mu_k(t)$ of the input $u(t)$ is given as follows

$$\mu_k(t) = \hat{\mu}_k \sin(\omega t + \phi_k), \qquad t \geq 0. \qquad \text{1-187}$$

Assume that all other components of the input are identically zero. Then the steady-state response of the i-th component $\eta_i(t)$ of the output $y(t)$ is given by

$$\eta_i(t) = |H_{ik}(j\omega)| \hat{\mu}_k \sin(\omega t + \phi_k + \psi_{ik}), \qquad \text{1-188}$$

where $H_{ik}(j\omega)$ is the (i, k)-th element of $H(j\omega)$ and

$$\psi_{ik} = \arg[H_{ik}(j\omega)]. \qquad \text{1-189}$$

A convenient manner of representing scalar frequency response functions is through asymptotic Bode plots (D'Azzo and Houpis, 1966). Melsa (1970) gives a FORTRAN computer program for plotting the modulus and the argument of a scalar frequency response function.

In conclusion, we remark that it follows from the results of this section that the steady-state response of an asymptotically stable system with frequency response matrix $H(j\omega)$ to a *constant* input

$$u(t) = u_m \qquad \text{1-190}$$

is given by

$$y(t) = H(0) u_m. \qquad \text{1-191}$$

Example 1.14. *Stirred tank.*

The stirred tank of Example 1.2 has the transfer matrix (Example 1.13)

$$H(s) = \begin{pmatrix} \dfrac{\frac{1}{2\theta}}{s + \frac{1}{2\theta}} & \dfrac{\frac{1}{2\theta}}{s + \frac{1}{2\theta}} \\ \dfrac{c_1 - c_0}{V_0} \dfrac{1}{s + \frac{1}{\theta}} & \dfrac{c_2 - c_0}{V_0} \dfrac{1}{s + \frac{1}{\theta}} \end{pmatrix}. \qquad \text{1-192}$$

The system is asymptotically stable so that it makes sense to consider the frequency response matrix. With the numerical data of Example 1.2, we have

$$H(j\omega) = \begin{pmatrix} \dfrac{0.01}{j\omega + 0.01} & \dfrac{0.01}{j\omega + 0.01} \\ \dfrac{-0.25}{j\omega + 0.02} & \dfrac{0.75}{j\omega + 0.02} \end{pmatrix}. \qquad \text{1-193}$$

1.5.3 Zeroes of Transfer Matrices

Let us consider the single-input single-output system

$$\dot{x}(t) = Ax(t) + b\mu(t),$$
$$\eta(t) = cx(t), \qquad \text{1-194}$$

where $\mu(t)$ and $\eta(t)$ are the scalar input and output variable, respectively, b is a column vector, and c a row vector. The transfer matrix of this system reduces to a transfer function which is given by

$$H(s) = c(sI - A)^{-1}b. \qquad \text{1-195}$$

Denote the characteristic polynomial of A as

$$\det(sI - A) = \phi(s). \qquad \text{1-196}$$

Then $H(s)$ can be written as

$$H(s) = \dfrac{\psi(s)}{\phi(s)}, \qquad \text{1-197}$$

where, if A is an $n \times n$ matrix, then $\phi(s)$ is a polynomial of degree n and $\psi(s)$ a polynomial of degree $n - 1$ or less. The roots of $\psi(s)$ we call the *zeroes of the system* **1-194**. Note that we determine the zeroes *before* cancelling any common factors of $\psi(s)$ and $\phi(s)$. The zeroes of $H(s)$ that remain *after* cancellation we call the *zeroes of the transfer function*.

In the case of a multiinput multioutput system, $H(s)$ is a matrix. Each entry of $H(s)$ is a transfer function which has its own zeroes. It is not obvious how to define "the zeroes of $H(s)$" in this case. In the remainder of this section we give a definition that is motivated by the results of Section 3.8. Only square transfer matrices are considered.

First we have the following result (Haley, 1967).

Theorem 1.19. *Consider the system*

$$\dot{x}(t) = Ax(t) + Bu(t),$$
$$y(t) = Cx(t), \qquad \text{1-198}$$

where the state x has dimension n and both the input u and the output variable y have dimension m. Let $H(s) = C(sI - A)^{-1}B$ be the transfer matrix of the system. Then

$$\det [H(s)] = \frac{\psi(s)}{\phi(s)}, \qquad \text{1-199}$$

where

$$\phi(s) = \det (sI - A), \qquad \text{1-200}$$

and $\psi(s)$ is a polynomial in s of degree $n - m$ or less.

Since this result is not generally known we shall prove it. We first state the following fact from matrix theory.

Lemma 1.1. *Let M and N be matrices of dimensions $m \times n$ and $n \times m$, respectively, and let I_m and I_n denote unit matrices of dimensions $m \times m$ and $n \times n$. Then*

(a) $$\det (I_m + MN) = \det (I_n + NM). \qquad \text{1-201}$$

(b) *Suppose* $\det (I_m + MN) \neq 0$; *then*

$$(I_m + MN)^{-1} = I_m - M(I_n + NM)^{-1}N. \qquad \text{1-202}$$

The proof of (a) follows from considerations involving the characteristic values of $I_m + MN$ (Plotkin, 1964; Sain, 1966). Part (b) is easily verified. It is not needed until later.

To prove Theorem 1.19 consider the expression

$$\det [\lambda I_m + C(sI_n - A)^{-1}B], \qquad \text{1-203}$$

where λ is a nonzero arbitrary scalar which later we let approach zero. Using part (a) of the lemma, we have

$$\det [\lambda I_m + C(sI_n - A)^{-1}B] = \det (\lambda I_m) \det \left[I_m + \frac{1}{\lambda} C(sI_n - A)^{-1}B\right]$$

$$= \lambda^m \det \left[I_n + \frac{1}{\lambda} (sI_n - A)^{-1}BC\right]$$

$$= \frac{\lambda^m \det \left[sI_n - A + \frac{1}{\lambda} BC\right]}{\det (sI_n - A)}. \qquad \text{1-204}$$

We see that the left-hand and the right-hand side of **1-204** are polynomials in λ that are equal for all nonzero λ; hence by letting $\lambda \to 0$ we obtain

$$\det [C(sI - A)^{-1}B] = \frac{\psi(s)}{\phi(s)}, \qquad \text{1-205}$$

where
$$\psi(s) = \lim_{\lambda \to 0} \lambda^m \det\left(sI_n - A + \frac{1}{\lambda}BC\right). \quad \text{1-206}$$

This immediately shows that $\psi(s)$ is a polynomial in s. We now consider the degree of this polynomial. For $|s| \to \infty$ we see from Theorem 1.18 that
$$\lim_{|s| \to \infty} s(sI - A)^{-1} = I. \quad \text{1-207}$$
Consequently,
$$\lim_{|s| \to \infty} \frac{s^m \psi(s)}{\phi(s)} = \lim_{|s| \to \infty} s^m \det[C(sI - A)^{-1}B]$$
$$= \lim_{|s| \to \infty} \det[Cs(sI - A)^{-1}B] = \det(CB). \quad \text{1-208}$$

This shows that the degree of $\phi(s)$ is greater than that of $\psi(s)$ by at least m, hence $\psi(s)$ has degree $n - m$ or less. If $\det(CB) \neq 0$, the degree of $\psi(s)$ is exactly $n - m$. This terminates the proof of Theorem 1.19.

We now introduce the following definition.

Definition 1.9. *The zeroes of the system*
$$\dot{x}(t) = Ax(t) + Bu(t),$$
$$y(t) = Cx(t), \quad \text{1-209}$$

where the state x has dimension n and both the input u and the output y have dimension m, are the zeroes of the polynomial $\psi(s)$, where
$$\det[H(s)] = \frac{\psi(s)}{\phi(s)}. \quad \text{1-210}$$

Here $H(s) = C(sI - A)^{-1}B$ is the transfer matrix and $\phi(s) = \det(sI - A)$ the characteristic polynomial of the system.

An n-dimensional system with m-dimensional input and output thus has at most $n - m$ zeroes. Note that for single-input single-output systems our definition of the zeroes of the system reduces to the conventional definition as described in the beginning of this section. In this case the system has at most $n - 1$ zeroes.

The numerical computation of the numerator polynomial for a system of some complexity presents problems. One possible way of going about this is to write the numerator polynomial as
$$\psi(s) = \phi(s) \det[H(s)], \quad \text{1-211}$$

where $\phi(s)$ is the characteristic polynomial of the system. The coefficients of $\psi(s)$ can then be found by substituting $n - m + 1$ suitable values for s into

42 Elements of Linear System Theory

the right-hand side of **1-211** and solving the resulting linear equations. Another, probably more practical, approach results from using the fact that from **1-206** we have

$$\psi(s) = \lim_{\lambda \to 0} \psi(s, \lambda),\qquad \text{1-212}$$

where

$$\psi(s, \lambda) = \lambda^m \det\left(sI - A + \frac{1}{\lambda}BC\right).\qquad \text{1-213}$$

Inspection shows that we can write

$$\psi(s, \lambda) = \sum_{i=0}^{m} \lambda^i \alpha_i(s),\qquad \text{1-214}$$

where $\alpha_i(s)$, $i = 0, 1, \cdots, m$, are polynomials in s. These polynomials can be computed by calculating $\psi(s, \lambda)$ for m different values of λ. The desired polynomial $\psi(s)$ is precisely $\alpha_0(s)$.

We illustrate the results of this section by the following example.

Example 1.15. *Stirred tank.*

The stirred tank of Example 1.2 (Section 1.2.3) has the transfer matrix

$$H(s) = \begin{pmatrix} \dfrac{\frac{1}{2\theta}}{s + \frac{1}{2\theta}} & \dfrac{\frac{1}{2\theta}}{s + \frac{1}{2\theta}} \\ \dfrac{c_1 - c_0}{V_0}\dfrac{1}{s + \frac{1}{\theta}} & \dfrac{c_2 - c_0}{V_0}\dfrac{1}{s + \frac{1}{\theta}} \end{pmatrix}.\qquad \text{1-215}$$

The characteristic polynomial of the system is

$$\phi(s) = \left(s + \frac{1}{2\theta}\right)\left(s + \frac{1}{\theta}\right).\qquad \text{1-216}$$

We find for the determinant of the transfer matrix

$$\det[H(s)] = \dfrac{\dfrac{1}{2\theta}\dfrac{c_2 - c_1}{V_0}}{\left(s + \dfrac{1}{2\theta}\right)\left(s + \dfrac{1}{\theta}\right)}.\qquad \text{1-217}$$

Apparently, the transfer matrix has no zeroes. This is according to expectation, since in this case $n - m = 0$ so that the degree of $\psi(s)$ is zero.

1.5.4 Interconnections of Linear Systems

In this section we discuss interconnections of linear systems. Two important examples of interconnected systems that we frequently encounter are the *series connection* of Fig. 1.5 and the *feedback configuration* or *closed-loop system* of Fig. 1.6.

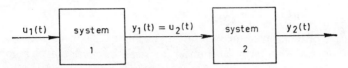

Fig. 1.5. Series connection.

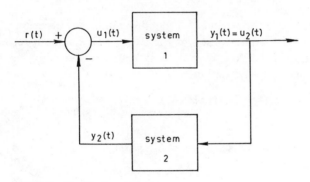

Fig. 1.6. Feedback connection.

We often describe interconnections of systems by the *state augmentation technique*. In the series connection of Fig. 1.5, let the individual systems be described by the state differential and output equations

$$\left.\begin{aligned}\dot{x}_1(t) &= A_1(t)x_1(t) + B_1(t)u_1(t)\\ y_1(t) &= C_1(t)x_1(t) + D_1(t)u_1(t)\end{aligned}\right\} \text{ system 1,} \quad \text{1-218}$$

$$\left.\begin{aligned}\dot{x}_2(t) &= A_2(t)x_2(t) + B_2(t)u_2(t)\\ y_2(t) &= C_2(t)x_2(t) + D_2(t)u_2(t)\end{aligned}\right\} \text{ system 2.}$$

Defining the *augmented state*

$$x(t) = \begin{pmatrix} x_1(t) \\ x_2(t) \end{pmatrix}, \quad \text{1-219}$$

the interconnected system is described by the state differential equation

$$\dot{x}(t) = \begin{pmatrix} A_1(t) & 0 \\ B_2(t)C_1(t) & A_2(t) \end{pmatrix} x(t) + \begin{pmatrix} B_1(t) \\ B_2(t)D_1(t) \end{pmatrix} u_1(t), \qquad \textbf{1-220}$$

where we have used the relation $u_2(t) = y_1(t)$. Taking $y_2(t)$ as the output of the interconnected system, we obtain for the output equation

$$y_2(t) = [D_2(t)C_1(t), C_2(t)]x(t) + D_2(t)D_1(t)u_1(t). \qquad \textbf{1-221}$$

In the case of time-invariant systems, it is sometimes convenient to describe an interconnection in terms of transfer matrices. Suppose that the individual transfer matrices of the systems 1 and 2 are given by $H_1(s)$ and $H_2(s)$, respectively. Then the overall transfer matrix is $H_2(s)H_1(s)$, as can be seen from

$$\mathbf{Y}_2(s) = H_2(s)\mathbf{U}_2(s) = H_2(s)H_1(s)\mathbf{U}_1(s). \qquad \textbf{1-222}$$

Note that the order of H_2 and H_1 generally cannot be interchanged.

In the feedback configuration of Fig. 1.6, $r(t)$ is the input to the overall system. Suppose that the individual systems are described by the state differential and output equations

$$\left.\begin{aligned}\dot{x}_1(t) &= A_1(t)x_1(t) + B_1(t)u_1(t) \\ y_1(t) &= C_1(t)x_1(t)\end{aligned}\right\} \text{ system 1},$$

$$\left.\begin{aligned}\dot{x}_2(t) &= A_2(t)x_2(t) + B_2(t)u_2(t) \\ y_2(t) &= C_2(t)x_2(t) + D_2(t)u_2(t)\end{aligned}\right\} \text{ system 2}. \qquad \textbf{1-223}$$

Note that we have taken system 1 without a direct link. This is to avoid implicit algebraic equations. In terms of the augmented state $x(t) = \text{col }[x_1(t), x_2(t)]$, the feedback connection can be described by the state differential equation

$$\dot{x}(t) = \begin{pmatrix} A_1(t) - B_1(t)D_2(t)C_1(t) & -B_1(t)C_2(t) \\ B_2(t)C_1(t) & A_2(t) \end{pmatrix} x(t) + \begin{pmatrix} B_1(t) \\ 0 \end{pmatrix} r(t), \qquad \textbf{1-224}$$

where we have used the relations $u_2(t) = y_1(t)$ and $u_1(t) = r(t) - y_2(t)$. If $y_1(t)$ is the overall output of the system, we have for the output equation

$$y_1(t) = [C_1(t), 0]x(t). \qquad \textbf{1-225}$$

Consider now the time-invariant case. Then we can write in terms of transfer matrices

$$\mathbf{Y}_1(s) = H_1(s)[\mathbf{R}(s) - H_2(s)\mathbf{Y}_1(s)], \qquad \textbf{1-226}$$

where $H_1(s)$ and $H_2(s)$ are the transfer matrices of the individual systems. Solving for $\mathbf{Y}_1(s)$, we find

$$\mathbf{Y}_1(s) = [I + H_1(s)H_2(s)]^{-1}H_1(s)\mathbf{R}(s). \qquad \text{1-227}$$

It is convenient to give the expression $I + H_1(s)H_2(s)$ a special name:

Definition 1.10. *Consider the feedback configuration of Fig. 1.6. and let the systems 1 and 2 be time-invariant systems with transfer matrices $H_1(s)$ and $H_2(s)$, respectively. Then the matrix function*

$$J(s) = I + H_1(s)H_2(s) \qquad \text{1-228}$$

is called the **return difference matrix.** *The matrix function*

$$L(s) = H_1(s)H_2(s) \qquad \text{1-229}$$

is called the **loop gain matrix.**

The term "return difference" can be clarified by Fig. 1.7. Here the loop is cut at the point indicated, and an external input variable $u_2(t)$ is connected.

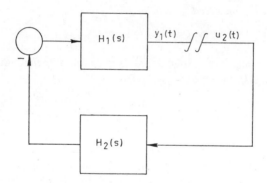

Fig. 1.7. Illustration of return difference.

This yields (putting $r(t) \equiv 0$)

$$\mathbf{Y}_1(s) = -H_1(s)H_2(s)\mathbf{U}_2(s). \qquad \text{1-230}$$

The difference between the "returned variable" $y_1(t)$ and the "injected variable" $u_2(t)$ is

$$\begin{aligned}\mathbf{U}_2(s) - \mathbf{Y}_1(s) &= [I + H_1(s)H_2(s)]\mathbf{U}_2(s) \\ &= J(s)\mathbf{U}_2(s).\end{aligned} \qquad \text{1-231}$$

Note that the loop can also be cut elsewhere, which will result in a different return difference matrix. We strictly adhere to the definition given above, however. The term "loop gain matrix" is self-explanatory.

A matter of great interest in control engineering is the stability of interconnections of systems. For series connections we have the following result, which immediately follows from a consideration of the characteristic polynomial of the augmented state differential equation **1-220**.

Theorem 1.20. *Consider the series connection of Fig. 1.5, where the systems 1 and 2 are time-invariant systems with characteristic polynomials $\phi_1(s)$ and $\phi_2(s)$, respectively. Then the interconnection has the characteristic polynomial $\phi_1(s)\phi_2(s)$. Hence the interconnected system is asymptotically stable if and only if both system 1 and system 2 are asymptotically stable.*

In terms of transfer matrices, the stability of the feedback configuration of Fig. 1.6 can be investigated through the following result (Chen, 1968a; Hsu and Chen, 1968).

Theorem 1.21. *Consider the feedback configuration of Fig. 1.6 in which the systems 1 and 2 are time-invariant linear systems with transfer matrices $H_1(s)$ and $H_2(s)$ and characteristic polynomials $\phi_1(s)$ and $\phi_2(s)$, respectively, and where system 1 does not have a direct link. Then the characteristic polynomial of the interconnected system is*

$$\phi_1(s)\phi_2(s) \det [I + H_1(s)H_2(s)]. \qquad \textbf{1-232}$$

Hence the interconnected system is stable if and only if the polynomial **1-232** *has zeroes with strictly negative real parts only.*

Before proving this result we remark the following. The expression $\det [I + H_1(s)H_2(s)]$ is a rational function in s. Unless cancellations take place, the denominator of this function is $\phi_1(s)\phi_2(s)$ so that the numerator of $\det [I + H_1(s)H_2(s)]$ is the characteristic polynomial of the interconnected system. We often refer to **1-232** as the *closed-loop characteristic polynomial*.

Theorem 1.21 can be proved as follows. In the time-invariant case, it follows from **1-224** for the state differential equation of the interconnected system

$$\dot{x}(t) = \begin{pmatrix} A_1 - B_1 D_2 C_1 & -B_1 C_2 \\ B_2 C_1 & A_2 \end{pmatrix} x(t) + \begin{pmatrix} B_1 \\ 0 \end{pmatrix} r(t). \qquad \textbf{1-233}$$

We show that the characteristic polynomial of this system is precisely **1-232**. For this we need the following result from matrix theory.

Lemma 1.2. *Let M be a square, partitioned matrix of the form*

$$M = \begin{pmatrix} M_1 & M_2 \\ M_3 & M_4 \end{pmatrix}. \qquad \textbf{1-234}$$

Then if $\det(M_1) \neq 0$,
$$\det(M) = \det(M_1)\det(M_4 - M_3 M_1^{-1} M_2). \qquad \text{1-235}$$

If $\det(M_4) \neq 0$,
$$\det(M) = \det(M_4)\det(M_1 - M_2 M_4^{-1} M_3). \qquad \text{1-236}$$

The lemma is easily proved by elementary row and column operations on M. With the aid of Lemmas 1.2 and 1.1 (Section 1.5.3), the characteristic polynomial of **1-233** can be written as follows.

$$\det\begin{pmatrix} sI - A_1 + B_1 D_2 C_1 & B_1 C_2 \\ -B_2 C_1 & sI - A_2 \end{pmatrix}$$
$$= \det(sI - A_2)\det[sI - A_1 + B_1 D_2 C_1 + B_1 C_2(sI - A_2)^{-1} B_2 C_1]$$
$$= \det(sI - A_2)\det(sI - A_1)$$
$$\cdot \det\{I + B_1[D_2 + C_2(sI - A_2)^{-1} B_2] C_1 (sI - A_1)^{-1}\}$$
$$= \det(sI - A_1)\det(sI - A_2)$$
$$\cdot \det\{I + C_1(sI - A_1)^{-1} B_1 [C_2(sI - A_2)^{-1} B_2 + D_2]\}. \qquad \text{1-237}$$

Since
$$\det(sI - A_1) = \phi_1(s),$$
$$\det(sI - A_2) = \phi_2(s), \qquad \text{1-238}$$
$$C_1(sI - A_1)^{-1} B_1 = H_1(s),$$
$$C_2(sI - A_2)^{-1} B_2 + D_2 = H_2(s),$$

1-237 can be rewritten as
$$\phi_1(s)\phi_2(s)\det[I + H_1(s)H_2(s)]. \qquad \text{1-239}$$

This shows that **1-232** is the characteristic polynomial of the interconnected system; thus the stability immediately follows from the roots of **1-232**.

This method for checking the stability of feedback systems is usually more convenient for single-input single-output systems than for multivariable systems. In the case of single-input single-output systems, we write

$$H_1(s) = \frac{\psi_1(s)}{\phi_1(s)}, \qquad H_2(s) = \frac{\psi_2(s)}{\phi_2(s)}, \qquad \text{1-240}$$

where $\psi_1(s)$ and $\psi_2(s)$ are the numerator polynomials of the systems. By Theorem 1.21 stability now follows from the roots of the polynomial

$$\phi_1(s)\phi_2(s)\left[1 + \frac{\psi_1(s)\psi_2(s)}{\phi_1(s)\phi_2(s)}\right] = \phi_1(s)\phi_2(s) + \psi_1(s)\psi_2(s). \qquad \text{1-241}$$

It often happens in designing linear feedback control systems that either in the feedback path or in the feedforward path a gain factor is left undetermined until a late stage in the design. Suppose by way of example that

$$H_1(s) = \rho \frac{\psi_1(s)}{\phi_1(s)}, \qquad \text{1-242}$$

where ρ is the undetermined gain factor. The characteristic values of the interconnected system are now the roots of

$$\phi_1(s)\phi_2(s) + \rho\psi_1(s)\psi_2(s). \qquad \text{1-243}$$

An interesting problem is to construct the loci of the roots of this polynomial as a function of the scalar parameter ρ. This is a special case of the more general problem of finding in the complex plane the loci of the roots of

$$\phi(s) + \rho\psi(s) \qquad \text{1-244}$$

as the parameter ρ varies, where $\phi(s)$ and $\psi(s)$ are arbitrary given polynomials. The rules for constructing such loci are reviewed in the next section.

Example 1.16. *Inverted pendulum*

Consider the inverted pendulum of Example 1.1 (Section 1.2.3) and suppose that we wish to stabilize it. It is clear that if the pendulum starts falling to the right the carriage must also move to the right. We therefore attempt a method of control whereby we apply a force $\mu(t)$ to the carriage which is proportional to the angle $\phi(t)$. This angle can be measured by a potentiometer at the pivot; the force $\mu(t)$ is exerted through a small servomotor. Thus we have

$$\mu(t) = k\phi(t), \qquad \text{1-245}$$

where k is a constant. It is easily found that the transfer function from $\mu(t)$ to $\phi(t)$ is given by

$$H_1(s) = \frac{-\dfrac{1}{L'M}s}{\left(s + \dfrac{F}{M}\right)\left(s^2 - \dfrac{g}{L'}\right)}. \qquad \text{1-246}$$

The transfer function of the feedback part of the system follows from **1-245**:

$$H_2(s) = -k. \qquad \text{1-247}$$

The characteristic polynomial of the pendulum positioning system is

$$\phi_1(s) = s\left(s + \frac{F}{M}\right)\left(s^2 - \frac{g}{L'}\right), \qquad \text{1-248}$$

while the characteristic polynomial of the feedback part is

$$\phi_2(s) = 1. \qquad \text{1-249}$$

It follows from **1-246** and **1-247** that in this case

$$1 + H_1(s)H_2(s) = \frac{s^3 + s^2 \frac{F}{M} + s\left(\frac{k}{L'M} - \frac{g}{L'}\right) - \frac{Fg}{ML'}}{\left(s + \frac{F}{M}\right)\left(s^2 - \frac{g}{L'}\right)}, \qquad \text{1-250}$$

while from **1-248** and **1-249** we obtain

$$\phi_1(s)\phi_2(s) = s\left(s + \frac{F}{M}\right)\left(s^2 - \frac{g}{L'}\right). \qquad \text{1-251}$$

We note that in this case the denominator of $1 + H_1(s)H_2(s)$ is not the product of the characteristic polynomials **1-251**, but that a factor s has been canceled. Therefore, the numerator of **1-250** is not the closed-loop characteristic polynomial. By multiplication of **1-250** and **1-251**, it follows that the characteristic polynomial of the feedback system is

$$s\left\{s^3 + s^2 \frac{F}{M} + s\left(\frac{k}{L'M} - \frac{g}{L'}\right) - \frac{Fg}{ML'}\right\}. \qquad \text{1-252}$$

We see that one of the closed-loop characteristic values is zero. Moreover, since the remaining factor contains a term with a negative coefficient, according to the well-known Routh–Hurwitz criterion (Schwarz and Friedland, 1965) there is at least one root with a positive real part. This means that the system cannot be stabilized in this manner. Example 2.6 (Section 2.4) presents a more sophisticated control scheme which succeeds in stabilizing the system.

Example 1.17. *Stirred tank*

Consider the stirred tank of Example 1.2 (Section 1.2.3). Suppose that it is desired to operate the system such that a constant flow $F(t)$ and a constant concentration $c(t)$ are maintained. One way of doing this is to use the main flow F_1 to regulate the flow F, and the minor flow F_2 to regulate the concentration c. Let us therefore choose μ_1 and μ_2 according to

$$\begin{aligned}\mu_1(t) &= -k_1\eta_1(t),\\ \mu_2(t) &= -k_2\eta_2(t).\end{aligned} \qquad \text{1-253}$$

This means that the system in the feedback loop has the transfer matrix

$$H_2(s) = \begin{pmatrix} k_1 & 0 \\ 0 & k_2 \end{pmatrix}. \qquad \text{1-254}$$

It is easily found with the numerical data of Example 1.2 that the transfer matrix of the system in the forward loop is given by

$$H_1(s) = \begin{pmatrix} \dfrac{0.01}{s+0.01} & \dfrac{0.01}{s+0.01} \\ \dfrac{-0.25}{s+0.02} & \dfrac{0.75}{s+0.02} \end{pmatrix}. \qquad \textbf{1-255}$$

With this the return difference matrix is

$$J(s) = I + H_1(s)H_2(s) = \begin{pmatrix} \dfrac{s+0.01k_1+0.01}{s+0.01} & \dfrac{0.01k_2}{s+0.01} \\ \dfrac{-0.25k_1}{s+0.02} & \dfrac{s+0.75k_2+0.02}{s+0.02} \end{pmatrix}. \qquad \textbf{1-256}$$

For the characteristic polynomials of the two systems, we have

$$\phi_1(s) = (s+0.01)(s+0.02),$$
$$\phi_2(s) = 1. \qquad \textbf{1-257}$$

It follows from **1-256** that

$$\det[J(s)] = \frac{(s+0.01k_1+0.01)(s+0.75k_2+0.02) + 0.0025k_1k_2}{(s+0.01)(s+0.02)}. \qquad \textbf{1-258}$$

Since the denominator of this expression is the product $\phi_1(s)\phi_2(s)$, its numerator is the closed-loop characteristic polynomial. Further evaluation yields for the closed-loop characteristic polynomial

$$s^2 + s(0.01k_1 + 0.75k_2 + 0.03) + (0.0002k_1 + 0.0075k_2 + 0.01k_1k_2 + 0.0002). \qquad \textbf{1-259}$$

This expression shows that for positive k_1 and k_2 the feedback system is stable. Let us choose for the gain coefficients $k_1 = 10$ and $k_2 = 0.1$. This gives for the characteristic polynomial

$$s^2 + 0.205s + 0.01295. \qquad \textbf{1-260}$$

The characteristic values are

$$-0.1025 \pm j0.04944. \qquad \textbf{1-261}$$

The effectiveness of such a control scheme **1-253** is investigated in Example 2.8 (Section 2.5.3).

1.5.5* Root Loci

In the preceding section we saw that sometimes it is of interest to find in the complex plane the loci of the roots of an expression of the form

$$\phi(s) + \rho\psi(s), \qquad \textbf{1-262}$$

where $\phi(s)$ and $\psi(s)$ are polynomials in s, as the scalar parameter ρ varies. In this section we give some rules pertaining to these loci, so as to allow us to determine some special points of the loci and, in particular, to determine the asymptotic behavior. These rules make it possible to sketch root loci quite easily for simple problems; for more complicated problems the assistance of a digital computer is usually indispensable. Melsa (1970) gives a FORTRAN computer program for computing root loci.

We shall assume the following forms for the polynomials $\phi(s)$ and $\psi(s)$:

$$\phi(s) = \prod_{i=1}^{n} (s - \pi_i),$$

$$\psi(s) = \prod_{i=1}^{m} (s - \nu_i). \qquad \textbf{1-263}$$

We refer to the π_i, $i = 1, 2, \cdots, n$, as the *open-loop poles*, and to the $\nu_i = 1, 2, \cdots, m$, as the *open-loop zeroes*. The roots of **1-262** will be called the *closed-loop poles*. This terminology stems from the significance that the polynomials $\phi(s)$ and $\psi(s)$ have in Section 1.5.4. We assume that $m \leq n$; this is no restriction since if $m > n$ the roles of $\phi(s)$ and $\psi(s)$ can be reversed by choosing $1/\rho$ as the parameter.

The most important properties of the root loci are the following.

(a) *Number of roots:* The number of roots of **1-262** is n. Each of the roots traces a continuous locus as ρ varies from $-\infty$ to ∞.

(b) *Origin of loci:* The loci originate for $\rho = 0$ at the poles π_i, $i = 1, 2, \ldots, n$. This is obvious, since for $\rho = 0$ the roots of **1-262** are the roots of $\phi(s)$.

(c) *Behavior of loci as* $\rho \to \pm\infty$: As $\rho \to \pm\infty$, m of the loci approach the zeroes ν_i, $i = 1, 2, \cdots, m$. The remaining $n - m$ loci go to infinity. This follows from the fact that the roots of **1-262** are also the roots of

$$\frac{1}{\rho}\phi(s) + \psi(s). \qquad \textbf{1-264}$$

(d) *Asymptotes of loci:* Those $n - m$ loci that go to infinity approach asymptotically $n - m$ straight lines which make angles

$$\frac{\pi + k2\pi}{n - m}, \quad k = 0, 1, \cdots, n - m - 1, \qquad \textbf{1-265}$$

with the positive real axis as $\rho \to +\infty$, and angles

$$\frac{k2\pi}{n-m}, \qquad k = 0, 1, \cdots, n-m-1, \qquad \textbf{1-266}$$

as $\rho \to -\infty$. The $n-m$ asymptotes intersect in one point on the real axis given by

$$\frac{\sum_{i=1}^{n} \pi_i - \sum_{i=1}^{m} \nu_i}{n-m}. \qquad \textbf{1-267}$$

These properties can be derived as follows. For large s we approximate **1-262** by

$$s^n + \rho s^m. \qquad \textbf{1-268}$$

The roots of this polynomial are

$$(-\rho)^{1/(n-m)}, \qquad \textbf{1-269}$$

which gives a first approximation for the faraway roots. A more refined analysis shows that a better approximation for the roots is given by

$$\frac{\sum_{i=1}^{n} \pi_i - \sum_{i=1}^{m} \nu_i}{n-m} + (-\rho)^{1/(n-m)}. \qquad \textbf{1-270}$$

This proves that the asymptotic behavior is as claimed.

(e) *Portions of root loci on real axis:* If ρ assumes only positive values, any portion of the real axis to the right of which an odd number of poles and zeroes lies on the real axis is part of a root locus. If ρ assumes only negative values, any portion of the real axis to the right of which an even number of poles and zeroes lies on the real axis is part of a root locus. This can be seen as follows. The roots of **1-262** can be found by solving

$$\frac{\phi(s)}{\psi(s)} = -\rho. \qquad \textbf{1-271}$$

If we assume ρ to be positive, **1-271** is equivalent to the real equations

$$\left| \frac{\phi(s)}{\psi(s)} \right| = \rho, \qquad \textbf{1-272}$$

$$\arg \frac{\phi(s)}{\psi(s)} = \pi + 2\pi k, \qquad \textbf{1-273}$$

where k is any integer. If s is real, there always exists a ρ for which **1-272** is satisfied. To satisfy **1-273** as well, there must be an odd number of zeroes and poles to the right of s. For negative ρ a similar argument holds.

Several other properties of root loci can be established (D'Azzo and Houpis, 1966) which are helpful in sketching root locus plots, but the rules listed above are sufficient for our purpose.

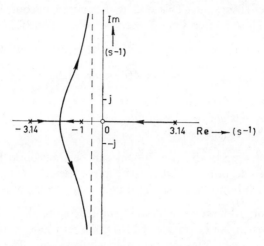

Fig. 1.8. Root locus for inverted pendulum. ×, open-loop poles; ○, open-loop zero.

Example 1.18. *Inverted pendulum*

Consider the proposed proportional feedback scheme of Example 1.16 where we found for the closed-loop characteristic polynomial

$$s\left(s + \frac{F}{M}\right)\left(s^2 - \frac{g}{L'}\right) + \frac{k}{L'M}s^2. \qquad \text{1-274}$$

Here k is varied from 0 to ∞. The poles are at 0, $-F/M$, $\sqrt{g/L'}$, and $-\sqrt{g/L'}$, while there is a double zero at 0. The asymptotes make angles of $\pi/2$ and $-\pi/2$ with the real axis as $k \to \infty$ since $n - m = 2$. The asymptotes intersect at $-\frac{1}{2}(F/M)$. The portions of the real axis between $\sqrt{g/L'}$ and 0, and between $-F/M$ and $-\sqrt{g/L'}$ belong to a locus. The pole at 0 coincides with a zero; this means that 0 is always one of the closed-loop poles. The loci of the remaining roots are sketched in Fig. 1.8 for the numerical values given in Example 1.1. It is seen that the closed-loop system is not stable for any k, as already concluded in Example 1.16.

1.6* CONTROLLABILITY

1.6.1* Definition of Controllability

For the solution of control problems, it is important to know whether or not a given system has the property that it may be steered from any given state

54   Elements of Linear System Theory

to any other given state. This leads to the concept of controllability (Kalman, 1960), which is discussed in this section. We give the following definition.

Definition 1.11.   *The linear system with state differential equation*

$$\dot{x}(t) = A(t)x(t) + B(t)u(t) \qquad \textbf{1-275}$$

is said to be **completely controllable** *if the state of the system can be transferred from the zero state at any initial time t_0 to any terminal state $x(t_1) = x_1$ within a finite time $t_1 - t_0$.*

Here, when we say that the system can be transferred from one state to another, we mean that there exists a piecewise continuous input $u(t)$, $t_0 \le t \le t_1$, which brings the system from one state to the other.

Definition 1.11 seems somewhat limited, since the only requirement is that the system can be transferred from the zero state to any other state. We shall see, however, that the definition implies more. The response from an arbitrary initial state is by **1-61** given by

$$x(t_1) = \Phi(t_1, t_0)x(t_0) + \int_{t_0}^{t_1} \Phi(t_1, \tau)B(\tau)u(\tau)\,d\tau, \qquad \textbf{1-276}$$

so that

$$x(t_1) - \Phi(t_1, t_0)x(t_0) = \int_{t_0}^{t_1} \Phi(t_1, \tau)B(\tau)u(\tau)\,d\tau. \qquad \textbf{1-277}$$

This shows that transferring the system from the state $x(t_0) = x_0$ to the state $x(t_1) = x_1$ is achieved by the same input that transfers $x(t_0) = 0$ to the state $x(t_1) = x_1 - \Phi(t_1, t_0)x_0$. This implies the following fact.

Theorem 1.22.   *The linear differential system*

$$\dot{x}(t) = A(t)x(t) + B(t)u(t) \qquad \textbf{1-278}$$

is completely controllable if and only if it can be transferred from any initial state x_0 at any initial time t_0 to any terminal state $x(t_1) = x_1$ within a finite time $t_1 - t_0$.

Example 1.19.   *Stirred tank*

Suppose that the feeds F_1 and F_2 of the stirred tank of Example 1.2 (Section 1.2.3) have equal concentrations $c_1 = c_2 = \bar{c}$. Then the steady-state concentration c_0 in the tank is also $\bar{c}$, and we find for the linearized state differential equation

$$\dot{x}(t) = \begin{pmatrix} -\dfrac{1}{2\theta} & 0 \\ 0 & -\dfrac{1}{\theta} \end{pmatrix} x(t) + \begin{pmatrix} 1 & 1 \\ 0 & 0 \end{pmatrix} u(t). \qquad \textbf{1-279}$$

It is clear from this equation that the second component of the state, which is the incremental concentration, cannot be controlled by manipulating the input, whose components are the incremental incoming flows. This is also clear physically, since the incoming feeds are assumed to have equal concentrations.

Therefore, the system obviously is not completely controllable if $c_1 = c_2$. If $c_1 \neq c_2$, the system is completely controllable, as we shall see in Example 1.21.

1.6.2* Controllability of Linear Time-Invariant Systems

In this section the controllability of linear time-invariant systems is studied. We first state the main result.

Theorem 1.23. *The n-dimensional linear time-invariant system*

$$\dot{x}(t) = Ax(t) + Bu(t) \qquad \text{1-280}$$

is completely controllable if and only if the column vectors of the **controllability matrix**

$$P = (B, AB, A^2B, \cdots, A^{n-1}B) \qquad \text{1-281}$$

span the n-dimensional space.

This result can be proved formally as follows. We write for the state at t_1, when at time t_0 the system is in the zero state,

$$x(t_1) = \int_{t_0}^{t_1} e^{A(t_1-\tau)} Bu(\tau)\, d\tau. \qquad \text{1-282}$$

The exponential may be represented in terms of its Taylor series; doing this we find

$$x(t_1) = B\int_{t_0}^{t_1} u(\tau)\, d\tau + AB\int_{t_0}^{t_1} (t_1 - \tau) u(\tau)\, d\tau$$
$$+ A^2 B \int_{t_0}^{t_1} \frac{(t_1 - \tau)^2}{2!} u(\tau)\, d\tau + \cdots. \qquad \text{1-283}$$

We see that the terminal state is in the linear subspace spanned by the column vectors of the infinite sequence of matrices $B, AB, A^2B, \cdots$. In this sequence there must eventually be a matrix, say $A^l B$, the column vectors of which are all linearly dependent upon the combined column vectors of the preceding matrices $B, AB, \cdots, A^{l-1}B$. There must be such a matrix since there cannot be more than n linearly independent vectors in n-dimensional space. This also implies that $l \leq n$.

56   Elements of Linear System Theory

Let us now consider $A^{l+1}B = A(A^l B)$. Since the column vectors of $A^l B$ depend linearly upon the combined column vectors of $B, AB, \cdots, A^{l-1}B$, we can write

$$A^l B = B\Lambda_0 + AB\Lambda_1 + \cdots + A^{l-1}B\Lambda_{l-1}, \qquad \text{1-284}$$

where the Λ_i, $i = 0, 1, \cdots, l-1$ are matrices which provide the correct coefficients to express each of the column vectors of $A^l B$ in terms of the column vectors of $B, AB, \cdots, A^{l-1}B$. Consequently, we write

$$A^{l+1}B = AB\Lambda_0 + A^2 B\Lambda_1 + \cdots + A^l B\Lambda_{l-1}, \qquad \text{1-285}$$

which very clearly shows that the columns of $A^{l+1}B$ also depend linearly upon the column vectors of $B, AB, \cdots, A^{l-1}B$. Similarly, it follows that the column vectors of all matrices $A^k B$ for $k \geq l$ depend linearly upon the column vectors of $B, AB, \cdots, A^{l-1}B$.

Returning now to **1-283**, we see that the terminal state $x(t_1)$ is in the linear subspace spanned by the column vectors of $B, AB, \cdots, A^{l-1}B$. Since $l \leq n$ we can just as well say that $x(t_1)$ is in the subspace spanned by the column vectors of $B, AB, \cdots, A^{n-1}B$. Now if these column vectors do *not* span the n-dimensional space, clearly only states in a linear subspace that is of smaller dimension than the entire n-dimensional space can be reached, hence the system is not completely controllable. This proves that if the system is completely controllable the column vectors of the controllability matrix P span the n-dimensional space.

To prove the other direction of the theorem, assume that the columns of P span the n-dimensional space. Then by a suitable choice of the input $u(\tau)$, $t_0 \leq \tau \leq t_1$ (e.g., involving orthogonal polynomials), the coefficient vectors

$$\int_{t_0}^{t_1} \frac{(t_1 - \tau)^i}{i!} u(\tau)\, d\tau \qquad \text{1-286}$$

in **1-283** can always be chosen so that the right-hand side of **1-283** equals any given vector in the space spanned by the columns of P. Since by assumption the columns of P span the entire n-dimensional space, this means that any terminal state can be reached, hence that the system is completely controllable. This terminates the proof of Theorem 1.23.

The controllability of the system **1-280** is of course completely determined by the matrices A and B. It is therefore convenient to introduce the following terminology.

Definition 1.12. *Let A be an $n \times n$ and B an $n \times k$ matrix. Then we say that the pair $\{A, B\}$ is completely controllable if the system*

$$\dot{x}(t) = Ax(t) + Bu(t) \qquad \text{1-287}$$

is completely controllable.

Example 1.20. *Inverted pendulum*

The inverted pendulum of Example 1.1 (Section 1.2.3) is a single-input system which is described by the state differential equation

$$\dot{x}(t) = \begin{pmatrix} 0 & 1 & 0 & 0 \\ 0 & -\dfrac{F}{M} & 0 & 0 \\ 0 & 0 & 0 & 1 \\ -\dfrac{g}{L'} & 0 & \dfrac{g}{L'} & 0 \end{pmatrix} x(t) + \begin{pmatrix} 0 \\ \dfrac{1}{M} \\ 0 \\ 0 \end{pmatrix} \mu(t). \qquad \text{1-288}$$

The controllability matrix of the system is

$$P = \begin{pmatrix} 0 & \dfrac{1}{M} & -\dfrac{F}{M}\dfrac{1}{M} & \left(\dfrac{F}{M}\right)^2 \dfrac{1}{M} \\ \dfrac{1}{M} & -\dfrac{F}{M}\dfrac{1}{M} & \left(\dfrac{F}{M}\right)^2 \dfrac{1}{M} & -\left(\dfrac{F}{M}\right)^3 \dfrac{1}{M} \\ 0 & 0 & 0 & -\dfrac{g}{L'}\dfrac{1}{M} \\ 0 & 0 & -\dfrac{g}{L'}\dfrac{1}{M} & \dfrac{g}{L'}\dfrac{F}{M}\dfrac{1}{M} \end{pmatrix}. \qquad \text{1-289}$$

It is easily seen that P has rank four for all values of the parameters, hence that the system is completely controllable.

1.6.3* The Controllable Subspace

In this section we analyze in some detail the structure of linear time-invariant systems that are not completely controllable. If a system is not completely controllable, clearly it is of interest to know what part of the state space can be reached. This motivates the following definition.

Definition 1.13. *The **controllable subspace** of the linear time-invariant system*

$$\dot{x}(t) = Ax(t) + Bu(t) \qquad \text{1-290}$$

is the linear subspace consisting of the states that can be reached from the zero state within a finite time.

In view of the role that the controllability matrix P plays, the following result is not surprising.

Theorem 1.24. *The controllable subspace of the n-dimensional linear time-invariant system*

$$\dot{x}(t) = Ax(t) + Bu(t) \qquad \text{1-291}$$

is the linear subspace spanned by the columns of the controllability matrix

$$P = (B, AB, \cdots, A^{n-1}B). \qquad \text{1-292}$$

This theorem immediately follows from the proof of Theorem 1.23 where we showed that any state that can be reached from the zero state is spanned by the columns of P, and any state not spanned by the columns of P cannot be reached. The controllable subspace possesses the following property.

Lemma 1.3. *The controllable subspace of the system $\dot{x}(t) = Ax(t) + Bu(t)$ is invariant under A, that is, if a vector x is in the controllable subspace, Ax is also in this subspace.*

The proof of this lemma follows along the lines of the proof of Theorem 1.23. The controllable subspace is spanned by the column vectors of $B, AB, \cdots, A^{n-1}B$. Thus the vector Ax, where x is in the controllable subspace, is in the linear subspace spanned by the column vectors of $AB, A^2B, \cdots, A^nB$. The column vectors of A^nB, however, depend linearly upon the column vectors of $B, AB, \cdots, A^{n-1}B$; therefore Ax is in the subspace spanned by the column vectors of $B, AB, \cdots, A^{n-1}B$, which means that Ax is in the controllable subspace. The controllable subspace is therefore invariant under A.

The concept of a controllable subspace can be further clarified by the following fact.

Theorem 1.25. *Consider the linear time-invariant system $\dot{x}(t) = Ax(t) + Bu(t)$. Then any initial state x_0 in the controllable subspace can be transferred to any terminal state x_1 in the controllable subspace within a finite time.*

We prove this result by writing for the state of the system at time t_1:

$$x(t_1) = e^{A(t_1-t_0)}x_0 + \int_{t_0}^{t_1} e^{A(t_1-\tau)}Bu(\tau)\,d\tau. \qquad \text{1-293}$$

Now if x_0 is in the controllable subspace, $\exp[A(t_1 - t_0)]x_0$ is also in the controllable subspace, since the controllable subspace is invariant under A and $\exp[A(t_1 - t_0)] = I + A(t_1 - t_0) + \frac{1}{2}A^2(t_1 - t_0) + \cdots$. Therefore, if x_1 is in the controllable subspace, $x_1 - \exp[A(t_1 - t_0)]x_0$ is also in the controllable subspace. Expression **1-293** shows that any input that transfers the zero state to the state $x_1 - \exp[A(t_1 - t_0)]x_0$ also transfers x_0 to x_1. Since $x_1 - \exp[A(t_1 - t_0)]x_0$ is in the controllable subspace, such an input exists; Theorem 1.25 is thus proved.

1.6 Controllability

We now find a state transformation that represents the system in a canonical form, which very clearly exhibits the controllability properties of the system. Let us suppose that P has rank $m \leq n$, that is, P possesses m linearly independent column vectors. This means that the controllable subspace of the system **1-290** has dimension m. Let us choose a basis $e_1, e_2, \cdots, e_m$ for the controllable subspace. Furthermore, let $e_{m+1}, e_{m+2}, \cdots, e_n$ be $n-m$ linearly independent vectors which together with $e_1, e_2, \cdots, e_m$ span the whole n-dimensional space. We now form the nonsingular transformation matrix

$$T = (T_1, T_2), \qquad \text{1-294}$$

where

$$T_1 = (e_1, e_2, \cdots, e_m), \qquad \text{1-295}$$

and

$$T_2 = (e_{m+1}, e_{m+2}, \cdots, e_n). \qquad \text{1-296}$$

Finally, we introduce a transformed state variable $x'(t)$ defined by

$$Tx'(t) = x(t). \qquad \text{1-297}$$

Substituting this into the state differential equation **1-290**, we obtain

$$T\dot{x}'(t) = ATx'(t) + Bu(t) \qquad \text{1-298}$$

or

$$\dot{x}'(t) = T^{-1}ATx'(t) + T^{-1}Bu(t). \qquad \text{1-299}$$

We partition T^{-1} as follows

$$T^{-1} = \begin{pmatrix} U_1 \\ U_2 \end{pmatrix}, \qquad \text{1-300}$$

where the partitioning corresponds to that of T in the sense that U_1 has m rows and U_2 has $n-m$ rows. With this partitioning it follows

$$T^{-1}T = \begin{pmatrix} U_1 \\ U_2 \end{pmatrix} (T_1, T_2) = \begin{pmatrix} U_1T_1 & U_1T_2 \\ U_2T_1 & U_2T_2 \end{pmatrix} = \begin{pmatrix} I_m & 0 \\ 0 & I_{n-m} \end{pmatrix}. \qquad \text{1-301}$$

From this we conclude that

$$U_2T_1 = 0. \qquad \text{1-302}$$

T_1 is composed of the vectors $e_1, e_2, \cdots, e_m$ which span the controllable subspace. This means that **1-302** implies that

$$U_2x = 0 \qquad \text{1-303}$$

for any vector x in the controllable subspace.

With the partitionings **1-294** and **1-300**, we write

$$T^{-1}AT = \begin{pmatrix} U_1 \\ U_2 \end{pmatrix} A(T_1, T_2) = \begin{pmatrix} U_1AT_1 & U_1AT_2 \\ U_2AT_1 & U_2AT_2 \end{pmatrix} \quad \textbf{1-304}$$

and

$$T^{-1}B = \begin{pmatrix} U_1 \\ U_2 \end{pmatrix} B = \begin{pmatrix} U_1B \\ U_2B \end{pmatrix}. \quad \textbf{1-305}$$

All the columns of T_1 are in the controllable subspace. This means that all the columns of AT_1 are also in the controllable subspace, since the controllable subspace is invariant under A (Lemma 1.3). However, then **1-303** implies that

$$U_2AT_1 = 0. \quad \textbf{1-306}$$

The columns of B are obviously all in the controllable subspace, since B is part of the controllability matrix. Therefore, we also have

$$U_2B = 0. \quad \textbf{1-307}$$

Our findings can be summarized as follows.

Theorem 1.26. *Consider the n-dimensional time-invariant system*

$$\dot{x}(t) = Ax(t) + Bu(t). \quad \textbf{1-308}$$

Form a nonsingular transformation matrix $T = (T_1, T_2)$ where the columns of T_1 form a basis for the m-dimensional ($m \leq n$) controllable subspace of **1-308** *and the column vectors of T_2 together with those of T_1 form a basis for the whole n-dimensional space. Define the transformed state*

$$x'(t) = T^{-1}x(t). \quad \textbf{1-309}$$

Then the state differential equation **1-308** *is transformed into the* **controllability canonical form**

$$\dot{x}'(t) = \begin{pmatrix} A'_{11} & A'_{12} \\ 0 & A'_{22} \end{pmatrix} x'(t) + \begin{pmatrix} B'_1 \\ 0 \end{pmatrix} u(t). \quad \textbf{1-310}$$

Here A'_{11} is an $m \times m$ matrix, and the pair $\{A'_{11}, B'_1\}$ is completely controllable.

Partitioning

$$x'(t) = \begin{pmatrix} x'_1(t) \\ x'_2(t) \end{pmatrix}, \quad \textbf{1-311}$$

where x'_1 has dimension m and x'_2 dimension $n - m$, we see from Theorem 1.26 that the transformed system can be represented as in Fig. 1.9. We note that

1.6 Controllability

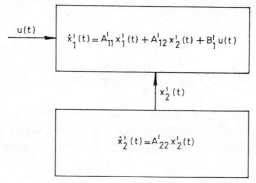

Fig. 1.9. The controllability canonical form of a linear time-invariant differential system.

x_2' behaves completely independently, while x_1' is influenced both by x_2' and the input u. The fact that $\{A_{11}', B_1'\}$ is completely controllable follows from the fact that any state of the form col $(x_0', 0)$ is in the controllable subspace of the system **1-310**. The proof is left as an exercise.

It should be noted that the controllability canonical form is not at all unique, since both T_1 and T_2 can to some extent be freely chosen. It is easily verified, however, that no matter how the transformation T is chosen the characteristic values of both A_{11}' and A_{22}' are always the same (Problem 1.5). Quite naturally, this leads us to refer to the characteristic values of A_{11}' as the *controllable poles* of the system, and to the characteristic values of A_{22}' as the *uncontrollable poles*. Let us now assume that all the characteristic values of the system **1-310** are distinct (this is not an essential restriction). Then it is not difficult to recognize (Problem 1.5) that *the controllable subspace of the system* **1-310** *is spanned by the characteristic vectors corresponding to the controllable poles of the system*. This statement is also true for the original representation **1-308** of the system. Then a natural definition for the *uncontrollable subspace* of the system, which we have so far avoided, is *the subspace spanned by the characteristic vectors corresponding to the uncontrollable poles of the system*.

Example 1.21. *Stirred tank*

The stirred tank of Example 1.2 (Section 1.2.3) is described by the state differential equation

$$\dot{x}(t) = \begin{pmatrix} -\dfrac{1}{2\theta} & 0 \\ 0 & -\dfrac{1}{\theta} \end{pmatrix} x(t) + \begin{pmatrix} 1 & 1 \\ \dfrac{c_1 - c_0}{V_0} & \dfrac{c_2 - c_0}{V_0} \end{pmatrix} u(t). \qquad \textbf{1-312}$$

62 Elements of Linear System Theory

The controllability matrix is

$$P = \begin{pmatrix} 1 & 1 & -\dfrac{1}{2\theta} & -\dfrac{1}{2\theta} \\ \dfrac{c_1-c_0}{V_0} & \dfrac{c_2-c_0}{V_0} & -\dfrac{1}{\theta}\dfrac{c_1-c_0}{V_0} & -\dfrac{1}{\theta}\dfrac{c_2-c_0}{V_0} \end{pmatrix}. \qquad \text{1-313}$$

P has rank two provided $c_1 \neq c_2$. The system is therefore completely controllable if $c_1 \neq c_2$.

If $c_1 = c_2 = \bar{c}$, then $c_0 = \bar{c}$ also and the controllability matrix takes the form

$$P = \begin{pmatrix} 1 & 1 & -\dfrac{1}{2\theta} & -\dfrac{1}{2\theta} \\ 0 & 0 & 0 & 0 \end{pmatrix}. \qquad \text{1-314}$$

The controllable subspace is therefore spanned by the vector col(1, 0). This means, as we saw in Example 1.19, that only the volume of fluid in the tank can be controlled but not the concentration.

We finally remark that if $c_1 = c_2 = c_0 = \bar{c}$ the state differential equation **1-312** takes the form **1-279**, which is already in controllability canonical form. The controllable pole of the system is $-1/(2\theta)$; the uncontrollable pole is $-1/\theta$.

1.6.4* Stabilizability

In this section we develop the notion of stabilizability (Galperin and Krasovski, 1963; Wonham 1968a). The terminology will be motivated in Section 3.2. In Section 1.4.3 we defined the stable and unstable subspaces for a time-invariant system. Any initial state $x(0)$ can be uniquely written as

$$x(0) = x_s(0) + x_u(0), \qquad \text{1-315}$$

where $x_s(0)$ is in the stable subspace and $x_u(0)$ in the unstable subspace. Clearly, in order to control the system properly, we must require that the unstable component can be completely controlled. This is the case if the unstable component $x_u(0)$ is in the controllable subspace. We thus state.

Definition 1.14. *The linear time-invariant system*

$$\dot{x}(t) = Ax(t) + Bu(t) \qquad \text{1-316}$$

is **stabilizable** *if its unstable subspace is contained in its controllable subspace, that is, any vector x in the unstable subspace is also in the controllable subspace.*

It is sometimes convenient to employ the following abbreviated terminology.

Definition 1.15. *The pair $\{A, B\}$ is stabilizable if the system*
$$\dot{x}(t) = Ax(t) + Bu(t) \qquad \text{1-317}$$
is stabilizable.

Obviously, we have the following result.

Theorem 1.27. *Any asymptotically stable time-invariant system is stabilizable. Any completely controllable system is stabilizable.*

The stabilizability of a system can conveniently be checked when the state differential equation is in controllability canonical form. This follows from the following fact.

Theorem 1.28. *Consider the time-invariant linear system*
$$\dot{x}(t) = Ax(t) + Bu(t). \qquad \text{1-318}$$
Suppose that it is transformed according to Theorem **1-26** *into the controllability canonical form*
$$\dot{x}'(t) = \begin{pmatrix} A'_{11} & A'_{12} \\ 0 & A'_{22} \end{pmatrix} x'(t) + \begin{pmatrix} B'_1 \\ 0 \end{pmatrix} u(t), \qquad \text{1-319}$$
where the pair $\{A'_{11}, B'_1\}$ is completely controllable. Then the system **1-318** *is stabilizable if and only if the matrix A'_{22} is asymptotically stable.*

This theorem can be summarized by stating that a system is stabilizable if and only if its uncontrollable poles are stable. We prove the theorem as follows.

(a) *Stabilizability implies A'_{22} asymptotically stable.* Suppose that the system **1-318** is stabilizable. Then the transformed system **1-319** is also stabilizable (Problem 1.6). Let us partition
$$x'(t) = \begin{pmatrix} x'_1(t) \\ x'_2(t) \end{pmatrix}, \qquad \text{1-320}$$
where the dimension m of $x'_1(t)$ is the dimension of the controllable subspace of the original system **1-318**. Suppose that A'_{22} is not stable. Choose an $(n - m)$-dimensional vector x'_2 in the unstable subspace of A'_{22}. Then obviously, the n-dimensional column vector col $(0, x'_2)$ is in the unstable subspace of **1-319**. This vector, however, is clearly not in the controllable subspace of **1-319**. This means that there is a vector that is in the unstable subspace of **1-319** but not in the controllable subspace. This contradicts the assumption of stabilizability. This proves that if the system **1-318** is stabilizable A'_{22} must be stable.

(b) *A'_{22} stable implies stabilizability:* Assume that A'_{22} is stable. Then any vector that is in the unstable subspace of **1-319** must be of the form col $(x_1', 0)$. However, since the pair $\{A_{11}', B_1'\}$ is completely controllable, this vector is also in the controllable subspace of **1-319**. This shows that any vector in the unstable subspace of **1-319** is also in the controllable subspace, hence that **1-319** is stabilizable. Consequently (Problem 1.6), the original system **1-318** is also stabilizable.

Example 1.22. *Stirred tank*

The stirred tank of Example 1.2 (Section 1.2.3) is described by the state differential equation

$$\dot{x}(t) = \begin{pmatrix} -\dfrac{1}{2\theta} & 0 \\ 0 & -\dfrac{1}{\theta} \end{pmatrix} x(t) + \begin{pmatrix} 1 & 1 \\ 0 & 0 \end{pmatrix} u(t), \qquad \text{1-321}$$

if we assume that $c_1 = c_2 = c_0 = \bar{c}$. As we have seen before, this system is not completely controllable. The state differential equation is already in the decomposed form for controllability. We see that the matrix A_{22}' has the characteristic value $-1/\theta$, which implies that the system is stabilizable. This means that even if the incremental concentration $\xi_2(t)$ initially has an incorrect value it will eventually approach zero.

1.6.5* Controllability of Time-Varying Linear Systems

The simple test for controllability of Theorem 1.24 does not apply to time-varying linear systems. For such systems we have the following result, which we shall not prove.

Theorem 1.29. *Consider the linear time-varying system with state differential equation*

$$\dot{x}(t) = A(t)x(t) + B(t)u(t). \qquad \text{1-322}$$

Define the nonnegative-definite symmetric matrix function

$$W(t_0, t) = \int_{t_0}^{t} \Phi(t, \tau) B(\tau) B^T(\tau) \Phi^T(t, \tau) \, d\tau, \qquad \text{1-323}$$

where $\Phi(t, t_0)$ is the transition matrix of the system. Then the system is completely controllable if and only if there exists for all t_0 a t_1 with $t_0 < t_1 < \infty$ such that $W(t_0, t_1)$ is nonsingular.

For a proof of this theorem, the reader is referred to Kalman, Falb, and Arbib (1969).

The matrix $W^{-1}(t_0, t_1)$ is related to the minimal "control energy" needed to transfer the system from one state to another when the "control energy"

is measured as
$$\int_{t_0}^{t_1} u^T(t)u(t)\,dt. \qquad \text{1-324}$$

A stronger form of controllability results if certain additional conditions are imposed upon the matrix $W(t_0, t)$ (Kalman, 1960):

Definition 1.16. *The time-varying system* **1-322** *is **uniformly completely controllable** if there exist positive constants* σ, α_0, α_1, β_0, *and* β_1 *such that*

(a) $\alpha_0 I \leq W(t_0, t_0 + \sigma) \leq \alpha_1 I \quad \text{for all } t_0;$ \qquad 1-325

(b) $\beta_0 I \leq \Phi(t_0, t_0 + \sigma)W(t_0, t_0 + \sigma)\Phi^T(t_0, t_0 + \sigma) \leq \beta_1 I \quad \text{for all } t_0,$

\qquad 1-326

where $W(t_0, t)$ *is the matrix* **1-323** *and* $\Phi(t, t_0)$ *is the transition matrix of the system.*

Uniform controllability implies not only that the system can be brought from any state to any other state but also that the control energy involved in this transfer and the transfer time are roughly independent of the initial time. In view of this remark, the following result for time-invariant systems is not surprising.

Theorem 1.30. *The time-invariant linear system*

$$\dot{x}(t) = Ax(t) + Bu(t) \qquad \text{1-327}$$

is uniformly completely controllable if and only if it is completely controllable.

1.7* RECONSTRUCTIBILITY

1.7.1* Definition of Reconstructibility

In Chapter 4 we discuss the problem of reconstructing the behavior of the state of the system from incomplete and possibly inaccurate observations. Before studying such problems it is important to know whether or not a given system has the property that it is at all possible to determine from the behavior of the output what the behavior of the state is. This leads to the concept of *reconstructibility* (Kalman, Falb, and Arbib, 1969), which is the subject of this section.

We first consider the following definition.

Definition 1.17. *Let* $y(t; t_0, x_0, u)$ *denote the response of the output variable* $y(t)$ *of the linear differential system*

$$\begin{aligned}\dot{x}(t) &= A(t)x(t) + B(t)u(t),\\ y(t) &= C(t)x(t),\end{aligned} \qquad \text{1-328}$$

66   Elements of Linear System Theory

to the initial state $x(t_0) = x_0$. Then the system is called **completely reconstructible** if for all t_1 there exists a t_0 with $-\infty < t_0 < t_1$ such that

$$y(t; t_0, x_0, u) = y(t; t_0, x_0', u), \qquad t_0 \leq t \leq t_1, \qquad \textbf{1-329}$$

for all $u(t)$, $t_0 \leq t \leq t_1$, implies $x_0 = x_0'$.

The definition implies that if a system is completely reconstructible, and the output variable is observed up to any time t_1, there always exists a time $t_0 < t_1$ at which the state of the system can be uniquely determined. If $x(t_0)$ is known, of course $x(t_1)$ can also be determined.

The following result shows that in order to study the reconstructibility of the system **1-328** we can confine ourselves to considering a simpler situation.

Theorem 1.31. *The system* **1-328** *is completely reconstructible if and only if for all t_1 there exists a t_0 with $-\infty < t_0 < t_1$ such that*

$$y(t; t_0, x_0, 0) = 0, \qquad t_0 \leq t \leq t_1, \qquad \textbf{1-330}$$

implies that $x_0 = 0$.

This result is not difficult to prove. Of course if the system **1-328** is completely reconstructible, it follows immediately from the definition that if **1-330** holds then $x_0 = 0$. This proves one direction of the theorem. However, since

$$y(t; t_0, x_0, u) = C(t)\left[\Phi(t, t_0)x_0 + \int_{t_0}^{t} \Phi(t, \tau)B(\tau)u(\tau)\,d\tau\right], \qquad \textbf{1-331}$$

the fact that

$$y(t; t_0, x_0, u) = y(t; t_0, x_0', u) \qquad \text{for } t_0 \leq t \leq t_1 \qquad \textbf{1-332}$$

implies and is implied by

$$C(t)\Phi(t, t_0)x_0 = C(t)\Phi(t, t_0)x_0' \qquad \text{for } t_0 \leq t \leq t_1. \qquad \textbf{1-333}$$

This in turn is equivalent to

$$C(t)\Phi(t, t_0)(x_0 - x_0') = 0 \qquad \text{for } t_0 \leq t \leq t_1. \qquad \textbf{1-334}$$

Evidently if **1-334** implies that $x_0 - x_0' = 0$, that is, $x_0 = x_0'$, the system is completely reconstructible. This finishes the proof of the other direction of Theorem 1.31.

The definition of reconstructibility is due to Kalman (Kalman, Falb, and Arbib, 1969). It should be pointed out that reconstructibility is complementary to *observability*. A system of the form **1-328** is said to be completely observable if for all t_0 there exists a $t_1 < \infty$ such that

$$y(t; t_0, x_0, u) = y(t; t_0, x_0', u), \qquad t_0 \leq t \leq t_1, \qquad \textbf{1-335}$$

for all $u(t)$, $t_0 \leq t \leq t_1$, implies that $x_0 = x_0'$. We note that observability means that is it possible to determine the state at time t_0 from the *future* output. In control and filtering problems, however, usually only past output values are available. It is therefore much more natural to consider reconstructibility, which regards the problem of determining the present state from *past* observations. It is easy to recognize that for time-invariant systems complete reconstructibility implies and is implied by complete observability.

Example 1.23. *Inverted pendulum*

Consider the inverted pendulum of Example 1.1 (Section 1.2.3) and take as the output variable the angle $\phi(t)$. Let us compare the states

$$\begin{pmatrix} 0 \\ 0 \\ 0 \\ 0 \end{pmatrix} \quad \text{and} \quad \begin{pmatrix} d_0 \\ 0 \\ d_0 \\ 0 \end{pmatrix} \qquad \text{1-336}$$

The second state differs from the zero state in that both carriage and pendulum are displaced over a distance d_0; otherwise, the system is at rest. If an input identical to zero is applied, the system stays in these positions, and $\phi(t) \equiv 0$ in both cases. It is clear that if only the angle $\phi(t)$ is observed it is impossible to decide at a given time whether the system is in one state or the other; as a result, the system is not completely reconstructible.

1.7.2* Reconstructibility of Linear Time-Invariant Systems

In this section the reconstructibility of linear time-invariant systems is discussed. The main result is the following.

Theorem 1.32. *The n-dimensional linear time-invariant system*

$$\begin{aligned} \dot{x}(t) &= Ax(t) + Bu(t), \\ y(t) &= Cx(t), \end{aligned} \qquad \text{1-337}$$

is completely reconstructible if and only if the row vectors of the **reconstructibility matrix**

$$Q = \begin{pmatrix} C \\ CA \\ CA^2 \\ \cdot \\ \cdot \\ \cdot \\ CA^{n-1} \end{pmatrix} \qquad \text{1-338}$$

span the n-dimensional space.

This can be proved as follows. Let us first assume that the system **1-337** is completely reconstructible. Then it follows from Theorem 1.31 that for all t_1 there exists a t_0 such that

$$Ce^{A(t-t_0)}x_0 = 0, \quad t_0 \leq t \leq t_1, \qquad \textbf{1-339}$$

implies that $x_0 = 0$. Expanding $\exp[A(t-t_0)]$ in terms of its Taylor series, **1-339** is equivalent to

$$\left[C + CA(t-t_0) + CA^2 \frac{(t-t_0)^2}{2!} + CA^3 \frac{(t-t_0)^3}{3!} + \cdots \right]x_0 = 0,$$

$$t_0 \leq t \leq t_1. \quad \textbf{1-340}$$

Now if the reconstructibility matrix Q does not have full rank, there exists a nonzero x_0 such that

$$Cx_0 = 0, \quad CAx_0 = 0, \quad \cdots, \quad CA^{n-1}x_0 = 0. \qquad \textbf{1-341}$$

By using the Cayley-Hamilton theorem, it is not difficult to see also that $CA^l x_0 = 0$ for $l \geq n$. Thus if Q does not have full rank there exists a nonzero x_0 such that **1-340** holds. Clearly, in this case **1-339** does not imply $x_0 = 0$, and the system is not completely reconstructible. This contradicts our assumption, which proves that Q must have full rank.

We now prove the other direction of Theorem 1.32. Asume that Q has full rank. Suppose that

$$y(t) = Ce^{A(t-t_0)}x_0 = 0 \quad \text{for} \quad t_0 \leq t \leq t_1. \qquad \textbf{1-342}$$

It follows by repeated differentiation of $y(t)$ that

$$\begin{aligned} y(t_0) &= Cx_0 &= 0, \\ y'(t_0) &= CAx_0 &= 0, \\ y''(t_0) &= CA^2 x_0 &= 0, \\ &\vdots \\ y^{(n-1)}(t_0) &= CA^{n-1}x_0 &= 0, \end{aligned} \qquad \textbf{1-343}$$

or

$$Qx_0 = 0. \qquad \textbf{1-344}$$

Since Q has full rank, **1-344** implies that $x_0 = 0$. Hence by Theorem 1.31 the system is completely reconstructible. This terminates the proof of Theorem 1.32.

Since the reconstructibility of the system **1-337** depends only on the matrices A and C, it is convenient to employ the following terminology.

1.7 Reconstructibility

Definition 1.18. *Let A be an $n \times n$ and C an $l \times n$ matrix. Then we call the pair $\{A, C\}$ completely reconstructible if the system*

$$\dot{x}(t) = Ax(t), \qquad \text{1-345}$$
$$y(t) = Cx(t), \qquad \text{1-346}$$

is completely reconstructible.

Example 1.24. *Inverted pendulum*

The inverted pendulum of Example 1.1 (Section 1.2.3) is described by the state differential equation

$$\dot{x}(t) = \begin{pmatrix} 0 & 1 & 0 & 0 \\ 0 & -\dfrac{F}{M} & 0 & 0 \\ 0 & 0 & 0 & 1 \\ -\dfrac{g}{L'} & 0 & \dfrac{g}{L'} & 0 \end{pmatrix} x(t) + \begin{pmatrix} 0 \\ \dfrac{1}{M} \\ 0 \\ 0 \end{pmatrix} \mu(t). \qquad \text{1-347}$$

If we take as the output variable $\eta(t)$ the angle $\phi(t)$, we have

$$\eta(t) = \left(-\dfrac{1}{L'},\ 0,\ \dfrac{1}{L'},\ 0\right) x(t). \qquad \text{1-348}$$

The reconstructibility matrix is

$$Q = \begin{pmatrix} -\dfrac{1}{L'} & 0 & \dfrac{1}{L'} & 0 \\ 0 & -\dfrac{1}{L'} & 0 & \dfrac{1}{L'} \\ -\dfrac{g}{L'}\dfrac{1}{L'} & \dfrac{F}{M}\dfrac{1}{L'} & \dfrac{g}{L'}\dfrac{1}{L'} & 0 \\ 0 & -\dfrac{g}{L'}\dfrac{1}{L'} - \left(\dfrac{F}{M}\right)^2 \dfrac{1}{L'} & 0 & \dfrac{g}{L'}\dfrac{1}{L'} \end{pmatrix}. \qquad \text{1-349}$$

This matrix has rank three; the system is therefore not completely reconstructible. This confirms the conclusion of Example 1.23. If we add as a second component of the output variable the displacement $s(t)$ of the carriage, we have

$$y(t) = \begin{pmatrix} -\dfrac{1}{L'} & 0 & \dfrac{1}{L'} & 0 \\ 1 & 0 & 0 & 0 \end{pmatrix} x(t). \qquad \text{1-350}$$

70   Elements of Linear System Theory

This yields for the reconstructibility matrix

$$Q = \begin{pmatrix} -\dfrac{1}{L'} & 0 & \dfrac{1}{L'} & 0 \\ 1 & 0 & 0 & 0 \\ 0 & -\dfrac{1}{L'} & 0 & \dfrac{1}{L'} \\ 0 & 1 & 0 & 0 \\ -\dfrac{g}{L'} & \dfrac{F}{M}\dfrac{1}{L'} & \dfrac{g}{L'}\dfrac{1}{L'} & 0 \\ 0 & -\dfrac{F}{M} & 0 & 0 \\ 0 & -\dfrac{g}{L'}\dfrac{1}{L'} - \left(\dfrac{F}{M}\right)^2 \dfrac{1}{L'} & 0 & \dfrac{g}{L'}\dfrac{1}{L'} \\ 0 & \left(\dfrac{F}{M}\right)^2 & 0 & 0 \end{pmatrix}. \quad \text{1-351}$$

With this output the system is completely reconstructible, since Q has rank four.

1.7.3* The Unreconstructible Subspace

In this section we analyze in some detail the structure of systems that are not completely reconstructible. If a system is not completely reconstructible, it is never possible to establish uniquely from the output what the state of the system is. Clearly, it is of interest to know exactly what uncertainty remains. This introduces the following definition.

Definition 1.19. *The **unreconstructible subspace** of the linear time-invariant system*

$$\dot{x}(t) = Ax(t) + Bu(t), \quad \text{1-352}$$
$$y(t) = Cx(t),$$

is the linear subspace consisting of the states x_0 for which

$$y(t; x_0, t_0, 0) = 0, \quad t \geq t_0. \quad \text{1-353}$$

The following theorem characterizes the unreconstructible subspace.

Theorem 1.33. *The unreconstructible subspace of the n-dimensional linear time-invariant system*

$$\dot{x}(t) = Ax(t) + Bu(t), \quad \text{1-354}$$
$$y(t) = Cx(t),$$

is the null space of the reconstructibility matrix

$$Q = \begin{pmatrix} C \\ CA \\ CA^2 \\ \cdot \\ \cdot \\ \cdot \\ CA^{n-1} \end{pmatrix}. \qquad \text{1-355}$$

The proof of this theorem immediately follows from the proof of Theorem 1.32 where we showed that any initial state in the null space of Q produces an output that is identical to zero in response to a zero input. Any initial state not in the null space of Q produces a nonzero response, which proves that the null space of Q is the unreconstructible subspace. The unreconstructible subspace possesses the following property.

Lemma 1.4. *The unreconstructible subspace of the system $\dot{x}(t) = Ax(t)$, $y(t) = Cx(t)$ is invariant under A.*

We leave the proof of this lemma as an exercise.

The concept of unreconstructible subspace can be clarified by the following fact.

Theorem 1.34. *Consider the time-invariant system*

$$\dot{x}(t) = Ax(t) + Bu(t),$$
$$y(t) = Cx(t). \qquad \text{1-356}$$

Suppose that the output $y(t)$ and the input $u(t)$ are known over an interval $t_0 \leq t \leq t_1$. Then the initial state of the system at time t_0 is determined within the addition of an arbitrary vector in the unreconstructible subspace. As a result, also the terminal state at time t_1 is determined within the addition of an arbitrary vector in the unreconstructible subspace.

To prove the first part of the theorem, we must show that if two initial states $x(t_0) = x_0$ and $x(t_0) = x_0'$ produce the same output $y(t)$, $t_0 \leq t \leq t_1$, for any input $u(t)$, $t_0 \leq t \leq t_1$, then $x_0 - x_0'$ lies in the unreconstructible subspace. This is obviously true since by the linearity of the system,

$$y(t; t_0, x_0, u) = y(t; t_0, x_0', u), \qquad t_0 \leq t \leq t_1, \qquad \text{1-357}$$

is equivalent to

$$y(t; t_0, x_0 - x_0', 0) = 0, \qquad t_0 \leq t \leq t_1, \qquad \text{1-358}$$

which shows that $x_0 - x_0'$ is in the unreconstructible subspace.

The second part of the theorem is proved as follows. The addition of an arbitrary vector x_0'' in the unreconstructible subspace to x_0 results in the addition of $\exp[A(t_1 - t_0)]x_0''$ to the terminal state. Since $\exp[A(t_1 - t_0)]$ can be expanded in powers of A, and the unreconstructible subspace is invariant under A, $\exp[A(t_1 - t_0)]x_0''$ is also in the unreconstructible subspace. Moreover, since $\exp[A(t_1 - t_0)]$ is nonsingular, this proves that also the terminal state is determined within the addition of an arbitrary vector in the unreconstructible subspace.

We now discuss a state transformation that represents the system in a *canonical form*, which clearly exhibits the reconstructibility properties of the system. Let us suppose that Q has rank $m \leq n$, that is, Q possesses m linearly independent row vectors. This means that the null space of Q, hence the unreconstructible subspace of the system, has dimension $n - m$. The row vectors of Q span an m-dimensional linear subspace; let the row vectors $f_1, f_2, \cdots, f_m$ be a basis for this subspace. An obvious choice for this basis is a set of m independent row vectors from Q. Furthermore, let $f_{m+1}, f_{m+2}, \cdots, f_n$ be $n - m$ linearly independent row vectors which together with $f_1, \cdots, f_m$ span the whole n-dimensional space. Now form the nonsingular transformation matrix

$$U = \begin{pmatrix} U_1 \\ U_2 \end{pmatrix}, \quad \text{1-359}$$

where

$$U_1 = \begin{pmatrix} f_1 \\ f_2 \\ \cdot \\ \cdot \\ \cdot \\ f_m \end{pmatrix} \quad \text{and} \quad U_2 = \begin{pmatrix} f_{m+1} \\ f_{m+2} \\ \cdot \\ \cdot \\ \cdot \\ f_n \end{pmatrix}. \quad \text{1-360}$$

Finally, introduce a transformed state variable $x'(t)$ as

$$x'(t) = Ux(t). \quad \text{1-361}$$

Substitution into **1-356** yields

$$\begin{aligned} U^{-1}\dot{x}'(t) &= AU^{-1}x'(t) + Bu(t), \\ y(t) &= CU^{-1}x'(t), \end{aligned} \quad \text{1-362}$$

or

$$\begin{aligned} \dot{x}'(t) &= UAU^{-1}x'(t) + UBu(t), \\ y(t) &= CU^{-1}x'(t). \end{aligned} \quad \text{1-363}$$

We partition U^{-1} as follows

$$U^{-1} = (T_1, T_2), \quad \text{1-364}$$

1.7 Reconstructibility

where the partitioning corresponds to that of U so that T_1 has m and T_2 $n - m$ columns. We have

$$UU^{-1} = \begin{pmatrix} U_1 \\ U_2 \end{pmatrix}(T_1, T_2) = \begin{pmatrix} U_1 T_1 & U_1 T_2 \\ U_2 T_1 & U_2 T_2 \end{pmatrix} = \begin{pmatrix} I_m & 0 \\ 0 & I_{n-m} \end{pmatrix}, \qquad \text{1-365}$$

from which we conclude that

$$U_1 T_2 = 0. \qquad \text{1-366}$$

The rows of U_1 are made up of linear combinations of the linearly independent rows of the reconstructibility matrix Q. This means that any vector x that satisfies $U_1 x = 0$ also satisfies $Qx = 0$, hence is in the unreconstructible subspace. Since

$$U_1 T_2 = 0, \qquad \text{1-367}$$

all column vectors of T_2 must be in the unreconstructible subspace. Because T_2 has $n - m$ linearly independent column vectors, and the unreconstrucible subspace has dimension $n - m$, the column vectors of T_2 form a basis for the subspace. With this it follows from **1-367** that $U_1 x = 0$ for any x in the subspace.

With the partitionings **1-359** and **1-364**, we have

$$UAU^{-1} = \begin{pmatrix} U_1 \\ U_2 \end{pmatrix} A(T_1, T_2) = \begin{pmatrix} U_1 A T_1 & U_1 A T_2 \\ U_2 A T_1 & U_2 A T_2 \end{pmatrix} \qquad \text{1-368}$$

and

$$CU^{-1} = (CT_1, CT_2). \qquad \text{1-369}$$

All column vectors of T_2 are in the unreconstructible subspace; because the subspace is invariant under A (Lemma 1.4), the columns of AT_2 are also in the subspace, and we have from **1-367**

$$U_1 A T_2 = 0. \qquad \text{1-370}$$

Since the rows of C are rows of the reconstructibility matrix Q, and the columns of T_2 are in the unreconstructible subspace, hence in the null space of Q, we must also have

$$CT_2 = 0. \qquad \text{1-371}$$

We summarize our results as follows.

Theorem 1.35. *Consider the n-th order time-invariant linear system*

$$\dot{x}(t) = Ax(t) + Bu(t),$$
$$y(t) = Cx(t). \qquad \text{1-372}$$

74   Elements of Linear System Theory

Form a nonsingular transformation matrix

$$U = \begin{pmatrix} U_1 \\ U_2 \end{pmatrix}, \qquad \text{1-373}$$

where the m rows of U_1 form a basis for the m-dimensional ($m \leq n$) subspace spanned by the rows of the reconstructibility matrix of the system. The $n - m$ rows of U_2 form together with the m rows of U_1 a basis for the whole n-dimensional space. Define a transformed state variable $x'(t)$ by

$$x'(t) = Ux(t). \qquad \text{1-374}$$

Then in terms of the transformed state variable the system is represented in the **reconstructibility canonical form**

$$\dot{x}'(t) = \begin{pmatrix} A'_{11} & 0 \\ A'_{21} & A'_{22} \end{pmatrix} x'(t) + \begin{pmatrix} B'_1 \\ B'_2 \end{pmatrix} u(t), \qquad \text{1-375}$$

$$y(t) = (C'_1, \ 0)x'(t).$$

Here A'_{11} is an $m \times m$ matrix, and the pair $\{A'_{11}, C'_1\}$ is completely reconstructible.

Partitioning

$$x'(t) = \begin{pmatrix} x'_1(t) \\ x'_2(t) \end{pmatrix}, \qquad \text{1-376}$$

where x'_1 has dimension m and x'_2 dimension $n - m$, we see from Theorem 1.35 that the system can be represented as in Fig. 1.10. We note that nothing about x'_2 can be inferred from observing the output y. The fact that the pair $\{A'_{11}, C'_1\}$ is completely reconstructible follows from the fact that if an initial

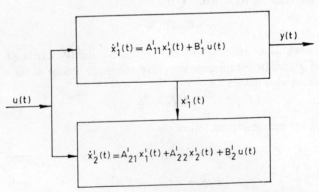

Fig. 1.10. Reconstructibility canonical form of a time-invariant linear differential system.

state $x'(t_0)$ produces a zero input response identical to zero, it must be of the form $x'(t_0) = \text{col}\,(0, x'_{20})$. The complete proof is left as an exercise.

We finally note that the reconstructibility canonical form is not unique because both U_1 and U_2 can to some extent be arbitrarily chosen. No matter how the transformation is performed, however, the characteristic values of A'_{11} and A'_{22} can be shown to be always the same. This leads us to refer to the characteristic values of A'_{11} as the *reconstructible poles*, and the characteristic values of A'_{22} as the *unreconstructible poles* of the system **1-372**. Let us assume for simplicity that all characteristic values of the system are distinct. Then it can be proved that *the unreconstructible subspace of the system is spanned by those characteristic vectors of the system that correspond to the unreconstructible poles*. This is true both for the transformed version **1-375** and the original representation **1-372** of the system. Quite naturally, we now define the *reconstructible subspace* of the system **1-372** as the *subspace spanned by the characteristic vectors of the system corresponding to the reconstructible poles*.

Example 1.25. *Inverted pendulum*

In Example 1.24 we saw that the inverted pendulum is not completely reconstructible if the angle $\phi(t)$ is chosen as the observed variable. We now determine the unreconstructible subspace and the reconstructibility canonical form. It is easy to see that the rows of the reconstructibility matrix Q as given by **1-349** are spanned by the row vectors

$$(-1, 0, 1, 0),\ (0, -1, 0, 1), \quad \text{and} \quad (0, 1, 0, 0). \qquad \textbf{1-377}$$

Any vector $x = \text{col}\,(\xi_1, \xi_2, \xi_3, \xi_4)$ in the null space of Q must therefore satisfy

$$-\xi_1 + \xi_3 = 0,$$
$$-\xi_2 + \xi_4 = 0, \qquad \textbf{1-378}$$
$$\xi_2 = 0.$$

This means that the unreconstructible subspace of the system is spanned by

$$\text{col}\,(1, 0, 1, 0). \qquad \textbf{1-379}$$

Any initial state proportional to this vector is indistinguishable from the zero state, as shown in Example 1.23.

To bring the system equations into reconstructibility canonical form, let us choose the row vectors **1-377** as the first three rows of the transformation matrix U. For the fourth row we select, rather arbitrarily, the row vector

$$(1, 0, 0, 0). \qquad \textbf{1-380}$$

With this we find for the transformation matrix U and its inverse

$$U = \begin{pmatrix} -1 & 0 & 1 & 0 \\ 0 & -1 & 0 & 1 \\ 0 & 1 & 0 & 0 \\ 1 & 0 & 0 & 0 \end{pmatrix}, \quad U^{-1} = \begin{pmatrix} 0 & 0 & 0 & 1 \\ 0 & 0 & 1 & 0 \\ 1 & 0 & 0 & 1 \\ 0 & 1 & 1 & 0 \end{pmatrix}. \quad \textbf{1-381}$$

It follows for the transformed representation

$$\dot{x}'(t) = \left(\begin{array}{cc|cc} 0 & 1 & 0 & 0 \\ \dfrac{g}{L'} & 0 & \dfrac{F}{M} & 0 \\ \hline 0 & 0 & -\dfrac{F}{M} & 0 \\ 0 & 0 & 1 & 0 \end{array}\right) x'(t) + \left(\begin{array}{c} 0 \\ -\dfrac{1}{M} \\ \hline \dfrac{1}{M} \\ 0 \end{array}\right) \mu(t), \quad \textbf{1-382}$$

$$\eta(t) = \left(\dfrac{1}{L'},\ 0,\ 0\ \Big|\ 0\right) x'(t).$$

The components of the transformed state are, from **1-24**,

$$\begin{aligned} \xi_1'(t) &= -\xi_1(t) + \xi_3(t) = L'\phi(t), \\ \xi_2'(t) &= -\xi_2(t) + \xi_4(t) = L'\dot\phi(t), \\ \xi_3'(t) &= \xi_2(t) = \dot s(t), \\ \xi_4'(t) &= \xi_1(t) = s(t). \end{aligned} \quad \textbf{1-383}$$

In this representation the position and velocity of the pendulum relative to the carriage, as well as the velocity of the carriage, can be reconstructed from the observed variable, but not the position of the carriage.

It is easily seen that the reconstructible poles of the system are $-F/M$ and $\pm\sqrt{g/L'}$. The unreconstructible pole is 0.

1.7.4* Detectablity

In the preceding section it was found that if the output variable of a not completely reconstructible system is observed there is always an uncertainty about the actual state of the system since to any possible state we can always add an arbitrary vector in the unreconstructible subspace (Theorem 1.34). The best we can hope for in such a situation is that any state in the unreconstructible subspace has the property that the zero input response of the system to this

1.7 Reconstructibility

state converges to zero. This is the case when any state in the unreconstructible subspace is also in the stable subspace of the system. Then, whatever we guess for the unreconstructible component of the state, the error will never grow indefinitely. A system with this property will be called detectable (Wonham, 1968a). We define this property as follows.

Definition 1.20. *The linear time-invariant system*

$$\dot{x}(t) = Ax(t) + Bu(t),$$
$$y(t) = Cx(t),$$
1-384

*is **detectable** if its unreconstructible subspace is contained in its stable subspace.*

It is convenient to employ the following abbreviated terminology.

Definition 1.21. *The pair $\{A, C\}$ is detectable if the system*

$$\dot{x}(t) = Ax(t),$$
$$y(t) = Cx(t),$$
1-385

is detectable.

The following result is an immediate consequence of the definition:

Theorem 1.36. *Any asymptotically stable system of the form* **1-384** *is detectable. Any completely reconstructible system of the form* **1-384** *is detectable.*

Detectable systems possess the following property.

Theorem 1.37. *Consider the linear time-invariant system*

$$\dot{x}(t) = Ax(t),$$
$$y(t) = Cx(t).$$
1-386

Suppose that it is transformed according to Theorem 1.35 into the form

$$\dot{x}'(t) = \begin{pmatrix} A'_{11} & 0 \\ A'_{21} & A'_{22} \end{pmatrix} x'(t),$$
$$y(t) = (C'_1, \ 0)x'(t),$$
1-387

where the pair $\{A'_{11}, C'_1\}$ is completely reconstructible. Then the system is detectable if and only if the matrix A'_{22} is asymptotically stable.

This theorem can be summarized by stating that a system is detectable if and only if its unreconstructible poles are stable. We prove the theorem as follows.

78 Elements of Linear System Theory

(a) *Detectability implies A'_{22} asymptotically stable:* Let us partition the transformed state variable as

$$x'(t) = \begin{pmatrix} x'_1(t) \\ x'_2(t) \end{pmatrix}, \qquad \text{1-388}$$

where the dimension m of $x'_1(t)$ is equal to the rank m of the reconstructibility matrix. The fact that the system is detectable implies that any initial state in the unreconstructible subspace gives a response that converges to zero. Any initial state in the unreconstructible subspace has in the transformed representation the form

$$x'(0) = \begin{pmatrix} 0 \\ x'_2(0) \end{pmatrix}. \qquad \text{1-389}$$

The response of the transformed state to this initial state is given by

$$x'(t) = \begin{pmatrix} 0 \\ e^{A'_{22}t} x'_2(0) \end{pmatrix}. \qquad \text{1-390}$$

Since this must give a response that converges to zero, A'_{22} must be stable.

(b) *A'_{22} asymptotically stable implies detectability:* Any initial state $x(0)$ in the unreconstructible subspace must in the transformed representation have the form

$$x'(0) = \begin{pmatrix} 0 \\ x'_2(0) \end{pmatrix}. \qquad \text{1-391}$$

The response to this initial state is

$$x'(t) = \begin{pmatrix} 0 \\ e^{A'_{22}t} x'_2(0) \end{pmatrix}. \qquad \text{1-392}$$

Since A'_{22} is stable, this response converges to zero, which shows that $x(0)$, which was assumed to be in the unreconstructible subspace, is also in the stable subspace. This implies that the system is detectable.

Example 1.26. *Inverted pendulum*

Consider the inverted pendulum in the transformed representation of Example 1.25. The matrix A'_{22} has the characteristic value 0, which implies that the system is not detectable. This means that if initially there is an uncertainty about the position of the carriage, the error made in guessing it will remain constant in time.

1.7.5* Reconstructibility of Time-Varying Linear Systems

The reconstructibility of time-varying linear systems can be ascertained by the following test.

Theorem 1.38. *Consider the linear time-varying system*

$$\dot{x}(t) = A(t)x(t) + B(t)u(t),$$
$$y(t) = C(t)x(t). \qquad \text{1-393}$$

Define the nonnegative-definite matrix function

$$M(t, t_1) = \int_t^{t_1} \Phi^T(\tau, t)C^T(\tau)C(\tau)\Phi(\tau, t)\, d\tau, \qquad \text{1-394}$$

where $\Phi(t, t_0)$ is the transition matrix of the system. Then the system is completely reconstructible if and only if for all t_1 there exists a t_0 with $-\infty < t_0 < t_1$ such that $M(t_0, t_1)$ is nonsingular.

For a proof we refer the reader to Bucy and Joseph (1968) and Kalman, Falb, and Arbib (1969). A stronger form of reconstructibility results by imposing further conditions on the matrix M (Kalman, 1960):

Definition 1.22. *The time-varying system* **1-393** *is* **uniformly completely reconstructible** *if there exist positive constants σ, α_0, α_1, β_0, and β_1 such that*

(a) $\qquad \alpha_0 I \leq M(t_1 - \sigma, t_1) \leq \alpha_1 I \qquad \text{for all } t_1;$ \hfill **1-395**

(b) $\beta_0 I \leq \Phi^T(t_1 - \sigma, t_1)M(t_1 - \sigma, t_1)\Phi(t_1 - \sigma, t_1) \leq \alpha_1 I \qquad \text{for all } t_1,$

\hfill **1-396**

where $M(t, t_1)$ is the matrix function **1-394**.

Uniform reconstructibility guarantees that identification of the state is always possible within roughly the same time. For time-invariant systems the following holds.

Theorem 1.39. *The time-invariant linear system*

$$\dot{x}(t) = Ax(t) + Bu(t),$$
$$y(t) = Cx(t), \qquad \text{1-397}$$

is uniformly completely reconstructible if and only if it is completely reconstructible.

1.8* DUALITY OF LINEAR SYSTEMS

In the discussion of controllability and reconstructibility, we have seen that there is a striking symmetry between these properties. This symmetry can be made explicit by introducing the idea of duality (Kalman, 1960; Kalman, Falb, and Arbib, 1969).

Definition 1.23. *Consider the linear time-varying system*

$$\dot{x}(t) = A(t)x(t) + B(t)u(t),$$
$$y(t) = C(t)x(t),$$
1-398

and also the system

$$\dot{x}^*(t) = A^T(t^* - t)x^*(t) + C^T(t^* - t)u^*(t),$$
$$y^*(t) = B^T(t^* - t)x^*(t),$$
1-399

where t^* is an arbitrary fixed time. Then **1-399** *is called the* **dual** *of the system* **1-398** *with respect to the time t^*.*

The purpose of introducing the dual system becomes apparent in Chapter 4 when we discuss the duality of linear optimal control problems and linear optimal observer problems. The following result is immediate.

Theorem 1.40. *The dual of the system* **1-399** *with respect to the time t^* is the original system* **1-398**.

There is a close connection between the reconstructibility and controllability of a system and its dual.

Theorem 1.41. *Consider the system* **1-398** *and its dual* **1-399** *where t^* is arbitrary.*
(a) *The system* **1-398** *is (uniformly) completely controllable if and only if its dual is (uniformly) completely reconstructible.*
(b) *The system* **1-398** *is (uniformly) completely reconstructible if and only if its dual is (uniformly) completely controllable.*
(c) *Assume that* **1-398** *is time-invariant. Then* **1-398** *is stabilizable if and only if its dual is detectable.*
(d) *Assume that* **1-398** *is time-invariant. Then* **1-398** *is detectable if and only if its dual is stabilizable.*

We give the proof only for time-invariant systems. The reconstructibility matrix of the dual system is given by

$$Q^* = \begin{pmatrix} B^T \\ B^T(A^T) \\ \vdots \\ B^T(A^T)^{n-1} \end{pmatrix} = P^T,$$
1-400

1.8 Duality of Linear Systems

where P is the controllability matrix of the original system. This immediately proves (a).

Part (b) of the theorem follows similarly. The controllability matrix of the dual system is given by

$$P^* = (C^T, A^T C^T, \cdots, (A^T)^{n-1} C^T) = Q^T, \qquad \text{1-401}$$

where Q is the reconstructibility matrix of the original system. This implies the validity of (b).

Part (c) can be proved as follows. The original system can be transformed by a transformation $x' = T^{-1}x$ according to Theorem 1.26 (Section 1.6.3) into the controllability canonical form

$$\dot{x}'(t) = \begin{pmatrix} A'_{11} & A'_{12} \\ 0 & A'_{22} \end{pmatrix} x'(t) + \begin{pmatrix} B'_1 \\ 0 \end{pmatrix} u(t), \qquad \text{1-402}$$

$$y(t) = (C'_1, \; C'_2) x'(t).$$

If **1-398** is stabilizable, the pair $\{A'_{11}, B'_1\}$ is completely controllable and A'_{22} is stable. The dual of the transformed system is

$$\dot{x}'^*(t) = \begin{pmatrix} A'^T_{11} & 0 \\ A'^T_{12} & A'^T_{22} \end{pmatrix} x'^*(t) + \begin{pmatrix} C'^T_1 \\ C'^T_2 \end{pmatrix} u^*(t), \qquad \text{1-403}$$

$$y^*(t) = (B'^T_1, \; 0) x'^*(t).$$

Since $\{A'_{11}, B'_{11}\}$ is completely controllable, $\{A'^T_{11}, B'^T_{11}\}$ is completely reconstructible [part (a)]. Since A'_{22} is stable, A'^T_{22} is also stable. This implies that the system **1-403** is detectable. By the transformation $T^T x^* = x'^*$ (see Problem 1.8), the system **1-403** is transformed into the dual of the original system. Therefore, since **1-403** is detectable, the dual of the original system is also detectable. By reversing the steps of the proof, the converse of Theorem 1.41(c) can also be proved. Part (d) can be proved completely analogously. The proofs of (a) and (b) for the time-varying case are left as an exercise for the reader.

We conclude this section with the following fact, relating the stability of a system and its dual.

Theorem 1.42. *The system* **1-398** *is exponentially stable if and only if its dual* **1-399** *is exponentially stable.*

This result is easily proved by first verifying that if the system **1-398** has the transition matrix $\Phi(t, t_0)$ its dual **1-399** has the transition matrix $\Phi^T(t^* - t_0, t^* - t)$, and then verifying Definition 1.5 (Section 1.4.1).

1.9* PHASE-VARIABLE CANONICAL FORMS

For single-input time-invariant linear systems, it is sometimes convenient to employ the so-called phase-variable canonical form.

Definition 1.24. *A single-input time-invariant linear system is in **phase-variable canonical form** if its system equations have the form*

$$\dot{x}(t) = \begin{pmatrix} 0 & 1 & 0 & \cdots\cdots & 0 \\ 0 & 0 & 1 & 0 & \cdots & 0 \\ \cdots\cdots\cdots\cdots\cdots\cdots\cdots\cdots \\ 0 & \cdots\cdots\cdots\cdots & 0 & 1 \\ -\alpha_0 & -\alpha_1 & \cdots\cdots & -\alpha_{n-1} \end{pmatrix} x(t) + \begin{pmatrix} 0 \\ 0 \\ \cdots \\ 0 \\ 1 \end{pmatrix} \mu(t) \quad , \quad \text{1-404}$$

$$y(t) = Cx(t).$$

Note that no special form is imposed upon the matrix C in this definition. It is not difficult to see that the numbers α_i, $i = 0, \cdots, n - 1$ are the coefficients of the characteristic polynomial

$$\sum_{i=0}^{n} \alpha_i s^i \qquad \text{1-405}$$

of the system, where $\alpha_n = 1$.

It is easily verified that the system **1-404** is always completely controllable. In fact, any completely controllable single-input system can be transformed into phase-variable canonical form.

Theorem 1.43. *Consider the completely controllable single-input time-invariant linear system*

$$\dot{x}(t) = Ax(t) + b\mu(t),$$
$$y(t) = Cx(t), \qquad \text{1-406}$$

where b is a column vector. Let P be the controllability matrix of the system,

$$P = (b, Ab, A^2b, \cdots, A^{n-1}b), \qquad \text{1-407}$$

and let

$$\det(sI - A) = \sum_{i=0}^{n} \alpha_i s^i, \qquad \text{1-408}$$

where $\alpha_n = 1$, be the characteristic polynomial of the matrix A. Then the system is transformed into phase-variable canonical form by a transformation

1.9 Phase-Variable Canonical Forms

$x(t) = Tx'(t)$. T is the nonsingular transformation matrix

$$T = PM,$$

where

$$M = \begin{pmatrix} \alpha_1 & \alpha_2 & \cdots\cdots & \alpha_n \\ \alpha_2 & \alpha_3 & \cdots & \alpha_n & 0 \\ \cdots\cdots\cdots\cdots\cdots\cdots \\ \alpha_{n-1} & \alpha_n & 0 & \cdots\cdots & 0 \\ \alpha_n & 0 & \cdots\cdots\cdots & 0 \end{pmatrix}. \qquad \textbf{1-409}$$

If the system **1-406** *is not completely controllable, no such transformation exists.*

This result can be proved as follows (Anderson and Luenberger, 1967). That the transformation matrix T is nonsingular is easily shown: P is nonsingular due to the assumption of complete controllability, and $\det(M) = 1$ because $\alpha_n = 1$. We now prove that T transforms the system into phase-variable canonical form. By postmultiplying P by M, it is easily seen that T can be written as

$$T = (t_1, t_2, \cdots, t_n), \qquad \textbf{1-410}$$

where the column vectors t_i of T are given by

$$\begin{aligned} t_1 &= \alpha_1 b + \alpha_2 Ab + \alpha_3 A^2 b + \cdots + \alpha_n A^{n-1} b, \\ t_2 &= \alpha_2 b + \alpha_3 Ab + \cdots + \alpha_n A^{n-2} b, \\ &\cdots \\ t_{n-1} &= \alpha_{n-1} b + \alpha_n Ab, \\ t_n &= \alpha_n b. \end{aligned} \qquad \textbf{1-411}$$

It is seen from **1-411** that

$$At_i = t_{i-1} - \alpha_{i-1} t_n, \quad i = 2, 3, \cdots, n, \qquad \textbf{1-412}$$

since $b = t_n$.

Now in terms of the new state variable, the state differential equation of the system is given by

$$\dot{x}'(t) = T^{-1}ATx'(t) + T^{-1}b\mu(t). \qquad \textbf{1-413}$$

Let us consider the matrix $T^{-1}AT$. To this end denote the rows of T^{-1} by r_i, $i = 1, 2, \cdots, n$. Then for $i = 1, 2, \cdots, n$ and $j = 2, 3, \cdots, n$, the (i,j)-th entry of $T^{-1}AT$ is given by

$$(T^{-1}AT)_{ij} = r_i(At_j) = r_i(t_{j-1} - \alpha_{j-1}t_n) = \begin{cases} 1 & \text{if } i = j-1, \\ -\alpha_{j-1} & \text{if } i = n, \\ 0 & \text{otherwise.} \end{cases} \qquad \textbf{1-414}$$

This proves that the last $n - 1$ columns of $T^{-1}AT$ have the form as required in the phase-variable canonical form. To determine the first column, we observe from **1-411** that

$$At_1 = (\alpha_1 A + \alpha_2 A^2 + \cdots + \alpha_n A^n)b = -\alpha_0 b = -\alpha_0 t_n, \qquad \textbf{1-415}$$

since according to the Cayley–Hamilton theorem

$$\alpha_0 I + \alpha_1 A + \alpha_2 A^2 + \cdots + \alpha_n A^n = 0. \qquad \textbf{1-416}$$

Thus we have for $i = 1, 2, \cdots, n$,

$$(T^{-1}AT)_{i1} = r_i(At_1) = -\alpha_0 r_i t_n = \begin{cases} -\alpha_0 & \text{if } i = n, \\ 0 & \text{otherwise.} \end{cases} \qquad \textbf{1-417}$$

Similarly, we can show that $T^{-1}b$ is in the form required, which terminates the proof of the first part of Theorem 1.43. The last statement of Theorem 1.43 is easily verified: if the system **1-406** is not completely controllable, no nonsingular transformation can bring the system into phase-variable canonical form, since nonsingular transformations preserve controllability properties (see Problem 1.6). An alternate method of finding the phase-variable canonical form is given by Ramaswami and Ramar (1968). Computational rules are described by Tuel (1966), Rane (1966), and Johnson and Wonham (1966).

For single-input systems represented in phase-variable canonical form, certain linear optimal control problems are much easier to solve than if the system is given in its general form (see, e.g., Section 3.2). Similarly, certain filtering problems involving the reconstruction of the state from observations of the output variable are more easily solved when the system is in the dual phase-variable canonical form.

Definition 1.25. *A single-output linear time-invariant system is in* **dual phase-variable canonical form** *if it is represented as follows:*

$$\dot{x}(t) = \begin{pmatrix} 0 & 0 & 0 & \cdots & 0 & -\alpha_0 \\ 1 & 0 & 0 & \cdots & 0 & -\alpha_1 \\ 0 & 1 & 0 & \cdots & 0 & -\alpha_2 \\ \cdots & \cdots & \cdots & \cdots & \cdots & \cdots \\ 0 & 0 & \cdots & \cdots & 0 & 1 & -\alpha_{n-1} \end{pmatrix} x(t) + Bu(t), \qquad \textbf{1-418}$$

$$\eta(t) = (0 \quad 0 \quad 0 \quad \cdots \quad 0 \quad 1) x(t).$$

It is noted that the definition imposes no special form on the matrix B. By "dualizing" Theorem 1.43, it is not difficult to establish a transformation to transform completely reconstructible systems into dual canonical form.

1.10 VECTOR STOCHASTIC PROCESSES

1.10.1 Definitions

In later chapters of this book we use stochastic processes as mathematical models for disturbances and noise phenomena. Often several disturbances and noise phenomena simultaneously influence a given system. This makes it necessary to introduce vector-valued stochastic processes, which constitute the topic of this section.

A stochastic process can be thought of as a family of time functions. Each time function we call a *realization* of the process. Suppose that $v_1(t)$, $v_2(t)$, $\cdots$, $v_n(t)$ are n scalar stochastic processes which are possibly mutually dependent. Then we call

$$v(t) = \text{col } [v_1(t), v_2(t), \cdots, v_n(t)] \qquad \text{1-419}$$

a *vector stochastic process*. We always assume that each of the components of $v(t)$ takes real values, and that $t \geq t_0$, with t_0 given.

A stochastic process can be characterized by specifying the joint probability distributions

$$P\{v(t_1) \leq v_1, v(t_2) \leq v_2, \cdots, v(t_m) \leq v_m\} \qquad \text{1-420}$$

for all real $v_1, v_2, \cdots, v_m$, for all $t_1, t_2, \cdots, t_m \geq t_0$ and for every natural number m. Here the vector inequality $v(t_i) \leq v_i$ is by definition satisfied if the inequalities

$$v_j(t_i) \leq v_{ij}, \qquad j = 1, 2, \cdots, n, \qquad \text{1-421}$$

are simultaneously satisfied. The v_{ij} are the components of v_i, that is, $v_i = \text{col } (v_{i1}, v_{i2}, \cdots, v_{in})$.

A special class of stochastic processes consists of those processes the statistical properties of which do not change with time. We define more precisely.

Definition 1.26. *A stochastic process $v(t)$ is **stationary** if*

$$P\{v(t_1) \leq v_1, \cdots, v(t_m) \leq v_m\}$$
$$= P\{v(t_1 + \theta) \leq v_1, \cdots, v(t_m + \theta) \leq v_m\} \qquad \text{1-422}$$

for all $t_1, t_2, \cdots, t_m$, for all $v_1, \cdots, v_m$, for every natural number m, and for all θ.

The joint probability distributions that characterize a stationary stochastic process are thus invariant with respect to a shift in the time origin.

In many cases we are interested only in the first and second-order properties of a stochastic process, namely, in the mean and covariance matrix or, equivalently, the second-order joint moment matrix. We define these notions as follows.

Definition 1.27. *Consider a vector-valued stochastic process $v(t)$. Then we call*

$$m(t) = E\{v(t)\} \qquad \text{1-423}$$

*the **mean** of the process,*

$$R_v(t_1, t_2) = E\{[v(t_1) - m(t_1)][v(t_2) - m(t_2)]^T\} \qquad \text{1-424}$$

*the **covariance matrix**, and*

$$C_v(t_1, t_2) = E\{v(t_1)v^T(t_2)\} \qquad \text{1-425}$$

*the **second-order joint moment matrix** of $v(t)$. $R_v(t, t) = Q(t)$ is termed the **variance matrix**, while $C_v(t, t) = Q'(t)$ is the **second-order moment matrix** of the process.*

Here E is the expectation operator. We shall often assume that the stochastic process under consideration has zero mean, that is, $m(t) = 0$ for all t; in this case the covariance matrix and the second-order joint moment matrix coincide. The joint moment matrix written out more explicitly is

$$C_v(t_1, t_2) = E\{v(t_1)v^T(t_2)\} = \begin{pmatrix} E\{v_1(t_1)v_1(t_2)\} & \cdots & E\{v_1(t_1)v_m(t_2)\} \\ E\{v_2(t_1)v_1(t_2)\} & \cdots & E\{v_2(t_1)v_m(t_2)\} \\ \cdots\cdots\cdots\cdots\cdots\cdots\cdots\cdots\cdots\cdots\cdots \\ E\{v_m(t_1)v_1(t_2)\} & \cdots & E\{v_m(t_1)v_m(t_2)\} \end{pmatrix}.$$

$$\text{1-426}$$

Each element of $C_v(t_1, t_2)$ is a scalar joint moment function. Similarly, each element of $R_v(t_1, t_2)$ is a scalar covariance function. It is not difficult to prove the following.

Theorem 1.44. *The covariance matrix $R_v(t_1, t_2)$ and the second-order joint moment matrix $C_v(t_1, t_2)$ of a vector-valued stochastic process $v(t)$ have the following properties.*

(a) $R_v(t_2, t_1) = R_v^T(t_1, t_2)$ *for all t_1, t_2, and* $\qquad$ 1-427

$\qquad C_v(t_2, t_1) = C_v^T(t_1, t_2)$ *for all t_1, t_2;* $\qquad$ 1-428

(b) $Q(t) = R_v(t, t) \geq 0$ *for all t, and* $\qquad$ 1-429

$\qquad Q'(t) = C_v(t, t) \geq 0$ *for all t;* $\qquad$ 1-430

(c) $C_v(t_1, t_2) = R_v(t_1, t_2) + m(t_1)m^T(t_2)$ *for all t_1, t_2,* $\qquad$ 1-431

where $m(t)$ is the mean of the process.

1.10 Vector Stochastic Processes

Here the notation $M \geq 0$, where M is a square symmetric real matrix, means that M is nonnegative-definite, that is,

$$x^T M x \geq 0 \quad \text{for all real } x. \qquad \text{1-432}$$

The theorem is easily proved from the definitions of $R_v(t_1, t_2)$ and $C_v(t_1, t_2)$. Since the second-order properties of the stochastic process are equally well characterized by the covariance matrix as by the joint moment matrix, we usually consider only the covariance matrix.

For stationary processes we have the following result.

Theorem 1.45. *Suppose that $v(t)$ is a stationary stochastic process. Then its mean $m(t)$ is constant and its covariance matrix $R_v(t_1, t_2)$ depends on $t_1 - t_2$ only.*

This is easily shown from the definition of stationarity.

It sometimes happens that a stochastic process has a constant mean and a covariance matrix that depends on $t_1 - t_2$ only, while its other statistical properties are not those of a stationary process. Since frequently we are interested only in the first- and second-order properties of a stochastic process, we introduce the following notion.

Definition 1.28. *The stochastic process $v(t)$ is called **wide-sense stationary** if its second-order moment matrix $C_v(t, t)$ is finite for all t, its mean $m(t)$ is constant, and its covariance matrix $R_v(t_1, t_2)$ depends on $t_1 - t_2$ only.*

Obviously, any stationary process with finite second-order moment matrix is also wide-sense stationary.

Let $v_1(t)$ and $v_2(t)$ be two vector stochastic processes. Then v_1 and v_2 are called *independent* processes if $\{v_1(t_1), v_1(t_2), \cdots, v_1(t_l)\}$ and $\{v_2(t_1'), v_2(t_2'), \cdots, v_2(t_m')\}$ are independent sets of stochastic variables for all $t_1, t_2, \cdots, t_l$, $t_1', t_2', \cdots, t_m' \geq t_0$ and for all natural numbers m and l. Furthermore, v_1 and v_2 are called *uncorrelated* stochastic processes if $v_1(t_1)$ and $v_2(t_2)$ are uncorrelated vector stochastic variables for all $t_1, t_2 \geq t_0$, that is,

$$E\{[v_1(t_1) - m_1(t_1)][v_2(t_2) - m_2(t_2)]^T\} = 0$$

for all t_1 and t_2, where m_1 is the mean of v_1 and m_2 that of v_2.

Example 1.27. *Gaussian stochastic process*

A Gaussian stochastic process v is a stochastic process where for each set of instants of time $t_1, t_2, \cdots, t_m \geq t_0$ the n-dimensional vector stochastic variables $v(t_1), v(t_2), \cdots, v(t_m)$ have a Gaussian joint probability distribution.

88 Elements of Linear System Theory

If the compound covariance matrix

$$R = \begin{pmatrix} R_v(t_1, t_1) & R_v(t_1, t_2) & \cdots & R_v(t_1, t_m) \\ R_v(t_2, t_1) & R_v(t_2, t_2) & \cdots & R_v(t_2, t_m) \\ \vdots & & & \\ R_v(t_m, t_1) & R_v(t_m, t_2) & \cdots & R_v(t_m, t_m) \end{pmatrix} \quad \text{1-433}$$

is nonsingular, the corresponding probability density function can be written as

$$p(v_1, v_2, \cdots, v_m) = \frac{1}{[(2\pi)^{mn} \det(R)]^{1/2}}$$
$$\cdot \exp\left\{-\tfrac{1}{2} \sum_{i=1}^{m} \sum_{j=1}^{m} [v_i - m(t_i)]^T \Lambda_{ij} [v_j - m(t_j)]\right\}. \quad \text{1-434}$$

The $n \times n$ matrices Λ_{ij} are obtained by partitioning $\Lambda = R^{-1}$ corresponding to the partitioning of R as follows:

$$\Lambda = \begin{pmatrix} \Lambda_{11} & \Lambda_{12} & \cdots & \Lambda_{1m} \\ \Lambda_{21} & \Lambda_{22} & \cdots & \Lambda_{2m} \\ \vdots & & & \\ \Lambda_{m1} & \Lambda_{m2} & \cdots & \Lambda_{mm} \end{pmatrix}. \quad \text{1-435}$$

Note that this process is completely characterized by its mean and covariance matrix; thus a Gaussian process is stationary if and only if it is wide-sense stationary.

Example 1.28. *Exponentially correlated noise*

A well-known type of wide-sense stationary stochastic process is the so-called exponentially correlated noise. This is a scalar stochastic process $v(t)$ with the covariance function

$$R_v(\tau) = \sigma^2 \exp\left(-\frac{|\tau|}{\theta}\right), \quad \text{1-436}$$

where σ^2 is the variance of the process and θ the "time constant." Many practical processes possess this covariance function.

Example 1.29. *Processes with uncorrelated increments*

A process $v(t)$, $t \geq t_0$, with uncorrelated increments can be defined as follows.

1. The initial value is given by

$$v(t_0) = 0. \quad \text{1-437}$$

1.10 Vector Stochastic Processes

2. For any sequence of instants t_1, t_2, t_3, and t_4, with $t_0 \leq t_1 \leq t_2 \leq t_3 \leq t_4$, the *increments* $v(t_2) - v(t_1)$ and $v(t_4) - v(t_3)$ have zero means and are uncorrelated, that is,

$$E\{v(t_2) - v(t_1)\} = E\{v(t_4) - v(t_3)\} = 0,$$
$$E\{[v(t_2) - v(t_1)][v(t_4) - v(t_3)]^T\} = 0. \qquad \text{1-438}$$

The mean of such a process is easily determined:

$$m(t) = E\{v(t)\} = E\{v(t) - v(t_0)\}$$
$$= 0, \qquad t \geq t_0. \qquad \text{1-439}$$

Suppose for the moment that $t_2 \geq t_1$. Then we have for the covariance matrix

$$R_v(t_1, t_2) = E\{v(t_1)v^T(t_2)\}$$
$$= E\{[v(t_1) - v(t_0)][v(t_2) - v(t_1) + v(t_1) - v(t_0)]^T\}$$
$$= E\{[v(t_1) - v(t_0)][v(t_1) - v(t_0)]^T\}$$
$$= E\{v(t_1)v^T(t_1)\}$$
$$= Q(t_1), \qquad t_2 \geq t_1 \geq t_0, \qquad \text{1-440}$$

where

$$Q(t) = E\{v(t)v^T(t)\} \qquad \text{1-441}$$

is the variance matrix of the process. Similarly,

$$R_v(t_1, t_2) = Q(t_2) \qquad \text{for } t_1 \geq t_2 \geq t_0. \qquad \text{1-442}$$

Clearly, a process with uncorrelated increments cannot be stationary or wide-sense stationary, except in the trivial case in which $Q(t) = 0$, $t \geq t_0$.

Let us now consider the variance matrix of the process. We can write for $t_2 \geq t_1 \geq t_0$:

$$Q(t_2) = E\{v(t_2)v^T(t_2)\}$$
$$= E\{[v(t_2) - v(t_1) + v(t_1) - v(t_0)][v(t_2) - v(t_1) + v(t_1) - v(t_0)]^T\}$$
$$= E\{[v(t_2) - v(t_1)][v(t_2) - v(t_1)]^T\} + Q(t_1). \qquad \text{1-443}$$

Obviously, $Q(t)$ is a monotonically nondecreasing matrix function of t in the sense that

$$Q(t_2) \geq Q(t_1) \qquad \text{for all } t_2 \geq t_1 \geq t_0. \qquad \text{1-444}$$

Here, if A and B are two symmetric real matrices, the notation

$$A \geq B \qquad \text{1-445}$$

implies that the matrix $A - B$ is nonnegative-definite. Let us now assume that the matrix function $Q(t)$ is absolutely continuous, that is, we can write

$$Q(t) = \int_0^t V(\tau) \, d\tau, \qquad \textbf{1-446}$$

where $V(t)$ is a nonnegative-definite symmetric matrix function. It then follows from **1-443** that the variance matrix of the increment $v(t_2) - v(t_1)$ is given by

$$E\{[v(t_2) - v(t_1)][v(t_2) - v(t_1)]^T\} = Q(t_2) - Q(t_1)$$
$$= \int_{t_1}^{t_2} V(\tau) \, d\tau. \qquad \textbf{1-447}$$

Combining **1-440** and **1-442**, we see that if **1-446** holds the covariance matrix of the process can be expressed as

$$R_v(t_1, t_2) = \int_{t_0}^{\min(t_1, t_2)} V(\tau) \, d\tau. \qquad \textbf{1-448}$$

One of the best-known processes with uncorrelated increments is the *Brownian motion process*, also known as the *Wiener process* or the *Wiener–Lévy process*. This is a process with uncorrelated increments where each of the increments $v(t_2) - v(t_1)$ is a Gaussian stochastic vector with zero mean and variance matrix $(t_2 - t_1)I$, where I is the unit matrix. A generalization of this process is obtained when it is assumed that each increment $v(t_2) - v(t_1)$ is a Gaussian stochastic vector with zero mean and variance matrix given in the form **1-447**. Since in the Brownian motion process the increments are uncorrelated and Gaussian, they are independent. Obviously, Brownian motion is a Gaussian process. It is an important tool in the theory of stochastic processes.

1.10.2 Power Spectral Density Matrices

For scalar wide-sense stationary stochastic processes, the power spectral density function is defined as the Fourier transform of the covariance function. Similarly, we define for vector stochastic processes:

Definition 1.29. *The* **power spectral density matrix** $\Sigma_v(\omega)$ *of a wide-sense stationary vector stochastic process is defined as the Fourier transform, if it exists, of the covariance matrix* $R_v(t_1 - t_2)$ *of the process, that is,*

$$\Sigma_v(\omega) = \int_{-\infty}^{\infty} e^{-j\omega\tau} R_v(\tau) \, d\tau. \qquad \textbf{1-449}$$

Note that we have allowed a slight inconsistency in the notation of the covariance matrix by replacing the two variables t_1 and t_2 by the single variable $t_1 - t_2$. The power spectral density matrix has the following properties.

1.10 Vector Stochastic Processes

Theorem 1.46. *Suppose that $\Sigma_v(\omega)$ is the spectral density matrix of a wide-sense stationary process $v(t)$. Then $\Sigma_v(\omega)$ is a complex matrix that has the properties:*

(a) $\Sigma_v(-\omega) = \Sigma_v^T(\omega)$ *for all ω;*   1-450

(b) $\Sigma_v^*(\omega) = \Sigma_v(\omega)$ *for all ω;*   1-451

(c) $\Sigma_v(\omega) \geq 0$ *for all ω.*   1-452

Here the asterisk denotes the complex conjugate transpose, while $M \geq 0$, where M is a complex matrix, indicates that M is a nonnegative-definite matrix, that is, $x^*Mx \geq 0$ for all complex x.

The proofs of parts (a) and (b) follow in a straightforward manner from the definition of $\Sigma_v(\omega)$ and Theorem 1.44. In order to prove part (c), one can extend the proof given by Davenport and Root (1958, Chapter 6) to the vector case. The reason for the term power spectral density matrix becomes apparent in Section 1.10.4.

Example 1.30. *Exponentially correlated noise*

In Example 1.28 we considered exponentially correlated noise, a scalar wide-sense stationary process $v(t)$ with covariance function

$$R_v(t_1 - t_2) = \sigma^2 \exp\left(-\frac{|t_1 - t_2|}{\theta}\right). \qquad 1\text{-}453$$

By Fourier transformation it easily follows that $v(t)$ has the power spectral density function

$$\Sigma_v(\omega) = \frac{2\sigma^2\theta}{1 + \omega^2\theta^2}, \qquad 1\text{-}454$$

provided $\theta > 0$.

1.10.3 The Response of Linear Systems to Stochastic Inputs

In this section we study the statistical properties of the response of a linear system if the input is a realization of a stochastic process. We have the following result.

Theorem 1.47. *Consider a linear system with impulse response matrix $K(t, \tau)$ which at time t_0 is in the zero state. Suppose that the input to the system is a realization of a stochastic process $u(t)$ with mean $m_u(t)$ and covariance matrix $R_u(t_1, t_2)$. Then the output of the system is a realization of a stochastic process $y(t)$ with mean*

$$m_y(t) = \int_{t_0}^{t} K(t, \tau) m_u(\tau) \, d\tau, \qquad 1\text{-}455$$

92 Elements of Linear System Theory

and covariance matrix

$$R_y(t_1, t_2) = \int_{t_0}^{t_1} d\tau_1 \int_{t_0}^{t_2} K(t_1, \tau_1) R_u(\tau_1, \tau_2) K^T(t_2, \tau_2) \, d\tau_2, \qquad \text{1-456}$$

provided the integrals exist.

We present a formal proof of these results. The output y, which is a stochastic process, is given by

$$y(t) = \int_{t_0}^{t} K(t, \tau) u(\tau) \, d\tau. \qquad \text{1-457}$$

Taking the expectation of both sides of **1-457**, interchanging the order of the integration and the expectation, one obtains **1-455**.

Similarly, we can write (assuming for simplicity that $m_u(t) = 0$)

$$\begin{aligned}
R_y(t_1, t_2) &= E\{y(t_1) y^T(t_2)\} \\
&= E\left\{ \left[\int_{t_0}^{t_1} K(t_1, \tau_1) u(\tau_1) \, d\tau_1 \right] \left[\int_{t_0}^{t_2} K(t_2, \tau_2) u(\tau_2) \, d\tau_2 \right]^T \right\} \\
&= E\left\{ \int_{t_0}^{t_1} d\tau_1 \int_{t_0}^{t_2} d\tau_2 \, K(t_1, \tau_1) u(\tau_1) u^T(\tau_2) K^T(t_2, \tau_2) \right\} \\
&= \int_{t_0}^{t_1} d\tau_1 \int_{t_0}^{t_2} d\tau_2 \, K(t_1, \tau_1) E\{u(\tau_1) u^T(\tau_2)\} K^T(t_2, \tau_2) \\
&= \int_{t_0}^{t_1} d\tau_1 \int_{t_0}^{t_2} d\tau_2 \, K(t_1, \tau_1) R_u(\tau_1, \tau_2) K^T(t_2, \tau_2). \qquad \text{1-458}
\end{aligned}$$

For a time-invariant system and a wide-sense stationary input process, we have the following result.

Theorem 1.48. *Suppose that the linear system of Theorem 1.47 is an asymptotically stable time-invariant system with impulse response matrix $K(t - \tau)$, and that the input stochastic process $u(t)$ is wide-sense stationary with covariance matrix $R_u(t_1 - t_2)$. Then if the input to the system is a realization of the process $u(t)$, which is applied from time $-\infty$ on, the output is a realization of a wide-sense stationary stochastic process $y(t)$ with covariance matrix*

$$R_y(t_1 - t_2) = \int_0^\infty d\tau_1 \int_0^\infty d\tau_2 \, K(\tau_1) R_u(t_1 - t_2 + \tau_2 - \tau_1) K^T(\tau_2). \qquad \text{1-459}$$

Note that we have introduced a slight inconsistency in the notation of the impulse response matrix K and the covariance matrix R_u. It is in Section 1.3.2 that we saw that the impulse response matrix of a time-invariant system depends on $t - \tau$ only. The result **1-459** can be found from **1-456** by letting $t_0 \to -\infty$ and making some simple substitutions.

For wide-sense stationary processes, it is of interest to consider the power density matrix.

Theorem 1.49. *Consider an asymptotically stable time-invariant linear system with transfer matrix $H(s)$. Suppose that the input is a realization of a wide-sense stationary stochastic process $u(t)$ with power spectral density matrix $\Sigma_u(\omega)$ which is applied from time $-\infty$ on. Then the output is a realization of a wide-sense stationary stochastic process $y(t)$ with power spectral density matrix*

$$\Sigma_y(\omega) = H(j\omega)\Sigma_u(\omega)H^T(-j\omega). \qquad \textbf{1-460}$$

This result follows easily by Fourier transforming **1-459** after replacing $t_1 - t_2$ with a variable τ, using the fact that $H(s)$ is the Laplace transform of $K(\tau)$.

Example 1.31. *Stirred tank*

Consider the stirred tank of Example 1.2 (Section 1.2.3) and assume that fluctuations occur in the concentrations c_1 and c_2 of the feeds. Let us therefore write

$$\begin{aligned} c_1(t) &= c_{10} + v_1(t), \\ c_2(t) &= c_{20} + v_2(t), \end{aligned} \qquad \textbf{1-461}$$

where c_{10} and c_{20} are the average concentrations and $v_1(t)$ and $v_2(t)$ fluctuations about the average. It is not difficult to show that the linearized system equations must be modified to the following:

$$\begin{aligned} \dot{x}(t) &= \begin{pmatrix} -\dfrac{1}{2\theta} & 0 \\ 0 & -\dfrac{1}{\theta} \end{pmatrix} x(t) + \begin{pmatrix} 1 & 1 \\ \dfrac{c_{10} - c_0}{V_0} & \dfrac{c_{20} - c_0}{V_0} \end{pmatrix} u(t) \\ &\quad + \begin{pmatrix} 0 & 0 \\ \dfrac{F_{10}}{V_0} & \dfrac{F_{20}}{V_0} \end{pmatrix} \begin{pmatrix} v_1(t) \\ v_2(t) \end{pmatrix}, \end{aligned} \qquad \textbf{1-462}$$

$$y(t) = \begin{pmatrix} \dfrac{1}{2\theta} & 0 \\ 0 & 1 \end{pmatrix} x(t).$$

If we take the input $u(t) \equiv 0$, the transfer matrix from the disturbances

$v(t) = \text{col}\,[v_1(t), v_2(t)]$ to the output variable $y(t)$ can be found to be

$$\begin{pmatrix} 0 & 0 \\ \dfrac{F_{10}/V_0}{s + \dfrac{1}{\theta}} & \dfrac{F_{20}/V_0}{s + \dfrac{1}{\theta}} \end{pmatrix}. \qquad \text{1-463}$$

Obviously, the disturbances affect only the second component of the output variable $\eta_2(t) = \xi_2(t)$. Let us assume that $v_1(t)$ and $v_2(t)$ are two independent exponentially correlated noise processes, so that we can write for the covariance matrix of $v(t)$

$$R_v(t_1 - t_2) = \begin{pmatrix} \sigma_1^2 \exp\left(-\dfrac{|t_1 - t_2|}{\theta_1}\right) & 0 \\ 0 & \sigma_2^2 \exp\left(-\dfrac{|t_1 - t_2|}{\theta_2}\right) \end{pmatrix}. \qquad \text{1-464}$$

With this we find for the power spectral density matrix of $v(t)$

$$\Sigma_v(\omega) = \begin{pmatrix} \dfrac{2\sigma_1^2 \theta_1}{1 + \omega^2 \theta_1^2} & 0 \\ 0 & \dfrac{2\sigma_2^2 \theta_2}{1 + \omega^2 \theta_2^2} \end{pmatrix}. \qquad \text{1-465}$$

It follows from **1-460** for the power spectral density matrix of the contribution of the disturbances $v(t)$ to the output variable $y(t)$

$$\Sigma_y(\omega) = \begin{pmatrix} 0 & 0 \\ 0 & \dfrac{1}{\omega^2 + \dfrac{1}{\theta^2}} \left[\dfrac{(F_{10}/V_0)^2 2\sigma_1^2 \theta_1}{1 + \omega^2 \theta_1^2} + \dfrac{(F_{20}/V_0)^2 2\sigma_2^2 \theta_2}{1 + \omega^2 \theta_2^2} \right] \end{pmatrix}. \qquad \text{1-466}$$

1.10.4 Quadratic Expressions

In later chapters of this book it will be convenient to use a measure for the mean square value of a stochastic process. For vector stochastic processes we introduce to this end quadratic expressions of the form

$$E\{v^T(t) W(t) v(t)\}, \qquad \text{1-467}$$

where $W(t)$ is a symmetric weighting matrix. If $v(t) = \text{col}\,[v_1(t), \cdots, v_n(t)]$

1.10 Vector Stochastic Processes

and W has elements W_{ij}, $i, j = 1, 2, \cdots, n$, **1-467** can be written as

$$E\{v^T(t)W(t)v(t)\} = E\left\{\sum_{i=1}^{n}\sum_{j=1}^{n} v_i(t)W_{ij}(t)v_j(t)\right\}, \quad \textbf{1-468}$$

which is the expectation of a quadratic expression in the components $v_i(t)$ of $v(t)$. Usually, $W(t)$ is chosen to be nonnegative-definite so that the expression assumes nonnegative values only.

It is helpful to develop expressions for quadratic expressions of this type in terms of the covariance matrix and power spectral density matrix of $v(t)$. We have the following result.

Theorem 1.50. *Let $v(t)$ be a vector-valued stochastic process. Then if $W(t)$ is a symmetric matrix,*

$$E\{v^T(t)W(t)v(t)\} = \operatorname{tr}[W(t)C_v(t, t)], \quad \textbf{1-469}$$

where $C_v(t_1, t_2)$ is the second-order joint moment matrix of $v(t)$. If $v(t)$ is wide-sense stationary with zero mean and covariance matrix $R_v(t_1 - t_2)$, and W is constant,

$$E\{v^T(t)Wv(t)\} = \operatorname{tr}[WR_v(0)]. \quad \textbf{1-470}$$

If $v(t)$ has zero mean and the power spectral density matrix $\Sigma_v(\omega)$,

$$E\{v^T(t)Wv(t)\} = \operatorname{tr}\left[\int_{-\infty}^{\infty} W\Sigma_v(\omega)\, df\right], \quad \textbf{1-471}$$

where

$$f = \omega/2\pi. \quad \textbf{1-472}$$

Furthermore,

$$R_v(0) = \int_{-\infty}^{\infty} \Sigma_v(\omega)\, df. \quad \textbf{1-473}$$

By $\operatorname{tr}(A)$ we mean the trace of the matrix A, that is,

$$\operatorname{tr}(A) = \sum_{i=1}^{n} \alpha_{ii}, \quad \textbf{1-474}$$

where α_{ii}, $i = 1, \cdots, n$ are the diagonal elements of the matrix. The first result of the theorem follows in an elementary manner:

$$\begin{aligned}
E\{v^T(t)W(t)v(t)\} &= E\left\{\sum_{i,j=1}^{n} v_i(t)W_{ij}(t)v_j(t)\right\} \\
&= \sum_{i,j=1}^{n} W_{ij}(t)E\{v_i(t)v_j(t)\} \\
&= \sum_{i,j=1}^{n} W_{ij}(t)C_{v,ij}(t, t) \\
&= \operatorname{tr}[W(t)C_v(t, t)], \quad \textbf{1-475}
\end{aligned}$$

where $C_{v,ij}(t, t)$ is the (i,j)-th element of $C_v(t, t)$. The second result, **1-470**, is immediate since under the assumptions stated $C_v(t, t) = R_v(0)$. The third result can be shown by recalling that the power spectral density matrix $\Sigma_v(\omega)$ is the Fourier transform of $R_v(\tau)$, and that consequently $R_v(\tau)$ is the inverse transform of $\Sigma_v(\omega)$:

$$R_v(\tau) = \int_{-\infty}^{\infty} \Sigma_v(\omega) e^{j\omega\tau} \, df. \qquad \textbf{1-476}$$

For $\tau = 0$ we immediately obtain **1-471** and **1-473**.

Equation **1-471** gives an interpretation of the term power spectral density matrix. Apparently, the total "power" $E\{v^T(t)Wv(t)\}$ of a zero-mean wide-sense stationary process $v(t)$ is obtained by integrating tr $[W\Sigma_v(\omega)]$ over all frequencies. Thus tr $[W\Sigma_v(\omega)]$ can be considered as a measure for the power "density" at the frequency ω. The weighting matrix W determines the contributions of the various components of $v(t)$ to the power.

Example 1.32. *Stirred tank*

We continue Example 1.31 where we computed the spectral density matrix of the output $y(t)$ due to disturbances $v(t)$ in the concentrations of the feeds of the stirred tank. Suppose we want to compute the mean square value of the fluctuations $\eta_2(t)$ in the concentration of the outgoing flow. This mean square value can be written as

$$E\{\eta_2^2(t)\} = E\{y^T(t)Wy(t)\}, \qquad \textbf{1-477}$$

where the weighting matrix W has the simple form

$$W = \begin{pmatrix} 0 & 0 \\ 0 & 1 \end{pmatrix}. \qquad \textbf{1-478}$$

Thus we find for the mean square error

$$\begin{aligned}
E\{y^T(t)Wy(t)\} &= \int_{-\infty}^{\infty} \text{tr}\,[W\Sigma_y(\omega)] \, df \\
&= \int_{-\infty}^{\infty} \frac{1}{\omega^2 + \dfrac{1}{\theta^2}} \left[\frac{(F_{10}/V_0)^2 2\sigma_1^2 \theta_1}{1 + \omega^2 \theta_1^2} + \frac{(F_{20}/V_0)^2 2\sigma_2^2 \theta_2}{1 + \omega^2 \theta_2^2} \right] df \\
&= \frac{(F_{10}/V_0)^2 \sigma_1^2 \theta_1 \theta^2}{\theta + \theta_1} + \frac{(F_{20}/V_0)^2 \sigma_2^2 \theta_2 \theta^2}{\theta + \theta_2}. \qquad \textbf{1-479}
\end{aligned}$$

Integrals of rational functions of the type appearing in **1-479** frequently occur in the computation of quadratic expressions as considered in this section. Tables of such integrals can be found in Newton, Gould, and Kaiser (1957, Appendix E) and Seifert and Steeg (1960, Appendix).

1.11 THE RESPONSE OF LINEAR DIFFERENTIAL SYSTEMS TO WHITE NOISE

1.11.1 White Noise

One frequently encounters in practice zero-mean scalar stochastic processes w with the property that $w(t_1)$ and $w(t_2)$ are uncorrelated even for values of $|t_2 - t_1|$ that are quite small, that is,

$$R_w(t_2, t_1) \simeq 0 \quad \text{for} \quad |t_2 - t_1| > \varepsilon, \qquad \text{1-480}$$

where ε is a "small" number. The covariance function of such stochastic processes can be idealized as follows.

$$R_w(t_2, t_1) = V(t_1)\, \delta(t_2 - t_1), \quad V(t_1) \geq 0. \qquad \text{1-481}$$

Here $\delta(t_2 - t_1)$ is a delta function and $V(t)$ is referred to as the *intensity* of the process at time t. Such processes are called *white noise processes* for reasons explained later. We can of course extend the notion of a white noise process to vector-valued processes:

Definition 1.30. *Let $w(t)$ be a zero mean vector-valued stochastic process with covariance matrix*

$$R_w(t_2, t_1) = V(t_1)\, \delta(t_2 - t_1), \qquad \text{1-482}$$

*where $V(t) \geq 0$. The process $w(t)$ is then said to be a **white noise** stochastic process with intensity $V(t)$.*

In the case in which the intensity of the white noise process is constant, the process is wide-sense stationary and we can introduce its power spectral density matrix. Formally, taking the Fourier transform of $V\delta(\tau)$, we see that wide-sense stationary white noise has the power spectral density matrix

$$\Sigma_w(\omega) = V. \qquad \text{1-483}$$

This shows that a wide-sense stationary white noise process has equal power density at *all* frequencies. This is why, in analogy with light, such processes are called white noise processes. This result also agrees with our physical intuition. A process with little correlation between two nearby values $w(t_1)$ and $w(t_2)$ is very irregular and thus contains power at quite high frequencies.

Unfortunately, when one computes the total power of a white noise process using Eq. **1-470** or **1-471**, one obtains an infinite value, which immediately points out that although white noise processes may be convenient to work with, they do not exist in the physical world. Also, from a strict mathematical viewpoint, white noise processes are not really well-defined. As we shall see in Example 1.33, white noise is the "derivative" of a process with uncorrelated increments; however, such a process can be shown to have no derivative. Once the white noise has passed at least one integration, however, we are again on a firm mathematical ground and the following integration rules, which are needed extensively, can be proved.

Theorem 1.51. *Let $w(t)$ be a vector-valued white noise process with intensity $V(t)$. Also, let $A_1(t)$, $A_2(t)$, and $A(t)$ be given time-varying matrices. Then*

(a) $\quad E\left\{\int_{t_1}^{t_2} A(t)w(t)\,dt\right\} = 0;$ **1-484**

(b) $\quad E\left\{\left[\int_{t_1}^{t_2} A_1(t)w(t)\,dt\right]^T W \left[\int_{t_3}^{t_4} A_2(t')w(t')\,dt'\right]\right\}$

$$= \int_I \operatorname{tr}\,[V(t)A_1^T(t)WA_2(t)]\,dt, \quad \textbf{1-485}$$

where I is the intersection of $[t_1, t_2]$ and $[t_3, t_4]$ and W is any weighting matrix;

(c) $\quad E\left\{\left[\int_{t_1}^{t_2} A_1(t)w(t)\,dt\right]\left[\int_{t_3}^{t_4} A_2(t')w(t')\,dt'\right]^T\right\}$

$$= \int_I A_1(t)V(t)A_2^T(t)\,dt, \quad \textbf{1-486}$$

where I is as defined before.

Formally, one can prove (a) by using the fact that $w(t)$ is a zero-mean process, while (b) can be made plausible as follows.

$$E\left\{\left[\int_{t_1}^{t_2} A_1(t)w(t)\,dt\right]^T W \left[\int_{t_3}^{t_4} A_2(t')w(t')\,dt'\right]\right\}$$

$$= E\left\{\int_{t_1}^{t_2} dt \int_{t_3}^{t_4} dt'\; w^T(t)A_1^T(t)WA_2(t')w(t')\right\} \quad \textbf{1-487a}$$

$$= E\left\{\int_{t_1}^{t_2} dt \int_{t_3}^{t_4} \operatorname{tr}\,[w(t')w^T(t)A_1^T(t)WA_2(t')]\,dt'\right\} \quad \textbf{1-487b}$$

$$= \int_{t_1}^{t_2} dt \int_{t_3}^{t_4} \operatorname{tr}\,[E\{w(t')w^T(t)\}A_1^T(t)WA_2(t')]\,dt' \quad \textbf{1-487c}$$

$$= \int_{t_1}^{t_2} dt \int_{t_3}^{t_4} \operatorname{tr}\,[V(t)A_1^T(t)WA_2(t')]\,\delta(t - t')\,dt' \quad \textbf{1-487d}$$

$$= \int_I \operatorname{tr}\,[V(t)A_1^T(t)WA_2(t)]\,dt. \quad \textbf{1-487e}$$

The transition from **1-487c** to **1-487d** uses **1-482**, and the transition from **1-487d** to **1-487e** follows from the properties of the delta function. We have

1.11 Response to White Noise

also used the fact that $\text{tr}(AB) = \text{tr}(BA)$ for any two matrices A and B of compatible dimensions.

The proof of (c) is similar to that of (b).

Example 1.33. *White noise as the derivative of a process with uncorrelated increments*

In Example 1.29 (Section 1.10.1) we considered processes $v(t)$, $t \geq t_0$, with uncorrelated increments, which we showed to be processes with zero means and covariance matrices of the form

$$R_v(t_1, t_2) = \begin{cases} Q(t_1) & \text{for } t_2 \geq t_1 \geq t_0, \\ Q(t_2) & \text{for } t_1 \geq t_2 \geq t_0, \end{cases} \quad \text{1-488}$$

Proceeding completely formally, let us show that the covariance matrix of the derivative process

$$\dot{v}(t) = \frac{dv(t)}{dt}, \quad t \geq t_0, \quad \text{1-489}$$

consists of a delta function. For the mean of the derivative process, we have

$$E\{\dot{v}(t)\} = \frac{d}{dt} E\{v(t)\} = 0, \quad t \geq t_0. \quad \text{1-490}$$

For the covariance matrix of the derivative process we write, completely formally,

$$\begin{aligned} R_{\dot{v}}(t_1, t_2) &= E\{\dot{v}(t_1)\dot{v}^T(t_2)\} \\ &= \frac{\partial^2}{\partial t_1 \, \partial t_2} E\{v(t_1)v^T(t_2)\} \\ &= \frac{\partial^2}{\partial t_1 \, \partial t_2} R_v(t_1, t_2), \quad t_1, t_2 \geq t_0. \end{aligned} \quad \text{1-491}$$

Now, successively carrying out the partial differentiations, we obtain

$$R_{\dot{v}}(t_1, t_2) = \dot{Q}(t_1) \, \delta(t_1 - t_2), \quad t_1, t_2 \geq t_0, \quad \text{1-492}$$

where

$$\dot{Q}(t) = \frac{dQ(t)}{dt}. \quad \text{1-493}$$

This shows that the derivative of a process with uncorrelated increments is a white noise process. When each increment $v(t_2) - v(t_1)$ of the process has a variance matrix that may be written in the form

$$\int_{t_1}^{t_2} V(t) \, dt, \quad \text{1-494}$$

the intensity of the white noise process that derives from the process with uncorrelated increments is $V(t)$, since (see Example 1.29)

$$Q(t) = \int_{t_0}^{t} V(\tau)\, d\tau. \qquad \textbf{1-495}$$

A special case that is of considerable interest occurs when the process $v(t)$ from which the white noise process derives is Brownian motion (see Example 1.29). The white noise process then obtained is often referred to as *Gaussian white noise*.

In the rigorous theory of white noise, the white noise process is never defined. Instead, the theory is developed in terms of increments of processes with uncorrelated increments. In particular, integrals of the type appearing in Theorem 1.51 are redefined in terms of such processes. Let us consider the integral

$$\int_{t_1}^{t_2} A(t) w(t)\, dt. \qquad \textbf{1-496}$$

This is replaced with

$$\int_{t_1}^{t_2} A(t)\, dv(t) = \lim_{\varepsilon \to 0} \sum_{0}^{n-1} A(\tau_i)[v(\tau_{i+1}) - v(\tau_i)], \qquad \textbf{1-497}$$

where $v(t)$ is the process with uncorrelated increments from which the white noise process $w(t)$ derives and where $t_1 = \tau_0 < \tau_1 < \cdots < \tau_n = t_2$, with

$$\varepsilon = \max_i |\tau_{i+1} - \tau_i|, \qquad \textbf{1-498}$$

is a partitioning of the interval $[t_1, t_2]$. The limit in **1-497** can be so defined that it is a proper stochastic variable, satisfying the properties of Theorem 1.51. For detailed treatments we refer the reader to Doob (1953), Gikhman and Skorokhod (1969), Åström (1970), and Kushner (1971). For an extensive and rigorous discussion of white noise, one should consult Hida (1970).

The material in this example is offered only for background. For our purposes, in the context of linear systems, it is sufficient to have Theorem 1.51 available.

1.11.2 Linear Differential Systems Driven by White Noise

It will turn out that a linear differential system driven by white noise is a very convenient model for formulating and solving linear control problems that involve disturbances and noise. In this section we obtain some of the statistical properties of the state of a linear differential system with a white noise process as input. In particular, we compute the mean, the covariance, joint moment, variance, and moment matrices of the state x.

1.11 Response to White Noise

Theorem 1.52. *Suppose that $x(t)$ is the solution of*

$$\dot{x}(t) = A(t)x(t) + B(t)w(t),$$
$$x(t_0) = x_0, \qquad \textbf{1-499}$$

where $w(t)$ is white noise with intensity $V(t)$ and x_0 is a stochastic variable independent of $w(t)$, with mean m_0 and $Q_0 = E\{(x_0 - m_0)(x_0 - m_0)^T\}$ as its variance matrix. Then $x(t)$ has mean

$$m_x(t) = \Phi(t, t_0)m_0, \qquad \textbf{1-500}$$

where $\Phi(t, t_0)$ is the transition matrix of the system **1-499**. *The covariance matrix of $x(t)$ is*

$$R_x(t_1, t_2) = \Phi(t_1, t_0)Q_0\Phi^T(t_2, t_0)$$
$$+ \int_{t_0}^{\min(t_1, t_2)} \Phi(t_1, \tau)B(\tau)V(\tau)B^T(\tau)\Phi^T(t_2, \tau)\, d\tau. \qquad \textbf{1-501}$$

The variance matrix $Q(t) = R_x(t, t)$ satisfies the matrix differential equation

$$\dot{Q}(t) = A(t)Q(t) + Q(t)A^T(t) + B(t)V(t)B^T(t), \qquad \textbf{1-502}$$
$$Q(t_0) = Q_0.$$

Furthermore,

$$R_x(t_1, t_2) = \begin{cases} Q(t_1)\Phi^T(t_2, t_1), & t_2 \geq t_1, \\ \Phi(t_1, t_2)Q(t_2), & t_1 \geq t_2. \end{cases} \qquad \textbf{1-503}$$

The second-order joint moment matrix of $x(t)$ is

$$C_x(t_1, t_2) = E\{x(t_1)x^T(t_2)\}$$
$$= \Phi(t_1, t_0)C_x(t_0, t_0)\Phi^T(t_2, t_0)$$
$$+ \int_{t_0}^{\min(t_1, t_2)} \Phi(t_1, \tau)B(\tau)V(\tau)B^T(\tau)\Phi^T(t_2, \tau)\, d\tau. \qquad \textbf{1-504}$$

The moment matrix $C_x(t, t) = Q'(t)$ satisfies the matrix differential equation

$$\dot{Q}'(t) = A(t)Q'(t) + Q'(t)A^T(t) + B(t)V(t)B^T(t), \qquad \textbf{1-505}$$
$$Q'(t_0) = E\{x_0 x_0^T\}. \qquad \textbf{1-506}$$

Finally,

$$C_x(t_1, t_2) = \begin{cases} Q'(t_1)\Phi^T(t_2, t_1), & t_2 \geq t_1, \\ \Phi(t_1, t_2)Q'(t_2), & t_1 \geq t_2. \end{cases} \qquad \textbf{1-507}$$

These results are easily proved by using the integration rules given in Theorem 1.51. Since

$$x(t) = \Phi(t, t_0)x_0 + \int_{t_0}^{t} \Phi(t, \tau)B(\tau)w(\tau)\, d\tau, \qquad \textbf{1-508}$$

it follows by **1-484** that $m_x(t)$ is given by **1-500**. To find the covariance and joint moment matrices, consider

$$E\{x(t_1)x^T(t_2)\} = \Phi(t_1, t_0)E\{x_0 x_0^T\}\Phi^T(t_2, t_0)$$
$$+ E\left\{[\Phi(t_1, t_0)x_0]\left[\int_{t_0}^{t_2}\Phi(t_2, \tau)B(\tau)w(\tau)\,d\tau\right]^T\right\}$$
$$+ E\left\{\left[\int_{t_0}^{t_1}\Phi(t_1, \tau)B(\tau)w(\tau)\,d\tau\right][\Phi(t_2, t_0)x_0]^T\right\}$$
$$+ E\left\{\left[\int_{t_0}^{t_1}\Phi(t_1, \tau)B(\tau)w(\tau)\,d\tau\right]\left[\int_{t_0}^{t_2}\Phi(t_2, \tau)B(\tau)w(\tau)\,d\tau\right]^T\right\}. \quad \text{1-509}$$

Because of the independence of x_0 and $w(t)$ and the fact that $w(t)$ has zero mean, the second and third terms of the right-hand side of **1-509** are zero. The fourth term is simplified by applying **1-486** so that **1-509** reduces to **1-504**. Similarly, **1-501** can be obtained. The variance $Q(t)$ is obtained by setting $t_1 = t_2 = t$ in **1-501**:

$$Q(t) = \Phi(t, t_0)Q_0\Phi^T(t, t_0) + \int_{t_0}^{t}\Phi(t, \tau)B(\tau)V(\tau)B^T(\tau)\Phi^T(t, \tau)\,d\tau. \quad \text{1-510}$$

The differential equation **1-502** is found by differentiating $Q(t)$ in **1-510** with respect to t. The initial condition **1-502** is obtained by setting $t = t_0$. The differential equation for $C_x(t, t) = Q'(t)$ follows similarly. Finally, **1-503** and **1-507** follow directly from **1-501** and **1-504**, respectively.

In passing, we remark that if x_0 is a Gaussian stochastic variable and the white noise $w(t)$ is Gaussian (see Example 1.33), then $x(t)$ is a Gaussian stochastic process. We finally note that in the analysis of linear systems it is often helpful to have a computer program available for the simulation of a linear differential system driven by white noise (see, e.g., Mehra, 1969).

Example 1.34. *A first-order differential system driven by white noise*
Consider the first-order stochastic differential equation

$$\dot{\xi}(t) = -\frac{1}{\theta}\xi(t) + \omega(t), \quad \text{1-511}$$

where $\omega(t)$ is scalar white noise with constant intensity μ. Let us suppose that $\xi(0) = \xi_0$, where ξ_0 is a scalar stochastic variable with mean zero and variance $E(\xi_0^2) = \sigma^2$. It is easily found that $\xi(t)$ has the covariance function

$$R_\xi(t_1, t_2) = \left(\sigma^2 - \frac{\mu\theta}{2}\right)e^{-(t_1+t_2)/\theta} + \frac{\mu\theta}{2}e^{-|t_1-t_2|/\theta}, \quad t_1, t_2 \geq 0. \quad \text{1-512}$$

The variance of the process is

$$Q(t) = \left(\sigma^2 - \frac{\mu\theta}{2}\right)e^{-2t/\theta} + \frac{\mu\theta}{2}, \qquad t \geq 0. \qquad \textbf{1-513}$$

1.11.3 The Steady-State Variance Matrix for the Time-Invariant Case

In the preceding section we found an expression [Eq. **1-510**] for the variance matrix of the state of a differential linear system driven by white noise. In this section we are interested in the asymptotic behavior of the variance matrix in the time-invariant case, that is, when A, B, and V are constant matrices. In this case **1-510** can be written as

$$Q(t) = e^{A(t-t_0)} Q_0 e^{A^T(t-t_0)} + \int_{t_0}^{t} e^{A(t-\tau)} BVB^T e^{A^T(t-\tau)} \, d\tau. \qquad \textbf{1-514}$$

It is not difficult to see that if, and only if, A is asymptotically stable, $Q(t)$ has the following limit for arbitrary Q_0:

$$\lim_{t \to \infty} Q(t) = \lim_{t_0 \to -\infty} Q(t) = \bar{Q} = \int_0^\infty e^{A\tau} BVB^T e^{A^T\tau} \, d\tau. \qquad \textbf{1-515}$$

Since $Q(t)$ is the solution of the differential equation **1-502**, its limit $\bar{Q}$ must also satisfy that differential equation, so that

$$A\bar{Q} + \bar{Q}A^T + BVB^T = 0. \qquad \textbf{1-516}$$

It is quite helpful to realize that this algebraic matrix equation in $\bar{Q}$ has a unique solution, which must then necessarily be given by **1-515**. This follows from the following result from matrix theory (Frame, 1964).

Lemma 1.5. *Let M_1, M_2, and M_3 be real $n \times n$, $m \times m$, and $n \times m$ matrices. Let λ_i, $i = 1, 2, \cdots, n$, and μ_j, $j = 1, 2, \cdots, m$ denote the characteristic values of M_1 and M_2, respectively. Then the matrix equation*

$$M_1 X + X M_2^T = M_3 \qquad \textbf{1-517}$$

has a unique $n \times m$ solution X if and only if for all i, j

$$\lambda_i + \mu_j \neq 0. \qquad \textbf{1-518}$$

In applying this lemma to **1-516**, we let $M_1 = A$, $M_2 = A^T$. It follows that $m = n$ and $\mu_j = \lambda_j$, $j = 1, 2, \cdots, m$. Since by assumption A is asymptotically stable, all characteristic values have strictly negative real parts, and necessarily

$$\lambda_i + \lambda_j \neq 0 \qquad \textbf{1-519}$$

for all i, j. Thus **1-516** has a unique solution.

104   Elements of Linear System Theory

We summarize as follows.

Theorem 1.53.   *Consider the stochastic differential equation*

$$\dot{x}(t) = Ax(t) + Bw(t),$$
$$x(t_0) = x_0,$$
  1-520

where A and B are constant and $w(t)$ is white noise with constant intensity V. Then if A is asymptotically stable and $t_0 \to -\infty$ or $t \to \infty$, the variance matrix of $x(t)$ tends to the constant nonnegative-definite matrix

$$\bar{Q} = \int_0^\infty e^{At} BVB^T e^{A^T t}\, dt,$$
  1-521

which is the unique solution of the matrix equation

$$0 = A\bar{Q} + \bar{Q}A^T + BVB^T.$$
  1-522

The matrix $\bar{Q}$ can thus be found as the limit of the solution of the differential equation **1-502**, with an arbitrary positive-semidefinite Q_0 as initial condition, from the integral **1-521** or from the algebraic equation **1-522**.

Matrix equations of the form **1-522** are also encountered in stability theory and are sometimes known as *Lyapunov equations*. Although the matrix equation **1-522** is linear in $\bar{Q}$, its solution cannot be directly obtained by simple matrix inversion. MacFarlane (1963) and Chen and Shieh (1968a) give useful suggestions for setting up linear equations from which $\bar{Q}$ can be solved. Barnett and Storey (1967), Davison and Man (1968), Smith (1968), Jameson (1968), Rome (1969), Kleinman (1970a), Müller (1970), Lu (1971), and Smith (1971) give alternative approaches. Hagander (1972) has made a comparison of various methods of solution, but his conclusions do not recommend one particular method. Also Barnett and Storey (1970) and Rothschild and Jameson (1970) review several methods of solution.

We remark that if A is asymptotically stable and $t_0 = -\infty$, the output of the differential system **1-499** is a wide-sense stationary process. The power spectral density of the state x is

$$\Sigma_x(\omega) = (j\omega I - A)^{-1} BVB^T (-j\omega I - A^T)^{-1}.$$
  1-523

Thus using **1-473** one can obtain yet another expression for $\bar{Q}$,

$$\bar{Q} = \int_{-\infty}^{\infty} (j\omega I - A)^{-1} BVB^T (-j\omega I - A^T)^{-1}\, df.$$
  1-524

The steady-state variance matrix $\bar{Q}$ has thus far been found in this section as the asymptotic solution of the variance differential equation for $t_0 \to -\infty$ or $t \to \infty$. Suppose now that we choose the steady-state variance matrix

$\bar{Q}$ as the initial variance at time t_0, that is, we set

$$Q_0 = \bar{Q}. \qquad \text{1-525}$$

By **1-502** this leads to

$$Q(t) = \bar{Q}, \qquad t \geq t_0. \qquad \text{1-526}$$

The process $x(t)$ thus obtained has all the properties of a wide-sense stationary process.

Example 1.35. *The steady-state covariance and variance functions of a first-order system*

Consider as in Example 1.34 the scalar first-order differential equation driven by white noise,

$$\dot{\xi}(t) = -\frac{1}{\theta}\xi(t) + \omega(t), \qquad \text{1-527}$$

where the scalar white noise $\omega(t)$ has intensity μ and $\theta > 0$. Denoting by $\bar{Q}$ the limit of $Q(t)$ as $t \to \infty$, one sees from **1-513** that

$$\bar{Q} = \frac{\mu\theta}{2}. \qquad \text{1-528}$$

The Lyapunov equation **1-522** reduces to

$$-\frac{2}{\theta}\bar{Q} + \mu = 0, \qquad \text{1-529}$$

which agrees with **1-528**. Also, **1-521** yields the same result:

$$\bar{Q} = \mu \int_0^\infty e^{-2t/\theta}\, dt = \frac{\mu\theta}{2}. \qquad \text{1-530}$$

Finally, one can also check that **1-524** yields:

$$\bar{Q} = \int_{-\infty}^{+\infty} \frac{\mu}{\left(\omega^2 + \frac{1}{\theta^2}\right)}\, df = \frac{\mu\theta}{2}. \qquad \text{1-531}$$

Note that the covariance function $R_\xi(t_1, t_2)$ given in **1-512** converges to

$$\frac{\mu\theta}{2} \exp\left(-\left|\frac{t_1 - t_2}{\theta}\right|\right) \qquad \text{1-532}$$

as $t_1 + t_2 \to \infty$ with $t_1 - t_2$ finite. $R_\xi(t_1, t_2)$ equals this limit at finite t_1 and t_2 if the variance of the initial state is

$$\sigma^2 = \frac{\mu\theta}{2}. \qquad \text{1-533}$$

Apparently, **1-527** represents exponentially correlated noise, provided $\xi(t_0)$ is a zero-mean stochastic variable with variance **1-533**.

1.11.4 Modeling of Stochastic Processes

In later chapters of this book we make almost exclusive use of linear differential systems driven by white noise to represent stochastic processes. This representation of a stochastic process $v(t)$ usually takes the following form. Suppose that $v(t)$ is given by

$$v(t) = C(t)x(t), \qquad \text{1-534}$$

with

$$\dot{x}(t) = A(t)x(t) + B(t)w(t), \qquad \text{1-535}$$

where $w(t)$ is white noise. Choosing such a representation for the stochastic process v, we call *modeling* of the stochastic process v. The use of such models can be justified as follows.

(a) Very often practical stochastic phenomena are generated by very fast fluctuations which act upon a much slower differential system. In this case the model of white noise acting upon a differential system is very appropriate. A typical example of this situation is thermal noise in an electronic circuit.

(b) As we shall see, in linear control theory almost always only the mean and covariance of the stochastic processes matter. Through the use of a linear model, it is always possible to approximate any experimentally obtained mean and covariance matrix arbitrarily closely.

(c) Sometimes the stochastic process to be modeled is a stationary process with known power spectral density matrix. Again, one can always generate a stochastic process by a linear differential equation driven by white noise so that its power spectral density matrix approximates arbitrarily closely the power spectral density matrix of the original stochastic process.

Examples 1.36 and 1.37, as well as Problem 1.11, illustrate the technique of modeling.

Example 1.36. *First-order differential system*

Suppose that the covariance function of a stochastic scalar process v, which is known to be stationary, has been measured and turns out to be the exponential function

$$R_v(t_1, t_2) = \sigma^2 e^{-|t_1 - t_2|/\theta}. \qquad \text{1-536}$$

One can model this process for $t \geq t_0$ as the state of a first-order differential system (see Example 1.35):

$$\dot{v}(t) = -\frac{1}{\theta} v(t) + \omega(t), \qquad \text{1-537}$$

with $\omega(t)$ white noise with intensity $2\sigma^2/\theta$ and where $v(t_0)$ is a stochastic variable with zero mean and variance σ^2.

Example 1.37. *Stirred tank*

Consider the stirred tank of Example 1.31 (Section 1.10.3) and suppose that we wish to compute the variance matrix of the output variable $y(t)$. In Example 1.31 the fluctuations in the concentrations in the feeds were assumed to be exponentially correlated noises and can thus be modeled as the solution of a first-order system driven by white noise. We now extend the state differential equation of the stirred tank with the models for the stochastic processes $v_1(t)$ and $v_2(t)$. Let us write

$$v_1(t) = \xi_3(t), \qquad \text{1-538}$$

where

$$\dot{\xi}_3(t) = -\frac{1}{\theta_1}\xi_3(t) + \omega_1(t). \qquad \text{1-539}$$

Here $\omega_1(t)$ is scalar white noise with intensity μ_1; to make the variance of $v_1(t)$ precisely σ_1^2, we take $\mu_1 = 2\sigma_1^2/\theta_1$. For $v_2(t) = \xi_4(t)$, we use a similar model. Thus we obtain the augmented system equation

$$\dot{x}(t) = \begin{pmatrix} -\dfrac{1}{2\theta} & 0 & 0 & 0 \\ 0 & -\dfrac{1}{\theta} & \dfrac{F_{10}}{V_0} & \dfrac{F_{20}}{V_0} \\ 0 & 0 & -\dfrac{1}{\theta_1} & 0 \\ 0 & 0 & 0 & -\dfrac{1}{\theta_2} \end{pmatrix} x(t)$$

$$+ \begin{pmatrix} 1 & 1 \\ \dfrac{c_{10}-c_0}{V_0} & \dfrac{c_{20}-c_0}{V_0} \\ 0 & 0 \\ 0 & 0 \end{pmatrix} u(t) + \begin{pmatrix} 0 & 0 \\ 0 & 0 \\ 1 & 0 \\ 0 & 1 \end{pmatrix} w(t), \qquad \text{1-540}$$

where $w(t) = \text{col}\,[\omega_1(t), \omega_2(t)]$. The two-dimensional white noise $w(t)$ has intensity

$$V = \begin{pmatrix} \dfrac{2\sigma_1^2}{\theta_1} & 0 \\ 0 & \dfrac{2\sigma_2^2}{\theta_2} \end{pmatrix}. \qquad \text{1-541}$$

Solution of **1-522** for the variance matrix $\bar{Q}$ yields, assuming that $u(t) \equiv 0$ in **1-540**,

$$\bar{Q} = \begin{pmatrix} 0 & 0 & 0 & 0 \\ 0 & q_{22} & q_{23} & q_{24} \\ 0 & q_{23} & \sigma_1^2 & 0 \\ 0 & q_{24} & 0 & \sigma_2^2 \end{pmatrix}, \qquad \text{1-542}$$

where

$$q_{22} = \frac{(F_{10}/V_0)^2 \sigma_1^2 \theta^2 \theta_1}{\theta + \theta_1} + \frac{(F_{20}/V_0)^2 \sigma_2^2 \theta^2 \theta_2}{\theta + \theta_2}, \qquad \text{1-543}$$

$$q_{23} = \frac{(F_{10}/V_0)^2 \sigma_1^2 \theta \theta_1}{\theta + \theta_1}, \qquad \text{1-544}$$

$$q_{24} = \frac{(F_{20}/V_0)^2 \sigma_2^2 \theta \theta_2}{\theta + \theta_2}. \qquad \text{1-545}$$

The variance of $\eta_2(t) = \xi_2(t)$ is q_{22}, which is in agreement with the result of Example 1.32 (Section 1.10.4).

1.11.5 Quadratic Integral Expressions

Consider the linear differential system

$$\dot{x}(t) = A(t)x(t) + B(t)w(t), \qquad \text{1-546}$$

where $w(t)$ is white noise with intensity $V(t)$ and where the initial state $x(t_0)$ is assumed to be a stochastic variable with second-order moment matrix

$$E\{x(t_0)x^T(t_0)\} = Q_0. \qquad \text{1-547}$$

In later chapters of this book we extensively employ quadratic integral expressions of the form

$$E\left\{ \int_{t_0}^{t_1} x^T(t)R(t)x(t)\,dt + x^T(t_1)P_1 x(t_1) \right\}, \qquad \text{1-548}$$

where $R(t)$ is a symmetric nonnegative-definite weighting matrix for all $t_0 \leq t \leq t_1$ and where P_1 is symmetric and nonnegative-definite. In this section formulas for such expressions are derived. These formulas of course are also applicable to the deterministic case, where $w(t) = 0$, $t \geq t_0$, $x(t_0)$ is a deterministic variable, and the expectation sign does not apply.

For the solution of the linear differential equation **1-546**, we write

$$x(t) = \Phi(t, t_0)x(t_0) + \int_{t_0}^{t} \Phi(t, \tau)B(\tau)w(\tau)\,d\tau, \qquad \text{1-549}$$

1.11 Response to White Noise

so that

$$\int_{t_0}^{t_1} x^T(t)R(t)x(t)\,dt + x^T(t_1)P_1 x(t_1)$$

$$= \int_{t_0}^{t_1}\left[x^T(t_0)\Phi^T(t,t_0) + \int_{t_0}^{t} w^T(\tau)B^T(\tau)\Phi^T(t,\tau)\,d\tau\right]$$

$$\cdot R(t)\left[\Phi(t,t_0)x(t_0) + \int_{t_0}^{t}\Phi(t,\tau)B(\tau)w(\tau)\,d\tau\right]dt$$

$$+ \left[x^T(t_0)\Phi^T(t_1,t_0) + \int_{t_0}^{t_1} w^T(\tau)B^T(\tau)\Phi^T(t_1,\tau)\,d\tau\right]$$

$$\cdot P_1\left[\Phi(t_1,t_0)x(t_0) + \int_{t_0}^{t_1}\Phi(t_1,\tau)B(\tau)w(\tau)\,d\tau\right]. \quad \textbf{1-550}$$

Taking the expectation of this expression and using the integration rules of Theorem 1.51, we obtain the result

$$E\left\{\int_{t_0}^{t_1} x^T(t)R(t)x(t)\,dt + x^T(t_1)P_1 x(t_1)\right\}$$

$$= \operatorname{tr}\left\{\left[\int_{t_0}^{t_1}\Phi^T(t,t_0)R(t)\Phi(t,t_0)\,dt + \Phi^T(t_1,t_0)P_1\Phi(t_1,t_0)\right]Q_0\right.$$

$$+ \int_{t_0}^{t_1}\left[\int_{t_0}^{t} V(\tau)B^T(\tau)\Phi^T(t,\tau)R(t)\Phi(t,\tau)B(\tau)\,d\tau\right]dt$$

$$\left. + \int_{t_0}^{t_1} V(\tau)B^T(\tau)\Phi^T(t_1,\tau)P_1\Phi(t_1,\tau)B(\tau)\,d\tau\right\}. \quad \textbf{1-551}$$

Now if M and N are arbitrary matrices of compatible dimensions, it is easily shown that $\operatorname{tr}(MN) = \operatorname{tr}(NM)$. Application of this fact to the last two terms of **1-551** and an interchange of the order of integration in the third term yields

$$\operatorname{tr}\left\{\int_{t_0}^{t_1}\left[\int_{t_0}^{t} V(\tau)B^T(\tau)\Phi^T(t,\tau)R(t)\Phi(t,\tau)B(\tau)\,d\tau\right]dt\right.$$

$$\left. + \int_{t_0}^{t_1} V(\tau)B^T(\tau)\Phi^T(t_1,\tau)P_1\Phi(t_1,\tau)B(\tau)\,d\tau\right\}$$

$$= \operatorname{tr}\left\{\int_{t_0}^{t_1}\left[\int_{t_0}^{t} B(\tau)V(\tau)B^T(\tau)\Phi^T(t,\tau)R(t)\Phi(t,\tau)\,d\tau\right]dt\right.$$

$$\left. + \int_{t_0}^{t_1} B(\tau)V(\tau)B^T(\tau)\Phi^T(t_1,\tau)P_1\Phi(t_1,\tau)\,d\tau\right\}$$

$$= \operatorname{tr}\left\{\int_{t_0}^{t_1} B(\tau)V(\tau)B^T(\tau)\left[\int_{\tau}^{t_1}\Phi^T(t,\tau)R(t)\Phi(t,\tau)\,dt\right.\right.$$

$$\left.\left. + \Phi^T(t_1,\tau)P_1\Phi(t_1,\tau)\right]d\tau\right\}. \quad \textbf{1-552}$$

Substitution of this into **1-551** shows that we can write

$$E\left\{\int_{t_0}^{t_1} x^T(t)R(t)x(t)\,dt + x^T(t_1)P_1 x(t_1)\right\}$$
$$= \text{tr}\left\{P(t_0)Q_0 + \int_{t_0}^{t_1} B(t)V(t)B^T(t)P(t)\,dt\right\}, \quad \textbf{1-553}$$

where the symmetric matrix $P(t)$ is given by

$$P(t) = \int_t^{t_1} \Phi^T(\tau, t)R(\tau)\Phi(\tau, t)\,d\tau + \Phi^T(t_1, t)P_1\Phi(t_1, t). \quad \textbf{1-554}$$

By using Theorem 1.2 (Section 1.3.1), it is easily shown by differentiation that $P(t)$ satisfies the matrix differential equation

$$-\dot{P}(t) = A^T(t)P(t) + P(t)A(t) + R(t). \quad \textbf{1-555}$$

Setting $t = t_1$ in **1-554** yields the terminal condition

$$P(t_1) = P_1. \quad \textbf{1-556}$$

We summarize these results as follows.

Theorem 1.54. *Consider the linear differential system*

$$\dot{x}(t) = A(t)x(t) + B(t)w(t), \quad \textbf{1-557}$$

where $w(t)$ is white noise with intensity $V(t)$ and where $x(t_0) = x_0$ is a stochastic variable with $E\{x_0 x_0^T\} = Q_0$. Let $R(t)$ be symmetric and nonnegative-definite for $t_0 \leq t \leq t_1$, and P_1 constant, symmetric, and nonnegative-definite. Then

$$E\left\{\int_{t_0}^{t_1} x^T(t)R(t)x(t)\,dt + x^T(t_1)P_1 x(t_1)\right\}$$
$$= \text{tr}\left\{P(t_0)Q_0 + \int_{t_0}^{t_1} B(t)V(t)B^T(t)P(t)\,dt\right\}, \quad \textbf{1-558}$$

where $P(t)$ is the symmetric nonnegative-definite matrix

$$P(t) = \int_t^{t_1} \Phi^T(\tau, t)R(\tau)\Phi(\tau, t)\,d\tau + \Phi^T(t_1, t)P_1\Phi(t_1, t). \quad \textbf{1-559}$$

$\Phi(t, t_0)$ *is the transition matrix of the system* **1-557**. *$P(t)$ satisfies the matrix differential equation*

$$-\dot{P}(t) = A^T(t)P(t) + P(t)A(t) + R(t) \quad \textbf{1-560}$$

with the terminal condition

$$P(t_1) = P_1. \quad \textbf{1-561}$$

1.11 Response to White Noise

In particular, if the differential system **1-557** reduces to an autonomous differential system:
$$\dot{x}(t) = A(t)x(t), \qquad \text{1-562}$$
that is, $V(t) = 0$ and $x(t_0)$ is deterministic, then
$$\int_{t_0}^{t_1} x^T(t)R(t)x(t)\,dt + x^T(t_1)P_1 x(t_1) = x^T(t_0)P(t_0)x(t_0). \qquad \text{1-563}$$

We conclude this section with a discussion of the asymptotic behavior of the matrix $P(t)$ as the terminal time t_1 goes to infinity. We limit ourselves to the time-invariant case where the matrices A, B, V, and R are constant, so that **1-559** reduces to:
$$P(t) = \int_t^{t_1} e^{A^T(\tau-t)} R e^{A(\tau-t)}\,d\tau + e^{A^T(t_1-t)} P_1 e^{A(t_1-t)}. \qquad \text{1-564}$$

If A is asymptotically stable, we obtain in the limit $t_1 \to \infty$:
$$P(t) \to \bar{P} = \int_t^{\infty} e^{A^T(\tau-t)} R e^{A(\tau-t)}\,d\tau. \qquad \text{1-565}$$

A change of integration variable shows that $\bar{P}$ can be written as
$$\bar{P} = \int_0^{\infty} e^{A^T t'} R e^{A t'}\,dt', \qquad \text{1-566}$$
which very clearly shows that $\bar{P}$ is a constant matrix. Since $\bar{P}$ satisfies the matrix differential equation **1-560**, we have
$$0 = A^T \bar{P} + \bar{P} A + R. \qquad \text{1-567}$$

Since by assumption A is asymptotically stable, Lemma 1.5 (Section 1.11.3) guarantees that this algebraic equation has a unique solution.

In the time-invariant case, it is not difficult to conjecture from **1-558** that for $t_1 \gg t_0$ we can approximate
$$E\left\{ \int_{t_0}^{t_1} x^T(t)R x(t)\,dt + x^T(t_1)P_1 x(t_1) \right\} \simeq \operatorname{tr}[\bar{P} Q_0 + (t_1 - t_0) B V B^T \bar{P}]. \qquad \text{1-568}$$

This shows that as $t_1 \to \infty$ the criterion **1-558** asymptotically increases with t_1 at the rate $\operatorname{tr}(BVB^T \bar{P})$.

Example 1.38. *Stirred tank*

Consider the stirred tank extended with the model for the disturbances of Example 1.37. Assume that $u(t) \equiv 0$ and suppose that we are interested in the integral expression
$$E\left\{ \int_{t_0}^{t_1} \xi_2^{\,2}(t)\,dt \right\}. \qquad \text{1-569}$$

This integral gives an indication of the average deviation of the concentration $\xi_2(t)$ from zero, where the average is taken both statistically and over time. This expression is of the general form **1-548** if we set

$$R = \begin{pmatrix} 0 & 0 & 0 & 0 \\ 0 & 1 & 0 & 0 \\ 0 & 0 & 0 & 0 \\ 0 & 0 & 0 & 0 \end{pmatrix}, \quad P_1 = 0. \qquad \text{1-570}$$

Solution of the algebraic equation

$$0 = A^T \bar{P} + \bar{P} A + R \qquad \text{1-571}$$

yields the steady-state solution

$$\bar{P} = \begin{pmatrix} 0 & 0 & 0 & 0 \\ 0 & p_{22} & p_{23} & p_{24} \\ 0 & p_{23} & p_{33} & p_{34} \\ 0 & p_{24} & p_{34} & p_{44} \end{pmatrix}, \qquad \text{1-572}$$

where

$$p_{22} = \frac{\theta}{2},$$

$$p_{23} = \frac{\dfrac{\theta}{2} \dfrac{F_{10}}{V_0}}{\dfrac{1}{\theta} + \dfrac{1}{\theta_1}},$$

$$p_{24} = \frac{\dfrac{\theta}{2} \dfrac{F_{20}}{V_0}}{\dfrac{1}{\theta} + \dfrac{1}{\theta_2}},$$

$$p_{33} = \frac{\left(\dfrac{F_{10}}{V_0}\right)^2 \dfrac{\theta \theta_1}{2}}{\dfrac{1}{\theta} + \dfrac{1}{\theta_1}}, \qquad \text{1-573}$$

$$p_{34} = \frac{\left(\dfrac{F_{10}}{V_0}\right)\left(\dfrac{F_{20}}{V_0}\right) \dfrac{\theta}{2}}{\dfrac{1}{\theta_1} + \dfrac{1}{\theta_2}} \left(\dfrac{1}{\dfrac{1}{\theta} + \dfrac{1}{\theta_1}} + \dfrac{1}{\dfrac{1}{\theta} + \dfrac{1}{\theta_2}} \right),$$

$$p_{44} = \frac{\left(\dfrac{F_{20}}{V_0}\right)^2 \dfrac{\theta \theta_2}{2}}{\dfrac{1}{\theta} + \dfrac{1}{\theta_2}}.$$

If we assume for V the form **1-541**, as we did in Example 1.37, we find for the rate at which the integral criterion **1-569** asymptotically increases with t_1 [see **1-568**]:

$$\operatorname{tr}(BVB^T \bar{P}) = \frac{\left(\frac{F_{10}}{V_0}\right)^2 \sigma_1^2 \theta}{\frac{1}{\theta} + \frac{1}{\theta_1}} + \frac{\left(\frac{F_{20}}{V_0}\right)^2 \sigma_2^2 \theta}{\frac{1}{\theta} + \frac{1}{\theta_2}}. \qquad \text{1-574}$$

Not unexpectedly, this is precisely the steady-state value of $E\{\xi_2^2(t)\}$ computed in Example 1.37.

1.12 PROBLEMS

1.1. Revolving satellite

Consider a satellite that revolves about its axis of symmetry (Fig. 1.11). The angular position of the satellite at time t is $\phi(t)$, while the satellite has a

Fig. 1.11. A revolving satellite.

constant moment of inertia J. By means of gas jets, a variable torque $\mu(t)$ can be exerted, which is considered the input variable to the system. The satellite experiences no friction.

(a) Choose as the components of the state the angular position $\phi(t)$ and the angular speed $\dot{\phi}(t)$. Let the output variable be $\eta(t) = \phi(t)$. Show that the state differential equation and the output equation of the system can be represented as

$$\dot{x}(t) = \begin{pmatrix} 0 & 1 \\ 0 & 0 \end{pmatrix} x(t) + \begin{pmatrix} 0 \\ \beta \end{pmatrix} \mu(t),$$

$$\eta(t) = (1, 0)x(t),$$

1-575

where $\beta = 1/J$.

(b) Compute the transition matrix, the impulse response function, and the step response function of the system. Sketch the impulse response and step response functions.

(c) Is the system stable in the sense of Lyapunov? Is it asymptotically stable?

(d) Determine the transfer function of the system.

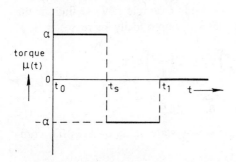

Fig. 1.12. Input torque for satellite repositioning.

(e) Consider the problem of rotating the satellite from one position in which it is at rest to another position, where it is at rest. In terms of the state, this means that the system must be transferred from the state $x(t_0) = \text{col}(\phi_0, 0)$ to the state $x(t_1) = \text{col}(\phi_1, 0)$, where ϕ_0 and ϕ_1 are given angles. Suppose that two gas jets are available; they produce torques in opposite directions such that the input variable assumes only the values $-\alpha$, 0, and $+\alpha$, where α is a fixed, given number. Show that the satellite can be rotated with an input of the form as sketched in Fig. 1.12. Calculate the switching time t_s and the terminal time t_1. Sketch the trajectory of the state in the state plane.

1.2. *Amplidyne*

An amplidyne is an electric machine used to control a large dc power through a small dc voltage. Figure 1.13 gives a simplified representation (D'Azzo and Houpis, 1966). The two armatures are rotated at a constant speed (in fact they are combined on a single shaft). The output voltage of each armature is proportional to the corresponding field current. Let L_1 and R_1 denote the inductance and resistance of the first field windings and L_2 and R_2 those of the first armature windings together with the second field windings.

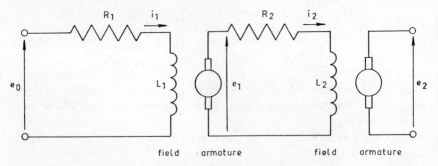

Fig. 1.13. Schematic representation of an amplidyne.

The induced voltages are given by

$$e_1 = k_1 i_1,$$
$$e_2 = k_2 i_2.$$ 1-576

The following numerical values are used:

$$R_1/L_1 = 10 \text{ s}^{-1}, \quad R_2/L_2 = 1 \text{ s}^{-1},$$
$$R_1 = 5 \, \Omega, \quad R_2 = 10 \, \Omega, \quad \text{1-577}$$
$$k_1 = 20 \text{ V/A}, \quad k_2 = 50 \text{ V/A}.$$

(a) Take as the components of the state $\xi_1(t) = i_1(t)$ and $\xi_2(t) = i_2(t)$ and show that the system equations are

$$\dot{x}(t) = \begin{pmatrix} -\dfrac{R_1}{L_1} & 0 \\ \dfrac{k_1}{L_2} & -\dfrac{R_2}{L_2} \end{pmatrix} x(t) + \begin{pmatrix} \dfrac{1}{L_1} \\ 0 \end{pmatrix} \mu(t),$$ 1-578

$$\eta(t) = (0, \ k_2) x(t),$$

where $\mu(t) = e_0(t)$ and $\eta(t) = e_2(t)$.

(b) Compute the transition matrix, the impulse response function, and the step response function of the system. Sketch for the numerical values given the impulse and step response functions.

(c) Is the system stable in the sense of Lyapunov? Is it asymptotically stable?

(d) Determine the transfer function of the system. For the numerical values given, sketch a Bode plot of the frequency response function of the system.

(e) Compute the modes of the system.

1.3. *Properties of time-invariant systems under state transformations*

Consider the linear time-invariant system

$$\dot{x}(t) = Ax(t) + Bu(t),$$
$$y(t) = Cx(t).$$ 1-579

We consider the effects of the state transformation $x' = Tx$.

(a) Show that the transition matrix $\Phi(t, t_0)$ of the system **1-579** and the transition matrix $\Phi'(t_1, t_0)$ of the transformed system are related by

$$\Phi'(t, t_0) = T\Phi(t, t_0)T^{-1}.$$ 1-580

(b) Show that the impulse response matrix and the step response matrix of the system do not change under a state transformation.

116 Elements of Linear System Theory

(c) Show that the characteristic values of the system do not change under a state transformation.

(d) Show that the transformed system is stable in the sense of Lyapunov if and only if the original system **1-579** is stable in the sense of Lyapunov. Similarly, prove that the transformed system is asymptotically stable if and only if the original system **1-579** is asymptotically stable.

(e) Show that the transfer matrix of the system does not change under a state transformation.

1.4. *Stability of amplidyne with feedback*

In an attempt to improve the performance of the amplidyne of Problem 1.2, the following simple proportional feedback scheme is considered.

$$\mu(t) = \lambda[\eta_r(t) - \eta(t)]. \quad \textbf{1-581}$$

Here $\eta_r(t)$ is an external reference voltage and λ a gain constant to be determined.

(a) Compute the transfer matrix of the amplidyne interconnected with the feedback scheme **1-581** from the reference voltage $\eta_r(t)$ to the output voltage $\eta(t)$.

(b) Determine the values of the gain constant λ for which the feedback system is asymptotically stable.

1.5*. *Structure of the controllable subspace*

Consider the controllability canonical form of Theorem 1.26 (Section 1.6.3).

(a) Prove that no matter how the transformation matrix T is chosen the characteristic values of A'_{11} and A'_{22} are always the same.

(b) Define the characteristic values of A'_{11} as the *controllable poles* and the characteristic values of A'_{22} as the *uncontrollable poles* of the system. Prove that the controllable subspace of the system **1-310** is spanned by the characteristic vectors and generalized characteristic vectors of the system that correspond to the controllable poles.

(c) Conclude that in the original representation **1-308** of the system the controllable subspace is similarly spanned by the characteristic vectors and generalized characteristic vectors corresponding to the controllable poles.

1.6*. *Controllability and stabilizability of a time-invariant system under a state transformation*

Consider the state transformation $x' = Tx$ for the linear time-invariant system

$$\dot{x}(t) = Ax(t) + Bu(t). \quad \textbf{1-582}$$

---

* See the preface for the significance of the problems marked with an asterisk.

(a) Prove that the transformed system is completely controllable if and only if the original system **1-582** is completely controllable.

(b) Prove directly (without using Theorem 1.26) that the transformed system is stabilizable if and only if the original system **1-582** is stabilizable.

1.7*. *Reconstructibility and detectability of a time-invariant system under a state transformation*

Consider the state transformation $x' = Tx$ for the time-invariant system

$$\dot{x}(t) = Ax(t), \qquad y(t) = Cx(t). \qquad \textbf{1-583}$$

(a) Prove that the transformed system is completely reconstructible if and only if the original system **1-583** is completely reconstructible.

(b) Prove directly (without using Theorem 1.35) that the transformed system is detectable if and only if the original system **1-583** is detectable.

1.8*. *Dual of a transformed system*

Consider the time-invariant system

$$\begin{aligned}\dot{x}(t) &= Ax(t) + Bu(t),\\ y(t) &= Cx(t).\end{aligned} \qquad \textbf{1-584}$$

Transform this system by defining $x'(t) = Tx(t)$ where T is a nonsingular transformation matrix. Show that the dual of the system **1-584** is transformed into the dual of the transformed system by the transformation $x^*(t) = T^T x'^*(t)$.

1.9. *"Damping" of stirred tank*

Consider the stirred tank with fluctuations in the concentrations c_1 and c_2 as described in Examples 1.31 and 1.32 (Sections 1.10.3 and 1.10.4). Assume that $u(t) \equiv 0$. The presence of the tank has the effect that the fluctuations in the concentrations c_1 and c_2 are reduced. Define the "damping factor" of the tank as the square root of the ratio of the mean square value of the fluctuations in the concentrations $c(t)$ of the outgoing flow and the mean square value of the fluctuations when the incoming feeds are mixed immediately without a tank ($V_0 = 0$). Compute the damping factor as a function of V_0. Assume $\sigma_1 = \sigma_2$, $\theta_1 = \theta_2 = 10$ s and use the numerical values of Example 1.2 (Section 1.2.3). Sketch a graph of the damping factor as a function of V_0.

1.10. *State of system driven by Gaussian white noise as a Markov process*

A stochastic process $v(t)$ is a Markov process if

$$P\{v(t_n) \leq v_n \,|\, v(t_1), v(t_2), \cdots, v(t_{n-1})\} = P\{v(t_n) \leq v_n \,|\, v(t_{n-1})\} \qquad \textbf{1-585}$$

for all n, all $t_1, t_2, \cdots, t_n$ with $t_n \geq t_{n-1} \geq t_{n-2} \geq \cdots \geq t_1$, and all v_n. Show that the state $x(t)$ of the system

$$\dot{x}(t) = A(t)x(t) + B(t)w(t),$$
$$x(t_0) = x_0,$$
1-586

where $w(t)$ is Gaussian white noise and x_0 a given stochastic variable, is a Markov process, provided x_0 is independent of $w(t)$, $t \geq t_0$.

1.11. *Modeling of second-order stochastic processes*

Consider the system

$$\dot{x}(t) = \begin{pmatrix} 0 & 1 \\ -\alpha_0 & -\alpha_1 \end{pmatrix} x(t) + \begin{pmatrix} 0 \\ 1 \end{pmatrix} \omega(t).$$
1-587

For convenience we have chosen the system to be in phase canonical form, but this is not essential. Let $\omega(t)$ be white noise with intensity 1. The output of the system is given by

$$v(t) = (\gamma_1, \gamma_2) x(t).$$
1-588

(a) Show that if **1-587** is asymptotically stable the power spectral density function of $v(t)$ is given by

$$\Sigma_v(\omega) = \left| \frac{\gamma_1 + (j\omega)\gamma_2}{(j\omega)^2 + \alpha_1(j\omega) + \alpha_0} \right|^2.$$
1-589

(b) Suppose that a stationary stochastic scalar process is given which has one of two following types of covariance functions:

$$R_v(\tau) = \beta_1 e^{-\sigma_1 |\tau|} + \beta_2 e^{-\sigma_2 |\tau|},$$
1-590

or

$$R_v(\tau) = \beta_1 e^{-\sigma_0 |\tau|} \cos(\omega_0 \tau) + \beta_2 e^{-\sigma_0 |\tau|} \cos(\omega_0 \tau),$$
1-591

where $\tau = t_1 - t_2$. Show that **1-587** and **1-588** can be used to model such a process. Express the constants occurring in **1-587** and **1-588** in terms of the constants occurring in **1-590** or **1-591**.

(c) Atmospheric turbulence manifests itself in the form of stochastically varying air speeds. The speed fluctuations in a direction perpendicular to the main flow can be represented as a scalar stochastic process with covariance function

$$R(\tau) = \sigma^2 e^{-|\tau|/\theta} \left(1 - \frac{1}{2} \frac{|\tau|}{\theta} \right),$$
1-592

where $\tau = t_1 - t_2$. Model this process.

2 ANALYSIS OF LINEAR CONTROL SYSTEMS

2.1 INTRODUCTION

In this introduction we give a brief description of control problems and of the contents of this chapter.

A control system is a dynamic system which as time evolves behaves in a certain prescribed way, generally without human interference. Control theory deals with the analysis and synthesis of control systems.

The essential components of a control system (Fig. 2.1) are: (1) the *plant*, which is the system to be controlled; (2) one or more *sensors*, which give information about the plant; and (3) the *controller*, the "heart" of the control system, which compares the measured values to their desired values and adjusts the input variables to the plant.

An example of a control system is a self-regulating home heating system, which maintains at all times a fairly constant temperature inside the home even though the outside temperature may vary considerably. The system operates without human intervention, except that the desired temperature must be set. In this control system the *plant* is the home and the heating equipment. The *sensor* generally consists of a temperature transducer inside the home, sometimes complemented by an outside temperature transducer. The *controller* is usually combined with the inside temperature sensor in the thermostat, which switches the heating equipment on and off as necessary.

Another example of a control system is a tracking antenna, which without human aid points at all times at a moving object, for example, a satellite. Here the plant is the antenna and the motor that drives it. The sensor consists of a potentiometer or other transducer which measures the antenna displacement, possibly augmented with a tachometer for measuring the angular velocity of the antenna shaft. The controller consists of electronic equipment which supplies the appropriate input voltage to the driving motor.

Although at first glance these two control problems seem different, upon further study they have much in common. First, in both cases the plant and the controller are described by differential equations. Consequently, the mathematical tool needed to analyze the behavior of the control system in

120   Analysis of Linear Control Systems

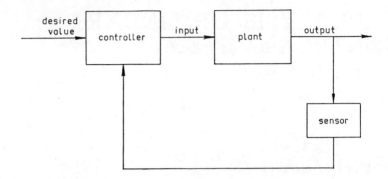

Fig. 2.1. Schematic representation of a control system.

both cases consists of the collection of methods usually referred to as system theory. Second, both control systems exhibit the feature of *feedback*, that is, the actual operation of the control system is compared to the desired operation and the input to the plant is adjusted on the basis of this comparison.

Feedback has several attractive properties. Since the actual operation is continuously compared to the desired operation, feedback control systems are able to operate satisfactorily despite adverse conditions, such as *disturbances* that act upon the system, or *variations in plant properties*. In a home heating system, disturbances are caused by fluctuations in the outside temperature and wind speed, and variations in plant properties may occur because the heating equipment in parts of the home may be connected or disconnected. In a tracking antenna disturbances in the form of wind gusts act upon the system, and plant variations occur because of different friction coefficients at different temperatures.

In this chapter we introduce control problems, describe possible solutions to these problems, analyze those solutions, and present basic design objectives. In the chapters that follow, we formulate control problems as mathematical optimization problems and use the results to synthesize control systems.

The basic design objectives discussed are stated mainly for time-invariant linear control systems. Usually, they are developed in terms of frequency domain characteristics, since in this domain the most acute insight can be gained. We also extensively discuss the time domain description of control systems via state equations, however, since numerical computations are often more conveniently performed in the time domain.

This chapter is organized as follows. In Section 2.2 a general description is given of tracking problems, regulator problems, and terminal control problems. In Section 2.3 closed-loop controllers are introduced. In the remaining sections various properties of control systems are discussed, such as stability, steady-state tracking properties, transient tracking properties, effects of disturbances and observation noise, and the influence of plant variations. Both single-input single-output and multivariable control systems are considered.

2.2 THE FORMULATION OF CONTROL PROBLEMS

2.2.1 Introduction

In this section the following two types of control problems are introduced: (1) *tracking problems* and, as special cases, *regulator problems;* and (2) *terminal control problems.*

In later sections we give detailed descriptions of possible control schemes and discuss at length how to analyze these schemes. In particular, the following topics are emphasized: root mean square (rms) tracking error, rms input, stability, transmission, transient behavior, disturbance suppression, observation noise suppression, and plant parameter variation compensation.

2.2.2 The Formulation of Tracking and Regulator Problems

We now describe in general terms an important class of control problems—*tracking problems.* Given is a system, usually called the *plant,* which cannot be altered by the designer, with the following variables associated with it (see Fig. 2.2).

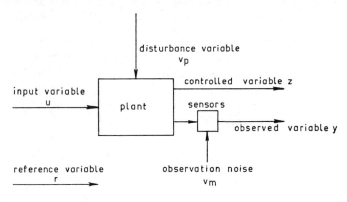

Fig. 2.2. The plant.

1. An *input variable* $u(t)$ which influences the plant and which can be manipulated;

2. A *disturbance variable* $v_p(t)$ which influences the plant but which cannot be manipulated;

122   Analysis of Linear Control Systems

3. An *observed variable* $y(t)$ which is measured by means of *sensors* and which is used to obtain information about the state of the plant; this observed variable is usually contaminated with *observation noise* $v_m(t)$;

4. A *controlled variable* $z(t)$ which is the variable we wish to control;

5. A *reference variable* $r(t)$ which represents the prescribed value of the controlled variable $z(t)$.

The *tracking problem* roughly is the following. For a given reference variable, find an appropriate input so that the controlled variable tracks the reference variable, that is,

$$z(t) \simeq r(t), \qquad t \geq t_0, \qquad \qquad \textbf{2-1}$$

where t_0 is the time at which control starts. Typically, the reference variable is not known in advance. A practical constraint is that the range of values over which the input $u(t)$ is allowed to vary is limited. Increasing this range usually involves replacement of the plant by a larger and thus more expensive one. As will be seen, this constraint is of major importance and prevents us from obtaining systems that track perfectly.

In designing tracking systems so as to satisfy the basic requirement **2-1**, the following aspects must be taken into account.

1. The disturbance influences the plant in an unpredictable way.
2. The plant parameters may not be known precisely and may vary.
3. The initial state of the plant may not be known.
4. The observed variable may not directly give information about the state of the plant and moreover may be contaminated with observation noise.

The input to the plant is to be generated by a piece of equipment that will be called the *controller*. We distinguish between two types of controllers: *open-loop* and *closed-loop* controllers. Open-loop controllers generate $u(t)$ on the basis of past and present values of the reference variable only (see Fig. 2.3), that is,

$$u(t) = f_{OL}[r(\tau), t_0 \leq \tau \leq t], \qquad t \geq t_0. \qquad \qquad \textbf{2-2}$$

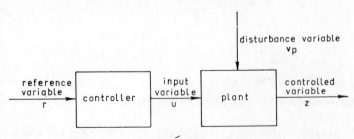

Fig. 2.3. An open-loop control system.

2.2 The Formulation of Control Problems

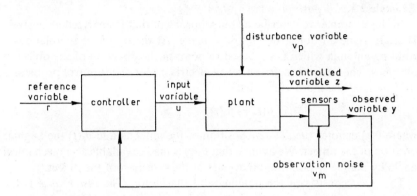

Fig. 2.4. A closed-loop control system.

Closed-loop controllers take advantage of the information about the plant that comes with the observed variable; this operation can be represented by (see Fig. 2.4)

$$u(t) = f_{CL}[r(\tau), t_0 \leq \tau \leq t; y(\tau), t_0 \leq \tau \leq t], \qquad t \geq t_0. \qquad \textbf{2-3}$$

Note that neither in **2-2** nor in **2-3** are future values of the reference variable or the observed variable used in generating the input variable since they are unknown. The plant and the controller will be referred to as the *control system*.

Already at this stage we note that closed-loop controllers are much more powerful than open-loop controllers. Closed-loop controllers can accumulate information about the plant during operation and thus are able to collect information about the initial state of the plant, reduce the effects of the disturbance, and compensate for plant parameter uncertainty and variations. Open-loop controllers obviously have no access to any information about the plant except for what is available before control starts. The fact that open-loop controllers are not afflicted by observation noise since they do not use the observed variable does not make up for this.

An important class of tracking problems consists of those problems where the reference variable is constant over long periods of time. In such cases it is customary to refer to the reference variable as the *set point* of the system and to speak of *regulator problems*. Here the main problem usually is to maintain the controlled variable at the set point in spite of disturbances that act upon the system. In this chapter tracking and regulator problems are dealt with simultaneously.

This section is concluded with two examples.

Example 2.1. *A position servo system*

In this example we describe a control problem that is analyzed extensively later. Imagine an object moving in a plane. At the origin of the plane is a rotating antenna which is supposed to point in the direction of the object at all times. The antenna is driven by an electric motor. The control problem is to command the motor such that

$$\theta(t) \simeq \theta_r(t), \qquad t \geq t_0, \qquad \textbf{2-4}$$

where $\theta(t)$ denotes the angular position of the antenna and $\theta_r(t)$ the angular position of the object. We assume that $\theta_r(t)$ is made available as a mechanical angle by manually pointing binoculars in the direction of the object.

The *plant* consists of the antenna and the motor. The *disturbance* is the torque exerted by wind on the antenna. The *observed variable* is the output of a potentiometer or other transducer mounted on the shaft of the antenna, given by

$$\eta(t) = \theta(t) + v(t), \qquad \textbf{2-5}$$

where $v(t)$ is the measurement noise. In this example the angle $\theta(t)$ is to be controlled and therefore is the *controlled variable*. The *reference variable* is the direction of the object $\theta_r(t)$. The *input* to the plant is the input voltage to the motor μ.

A possible method of forcing the antenna to point toward the object is as follows. Both the angle of the antenna $\theta(t)$ and the angle of the object $\theta_r(t)$ are converted to electrical variables using potentiometers or other transducers mounted on the shafts of the antenna and the binoculars. Then $\theta(t)$ is subtracted from $\theta_r(t)$; the difference is amplified and serves as the input voltage to the motor. As a result, when $\theta_r(t) - \theta(t)$ is positive, a positive input voltage is produced that makes the antenna rotate in a positive direction so that the difference between $\theta_r(t)$ and $\theta(t)$ is reduced. Figure 2.5 gives a representation of this control scheme.

This scheme obviously represents a closed-loop controller. An open-loop controller would generate the driving voltage $\mu(t)$ on the basis of the reference angle $\theta_r(t)$ alone. Intuitively, we immediately see that such a controller has no way to compensate for external disturbances such as wind torques, or plant parameter variations such as different friction coefficients at different temperatures. As we shall see, the closed-loop controller does offer protection against such phenomena.

This problem is a typical *tracking* problem.

Example 2.2. *A stirred tank regulator system*

The preceding example is relatively simple since the plant has only a single input and a single controlled variable. *Multivariable* control problems, where the plant has several inputs and several controlled variables, are usually

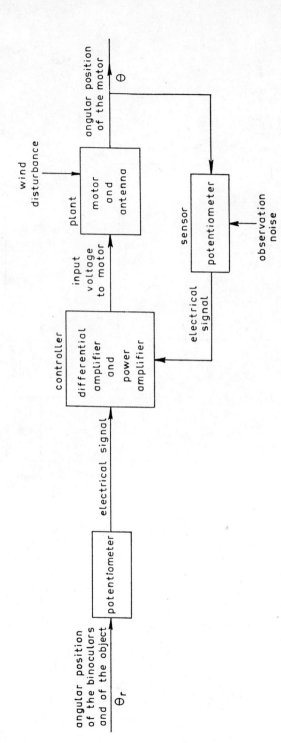

Fig. 2.5. A position servo system.

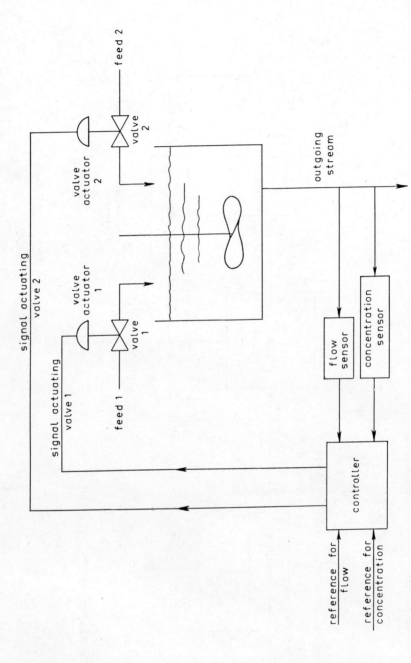

Fig. 2.6. The stirred-tank control system.

much more difficult to deal with. As an example of a multivariable problem, we consider the stirred tank of Example 1.2 (Section 1.2.3). The tank has two feeds; their flows can be adjusted by valves. The concentration of the material dissolved in each of the feeds is fixed and cannot be manipulated. The tank has one outlet and the control problem is to design equipment that automatically adjusts the feed valves so as to maintain both the outgoing flow and the concentration of the outgoing stream constant at given reference values (see Fig. 2.6).

This is a typical *regulator* problem. The components of the input variable are the flows of the incoming feeds. The components of the controlled variable are the outgoing flow and the concentration of the outgoing stream. The set point also has two components: the desired outgoing flow and the desired outgoing concentration. The following disturbances may occur: fluctuations in the incoming concentrations, fluctuations in the incoming flows resulting from pressure fluctuations before the valves, loss of fluid because of leaks and evaporation, and so on. In order to control the system well, both the outgoing flow and concentration should be measured; these then are the components of the observed variable. A closed-loop controller uses these measurements as well as the set points to produce a pneumatic or electric signal which adjusts the valves.

2.2.3 The Formulation of Terminal Control Problems

The framework of terminal control problems is similar to that of tracking and regulator problems, but a somewhat different goal is set. Given is a plant with input variable u, disturbance variable v_p, observed variable y, and controlled variable z, as in the preceding section. Then a typical terminal control problem is roughly the following. Find $u(t)$, $t_0 \leq t \leq t_1$, so that $z(t_1) \simeq r$, where r is a given vector and where the terminal time t_1 may or may not be specified. A practical restriction is that the range of possible input amplitudes is limited. The input is to be produced by a controller, which again can be of the closed-loop or the open-loop type.

In this book we do not elaborate on these problems, and we confine ourselves to giving the following example.

Example 2.3. *Position control as a terminal control problem*

Consider the antenna positioning problem of Example 2.1. Suppose that at a certain time t_0 the antenna is at rest at an angle θ_0. Then the problem of repositioning the antenna at an angle θ_1, where it is to be at rest, in as short a time as possible without overloading the motor is an example of a terminal control problem.

2.3 CLOSED-LOOP CONTROLLERS; THE BASIC DESIGN OBJECTIVE

In this section we present detailed descriptions of the plant and of closed-loop controllers. These descriptions constitute the framework for the discussion of the remainder of this chapter. Furthermore, we define the mean square tracking error and the mean square input and show how these quantities can be computed.

Throughout this chapter and, indeed, throughout most of this book, it is assumed that the plant can be described as a linear differential system with some of its inputs stochastic processes. The state differential equation of the system is

$$\dot{x}(t) = A(t)x(t) + B(t)u(t) + v_p(t),$$
$$x(t_0) = x_0.$$
2-6

Here $x(t)$ is the *state* of the plant and $u(t)$ the *input variable*. The *initial state* x_0 is a stochastic variable, and the *disturbance variable* $v_p(t)$ is assumed to be a stochastic process. The *observed variable* $y(t)$ is given by

$$y(t) = C(t)x(t) + v_m(t),$$
2-7

where the *observation noise* $v_m(t)$ is also assumed to be a stochastic process. The *controlled variable* is

$$z(t) = D(t)x(t).$$
2-8

Finally, the *reference variable* $r(t)$ is assumed to be a stochastic process of the same dimension as the controlled variable $z(t)$.

The general *closed-loop controller* will also be taken to be a linear differential system, with the reference variable $r(t)$ and the observed variable $y(t)$ as inputs, and the plant input $u(t)$ as output. The state differential equation of the closed-loop controller will have the form

$$\dot{q}(t) = L(t)q(t) + K_r(t)r(t) - K_f(t)y(t),$$
$$q(t_0) = q_0,$$
2-9

while the output equation of the controller is of the form

$$u(t) = F(t)q(t) + H_r(t)r(t) - H_f(t)y(t).$$
2-10

Here the index r refers to the reference variable and the index f to feedback. The quantity $q(t)$ is the state of the controller. The initial state q_0 is either a given vector or a stochastic variable. Figure 2.7 clarifies the interconnection of plant and controller, which is referred to as the *control system*. If $K_f(t) \equiv 0$ and $H_f(t) \equiv 0$, the closed-loop controller reduces to an *open-loop controller* (see Fig. 2.8). We refer to a control system with a closed-loop

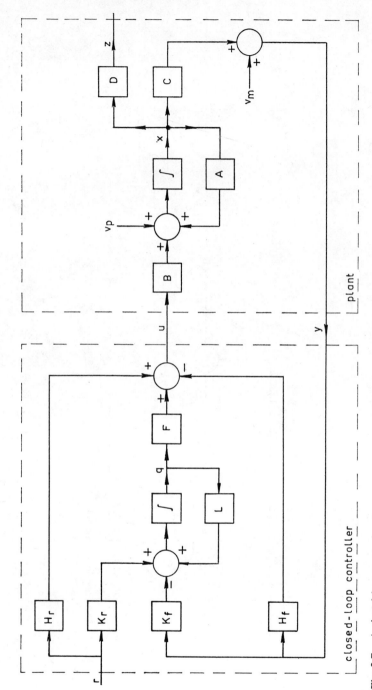

Fig. 2.7. A closed-loop control system.

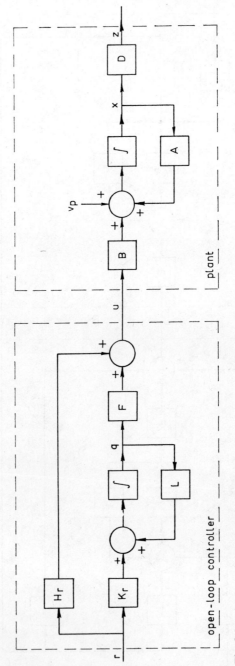

Fig. 2.8. An open-loop control system.

controller as a *closed-loop control system*, and to a control system with an open-loop controller as an *open-loop control system*.

We now define two measures of control system performance that will serve as our main tools in evaluating how well a control system performs its task:

Definition 2.1. *The **mean square tracking error** $C_e(t)$ and the **mean square input** $C_u(t)$ are defined as:*

$$C_e(t) = E\{e^T(t)W_e(t)e(t)\}, \qquad t \geq t_0,$$
$$C_u(t) = E\{u^T(t)W_u(t)u(t)\}, \qquad t \geq t_0. \qquad \textbf{2-11}$$

*Here the **tracking error** $e(t)$ is given by*

$$e(t) = z(t) - r(t), \qquad t \geq t_0, \qquad \textbf{2-12}$$

and $W_e(t)$ and $W_u(t)$, $t \geq t_0$, are given nonnegative-definite symmetric weighting matrices.

When $W_e(t)$ is diagonal, as it usually is, $C_e(t)$ is the weighted sum of the mean square tracking errors of each of the components of the controlled variable. When the error $e(t)$ is a scalar variable, and $W_e = 1$, then $\sqrt{C_e(t)}$ is the *rms tracking error*. Similarly, when the input $u(t)$ is scalar, and $W_u = 1$, then $\sqrt{C_u(t)}$ is the *rms input*.

Our aim in designing a control system is to reduce the mean square tracking error $C_e(t)$ as much as possible. Decreasing $C_e(t)$ usually implies increasing the mean square input $C_u(t)$. Since the maximally permissible value of the mean square input is determined by the capacity of the plant, a compromise must be found between the requirement of a small mean square tracking error and the need to keep the mean square input down to a reasonable level. We are thus led to the following statement.

Basic Design Objective. *In the design of control systems, the lowest possible mean square tracking error should be achieved without letting the mean square input exceed its maximally permissible value.*

In later sections we derive from the basic design objective more specific design rules, in particular for time-invariant control systems.

We now describe how the mean square tracking error $C_e(t)$ and the mean square input $C_u(t)$ can be computed. First, we use the state augmentation technique of Section 1.5.4 to obtain the state differential equation of the control system. Combining the various state and output equations we find

$$\begin{pmatrix} \dot{x}(t) \\ \dot{q}(t) \end{pmatrix} = \begin{pmatrix} A(t) - B(t)H_f(t)C(t) & B(t)F(t) \\ -K_f(t)C(t) & L(t) \end{pmatrix} \begin{pmatrix} x(t) \\ q(t) \end{pmatrix}$$
$$+ \begin{pmatrix} B(t)H_r(t) \\ K_r(t) \end{pmatrix} r(t) + \begin{pmatrix} I & -B(t)H_f(t) \\ 0 & -K_f(t) \end{pmatrix} \begin{pmatrix} v_p(t) \\ v_m(t) \end{pmatrix}.$$

2-13

For the tracking error and the input we write

$$e(t) = [D(t), 0] \begin{pmatrix} x(t) \\ q(t) \end{pmatrix} - r(t),$$

2-14

$$u(t) = [-H_f(t)C(t), F(t)] \begin{pmatrix} x(t) \\ q(t) \end{pmatrix} + H_r(t)r(t) - H_f(t)v_m(t).$$

The computation of $C_e(t)$ and $C_u(t)$ is performed in two stages. First, we determine the *mean* or *deterministic part* of $e(t)$ and $u(t)$, denoted by

$$\bar{e}(t) = E\{e(t)\}, \quad \bar{u}(t) = E\{u(t)\}, \quad t \geq t_0.$$

2-15

These means are computed by using the augmented state equation **2-13** and the output relations **2-14** where the stochastic processes $r(t)$, $v_p(t)$, and $v_m(t)$ are replaced with their means, and the initial state is taken as the mean of col $[x(t_0), q(t_0)]$.

Next we denote by $\tilde{x}(t)$, $\tilde{q}(t)$, and so on, the variables $x(t)$, $q(t)$, and so on, with their means $\bar{x}(t)$, $\bar{q}(t)$, and so on, subtracted:

$$\tilde{x}(t) = x(t) - \bar{x}(t), \quad \tilde{q}(t) = q(t) - \bar{q}(t), \text{ and so on}, \quad t \geq t_0. \quad \text{2-16}$$

With this notation we write for the mean square tracking error and the mean square input

$$C_e(t) = E\{e^T(t)W_e(t)e(t)\} = \bar{e}^T(t)W_e(t)\bar{e}(t) + E\{\tilde{e}^T(t)W_e(t)\tilde{e}(t)\},$$
$$C_u(t) = E\{u^T(t)W_u(t)u(t)\} = \bar{u}^T(t)W_u(t)\bar{u}(t) + E\{\tilde{u}^T(t)W_u(t)\tilde{u}(t)\}.$$

2-17

The terms $E\{\tilde{e}^T(t)W_e(t)\tilde{e}(t)\}$ and $E\{\tilde{u}^T(t)W_u(t)\tilde{u}(t)\}$ can easily be found when the variance matrix of col $[\tilde{x}(t), \tilde{q}(t)]$ is known. In order to determine this variance matrix, we must model the zero mean parts of $r(t)$, $v_p(t)$, and $v_m(t)$ as output variables of linear differential systems driven by white noise (see Section 1.11.4). Then col $[\tilde{x}(t), \tilde{q}(t)]$ is augmented with the state of the models generating the various stochastic processes, and the variance matrix of the resulting augmented state can be computed using the differential equation for the variance matrix of Section 1.11.2. The entire procedure is illustrated in the examples.

2.3 Closed-Loop Controllers

Example 2.4. *The position servo with three different controllers*

We continue Example 2.1 (Section 2.2.2). The motion of the antenna can be described by the differential equation

$$J\ddot{\theta}(t) + B\dot{\theta}(t) = \tau(t) + \tau_d(t). \qquad 2\text{-}18$$

Here J is the moment of inertia of all the rotating parts, including the antenna. Furthermore, B is the coefficient of viscous friction, $\tau(t)$ is the torque applied by the motor, and $\tau_d(t)$ is the disturbing torque caused by the wind. The motor torque is assumed to be proportional to $\mu(t)$, the input voltage to the motor, so that

$$\tau(t) = k\mu(t).$$

Defining the state variables $\xi_1(t) = \theta(t)$ and $\xi_2(t) = \dot{\theta}(t)$, the state differential equation of the system is

$$\dot{x}(t) = \begin{pmatrix} 0 & 1 \\ 0 & -\alpha \end{pmatrix} x(t) + \begin{pmatrix} 0 \\ \kappa \end{pmatrix} \mu(t) + \begin{pmatrix} 0 \\ \gamma \end{pmatrix} \tau_d(t), \qquad 2\text{-}19$$

where

$$x(t) = \text{col}\,[\xi_1(t), \xi_2(t)], \qquad \alpha = \frac{B}{J}, \qquad \kappa = \frac{k}{J}, \qquad \gamma = \frac{1}{J}. \qquad 2\text{-}20$$

The controlled variable $\zeta(t)$ is the angular position of the antenna:

$$\zeta(t) = (1, 0)x(t). \qquad 2\text{-}21$$

When appropriate, the following numerical values are used:

$$\begin{aligned} \alpha &= 4.6 \text{ s}^{-1}, \\ \kappa &= 0.787 \text{ rad}/(\text{V s}^2), \qquad 2\text{-}22 \\ J &= 10 \text{ kg m}^2. \end{aligned}$$

Design I. *Position feedback via a zero-order controller*

In a first attempt to design a control system, we consider the control scheme outlined in Example 2.1. The only variable measured is the angular position $\theta(t)$, so that we write for the observed variable

$$\eta(t) = (1, 0)x(t) + \nu(t), \qquad 2\text{-}23$$

where $\nu(t)$ is the measurement noise. The controller proposed can be described by the relation

$$\mu(t) = \lambda[\theta_r(t) - \eta(t)], \qquad 2\text{-}24$$

where $\theta_r(t)$ is the reference angle and λ a gain constant. Figure 2.9 gives a simplified block diagram of the control scheme. Here it is seen how an input voltage to the motor is generated that is proportional to the difference between the reference angle $\theta_r(t)$ and the observed angular position $\eta(t)$.

134   Analysis of Linear Control Systems

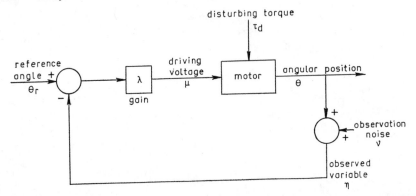

Fig. 2.9. Simplified block diagram of a position feedback control system via a zero-order controller.

The signs are so chosen that a positive value of $\theta_r(t) - \eta(t)$ results in a positive torque upon the shaft of the antenna. The question what to choose for λ is left open for the time being; we return to it in the examples of later sections.

The state differential equation of the closed-loop system is obtained from **2-19**, **2-23**, and **2-24**:

$$\dot{x}(t) = \begin{pmatrix} 0 & 1 \\ -\kappa\lambda & -\alpha \end{pmatrix} x(t) + \begin{pmatrix} 0 \\ \kappa\lambda \end{pmatrix} \theta_r(t) + \begin{pmatrix} 0 \\ \gamma \end{pmatrix} \tau_d(t) + \begin{pmatrix} 0 \\ -\kappa\lambda \end{pmatrix} v(t). \qquad \textbf{2-25}$$

We note that the controller **2-24** does not increase the dimension of the closed-loop system as compared to the plant, since it does not contain any dynamics. We refer to controllers of this type as *zero-order controllers*.

In later examples it is seen how the mean square tracking error and the mean square input can be computed when specific models are assumed for the stochastic processes $\theta_r(t)$, $\tau_d(t)$, and $v(t)$ entering into the closed-loop system equation.

Design II. *Position and velocity feedback via a zero-order controller*

As we shall see in considerable detail in later chapters, the more information the control system has about the state of the system the better it can be made to perform. Let us therefore introduce, in addition to the potentiometer that measures the angular position, a tachometer, mounted on the shaft of the antenna, which measures the angular velocity. Thus we observe the complete state, although contaminated with observation noise, of course. We write for the observed variable

$$y(t) = \begin{pmatrix} 1 & 0 \\ 0 & 1 \end{pmatrix} x(t) + v(t), \qquad \textbf{2-26}$$

2.3 Closed-Loop Controllers

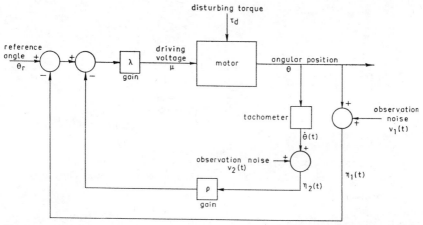

Fig. 2.10. Simplified block diagram of a position and velocity feedback control system via a zero-order controller.

where $y(t) = \text{col } [\eta_1(t), \eta_2(t)]$ and where $v(t) = \text{col } [v_1(t), v_2(t)]$ is the observation noise.

We now suggest the following simple control scheme (see Fig. 2.10):

$$\mu(t) = \lambda[\theta_r(t) - \eta_1(t)] - \lambda\rho\eta_2(t). \qquad \text{2-27}$$

This time the motor receives as input a voltage that is not only proportional to the tracking error $\theta_r(t) - \theta(t)$ but which also contains a contribution proportional to the angular velocity $\dot{\theta}(t)$. This serves the following purpose. Let us assume that at a given instant $\theta_r(t) - \theta(t)$ is positive, and that $\dot{\theta}(t)$ is positive and large. This means that the antenna moves in the right direction but with great speed. Therefore it is probably advisable not to continue driving the antenna but to start decelerating and thus avoid "overshooting" the desired position. When ρ is correctly chosen, the scheme **2-27** can accomplish this, in contrast to the scheme **2-24**. We see later that the present scheme can achieve much better performance than that of Design I.

Design III. *Position feedback via a first-order controller*

In this design approach it is assumed, as in Design I, that only the angular position $\theta(t)$ is measured. If the observation did not contain any noise, we could use a differentiator to obtain $\dot{\theta}(t)$ from $\theta(t)$ and continue as in Design II. Since observation noise is always present, however, we cannot differentiate since this greatly increases the noise level. We therefore attempt to use an approximate differentiator (see Fig. 2.11), which has the property of "filtering" the noise to some extent. Such an approximate differentiator can be realized as a system with transfer function

$$\frac{s}{T_d s + 1}, \qquad \text{2-28}$$

136 Analysis of Linear Control Systems

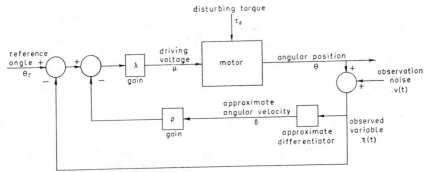

Fig. 2.11. Simplified block diagram of a position feedback control system using a first-order controller.

where T_d is a (small) positive time constant. The larger T_d is the less accurate the differentiator is, but the less the noise is amplified.

The input to the plant can now be represented as

$$\mu(t) = \lambda[\theta_r(t) - \eta(t)] - \lambda\rho\delta(t), \qquad \textbf{2-29}$$

where $\eta(t)$ is the observed angular position as in **2-23** and where $\delta(t)$ is the "approximate derivative," that is, $\delta(t)$ satisfies the differential equation

$$T_d\dot{\delta}(t) + \delta(t) = \dot{\eta}(t). \qquad \textbf{2-30}$$

This time the controller is dynamic, of order one. Again, we defer to later sections the detailed analysis of the performance of this control system; this leads to a proper choice of the time constant T_d and the gains λ and ρ. As we shall see, the performance of this design is in between those of Design I and Design II; better performance can be achieved than with Design I, although not as good as with Design II.

2.4 THE STABILITY OF CONTROL SYSTEMS

In the preceding section we introduced the control system performance measures $C_e(t)$ and $C_u(t)$. Since generally we expect that the control system will operate over long periods of time, the least we require is that both $C_e(t)$ and $C_u(t)$ remain bounded as t increases. This leads us directly to an investigation of the stability of the control system.

If the control system is not stable, sooner or later some variables will start to grow indefinitely, which is of course unacceptable in any control system that operates for some length of time (i.e., during a period larger than the time constant of the growing exponential). If the control system is

2.4 The Stability of Control Systems

unstable, usually $C_e(t)$ or $C_u(t)$, or both, will also grow indefinitely. We thus arrive at the following design objective.

Design Objective 2.1. *The control system should be asymptotically stable.*

Under the assumption that the control system is time-invariant, Design Objective 2.1 is equivalent to the requirement that all characteristic values of the augmented system **2-13**, that is, the characteristic values of the matrix

$$\begin{pmatrix} A - BH_fC & BF \\ -K_fC & L \end{pmatrix}, \qquad \text{2-31}$$

have strictly negative real parts. By referring back to Section 1.5.4, Theorem 1.21, the characteristic polynomial of **2-31** can be written as

$$\det(sI - A) \det(sI - L) \det[I + H(s)G(s)], \qquad \text{2-32}$$

where we have denoted by

$$H(s) = C(sI - A)^{-1}B \qquad \text{2-33}$$

the transfer matrix of the plant from the input u to be the observed variable y, and by

$$G(s) = F(sI - L)^{-1}K_f + H_f \qquad \text{2-34}$$

the transfer matrix of the controller from y to $-u$.

One of the functions of the controller is to move the poles of the plant to better locations in the left-hand complex plane so as to achieve an improved system performance. If the plant by itself is unstable, *stabilizing* the system by moving the closed-loop poles to proper locations in the left-half complex plane is the *main* function of the controller (see Example 2.6).

Example 2.5. *Position servo*

Let us analyze the stability of the zero-order position feedback control system proposed for the antenna drive system of Example 2.4, Design I. The plant transfer function (the transfer function from the driving voltage to the antenna position) is given by

$$H(s) = \frac{\kappa}{s(s + \alpha)}. \qquad \text{2-35}$$

The controller transfer function is

$$G(s) = \lambda. \qquad \text{2-36}$$

Thus by **2-32** the closed-loop poles are the roots of

$$s(s + \alpha)\left[1 + \frac{\kappa\lambda}{s(s + \alpha)}\right] = s^2 + \alpha s + \kappa\lambda. \qquad \text{2-37}$$

138 Analysis of Linear Control Systems

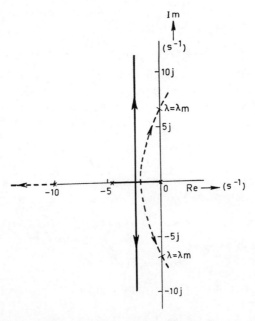

Fig. 2.12. Root loci for position servo. Solid lines, loci for second-order system; dashed lines, modifications of loci due to the presence of the pole at -10 s^{-1}.

Figure 2.12 shows the loci of the closed-loop poles with λ as a parameter for the numerical values **2-22**.

It is seen that ideally the control system is stable for all positive values of λ. In practice, however, the system becomes unstable for large λ. The reason is that, among other things, we have neglected the electrical time constant T_e of the motor. Taking this into account, the transfer function of motor plus antenna is

$$H(s) = \frac{\kappa}{s(s+\alpha)(sT_e + 1)}. \qquad \text{2-38}$$

As a result, the closed-loop characteristic polynomial is

$$s(s+\alpha)\left(s + \frac{1}{T_e}\right) + \frac{\kappa\lambda}{T_e}. \qquad \text{2-39}$$

Figure 2.12 shows the modification of the root loci that results for

$$T_e = 0.1 \text{ s}. \qquad \text{2-40}$$

For $\lambda \geq \lambda_m$, where

$$\lambda_m = \frac{\alpha}{\kappa}\left(\alpha + \frac{1}{T_e}\right), \qquad \text{2-41}$$

the closed-loop system is unstable. In the present case $\lambda_m = 85.3$ V/rad.

Example 2.6. *The stabilization of the inverted pendulum*

As an example of an unstable plant, we consider the inverted pendulum of Example 1.1 (Section 1.2.3). In Example 1.16 (Section 1.5.4), we saw that by feeding back the angle $\phi(t)$ via a zero-order controller of the form

$$\mu(t) = \lambda\phi(t) \qquad \text{2-42}$$

it is not possible to stabilize the system for any value of the gain λ. It is possible, however, to stabilize the system by feeding back the complete state $x(t)$ as follows

$$\mu(t) = -kx(t). \qquad \text{2-43}$$

Here k is a constant row vector to be determined. We note that implementation of this controller requires measurement of all four state variables.

In Example 1.1 we gave the linearized state differential equation of the system, which is of the form

$$\dot{x}(t) = Ax(t) + b\mu(t), \qquad \text{2-44}$$

where b is a column vector. Substitution of **2-43** yields

$$\dot{x}(t) = Ax(t) - bkx(t), \qquad \text{2-45}$$

or

$$\dot{x}(t) = (A - bk)x(t). \qquad \text{2-46}$$

The stability of this system is determined by the characteristic values of the matrix $A - bk$. In Chapter 3 we discuss methods for determining *optimal* controllers of the form **2-43** that stabilize the system. By using those methods, and using the numerical values of Example 1.1, it can be found, for example, that

$$k = (86.81, 12.21, -118.4, -33.44) \qquad \text{2-47}$$

stabilizes the linearized system. With this value for k, the closed-loop characteristic values are $-4.706 \pm j1.382$ and $-1.902 \pm j3.420$.

To determine the stability of the actual (nonlinear) closed-loop system, we consider the nonlinear state differential equation

$$\dot{\xi}_1(t) = \xi_2(t),$$

$$\dot{\xi}_2(t) = \frac{1}{M}\mu(t) - \frac{F}{M}\xi_2(t),$$

$$\dot{\xi}_3(t) = \xi_4(t),$$

$$\dot{\xi}_4(t) = g \sin\left[\frac{\xi_3(t) - \xi_1(t)}{L'}\right] \qquad \text{2-48}$$

$$+ \left[\frac{\mu(t) - F\xi_2(t)}{M}\right]\left[1 - \cos\frac{\xi_3(t) - \xi_1(t)}{L'}\right],$$

140   Analysis of Linear Control Systems

where the definitions of the components ξ_1, ξ_2, ξ_3, and ξ_4 are the same as for the linearized equations. Substitution of the expression **2-43** for $\mu(t)$ into **2-48** yields the closed-loop state differential equation. Figure 2.13 gives the closed-loop response of the angle $\phi(t)$ for different initial values $\phi(0)$ while all other initial conditions are zero. For $\phi(0) = 10°$ the motion is indistinguishable from the motion that would be found for the linearized system. For $\phi(0) = 20°$ some deviations occur, while for $\phi(0) = 30°$ the system is no longer stabilized by **2-47**.

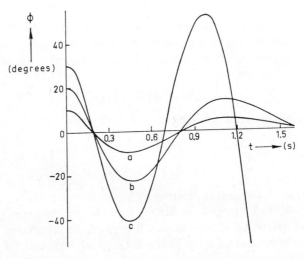

Fig. 2.13. The behavior of the angle $\phi(t)$ for the stabilized inverted pendulum: (a) $\phi(0) = 10°$; (b) $\phi(0) = 20°$; (c) $\phi(0) = 30°$.

This example also illustrates Theorem 1.16 (Section 1.4.4), where it is stated that when a linearized system is asymptotically stable the nonlinear system from which it is derived is also asymptotically stable. We see that in the present case the range over which linearization gives useful results is quite large.

2.5 THE STEADY-STATE ANALYSIS OF THE TRACKING PROPERTIES

2.5.1 The Steady-State Mean Square Tracking Error and Input

In Section 2.3 we introduced the mean square tracking error C_e and the mean square input C_u. From the control system equations **2-13** and **2-14**, it can be seen that all three processes $r(t)$, $v_p(t)$, and $v_m(t)$, that is, the reference

2.5 Steady-State Tracking Properties

variable, the disturbance variable, and the observation noise, have an effect on C_e and C_u. From now until the end of the chapter, we assume that $r(t)$, $v_p(t)$, and $v_m(t)$ are *statistically uncorrelated* stochastic processes so that their contributions to C_e and C_u can be investigated separately. In the present and the following section, we consider the contribution of the reference variable $r(t)$ to $C_e(t)$ and $C_u(t)$ alone. The effect of the disturbance and the observation noise are investigated in later sections.

We divide the duration of a control process into two periods: the *transient* and the *steady-state* period. These two periods can be characterized as follows. The transient period starts at the beginning of the process and terminates when the quantities we are interested in (usually the mean square tracking error and input) approximately reach their steady-state values. From that time on we say that the process is in its steady-state period. We assume, of course, that the quantities of interest converge to a certain limit as time increases. The duration of the transient period will be referred to as the *settling time*.

In the design of control systems, we must take into account the performance of the system during both the transient period and the steady-state period. The present section is devoted to the analysis of the steady-state properties of tracking systems. In the next section the transient analysis is discussed. In this section and the next, the following assumptions are made.

1. *Design Objective 2.1 is satisfied, that is, the control system is asymptotically stable;*
2. *The control system is time-invariant and the weighting matrices W_e and W_u are constant;*
3. *The disturbance $v_p(t)$ and the observation noise $v_m(t)$ are identical to zero;*
4. *The reference variable $r(t)$ can be represented as*

$$r(t) = r_0 + r_v(t), \qquad \text{2-49}$$

where r_0 is a stochastic vector and $r_v(t)$ is a zero-mean wide-sense stationary vector stochastic process, uncorrelated with r_0.

Here the stochastic vector r_0 is the *constant part* of the reference variable and is in fact the *set point* for the controlled variable. The zero-mean process $r_v(t)$ is the *variable part* of the reference variable. We assume that the second-order moment matrix of r_0 is given by

$$E\{r_0 r_0^T\} = R_0, \qquad \text{2-50}$$

while the variable part $r_v(t)$ will be assumed to have the power spectral density matrix $\Sigma_r(\omega)$.

Under the assumptions stated the mean square tracking error and the mean square input converge to constant values as t increases. We thus define the *steady-state mean square tracking error*

$$C_{e\infty} = \lim_{t \to \infty} C_e(t), \qquad 2\text{-}51$$

and the *steady-state mean square input*

$$C_{u\infty} = \lim_{t \to \infty} C_u(t). \qquad 2\text{-}52$$

In order to compute $C_{e\infty}$ and $C_{u\infty}$, let us denote by $T(s)$ the *transmission* of the closed-loop control system, that is, the transfer matrix from the reference variable r to the controlled variable z. We furthermore denote by $N(s)$ the transfer matrix of the closed-loop system from the reference variable r to the input variable u.

In order to derive expressions for the steady-state mean square tracking error and input, we consider the contributions of the constant part r_0 and the variable part $r_v(t)$ of the reference variable separately. The constant part of the reference variable yields a steady-state response of the controlled variable and input as follows

$$\lim_{t \to \infty} z(t) = T(0)r_0 \qquad 2\text{-}53$$

and

$$\lim_{t \to \infty} u(t) = N(0)r_0, \qquad 2\text{-}54$$

respectively. The corresponding contributions to the steady-state square tracking error and input are

$$[T(0)r_0 - r_0]^T W_e[T(0)r_0 - r_0] = \operatorname{tr}\{r_0 r_0^T [T(0) - I]^T W_e[T(0) - I]\} \qquad 2\text{-}55$$

and

$$[N(0)r_0]^T W_u[N(0)r_0] = \operatorname{tr}[r_0 r_0^T N^T(0) W_u N(0)]. \qquad 2\text{-}56$$

It follows that the contributions of the *constant* part of the reference variable to the steady-state mean square tracking error and input, respectively, are

$$\operatorname{tr}\{R_0[T(0) - I]^T W_e[T(0) - I]\} \quad \text{and} \quad \operatorname{tr}[R_0 N^T(0) W_u N(0)]. \qquad 2\text{-}57$$

The contributions of the *variable* part of the reference variable to the steady-state mean square tracking error and input are easily found by using the results of Section 1.10.4 and Section 1.10.3. The steady-state mean square tracking error turns out to be

$$C_{e\infty} = \operatorname{tr}\left\{R_0[T(0) - I]^T W_e[T(0) - I] \right. \\ \left. + \int_{-\infty}^{\infty} \Sigma_r(\omega)[T(-j\omega) - I]^T W_e[T(j\omega) - I]\, df\right\}, \qquad 2\text{-}58$$

2.5 Steady-State Tracking Properties

while the steady-state mean square input is

$$C_{u\infty} = \text{tr}\left\{R_0 N^T(0) W_u N(0) + \int_{-\infty}^{\infty} \Sigma_r(\omega) N^T(-j\omega) W_u N(j\omega) \, df\right\}. \quad \text{2-59}$$

These formulas are the starting point for deriving specific design objectives. In the next subsection we confine ourselves to the single-input single-output case, where both the input u and the controlled variable z are scalar and where the interpretation of the formulas **2-58** and **2-59** is straightforward. In Section 2.5.3 we turn to the more general multiinput multioutput case.

In conclusion we obtain expressions for $T(s)$ and $N(s)$ in terms of the various transfer matrices of the plant and the controller. Let us denote the transfer matrix of the plant **2-6–2-8** (now assumed to be time-invariant) from the input u to the controlled variable z by $K(s)$ and that from the input u to the observed variable y by $H(s)$. Also, let us denote the transfer matrix of the controller **2-9**, **2-10** (also time-invariant) from the reference variable r to u by $P(s)$, and from the plant observed variable y to $-u$ by $G(s)$. Thus we have:

$$K(s) = D(sI - A)^{-1}B, \qquad H(s) = C(sI - A)^{-1}B,$$
$$P(s) = F(sI - L)^{-1}K_r + H_r, \qquad G(s) = F(sI - L)^{-1}K_f + H_f. \quad \text{2-60}$$

The block diagram of Fig. 2.14 gives the relations between the several system variables in terms of transfer matrices. From this diagram we see that, if

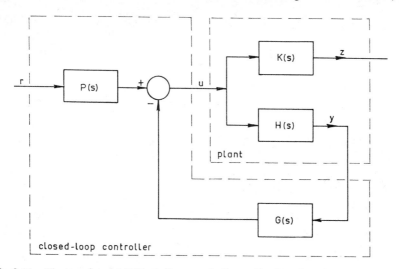

Fig. 2.14. The transfer matrix block diagram of a linear time-invariant closed-loop control system.

$r(t)$ has a Laplace transform $\mathbf{R}(s)$, in terms of Laplace transforms the several variables are related by

$$\mathbf{U}(s) = P(s)\mathbf{R}(s) - G(s)\mathbf{Y}(s),$$
$$\mathbf{Y}(s) = H(s)\mathbf{U}(s), \qquad \qquad \text{2-61}$$
$$\mathbf{Z}(s) = K(s)\mathbf{U}(s).$$

Elimination of the appropriate variables yields

$$\mathbf{Z}(s) = T(s)\mathbf{R}(s),$$
$$\mathbf{U}(s) = N(s)\mathbf{R}(s), \qquad \qquad \text{2-62}$$

where

$$T(s) = K(s)[I + G(s)H(s)]^{-1}P(s),$$
$$N(s) = [I + G(s)H(s)]^{-1}P(s). \qquad \text{2-63}$$

$T(s)$ and $N(s)$ are of course related by

$$T(s) = K(s)N(s). \qquad \qquad \text{2-64}$$

2.5.2 The Single-Input Single-Output Case

In this section it is assumed that both the input u and the controlled variable z, and therefore also the reference variable r, are scalar variables. Without loss of generality we take both $W_e = 1$ and $W_u = 1$. As a result, the steady-state mean square tracking error and the steady-state mean square input can be expressed as

$$C_{e\infty} = R_0 \, |T(0) - 1|^2 + \int_{-\infty}^{\infty} \Sigma_r(\omega) \, |T(j\omega) - 1|^2 \, df, \qquad \text{2-65a}$$

$$C_{u\infty} = R_0 \, |N(0)|^2 + \int_{-\infty}^{\infty} \Sigma_r(\omega) \, |N(j\omega)|^2 \, df. \qquad \text{2-65b}$$

From the first of these expressions, we see that since we wish to design tracking systems with a small steady-state mean square tracking error the following advice must be given.

Design Objective 2.2. *In order to obtain a small steady-state mean square tracking error, the transmission $T(s)$ of a time-invariant linear control system should be designed such that*

$$\Sigma_r(\omega) \, |T(j\omega) - 1|^2 \qquad \qquad \text{2-66}$$

is small for all real ω. In particular, when nonzero set points are likely to occur, $T(0)$ should be made close to 1.

The remark about $T(0)$ can be clarified as follows. In certain applications it is important that the set point of the control system be maintained very accurately. In particular, this is the case in regulator problems, where the

2.5 Steady-State Tracking Properties

variable part of the reference variable is altogether absent. In such a case it may be necessary that $T(0)$ very precisely equal 1.

We now examine the contributions to the integral in **2-65a** from various frequency regions. Typically, as ω increases, $\Sigma_r(\omega)$ decreases to zero. It thus follows from **2-65a** that it is sufficient to make $|T(j\omega) - 1|$ small for those frequencies where $\Sigma_r(\omega)$ assumes significant values.

In order to emphasize these remarks, we introduce two notions: the *frequency band of the control system* and the *frequency band of the reference variable*. The frequency band of the control system is roughly the range of frequencies over which $T(j\omega)$ is "close" to 1:

Definition 2.2. *Let $T(s)$ be the scalar transmission of an asymptotically stable time-invariant linear control system with scalar input and scalar controlled variable. Then the **frequency band** of the control system is defined as the set of frequencies ω, $\omega \geq 0$, for which*

$$|T(j\omega) - 1| \leq \varepsilon, \qquad \textbf{2-67}$$

*where ε is a given number that is small with respect to 1. If the frequency band is an interval $[\omega_1, \omega_2]$, we call $\omega_2 - \omega_1$ the **bandwidth** of the control system. If the frequency band is an interval $[0, \omega_c]$, we refer to ω_c as the **cutoff frequency** of the system.*

Figure 2.15 illustrates the notions of frequency band, bandwidth, and cutoff frequency.

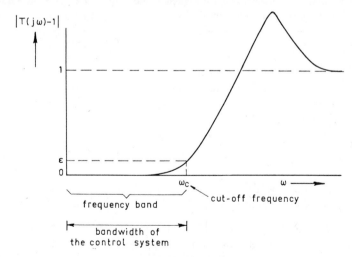

Fig. 2.15. Illustration of the definition of the frequency band, bandwidth, and cutoff frequency of a single-input single-output time-invariant control system. It is assumed that $T(j\omega) \to 0$ as $\omega \to \infty$.

In this book we usually deal with *low-pass* transmissions where the frequency band is the interval from the zero frequency to the cutoff frequency ω_c. The precise value of the cutoff frequency is of course very much dependent upon the number ε. When $\varepsilon = 0.01$, we refer to ω_c as the 1% *cutoff frequency*. We use a similar terminology for different values of ε. Frequently, however, we find it convenient to speak of the *break frequency* of the control system, which we define as that corner frequency where the asymptotic Bode plot of $|T(j\omega)|$ breaks away from unity. Thus the break frequency of the first-order transmission

$$T(s) = \frac{\alpha}{s + \alpha} \qquad \text{2-68}$$

is α, while the break frequency of the second-order transmission

$$T(s) = \frac{\omega_0^2}{s^2 + 2\zeta\omega_0 s + \omega_0^2} \qquad \text{2-69}$$

is ω_0. Note, however, that in both cases the cutoff frequency is considerably *smaller* than the break frequency, dependent upon ε, and, in the second-order case, dependent upon the relative damping ζ. Table 2.1 lists the 1% and 10% cut-off frequencies for various cases.

Table 2.1 Relation between Break Frequency and Cutoff Frequency for First- and Second-Order Scalar Transmissions

	First-order system with break frequency α	Second-order system with break frequency ω_0		
		$\zeta = 0.4$	$\zeta = 0.707$	$\zeta = 1.5$
1% cutoff freq.	0.01α	$0.012\omega_0$	$0.0071\omega_0$	$0.0033\omega_0$
10% cutoff freq.	0.1α	$0.12\omega_0$	$0.071\omega_0$	$0.033\omega_0$

Next we define the frequency band of the reference variable, which is the range of frequencies over which $\Sigma_r(\omega)$ is significantly different from zero:

Definition 2.3. *Let r be a scalar wide-sense stationary stochastic process with power spectral density function $\Sigma_r(\omega)$. The **frequency band** Ω of $r(t)$ is defined as the set of frequencies ω, $\omega \geq 0$, for which*

$$\Sigma_r(\omega) \geq \alpha. \qquad \text{2-70}$$

2.5 Steady-State Tracking Properties

Here α is so chosen that the frequency band contains a given fraction $1 - \varepsilon$ *where ε is small with respect to 1, of half of the power of the process, that is*

$$\int_{\omega \varepsilon \Omega} \Sigma_r(\omega) \, df = (1 - \varepsilon) \int_{\omega > 0} \Sigma_r(\omega) \, df. \qquad \text{2-71}$$

If the frequency band is an interval $[\omega_1, \omega_2]$, *we define* $\omega_2 - \omega_1$ *as the **bandwidth** of the process. If the frequency band is an interval* $[0, \omega_c]$, *we refer to* ω_c *as the **cutoff frequency** of the process.*

Figure 2.16 illustrates the notions of frequency band, bandwidth, and cutoff frequency of a stochastic process.

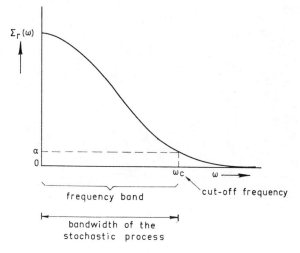

Fig. 2.16. Illustration of the definition of the frequency band, bandwidth, and cutoff frequency of a scalar stochastic process r.

Usually we deal with low-pass-type stochastic processes that have an interval of the form $[0, \omega_c]$ as a frequency band. The precise value of the cutoff frequency is of course very much dependent upon the value of ε. When $\varepsilon = 0.01$, we speak of the *1 % cutoff frequency*, which means that the interval $[0, \omega_c]$ contains 99% of half the power of the process. A similar terminology is used for other values of ε. Often, however, we find it convenient to speak of the *break frequency* of the process, which we define as the corner frequency where the asymptotic Bode plot of $\Sigma_r(\omega)$ breaks away from its low-frequency asymptote, that is, from $\Sigma_r(0)$. Let us take as an example exponentially correlated noise with rms value σ and time constant θ. This

148 Analysis of Linear Control Systems

process has the power spectral density function

$$\frac{2\sigma^2\theta}{1 + \omega^2\theta^2}, \qquad \text{2-72}$$

so that its break frequency is $1/\theta$. Since this power spectral density function decreases very slowly with ω, the 1 and 10% cutoff frequencies are much *larger* than $1/\theta$; in fact, they are $63.66/\theta$ and $6.314/\theta$, respectively.

Let us now reconsider the integral in **2-65a**. Using the notions just introduced, we see that the main contribution to this integral comes from those frequencies which are in the frequency band of the reference variable but *not* in the frequency band of the system (see Fig. 2.17). We thus rephrase Design Objective 2.2 as follows.

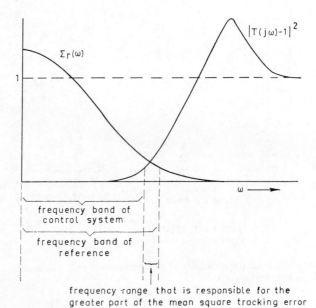

Fig. 2.17. Illustration of Design Objective 2.2.A.

Design Objective 2.2A. *In order to obtain a small steady-state mean square tracking error, the frequency band of the control system should contain as much as possible of the frequency band of the variable part of the reference variable. If nonzero set points are likely to occur, $T(0)$ should be made close to 1.*

An important aspect of this design rule is that it is also useful when very little is known about the reference variable except for a rough idea of its frequency band.

2.5 Steady-State Tracking Properties

Let us now consider the second aspect of the design—the steady-state mean square input. A consideration of **2-65b** leads us to formulate our next design objective.

Design Objective 2.3. *In order to obtain a small steady-state mean square input in an asymptotically stable single-input single-output time-invariant linear control system,*

$$\Sigma_r(\omega) |N(j\omega)|^2 \qquad \text{2-73}$$

should be made small for all real ω. This can be achieved by making $|N(j\omega)|$ sufficiently small over the frequency band of the reference variable.

It should be noted that this objective does not contain the advice to keep $N(0)$ small, such as would follow from considering the first term of **2-65b**. This term represents the contribution of the constant part of the reference variable, that is, the set point, to the input. The set point determines the desired level of the controlled variable and therefore also that of the input. It must be assumed that the plant is so designed that it is capable of sustaining this level. The second term in **2-65b** is important for the dynamic range of the input, that is, the variations in the input about the set point that are permissible. Since this dynamic range is restricted, the magnitude of the second term in **2-65b** must be limited.

It is not difficult to design a control system so that *one* of the Design Objectives 2.2A or 2.3 is completely satisfied. Since $T(s)$ and $N(s)$ are related by

$$T(s) = K(s)N(s), \qquad \text{2-74}$$

however, the design of $T(s)$ affects $N(s)$, and vice-versa. We elaborate a little on this point and show how Objectives 2.2 and 2.3 may conflict. The plant frequency response function $|K(j\omega)|$ usually decreases beyond a certain frequency, say ω_p. If $|T(j\omega)|$ is to stay close to 1 beyond this frequency, it is seen from **2-74** that $|N(j\omega)|$ must *increase* beyond ω_p. The fact that $|T(j\omega)|$ is not allowed to decrease beyond ω_p implies that the reference variable frequency band extends beyond ω_p. As a result, $|N(j\omega)|$ will be large over a frequency range where $\Sigma_r(\omega)$ is not small, which may mean an important contribution to the mean square input. If this results in overloading the plant, either the bandwidth of the control system must be reduced (at the expense of a larger tracking error), or the plant must be replaced by a more powerful one.

The designer must find a technically sound compromise between the requirements of a small mean square tracking error and a mean square input that matches the dynamic range of the plant. This compromise should be based on the specifications of the control system such as the maximal

150 Analysis of Linear Control Systems

allowable rms tracking error or the maximal power of the plant. In later chapters, where we are concerned with the synthesis problem, *optimal* compromises to this dilemma are found.

At this point a brief comment on computational aspects is in order. In Section 2.3 we outlined how time domain methods can be used to calculate the mean square tracking error and mean square input. In the time-invariant case, the integral expressions **2-65a** and **2-65b** offer an alternative computational approach. Explicit solutions of the resulting integrals have been tabulated for low-order cases (see, e.g., Newton, Gould, and Kaiser (1957), Appendix E; Seifert and Steeg (1960), Appendix). For numerical computations we usually prefer the time-domain approach, however, since this is better suited for digital computation. Nevertheless, the frequency domain expressions as given are extremely important since they allow us to formulate design objectives that cannot be easily seen, if at all, from the time domain approach.

Example 2.7. *The tracking properties of the position servo*

Let us consider the position servo problem of Examples 2.1 (Section 2.2.2) and 2.4 (Section 2.3), and let us assume that the reference variable is adequately represented as zero-mean exponentially correlated noise with rms value σ and time constant T_r. We use the numerical values

$$\sigma = 1 \text{ rad},$$
$$T_r = 10 \text{ s}. \qquad \textbf{2-75}$$

It follows from the value of the time constant and from **2-72** that the reference variable break frequency is 0.1 rad/s, its 10% cutoff frequency 0.63 rad/s, and its 1% cutoff frequency 6.4 rad/s.

Design I. Let us first consider Design I of Example 2.4, where zero-order feedback of the position has been assumed. It is easily found that the transmission $T(s)$ and the transfer function $N(s)$ are given by

$$T(s) = \frac{\kappa\lambda}{s^2 + \alpha s + \kappa\lambda},$$
$$N(s) = \frac{\lambda s(s + \alpha)}{s^2 + \alpha s + \kappa\lambda}. \qquad \textbf{2-76}$$

We rewrite the transmission as

$$T(s) = \frac{\omega_0^2}{s^2 + 2\zeta\omega_0 s + \omega_0^2}, \qquad \textbf{2-77}$$

where

$$\omega_0 = \sqrt{\kappa\lambda} \qquad \textbf{2-78}$$

is the undamped natural frequency, and

$$\zeta = \frac{\alpha}{2\sqrt{\kappa\lambda}} \qquad 2\text{-}79$$

the relative damping. In Fig. 2.18 we plot $|T(j\omega)|$ as a function of ω for various values of the gain λ. Following Design Objective 2.2A the gain λ should probably not be chosen less than about 15 V/rad, since otherwise the cutoff frequency of the control system would be too small as compared to the 1% cutoff frequency of the reference variable of 6.4 rad/s. However, the cutoff

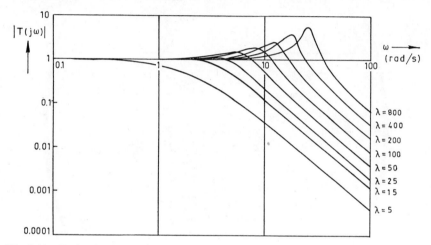

Fig. 2.18. Bode plots of the transmission of the position control system, Design I, for various values of the gain λ.

frequency does not seem to increase further with the gain, due to the peaking effect which becomes more and more pronounced. The value of 15 V/rad for the gain corresponds to the case where the relative damping ζ is about 0.7.

It remains to be seen whether or not this gain leads to acceptable values of the rms tracking error and the rms input voltage. To this end we compute both. The reference variable can be modeled as follows

$$\dot{\theta}_r(t) = -\frac{1}{T_r}\theta_r(t) + w(t), \qquad 2\text{-}80$$

where $w(t)$ is white noise with intensity $2\sigma^2/T_r$. The combined state equations

152   Analysis of Linear Control Systems

of the control system and the reference variable are from **2-19, 2-24,** and **2-80**:

$$\begin{pmatrix} \dot{\xi}_1(t) \\ \dot{\xi}_2(t) \\ \dot{\theta}_r(t) \end{pmatrix} = \begin{pmatrix} 0 & 1 & 0 \\ -\kappa\lambda & -\alpha & \kappa\lambda \\ 0 & 0 & -\dfrac{1}{T_r} \end{pmatrix} \begin{pmatrix} \xi_1(t) \\ \xi_2(t) \\ \theta_r(t) \end{pmatrix} + \begin{pmatrix} 0 \\ 0 \\ 1 \end{pmatrix} w(t). \qquad \text{2-81}$$

With this equation as a starting point, it is easy to set up and solve the Lyapunov equation for the steady-state variance matrix $\bar{Q}$ of the augmented state col $[\xi_1(t), \xi_2(t), \theta_r(t)]$ (Theorem 1.53, Section 1.11.3). The result is

$$\bar{Q} = \begin{pmatrix} \dfrac{\kappa\lambda\left(\dfrac{1}{\alpha}+T_r\right)}{\alpha+\dfrac{1}{T_r}+\kappa\lambda T_r}\sigma^2 & 0 & \dfrac{\kappa\lambda T_r}{\alpha+\dfrac{1}{T_r}+\kappa\lambda T_r}\sigma^2 \\ 0 & \dfrac{(\kappa\lambda)^2}{\alpha+\dfrac{1}{T_r}+\kappa\lambda T_r}\sigma^2 & \dfrac{\kappa\lambda}{\alpha+\dfrac{1}{T_r}+\kappa\lambda T_r}\sigma^2 \\ \dfrac{\kappa\lambda T_r}{\alpha+\dfrac{1}{T_r}+\kappa\lambda T_r}\sigma^2 & \dfrac{\kappa\lambda}{\alpha+\dfrac{1}{T_r}+\kappa\lambda T_r}\sigma^2 & \sigma^2 \end{pmatrix}. \qquad \text{2-82}$$

As a result, we obtain for the steady-state mean square tracking error:

$$C_{e\infty} = \lim_{t\to\infty} E\{[\theta(t)-\theta_r(t)]^2\} = \bar{q}_{11} - 2\bar{q}_{13} + \bar{q}_{33}$$

$$= \dfrac{\alpha+\dfrac{1}{T_r}+\dfrac{\kappa\lambda}{\alpha}}{\alpha+\dfrac{1}{T_r}+\kappa\lambda T_r}\sigma^2, \qquad \text{2-83}$$

where the $\bar{q}_{ij}$ are the entries of $\bar{Q}$. A plot of the steady-state rms tracking error is given in Fig. 2.19. We note that increasing λ beyond 15–25 V/rad decreases the rms tracking error only very little. The fact that $C_{e\infty}$ does not decrease to zero as $\lambda \to \infty$ is attributable to the peaking effect in the transmission which becomes more and more pronounced as λ becomes larger.

The steady-state rms input voltage can be found to be given by

$$C_{u\infty} = E\{\mu^2(t)\} = E\{\lambda^2[\theta(t)-\theta_r(t)]^2\} = \lambda^2 C_{e\infty}. \qquad \text{2-84}$$

2.5 Steady-State Tracking Properties

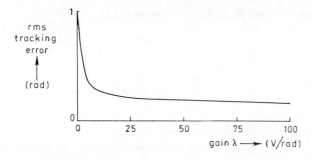

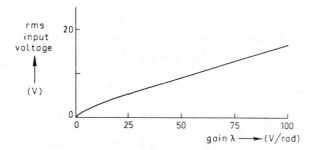

Fig. 2.19. Rms tracking error and rms input voltage as functions of the gain λ for the position servo, Design I.

Figure 2.19 shows that, according to what one would intuitively feel, the rms input keeps increasing with the gain λ. Comparing the behavior of the rms tracking error and the rms input voltage confirms the opinion that there is very little point in increasing the gain beyond 15–25 V/rad, since the increase in rms input voltage does not result in any appreciable reduction in the rms tracking error. We observe, however, that the resulting design is not very good, since the rms tracking error achieved is about 0.2 rad, which is not very small as compared to the rms value of the reference variable of 1 rad.

Design II. The second design suggested in Example 2.4 gives better results, since in this case the tachometer feedback gain factor ρ can be so chosen that the closed-loop system is well-damped for each desired bandwidth, which eliminates the peaking effect. In this design we find for the transmission

$$T(s) = \frac{\kappa\lambda}{s^2 + (\alpha + \kappa\lambda\rho)s + \kappa\lambda}, \qquad \textbf{2-85}$$

which is similar to **2-76** except that α is replaced with $\alpha + \kappa\lambda\rho$. As a result, the undamped natural frequency of the system is

$$\omega_0 = \sqrt{\kappa\lambda} \qquad \text{2-86}$$

and the relative damping

$$\zeta = \frac{\alpha + \kappa\lambda\rho}{2\sqrt{\kappa\lambda}}. \qquad \text{2-87}$$

The break frequency of the system is ω_0, which can be made arbitrarily large by choosing λ large enough. By choosing ρ such that the relative damping is in the neighborhood of 0.7, the cutoff frequency of the control system can be made correspondingly large. The steady-state rms tracking error is

$$C_{e\infty} = \frac{(\alpha + \kappa\lambda\rho)^2 + (\alpha + \kappa\lambda\rho)\dfrac{1}{T_r} + \kappa\lambda}{(\alpha + \kappa\lambda\rho)\left(\alpha + \kappa\lambda\rho + \dfrac{1}{T_r} + \kappa\lambda T_r\right)}\sigma^2, \qquad \text{2-88}$$

while the steady-state mean square input voltage is given by

$$C_{u\infty} = \lambda^2 \frac{\alpha^2 + \dfrac{\alpha}{T_r} + \dfrac{\kappa\lambda\rho}{T_r} + \kappa\lambda}{\left(\alpha + \kappa\lambda\rho + \dfrac{1}{T_r} + \kappa\lambda T_r\right)(\alpha + \kappa\lambda\rho)}\sigma^2. \qquad \text{2-89}$$

$C_{e\infty}$ can be made arbitrarily small by choosing λ and ρ large enough. For a given rms input voltage, it is possible to achieve an rms tracking error that is less than for Design I. The problem of how to choose the gains λ and ρ such that for a given rms input a minimal rms tracking error is obtained is a mathematical optimization problem.

In Chapter 3 we see how this optimization problem can be solved. At present we confine ourselves to an intuitive argument as follows. Let us suppose that for each value of λ the tachometer gain ρ is so chosen that the relative damping ζ is 0.7. Let us furthermore suppose that it is given that the steady-state rms input voltage should not exceed 30 V. Then by trial and error it can be found, using the formulas **2-88** and **2-89**, that for

$$\lambda = 500 \text{ V/rad}, \qquad \rho = 0.06 \text{ s}, \qquad \text{2-90}$$

the steady-state rms tracking error is 0.1031 rad, while the steady-state rms input voltage is 30.64 V. These values of the gain yield a near-minimal rms tracking error for the given rms input. We observe that this design is better than Design I, where we achieved an rms tracking error of about 0.2 rad. Still Design II is not very good, since the rms tracking error of 0.1 rad is not very small as compared to the rms value of the reference variable of 1 rad.

2.5 Steady-State Tracking Properties

This situation can be remedied by either replacing the motor by a more powerful one, or by lowering the bandwidth of the reference variable. The 10% cutoff frequency of the present closed-loop design is $0.071\omega_0 = 0.071\sqrt{\kappa\lambda} \simeq 1.41$ rad/s, where ω_0 is the break frequency of the system (see Table 2.1). This cutoff frequency is not large enough compared to the 1% cutoff frequency of 6.4 rad/s of the reference variable.

Design III. The third design proposed in Example 2.4 is an intermediate design: for $T_d = 0$ it reduces to Design II and for $T_d = \infty$ to Design I. For a given value of T_d, we expect its performance to lie in between that of the two other designs, which means that for a given rms input voltage an rms tracking error may be achieved that is less than that for Design I but larger than that for Design II.

From the point of view of tracking performance, T_d should of course be chosen as small as possible. A too small value of T_d, however, will unduly enhance the effect of the observation noise. In Example 2.11 (Section 2.8), which concludes the section on the effect of observation noise in the control system, we determine the most suitable value of T_d.

2.5.3 The Multiinput Multioutput Case

In this section we return to the case where the plant input, the controlled variable, and the reference variable are multidimensional variables, for which we rephrase the design objectives of Section 2.5.2.

When we first consider the steady-state mean square tracking error as given by **2-58**, we see that Design Objective 2.2 should be modified in the sense that

$$\text{tr}\,\{\Sigma_r(\omega)[T(-j\omega) - I]^T W_e[T(j\omega) - I]\} \qquad \textbf{2-91}$$

is to be made small for all real $\omega \geq 0$, and that when nonzero set points are likely to occur,

$$\text{tr}\,\{R_0[T(0) - I]^T W_e[T(0) - I]\} \qquad \textbf{2-92}$$

must be made small. Obviously, this objective is achieved when $T(j\omega)$ equals the unit matrix for all frequencies. It clearly is *sufficient*, however, that $T(j\omega)$ be close to the unit matrix for all frequencies for which $\Sigma_r(\omega)$ is significantly different from zero. In order to make this statement more precise, the following assumptions are made.

1. *The variable part of the reference variable is a stochastic process with uncorrelated components, so that its power spectral density matrix can be expressed as*

$$\Sigma_r(\omega) = \text{diag}\,[\Sigma_{r,1}(\omega), \Sigma_{r,2}(\omega), \cdots, \Sigma_{r,m}(\omega)]. \qquad \textbf{2-93}$$

156   Analysis of Linear Control Systems

2. *The constant part of the reference variable is a stochastic variable with uncorrelated components, so that its second-order moment matrix can be expressed as*

$$R_0 = \text{diag}(R_{0,1}, R_{0,2}, \cdots, R_{0,m}). \qquad \textbf{2-94}$$

From a practical point of view, these assumptions are not very restrictive. By using **2-93** and **2-94**, it is easily found that the steady-state mean square tracking error can be expressed as

$$C_{e\infty} = \sum_{i=1}^{m} R_{0,i}\{[T(0) - I]^T W_e[T(0) - I]\}_{ii}$$

$$+ \sum_{i=1}^{m} \int_{-\infty}^{\infty} \Sigma_{r,i}(\omega)\{[T(-j\omega) - I]^T W_e[T(j\omega) - I]\}_{ii}\, df, \qquad \textbf{2-95}$$

where

$$\{[T(-j\omega) - I]^T W_e[T(j\omega) - I]\}_{ii} \qquad \textbf{2-96}$$

denotes the i-th diagonal element of the matrix $[T(-j\omega) - I]^T W_e[T(j\omega) - I]$.

Let us now consider one of the terms on the right-hand side of **2-95**:

$$\int_{-\infty}^{\infty} \Sigma_{r,i}(\omega)\{[T(-j\omega) - I]^T W_e[T(j\omega) - I]\}_{ii}\, df. \qquad \textbf{2-97}$$

This expression describes the contribution of the i-th component of the reference variable to the tracking error as transmitted through the system. It is therefore appropriate to introduce the following notion.

Definition 2.4. *Let $T(s)$ be the $m \times m$ transmission of an asymptotically stable time-invariant linear control system. Then we define the **frequency band of the i-th link** of the control system as the set of frequencies ω, $\omega \geq 0$, for which*

$$\{[T(-j\omega) - I]^T W_e[T(j\omega) - I]\}_{ii} \leq \varepsilon^2 W_{e,ii}. \qquad \textbf{2-98}$$

Here ε is a given number which is small with respect to 1, W_e is the weighting matrix for the mean square tracking error, and $W_{e,ii}$ denotes the i-th diagonal element of W_e.

Once the frequency band of the i-th link is established, we can of course define the *bandwidth* and the *cutoff frequency* of the i-th link, if they exist, as in Definition 2.2. It is noted that Definition 2.4 also holds for nondiagonal weighting matrices W_e. The reason that the magnitude of

$$\{[T(-j\omega) - I]^T W_e[T(j\omega) - I]\}_{ii}$$

is compared to $W_{e,ii}$ is that it is reasonable to compare the contribution **2-97** of the i-th component of the reference variable to the mean square tracking error to its contribution when no control is present, that is, when

$T(s) = 0$. This latter contribution is given by

$$\int_{-\infty}^{\infty} \Sigma_{r,i}(\omega) W_{e,ii}\, df. \qquad \textbf{2-99}$$

We refer to the normalized function $\{[T(-j\omega) - I]^T W_e[T(j\omega) - I]\}_{ii}/W_{e,ii}$ as the *difference function* of the i-th link. In the single-input single-output case, this function is $|T(j\omega) - 1|^2$.

We are now in a position to extend Design Objective 2.2A as follows.

Design Objective 2.2B. *Let $T(s)$ be the $m \times m$ transmission of an asymptotically stable time-invariant linear control system for which both the constant part and the variable part of the reference variable have uncorrelated components. Then in order to obtain a small steady-state mean square tracking error, the frequency band of each of the m links should contain as much as possible of the frequency band of the corresponding component of the reference variable. If the i-th component, $i = 1, 2, \cdots, m$, of the reference variable is likely to have a nonzero set point, $\{[T(0) - I]^T W_e[T(0) - I]\}_{ii}$ should be made small as compared to $W_{e,ii}$.*

As an amendment to this rule, we observe that if the contribution to $C_{e\infty}$ of one particular term in the expression **2-95** is much larger than those of the remaining terms, then the advice of the objective should be applied more severely to the corresponding link than to the other links.

In view of the assumptions 1 and 2, it is not unreasonable to suppose that the weighting matrix W_e is diagonal, that is,

$$W_e = \mathrm{diag}\,(W_{e,11}, W_{e,22}, \cdots, W_{e,mm}). \qquad \textbf{2-100}$$

Then we can write

$$\{[T(-j\omega) - I]^T W_e[T(j\omega) - I]\}_{ii}$$
$$= \sum_{l=1}^{m} |\{T(j\omega) - I\}_{li}|^2 W_{e,ll}, \qquad i = 1, 2, \cdots, m, \qquad \textbf{2-101}$$

where $\{T(j\omega) - I\}_{li}$ denotes the (l, i)-th element of $T(j\omega) - I$. This shows that the frequency band of the i-th link is determined by the i-th column of the transmission $T(s)$.

It is easy to see, especially in the case where W_e is diagonal, that the design objective forces the diagonal elements of the transmission $T(j\omega)$ to be close to 1 over suitable frequency bands, while the off-diagonal elements are to be small in an appropriate sense. If all off-diagonal elements of $T(j\omega)$ are zero, that is, $T(j\omega)$ is diagonal, we say that the control system is completely *decoupled*. A control system that is not completely decoupled is said to exhibit *interaction*. A well-designed control system shows little interaction. A control system for which $T(0)$ is diagonal will be called *statically decoupled*.

We consider finally the steady-state mean square input. If the components

158   Analysis of Linear Control Systems

of the reference variable are uncorrelated (assumptions 1 and 2), we can write

$$C_{u\infty} = \sum_{i=1}^{m} R_{0,i}\{N^T(0)W_uN(0)\}_{ii} + \sum_{i=1}^{m}\int_{-\infty}^{\infty}\Sigma_{r,i}(\omega)\{N^T(-j\omega)W_uN(j\omega)\}_{ii}\,df,$$

2-102

where $\{N^T(-j\omega)W_uN(j\omega)\}_{ii}$ is the i-th diagonal element of $N^T(-j\omega) \cdot W_uN(j\omega)$. This immediately leads to the following design objective.

Design Objective 2.3A. *In order to obtain a small steady-state mean square input in an asymptotically stable time-invariant linear control system with an m-dimensional reference variable with uncorrelated components,*

$$\{N^T(-j\omega)W_uN(j\omega)\}_{ii}$$
2-103

should be made small over the frequency band of the i-th component of the reference variable, for $i = 1, 2, \cdots, m$.

Again, as in Objective 2.3, we impose no special restrictions on $\{N^T(0)W_uN(0)\}_{ii}$ even if the i-th component of the reference variable is likely to have a nonzero set point, since only the fluctuations *about* the set point of the input need be restricted.

Example 2.8. *The control of a stirred tank*

Let us take up the problem of controlling a stirred tank, as described in Example 2.2 (Section 2.2.2). The linearized state differential equation is given in Example 1.2 (Section 1.2.3); it is

$$\dot{x}(t) = \begin{pmatrix} -0.01 & 0 \\ 0 & -0.02 \end{pmatrix}x(t) + \begin{pmatrix} 1 & 1 \\ -0.25 & 0.75 \end{pmatrix}u(t).$$

2-104

As the components of the controlled variable $z(t)$ we choose the outgoing flow and the outgoing concentration so that we write

$$z(t) = \begin{pmatrix} \zeta_1(t) \\ \zeta_2(t) \end{pmatrix} = \begin{pmatrix} 0.01 & 0 \\ 0 & 1 \end{pmatrix}x(t).$$

2-105

The reference variable $r(t)$ thus has as its components $\rho_1(t)$ and $\rho_2(t)$, the desired outgoing flow and the desired outgoing concentration, respectively.

We now propose the following simple controller. If the outgoing flow is too small, we adjust the flow of feed 1 proportionally to the difference between the actual flow and the desired flow; thus we let

$$\mu_1(t) = k_1[\rho_1(t) - \zeta_1(t)].$$

2-106

However, if the outgoing concentration differs from the desired value, the flow of feed 2 is adjusted as follows:

$$\mu_2(t) = k_2[\rho_2(t) - \zeta_2(t)].$$

2-107

Figure 2.20 gives a block diagram of this control scheme. The reason that

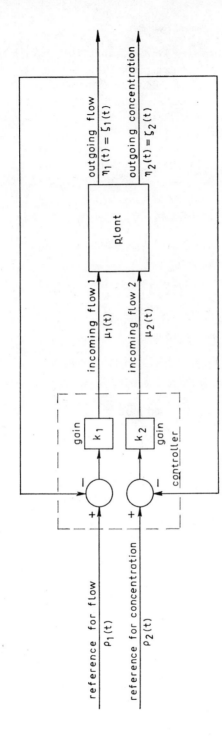

Fig. 2.20. A closed-loop control scheme for the stirred tank.

this simple scheme is expected to work is that feed 2 has a higher concentration than feed 1; thus the concentration is more sensitive to adjustments of the second flow. As a result, the first flow is more suitable for regulating the outgoing flow. However, since the second flow also affects the outgoing flow, and the first flow the concentration, a certain amount of interaction seems unavoidable in this scheme.

For this control system the various transfer matrices occurring in Fig. 2.14 can be expressed as follows:

$$K(s) = H(s) = \begin{pmatrix} \dfrac{0.01}{s+0.01} & \dfrac{0.01}{s+0.01} \\ \dfrac{-0.25}{s+0.02} & \dfrac{0.75}{s+0.02} \end{pmatrix}, \quad \text{2-108}$$

$$P(s) = G(s) = \begin{pmatrix} k_1 & 0 \\ 0 & k_2 \end{pmatrix}.$$

In Example 1.17 (Section 1.5.4), we found that the characteristic polynomial of the closed-loop system is given by

$$\phi_c(s) = s^2 + s(0.01k_1 + 0.75k_2 + 0.03) \\ + (0.0002k_1 + 0.0075k_2 + 0.01k_1k_2 + 0.0002), \quad \text{2-109}$$

from which we see that the closed-loop system is asymptotically stable for all positive values of the gains k_1 and k_2.

It can be found that the transmission of the system is given by

$$T(s) = K(s)[I + G(s)H(s)]^{-1}P(s) \\ = \dfrac{1}{\phi_c(s)} \begin{pmatrix} 0.01k_1(s + k_2 + 0.02) & 0.01k_2(s + 0.02) \\ -0.25k_1(s + 0.01) & k_2(0.75s + 0.01k_1 + 0.0075) \end{pmatrix}. \quad \text{2-110}$$

As a result, we find that

$$T(s) - I \\ = \dfrac{1}{\phi_c(s)} \begin{pmatrix} -[s^2 + s(0.75k_2 + 0.03) \\ + 0.0075k_2 + 0.0002] & 0.01k_2(s + 0.02) \\ -0.25k_1(s + 0.01) & -[s^2 + s(0.01k_1 + 0.03) \\ + 0.0002k_1 + 0.0002] \end{pmatrix}. \\ \text{2-111}$$

It is easy to see that if k_1 and k_2 simultaneously approach infinity then $[T(s) - I] \to 0$ so that perfect tracking is obtained.

2.5 Steady-State Tracking Properties

The transfer matrix $N(s)$ can be found to be

$$N(s) = [I + G(s)H(s)]^{-1}P(s)$$

$$= \frac{1}{\phi_c(s)} \begin{pmatrix} k_1[s^2 + s(0.75k_2 + 0.03) \\ + 0.0075k_2 + 0.0002] & -0.01k_1k_2(s + 0.02) \\ 0.25k_1k_2(s + 0.01) & k_2[s^2 + s(0.01k_1 + 0.03) \\ + 0.0002k_1 + 0.0002] \end{pmatrix}.$$

2-112

When k_1 and k_2 simultaneously approach infinity,

$$N(s) \to \begin{pmatrix} 75(s + 0.01) & -(s + 0.02) \\ 25(s + 0.01) & s + 0.02 \end{pmatrix},$$

2-113

which means that the steady-state mean square input $C_{u\infty}$ will be infinite unless the entries of $\Sigma_r(\omega)$ decrease fast enough with ω.

In order to find suitable values for the gains k_1 and k_2, we now apply Design Objective 2.2B and determine k_1 and k_2 so that the frequency bands of the two links of the system contain the frequency bands of the components of the reference variable. This is a complicated problem, however, and therefore we prefer to use a trial-and-error approach that is quite typical of the way multivariable control problems are commonly solved. This approach is as follows. To determine k_1 we assume that the second feedback link has not yet been connected. Similarly, in order to determine k_2, we assume that the first feedback link is disconnected. Thus we obtain two single-input single-output problems which are much easier to solve. Finally, the control system with both feedback links connected is analyzed and if necessary the design is revised.

When the second feedback link is disconnected, the transfer function from the first input to the first controlled variable is

$$H_{11}(s) = \frac{0.01}{s + 0.01}.$$

2-114

Proportional feedback according to **2-106** results in the following closed-loop transfer function from $\rho_1(t)$ to $\zeta_1(t)$:

$$\frac{0.01k_1}{s + 0.01k_1 + 0.01}.$$

2-115

We immediately observe that the zero-frequency transmission is different from 1; this can be remedied by inserting an extra gain f_1 into the connection from the first component of the reference variable as follows:

$$\mu_1(t) = k_1[f_1\rho_1(t) - \zeta_1(t)].$$

2-116

With this **2-115** is modified to

$$\frac{0.01k_1f_1}{s + 0.01k_1 + 0.01}. \qquad \textbf{2-117}$$

For each value of k_1, it is possible to choose f_1 so that the zero-frequency transmission is 1. Now the value of k_1 depends upon the cutoff frequency desired. For $k_1 = 10$ the 10% cutoff frequency is 0.011 rad/s (see Table 2.1). Let us assume that this is sufficient for the purpose of the control system. The corresponding value that should be chosen for f_1 is 1.1.

When studying the second link in a similar manner, it can be found that the feedback scheme

$$\mu_2(t) = k_2[f_2\rho_2(t) - \zeta_2(t)] \qquad \textbf{2-118}$$

results in the following closed-loop transfer function from $\rho_2(t)$ to $\zeta_2(t)$ (assuming that the first feedback link is disconnected):

$$\frac{0.75k_2f_2}{s + 0.75k_2 + 0.02}. \qquad \textbf{2-119}$$

For $k_2 = 0.1$ and $f_2 = 1.267$, the zero-frequency transmission is 1 and the 10% cutoff frequency 0.0095 rad/s.

Let us now investigate how the multivariable control system with

$$G(s) = \begin{pmatrix} k_1 & 0 \\ 0 & k_2 \end{pmatrix} = \begin{pmatrix} 10 & 0 \\ 0 & 0.1 \end{pmatrix} \qquad \textbf{2-120}$$

and

$$P(s) = \begin{pmatrix} k_1f_1 & 0 \\ 0 & k_2f_2 \end{pmatrix} = \begin{pmatrix} 11 & 0 \\ 0 & 0.1267 \end{pmatrix}, \qquad \textbf{2-121}$$

performs. It can be found that the control system transmission is given by

$$T(s) = \frac{1}{s^2 + 0.205s + 0.01295} \begin{pmatrix} 0.11s + 0.0132 & 0.001267s + 0.00002534 \\ -2.75s - 0.0275 & 0.09502s + 0.01362 \end{pmatrix}, \qquad \textbf{2-122}$$

hence that

$$T(s) - I = \frac{1}{s^2 + 0.205s + 0.01295} \cdot \begin{pmatrix} -s^2 - 0.095s + 0.00025 & 0.001267s + 0.00002534 \\ -2.75s - 0.0275 & -s^2 - 0.1100s + 0.00067 \end{pmatrix}. \qquad \textbf{2-123}$$

2.5 Steady-State Tracking Properties

Now in order to determine the frequency bands of the two links of the control system, we must first choose the weighting matrix W_e. The two controlled variables are the outgoing flow and the outgoing concentration. The flow has the constant nominal value 0.02 m³/s, while the concentration has the constant nominal value 1.25 kmol/m³. A 10% change in the flow therefore corresponds to 0.002 m³/s, while a 10% change in the concentration is about 0.1 kmol/m³. Now let us suppose that we make the weighting matrix W_e diagonal, with diagonal entries $W_{e,1}$ and $W_{e,2}$. Let us also assume that 10% changes in either the flow or the concentration make equal contributions to the mean square tracking error. Then we have

$$(0.002)^2 W_{e,1} = (0.1)^2 W_{e,2}, \qquad \text{2-124}$$

or

$$\frac{W_{e,1}}{W_{e,2}} = 2500. \qquad \text{2-125}$$

Let us therefore choose

$$W_e = \text{diag}(50, 0.02). \qquad \text{2-126}$$

Since W_e is diagonal, we can use **2-101** to determine the frequency band of the i-th link. The frequency band of the *first* link (the flow link) thus follows from considering the inequality

$$50 \left| \frac{(j\omega)^2 + 0.095(j\omega) - 0.00025}{(j\omega)^2 + 0.205(j\omega) + 0.01295} \right|^2 + 0.02 \left| \frac{2.75(j\omega) + 0.0275}{(j\omega)^2 + 0.205(j\omega) + 0.01295} \right|^2$$
$$\leq 50\varepsilon^2. \qquad \text{2-127}$$

Dividing by 50 and rearranging, we obtain

$$\frac{|(j\omega)^2 + 0.095(j\omega) - 0.00025|^2 + 0.0004 |2.75(j\omega) + 0.0275|^2}{|(j\omega)^2 + 0.205(j\omega) + 0.01295|^2} \leq \varepsilon^2. \qquad \text{2-128}$$

Figure 2.21 shows a Bode plot of the left-hand side of this inequality, which is precisely the difference function of the first link. It is seen that ε cannot be

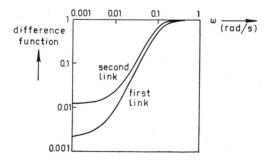

Fig. 2.21. Difference functions of the first and the second link of the stirred-tank control system.

chosen arbitrarily small since the left-hand side of **2-128** is bounded from below. For $\varepsilon = 0.1$ the cutoff frequency is about 0.01 rad/s. The horizontal part of the curve at low frequencies is mainly attributable to the second term in the numerator of **2-128**, which originates from the off-diagonal entry in the first column of $T(j\omega) - I$. This entry represents part of the interaction present in the system.

We now consider the *second* link (the concentration link). Its frequency band follows from the inequality

$$50 \left| \frac{0.001267(j\omega) + 0.00002534}{(j\omega)^2 + 0.205(j\omega) + 0.01295} \right|^2 + 0.02 \left| \frac{(j\omega)^2 + 0.1100(j\omega) - 0.00067}{(j\omega)^2 + 0.205(j\omega) + 0.01295} \right|^2$$
$$\leq 0.02\varepsilon^2. \quad \textbf{2-129}$$

By dividing by 0.02 and rearranging, it follows for this inequality,

$$\frac{|(j\omega)^2 + 0.1100(j\omega) - 0.00067|^2 + 2500\,|0.001267(j\omega) + 0.00002534|^2}{|(j\omega)^2 + 0.205(j\omega) + 0.01295|^2} \leq \varepsilon^2.$$

2-130

The Bode plot of the left-hand side of this inequality, which is the difference function of the second link, is also shown in Fig. 2.21. In this case as well, the horizontal part of the curve at low frequencies is caused by the interaction in the system. If the requirements on ε are not too severe, the cutoff frequency of the second link is somewhere near 0.01 rad/s.

The cutoff frequencies obtained are reasonably close to the 10% cutoff frequencies of 0.011 rad/s and 0.0095 rad/s of the single-loop designs. Moreover, the interaction in the system seems to be limited. In conclusion, Fig. 2.22 pictures the step response matrix of the control system. The plots confirm that the control system exhibits moderate interaction (both dynamic and static). Each link has the step response of a first-order system with a time constant of approximately 10 s.

A rough idea of the resulting input amplitudes can be obtained as follows. From **2-116** we see that a step of 0.002 m³/s in the flow (assuming that this is a typical value) results in an initial flow change in feed 1 of $k_1 f_1 0.002 = 0.022$ m³/s. Similarly, a step of 0.1 kmol/m³ in the concentration results in an initial flow change in feed 2 of $k_2 f_2 0.1 = 0.01267$ m³/s. Compared to the nominal values of the incoming flows (0.015 m³/s and 0.005 m³/s, respectively), these values are far too large, which means that either smaller step input amplitudes must be chosen or the desired transition must be made more gradually. The latter can be achieved by redesigning the control system with smaller bandwidths.

In Problem 2.2 a more sophisticated design of a controller for the stirred tank is considered.

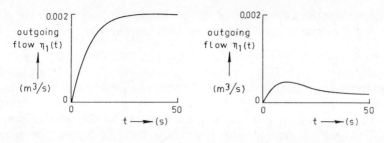

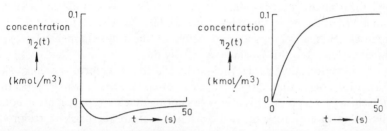

Fig. 2.22. Step response matrix of the stirred-tank control system. Left column: Responses of the outgoing flow and concentration to a step of 0.002 m³/s in the set point of the flow. Right column: Responses of the outgoing flow and concentration to a step of 0.1 kmol/m³ in the set point of the concentration.

2.6 THE TRANSIENT ANALYSIS OF THE TRACKING PROPERTIES

In the previous section we quite extensively discussed the steady-state properties of tracking systems. This section is devoted to the *transient* behavior of tracking systems, in particular that of the mean square tracking error and the mean square input. We define the *settling time* of a certain quantity (be it the mean square tracking error, the mean square input, or any other variable) as the time it takes the variable to reach its steady-state value to within a specified accuracy. When this accuracy is, say, 1 % of the maximal deviation from the steady-state value, we speak of the 1 % *settling time*. For other percentages similar terminology is used.

Usually, when a control system is started the initial tracking error, and as a result the initial input also, is large. Obviously, it is desirable that the mean square tracking error settles down to its steady-state value as quickly as possible after starting up or after upsets. We thus formulate the following directive.

Design Objective 2.4. *A control system should be so designed that the settling time of the mean square tracking error is as short as possible.*

As we have seen in Section 2.5.1, the mean square tracking error attributable to the reference variable consists of two contributions. One originates from the constant part of the reference variable and the other from the variable part. The transient behavior of the contribution of the variable part must be found by solving the matrix differential equation for the variance matrix of the state of the control system, which is fairly laborious. The transient behavior of the contribution of the constant part of the reference variable to the mean square tracking error is much simpler to find; this can be done simply by evaluating the response of the control system to nonzero initial conditions and to steps in the reference variable. As a rule, computing these responses gives a very good impression of the transient behavior of the control system, and this is what we usually do.

For asymptotically stable time-invariant linear control systems, some information concerning settling times can often be derived from the locations of the closed-loop poles. This follows by noting that *all* responses are exponentially damped motions with time constants that are the negative reciprocals of the real parts of the closed-loop characteristic values of the system. Since the 1% settling time of

$$e^{-t/\theta}, t \geq 0,\qquad\qquad \textbf{2-131}$$

is 4.6θ, a bound for the 1% settling time t_s of any variable is

$$t_s \leq 4.6 \max_i \left\{ \frac{1}{|\text{Re}(\lambda_i)|} \right\},\qquad\qquad \textbf{2-132}$$

where λ_i, $i = 1, 2, \cdots, n$, are the closed-loop characteristic values. Note that for *squared* variables such as the mean square tracking error and the mean square input, the settling time is half that of the variable itself.

The bound **2-132** sometimes gives misleading results, since it may easily happen that the response of a given variable does not depend upon certain characteristic values. Later (Section 3.8) we meet instances, for example, where the settling time of the rms tracking error is determined by the closed-loop poles furthest from the origin and not by the nearby poles, while the settling time of the rms input derives from the nearby closed-loop poles.

Example 2.9. *The settling time of the tracking error of the position servo*

Let us consider Design I of Example 2.4 (Section 2.3) for the position servo. From the steady-state analysis in Example 2.7 (Section 2.5.2), we learned that as the gain λ increases the rms steady-state tracking error keeps decreasing, although beyond a certain value (15–25 V/rad) very little improvement in the rms tracking error is obtained, while the rms input voltage becomes larger and larger. We now consider the settling time of the tracking error. To this end, in Fig. 2.23 the response of the controlled variable to a step in the reference variable is plotted for various values of λ, from zero initial conditions. As can be seen, the settling time of the step response (hence also that of the tracking error) first decreases rapidly as λ increases, but beyond a value of λ of about 15 V/rad the settling time fails to improve because of the increasingly oscillatory behavior of the response. In this case

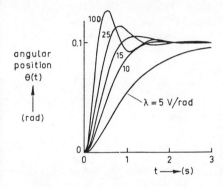

Fig. 2.23. Response of Design I of the position servo to a step of 0.1 rad in the reference variable for various values of the gain λ.

as well, the most favorable value of λ seems to be about 15 V/rad, which corresponds to a relative damping ζ (see Example 2.7) of about 0.7. From the plots of $|T(j\omega)|$ of Fig. 2.18, we see that for this value of the gain the largest bandwidth is achieved without undesirable peaking of the transmission.

2.7 THE EFFECTS OF DISTURBANCES IN THE SINGLE-INPUT SINGLE-OUTPUT CASE

In Section 2.3 we saw that very often disturbances act upon a control system, adversely affecting its tracking or regulating performance. In this section we derive expressions for the increases in the steady-state mean square tracking error and the steady-state mean square input attributable to disturbances, and formulate design objectives which may serve as a guide in designing control systems capable of counteracting disturbances.

Throughout this section the following assumptions are made.

1. *The disturbance variable $v_p(t)$ is a stochastic process that is uncorrelated with the reference variable $r(t)$ and the observation noise $v_m(t)$.*
As a result, we can obtain the increase in the mean square tracking error and the mean square input simply by setting $r(t)$ and $v_m(t)$ identical to zero.

2. *The controlled variable is also the observed variable, that is, $C = D$.*

168   Analysis of Linear Control Systems

This means that we can write

$$y(t) = z(t) + v_m(t), \qquad 2\text{-}133$$

and that in the time-invariant case

$$H(s) = K(s). \qquad 2\text{-}134$$

The assumption that the controlled variable is also the observed variable is quite reasonable, since it is intuitively clear that feedback is most effective when the controlled variable itself is directly fed back.

3. *The control system is asymptotically stable and time-invariant.*
4. *The input variable and the controlled variable, hence also the reference variable, are scalars. W_e and W_u are both 1.*

The analysis of this section can be extended to multivariable systems but doing so adds very little to the conclusions of this and the following sections.

5. *The disturbance variable $v_p(t)$ can be written as*

$$v_p(t) = v_{p0} + v_{pv}(t), \qquad 2\text{-}135$$

where the constant part v_{p0} of the disturbance variable is a stochastic vector with given second-order moment matrix, and where the variable part $v_{pv}(t)$ of the disturbance variable is a wide-sense stationary zero mean stochastic process with power spectral density matrix $\Sigma_{vp}(\omega)$, uncorrelated with v_{p0}.

The transfer matrix from the disturbance variable $v_p(t)$ to the controlled variable $z(t)$ can be found from the relation (see Fig. 2.24)

$$\mathbf{Z}(s) = -H(s)G(s)\mathbf{Z}(s) + D(sI - A)^{-1}\mathbf{V}_p(s), \qquad 2\text{-}136$$

where $\mathbf{Z}(s)$ and $\mathbf{V}_p(s)$ denote the Laplace transforms of $z(t)$ and $v_p(t)$,

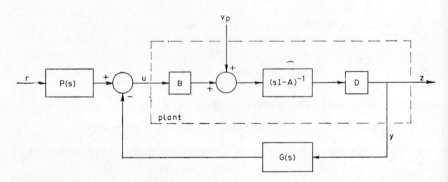

Fig. 2.24. Transfer matrix block diagram of a closed-loop control system with plant disturbance v_p.

respectively, so that

$$\mathbf{Z}(s) = \frac{1}{1 + H(s)G(s)} D(sI - A)^{-1}\mathbf{V}_p(s). \qquad 2\text{-}137$$

Here we have used the fact that the controlled variable is a scalar so that $1 + H(s)G(s)$ is also a scalar function. We now introduce the function

$$S(s) = \frac{1}{1 + H(s)G(s)}, \qquad 2\text{-}138$$

which we call the *sensitivity function* of the control system for reasons to be explained later.

We compute the contribution of the disturbance variable to the steady-state mean square tracking error as the sum of two terms, one originating from the constant part and one from the variable part of the disturbance. Since

$$\mathbf{Z}(s) = S(s)D(sI - A)^{-1}\mathbf{V}_p(s), \qquad 2\text{-}139$$

the steady-state response of the controlled variable to the constant part of the disturbance is given by

$$\lim_{t \to \infty} z(t) = S(0)D(-A)^{-1}v_{p0} = S(0)v_{00}. \qquad 2\text{-}140$$

Here we have assumed that the matrix A is nonsingular—the case where A is singular is treated in Problem 2.4. Furthermore, we have abbreviated

$$v_{00} = D(-A)^{-1}v_{p0}. \qquad 2\text{-}141$$

As a result of **2-140**, the contribution of the constant part of the disturbance to the steady-state mean square tracking error is

$$E\{|S(0)v_{00}|^2\} = |S(0)|^2 \, V_0, \qquad 2\text{-}142$$

where V_0 is the second-order moment of v_{00}, that is, $V_0 = E\{v_{00}^2\}$. Furthermore it follows from **2-139** with the methods of Sections 1.10.4 and 1.10.3 that the contribution of the variable part of the disturbance to the steady-state mean square tracking error can be expressed as

$$\int_{-\infty}^{\infty} |S(j\omega)|^2 \, D(j\omega I - A)^{-1}\Sigma_{vp}(\omega)(-j\omega I - A^T)^{-1}D^T \, df$$

$$= \int_{-\infty}^{\infty} |S(j\omega)|^2 \Sigma_{v0}(\omega) \, df. \qquad 2\text{-}143$$

Here we have abbreviated

$$\Sigma_{v0}(\omega) = D(j\omega I - A)^{-1}\Sigma_{vp}(\omega)(-j\omega I - A^T)^{-1}D^T. \qquad 2\text{-}144$$

Consequently, the increase in the steady-state mean square tracking error

170   Analysis of Linear Control Systems

attributable to the disturbance is given by

$C_{e\infty}$ (with disturbance) $- C_{e\infty}$ (without disturbance)

$$= |S(0)|^2 V_0 + \int_{-\infty}^{\infty} |S(j\omega)|^2 \Sigma_{v0}(\omega) \, df. \quad \text{2-145}$$

Before discussing how to make this expression small, we give an interpretation. Consider the situation of Fig. 2.25 where a variable $v_0(t)$ acts upon the closed-loop system. This variable is added to the controlled variable. It is

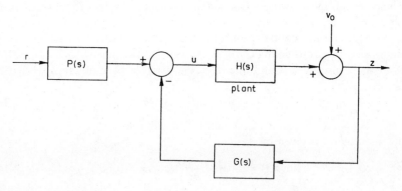

Fig. 2.25. Transfer matrix block diagram of a closed-loop control system with the equivalent disturbance v_0 at the controlled variable.

easily found that in terms of Laplace transforms with the reference variable and the initial conditions identical to zero the controlled variable is given by

$$\mathbf{Z}(s) = S(s)\mathbf{V}_0(s), \quad \text{2-146}$$

where $\mathbf{V}_0(s)$ denotes the Laplace transform of $v_0(t)$. We immediately see that if $v_0(t)$ is a stochastic process with as constant part a stochastic variable with second-order moment V_0 and as variable part a zero-mean wide-sense stationary stochastic process with power spectral density $\Sigma_{v0}(\omega)$, the increase in the steady-state mean square tracking error is exactly given by **2-145**. We therefore call the process $v_0(t)$ with these properties the *equivalent disturbance at the controlled variable*.

An examination of **2-145** leads to the following design rule.

Design Objective 2.5. *In order to reduce the increase of the steady-state mean square tracking error attributable to disturbances in an asymptotically stable linear time-invariant control system with a scalar controlled variable, which is also the observed variable, the absolute value of the sensitivity function $S(j\omega)$ should be made small over the frequency band of the equivalent disturbance at* the controlled variable. *If constant errors are of special concern, $S(0)$ should be made small, preferably zero.*

The last sentence of this design rule is not valid without further qualification for control systems where the matrix A of the plant is singular; this case is discussed in Problem 2.4. It is noted that since $S(j\omega)$ is given by

$$S(j\omega) = \frac{1}{1 + H(j\omega)G(j\omega)}, \qquad \text{2-147}$$

a small $S(j\omega)$ generally must be achieved by making the loop gain $H(j\omega)G(j\omega)$ of the control system large over a suitable frequency range. This easily conflicts with Design Objective 2.1 (Section 2.4) concerning the stability of the control system (see Example 2.5, Section 2.4), and with Objective 2.3 (Section 2.5.2) concerning the mean square input. A compromise must be found.

Reduction of constant errors is of special importance for regulator and tracking systems where the set point of the controlled variable must be maintained with great precision. Constant disturbances occur very easily in control systems, especially because of errors made in establishing the nominal input. Constant errors can often be completely eliminated by making $S(0) = 0$, which is usually achieved by introducing *integrating action*, that is, by letting the controller transfer function $G(s)$ have a pole at the origin (see Problem 2.3).

Let us now turn to a consideration of the steady-state mean square input. It is easily found that in terms of Laplace transforms we can write (see Fig. 2.24)

$$\mathbf{U}(s) = \frac{-G(s)}{1 + H(s)G(s)} D(sI - A)^{-1}\mathbf{V}_p(s), \qquad \text{2-148}$$

where $\mathbf{U}(s)$ is the Laplace transform of $u(t)$. It follows for the increase in the steady-state mean square input, using the notation introduced earlier in this section,

$C_{u\infty}$ (with disturbance) $- C_{u\infty}$ (without disturbance)

$$= \left| \frac{G(0)}{1 + H(0)G(0)} \right|^2 V_0 + \int_{-\infty}^{\infty} \left| \frac{G(j\omega)}{1 + H(j\omega)G(j\omega)} \right|^2 \Sigma_{v0}(\omega) \, df. \qquad \text{2-149}$$

This expression results in the following directive.

Design Objective 2.6. *In order to obtain a small increase in the steady-state mean square input attributable to the disturbance in an asymptotically stable linear time-invariant control system with a scalar controlled variable that is* also the observed variable and a scalar input,

$$\left| \frac{G(j\omega)}{1 + H(j\omega)G(j\omega)} \right| \qquad \textbf{2-150}$$

should be made small over the frequency band of the equivalent disturbance at the controlled variable.

In this directive no attention is paid to the constant part of the input since, as assumed in the discussion of Objective 2.3, the plant must be able to sustain these constant deviations.

Design Objective 2.6 conflicts with Objective 2.5. Making the loop gain $H(j\omega)G(j\omega)$ large, as required by Objective 2.5, usually does not result in small values of **2-150**. Again a compromise must be found.

Example 2.10. *The effect of disturbances on the position servo*

In this example we study the effect of disturbances on Design I of the position servo of Example 2.4 (Section 2.3). It is easily found that the sensitivity function of the control system as proposed is given by

$$S(s) = \frac{s(s + \alpha)}{s^2 + \alpha s + \kappa \lambda}. \qquad \textbf{2-151}$$

In Fig. 2.26 Bode plots of $|S(j\omega)|$ are given for several values of the gain λ. It is seen that by choosing λ larger the frequency band over which disturbance suppression is obtained also becomes larger. If the equivalent disturbance at the controlled variable, however, has much power near the frequency where $|S(j\omega)|$ has its peak, then perhaps a smaller gain is advisable.

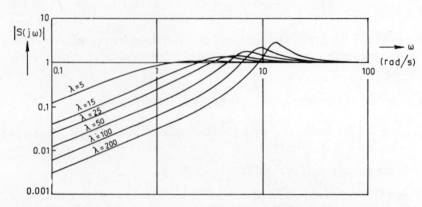

Fig. 2.26. Bode plots of the sensitivity function of the position control system, Design I, as a function of the gain λ.

2.7 Effects of Disturbances

In Example 2.4 we assumed that the disturbance enters as a disturbing torque $\tau_d(t)$ acting on the shaft of the motor. If the variable part of this disturbing torque has the power spectral density function $\Sigma_{rd}(\omega)$, the variable part of the equivalent disturbance at the controlled variable has the power spectral density function

$$\left| \frac{\gamma}{j\omega(j\omega + \alpha)} \right|^2 \Sigma_{rd}(\omega). \qquad \text{2-152}$$

The power spectral density of the contribution of the disturbing torque to the controlled variable is found by multiplying **2-152** by $|S(j\omega)|^2$ and thus is given by

$$\left| \frac{\gamma}{(j\omega)^2 + \alpha(j\omega) + \kappa\lambda} \right|^2 \Sigma_{rd}(\omega). \qquad \text{2-153}$$

Let us suppose that the variable part of the disturbing torque can be represented as exponentially correlated noise with rms value σ_{rd} and time constant T_{rd} so that

$$\Sigma_{rd}(\omega) = \frac{2\sigma_{rd}^2 T_{rd}}{1 + \omega^2 T_{rd}^2}. \qquad \text{2-154}$$

The increase in the steady-state mean square tracking error attributable to the disturbing torque can be computed by integrating **2-153**, or by modeling the disturbance, augmenting the state differential equation, and solving for the steady-state variance matrix of the augmented state. Either way we find

$C_{e\infty}$ (with disturbing torque) $- C_{e\infty}$ (without disturbing torque)

$$= \frac{1 + \alpha T_{rd}}{1 + \alpha T_{rd} + \kappa\lambda T_{rd}^2} \frac{\gamma^2 T_{rd}}{\alpha\kappa\lambda} \sigma_{rd}^2. \qquad \text{2-155}$$

From this we see that the addition to $C_{e\infty}$ monotonically decreases to zero with increasing λ. Thus the larger λ the less the disturbing torque affects the tracking properties.

In the absence of the reference variable, we have $\mu(t) = -\lambda\eta(t)$ so that the increase in the mean square input voltage attributable to the disturbing torque is λ^2 times the increase in the mean square tracking error:

$C_{u\infty}$ (with disturbing torque) $- C_{u\infty}$ (without disturbing torque)

$$= \frac{(1 + \alpha T_{rd})\lambda}{1 + \alpha T_{rd} + \kappa\lambda T_{rd}^2} \frac{\gamma^2 T_{rd}}{\alpha\kappa} \sigma_{rd}^2. \qquad \text{2-156}$$

For $\lambda \to \infty$, $C_{u\infty}$ monotonically increases to

$$\frac{(1 + \alpha T_{rd})\gamma^2}{\alpha T_{rd}\kappa^2} \sigma_{rd}^2. \qquad \text{2-157}$$

174 Analysis of Linear Control Systems

It is easily found from **2-25** that a *constant* disturbing torque τ_0 results in a steady-state displacement of the controlled variable of

$$\frac{\gamma \tau_0}{\kappa \lambda}. \qquad 2\text{-}158$$

Clearly, this displacement can also be made arbitrarily small by making the gain λ sufficiently large.

2.8 THE EFFECTS OF OBSERVATION NOISE IN THE SINGLE-INPUT SINGLE-OUTPUT CASE

In any closed-loop scheme, the effect of observation noise is to some extent felt. In this section the contribution of the observation noise to the mean square tracking error and the mean square input is analyzed. To this end, the following assumptions are made.

1. *The observation noise $v_m(t)$ is a stochastic process which is uncorrelated with the reference variable $r(t)$ and the plant disturbance $v_p(t)$.*
As a result, the increase in the mean square tracking error and the mean square input attributable to the observation noise may be computed simply by setting $r(t)$ and $v_p(t)$ identical to zero.

2. *The controller variable is also the observed variable, that is, $C = D$, so that*

$$y(t) = z(t) + v_m(t), \qquad 2\text{-}159$$

and, in the time-invariant case,

$$H(s) = K(s). \qquad 2\text{-}160$$

3. *The control system is asymptotically stable and time-invariant.*
4. *The input variable and the controlled variable, hence also the reference variable, are scalars. W_e and W_u are both 1.*
Here also the analysis can be extended to multivariable systems but again very little additional insight is gained.

5. *The observation noise is a zero-mean wide-sense stationary stochastic process with power spectral density function $\Sigma_{vm}(\omega)$.*

Figure 2.27 gives a transfer function block diagram of the situation that results from these assumptions. It is seen that in terms of Laplace transforms

$$\mathbf{Z}(s) = -H(s)G(s)[\mathbf{V}_m(s) + \mathbf{Z}(s)], \qquad 2\text{-}161$$

so that

$$\mathbf{Z}(s) = -\frac{H(s)G(s)}{1 + H(s)G(s)} \mathbf{V}_m(s). \qquad 2\text{-}162$$

2.8 Effects of Observation Noise

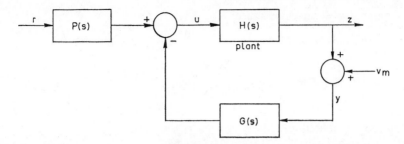

Fig. 2.27. Transfer matrix block diagram of a closed-loop control system with observation noise.

Consequently, the increase in the steady-state mean square tracking error attributable to the observation noise can be written as

$C_{e\infty}$ (with observation noise) $- C_{e\infty}$ (without observation noise)

$$= \int_{-\infty}^{\infty} \left| \frac{H(j\omega)G(j\omega)}{1 + H(j\omega)G(j\omega)} \right|^2 \Sigma_{vm}(\omega) \, df. \qquad \text{2-163}$$

Our next design objective can thus be formulated as follows.

Design Objective 2.7. *In order to reduce the increase in the steady-state mean square tracking error attributable to observation noise in an asymptotically stable linear time-invariant control system with a scalar controlled variable that is also the observed variable, the system should be designed so that*

$$\left| \frac{H(j\omega)G(j\omega)}{1 + H(j\omega)G(j\omega)} \right| \qquad \text{2-164}$$

is small over the frequency band of the observation noise.

Obviously, this objective is in conflict with Objective 2.5, since making the loop gain $H(j\omega)G(j\omega)$ large, as required by Objective 2.5, results in a value of **2-164** that is near 1, which means that the observation noise appears unattenuated in the tracking error. This is a result of the fact that if a large loop gain $H(j\omega)G(j\omega)$ is used the system is so controlled that $z(t) + v_m(t)$ instead of $z(t)$ tracks the reference variable.

A simple computation shows that the transfer function from the observation noise to the plant input is given by

$$U(s) = -\frac{G(s)}{1 + G(s)H(s)} V_m(s), \qquad \text{2-165}$$

which results in the following increase in the steady-state mean square input

176   Analysis of Linear Control Systems

attributable to observation noise:

$C_{u\infty}$ (with observation noise) $-$ $C_{u\infty}$ (without observation noise)

$$= \int_{-\infty}^{\infty} \left| \frac{G(j\omega)}{1 + G(j\omega)H(j\omega)} \right|^2 \Sigma_{vm}(\omega) \, df. \qquad \textbf{2-166}$$

This yields the design rule that to make the increase in the steady-state mean square input attributable to the observation noise small,

$$\left| \frac{G(j\omega)}{1 + G(j\omega)H(j\omega)} \right| \qquad \textbf{2-167}$$

should be made small over the frequency band of the observation noise. Clearly, this rule is also in conflict with Objective 2.5.

Example 2.11. *The position servo with position feedback only*

Let us once again consider the position servo of Example 2.4 (Section 2.3) with the three different designs proposed. In Examples 2.7 (Section 2.5.2) and 2.9 (Section 2.6), we analyzed Design I and chose $\lambda = 15$ V/rad as the best value of the gain. In Example 2.7 it was found that Design II gives better performance because of the additional feedback link from the angular velocity. Let us now suppose, however, that for some reason (financial or technical) a tachometer cannot be installed. We then resort to Design III, which attempts to approximate Design II by using an approximate differentiator with time constant T_d. If no observation noise were present, we could choose $T_d = 0$ and Design III would reduce to Design II. Let us suppose that observation noise is present, however, and that is can be represented as exponentially correlated noise with time constant

$$T_m = 0.02 \text{ s} \qquad \textbf{2-168}$$

and rms value

$$\sigma_m = 0.001 \text{ rad}. \qquad \textbf{2-169}$$

The presence of the observation noise forces us to choose $T_d > 0$. In order to determine a suitable value of T_d, we first assume that T_d will turn out to be small enough so that the gains ρ and λ can be chosen as in Design II. Then we see how large T_d can be made without spoiling the performance of Design II, while at the same time sufficiently reducing the effect of the observation noise.

It is easily found that the transmission of the control system according to Design III is given by

$$T(s) = \frac{\kappa\lambda(T_d s + 1)}{T_d s^3 + (\alpha T_d + 1)s^2 + (\alpha + \kappa\lambda T_d + \rho\lambda\kappa)s + \lambda\kappa}. \qquad \textbf{2-170}$$

2.8 Effects of Observation Noise

To determine a suitably small value of T_d, we argue as follows. The closed-loop system according to Design II, with the numerical values obtained in Example 2.7 for λ and ρ, has an undamped natural frequency ω_0 of about 20 rad/s with a relative damping of 0.707. Now in order not to impede the behavior of the system, the time constant T_d of the differentiator should be chosen small with respect to the inverse natural frequency, that is, small with respect to 0.05 s. In Fig. 2.28 we have plotted the transmission **2-170** for

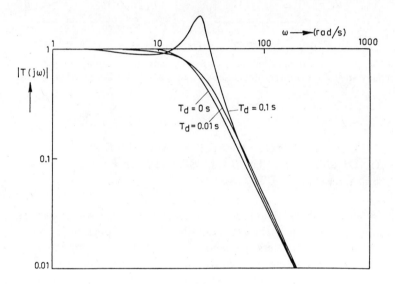

Fig. 2.28. The effect of T_d on the transmission of Design III of the position servo.

various values of T_d. It is seen that for $T_d = 0.01$ s the transmission is hardly affected by the approximate derivative operation, but that for $T_d = 0.1$ s discrepancies occur.

Let us now consider the effect of the observation noise. Modeling $v_m(t)$ in the usual way, the additions to the steady-state mean square tracking error and input attributable to the observation noise can be computed from the variance matrix of the augmented state. The numerical results are plotted in Fig. 2.29. These plots show that for small T_d the steady-state mean square input is greatly increased. An acceptable value of T_d seems to be about 0.01 s. For this value the square root of the increase in the steady-state mean square input is only about 2 V, the square root of the increase in the steady-state mean square tracking error of about 0.0008 rad is very small, and the transmission of the control system is hardly affected.

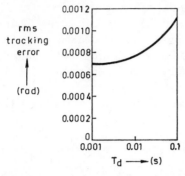
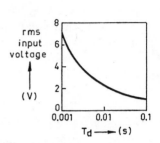

Fig. 2.29. The square roots of the additions to the steady-state mean square tracking error and input voltage due to observation noise as a function of T_d for Design III of the position servo.

2.9 THE EFFECT OF PLANT PARAMETER UNCERTAINTY IN THE SINGLE-INPUT SINGLE-OUTPUT CASE

Quite often a control system must be designed for a plant whose parameters are not exactly known to the designer. Also, it may happen in practice that changes of plant parameters frequently occur and that it is too laborious to measure the plant parameters each time and adjust the controller.

We shall see that closed-loop controllers can be designed so that the performance of the control system deteriorates very little even though there may be quite a large discrepancy between the actual plant parameters and the *nominal* plant parameters, that is, the parameter values that have been used while designing the controller. To this end we investigate the addition to the steady-state mean square tracking error attributable to parameter deviations.

In this section we work with the following assumptions.

1. *The control system is time-invariant and asymptotically stable.*
2. *The controlled variable is also the observed variable, that is, $C = D$, hence $K(s) = H(s)$.*
3. *The input variable and the controlled variable, hence also the reference variable, are scalar. W_e and W_u are both 1.*
Extension to the multivariable case is possible, but does not give much additional insight.
4. *Only the effect of parameter changes on the tracking properties is considered and not that on the disturbance suppression or noise reduction properties.*
5. *The reference variable has a constant part r_0, which is a stochastic vector,*

2.9 The Effect of Plant Parameter Uncertainty

with second-order moment R_0 and as variable part a zero-mean wide-sense stationary stochastic process with power spectral density function $\Sigma_r(\omega)$.

We denote by $H_0(s)$ the *nominal* transfer function of the plant, and by $H_1(s)$ the *actual* transfer function. Similarly, we write $T_0(s)$ for the transmission of the control system with the nominal plant transfer function and $T_1(s)$ for the transmission with the actual plant transfer function. We assume that the transfer function $G(s)$ in the feedback link and the transfer function $P(s)$ in the link from the reference variable (see the block diagram of Fig. 2.14, Section 2.5.1) are precisely known and not subject to change.

Using **2-63**, we obtain for the nominal transmission

$$T_0(s) = \frac{H_0(s)P(s)}{1 + G(s)H_0(s)}, \qquad \textbf{2-171}$$

and for the actual transmission

$$T_1(s) = \frac{H_1(s)P(s)}{1 + G(s)H_1(s)}. \qquad \textbf{2-172}$$

For the actual control system, the steady-state mean square tracking error is given by

$$C_{e\infty} = |T_1(0) - 1|^2 R_0 + \int_{-\infty}^{\infty} |T_1(j\omega) - 1|^2 \Sigma_r(\omega)\, df. \qquad \textbf{2-173}$$

We now make an estimate of the increase in the mean square tracking error attributable to a change in the transmission. Let us denote

$$\Delta T(s) = T_1(s) - T_0(s). \qquad \textbf{2-174}$$

Inserting $T_1(s) = T_0(s) + \Delta T(s)$ into **2-173**, we obtain

$$\begin{aligned} C_{e\infty} &= |T_0(0) - 1|^2 R_0 + \int_{-\infty}^{\infty} |T_0(j\omega) - 1|^2 \Sigma_r(\omega)\, df \\ &+ 2[T_0(0) - 1]\Delta T(0) R_0 + 2\text{Re}\left\{\int_{-\infty}^{\infty} [T_0(j\omega) - 1]\Delta T(-j\omega)\Sigma_r(\omega)\, df\right\} \\ &+ |\Delta T(0)|^2 R_0 + \int_{-\infty}^{\infty} |\Delta T(j\omega)|^2 \Sigma_r(\omega)\, df. \end{aligned} \qquad \textbf{2-175}$$

We now proceed by assuming that the nominal control system is well-designed so that the transmission $T_0(j\omega)$ is very close to 1 over the frequency band of the reference variable. In this case we can neglect the first four terms of **2-175** and we approximate

$$C_{e\infty} \simeq |\Delta T(0)|^2 R_0 + \int_{-\infty}^{\infty} |\Delta T(j\omega)|^2 \Sigma_r(\omega)\, df. \qquad \textbf{2-176}$$

180   Analysis of Linear Control Systems

This approximation amounts to the assumption that

$$|T_0(j\omega) - 1| \ll |\Delta T(j\omega)| \qquad \text{2-177}$$

for all ω in the frequency band of the reference variable.

Our next step is to express $\Delta T(s)$ in terms of $\Delta H(s)$, where

$$\Delta H(s) = H_1(s) - H_0(s). \qquad \text{2-178}$$

We obtain:

$$\Delta T(s) = \frac{H_1(s)P(s)}{1 + G(s)H_1(s)} - \frac{H_0(s)P(s)}{1 + G(s)H_0(s)}$$

$$= \frac{\Delta H(s)P(s)}{[1 + H_1(s)G(s)][1 + G(s)H_0(s)]}$$

$$= S_1(s)\,\Delta H(s) N_0(s), \qquad \text{2-179}$$

where

$$S_1(s) = \frac{1}{1 + H_1(s)G(s)} \qquad \text{2-180}$$

is the sensitivity function of the actual control system, and where

$$N_0(s) = \frac{P(s)}{1 + G(s)H_0(s)} \qquad \text{2-181}$$

is the transfer function of the nominal control system from the reference variable r to the input variable u. Now with the further approximation

$$S_1(j\omega) \simeq S_0(j\omega), \qquad \text{2-182}$$

where

$$S_0(s) = \frac{1}{1 + H_0(s)G(s)} \qquad \text{2-183}$$

is the sensitivity function of the nominal control system (which is known), we write for the steady-state mean square tracking error

$$C_{e\infty} \simeq |S_0(0)|^2 |\Delta H(0) N_0(0)|^2 R_0 + \int_{-\infty}^{\infty} |S_0(j\omega)|^2 |\Delta H(j\omega) N_0(j\omega)|^2 \Sigma_r(\omega)\, df.$$

$$\text{2-184}$$

We immediately conclude the following design objective.

Design Objective 2.8. *Consider a time-invariant asymptotically stable linear closed-loop control system with a scalar controlled variable that is also the observed variable. Then in order to reduce the steady-state mean square tracking error attributable to a variation $\Delta H(s)$ in the plant transfer function $H(s)$, the control system sensitivity function $S_0(j\omega)$ should be made small over*

2.9 The Effect of Plant Parameter Uncertainty

the frequency band of $|\Delta H(j\omega)N_0(j\omega)|^2 \Sigma_r(\omega)$. *If constant errors are of special concern, $S_0(0)$ should be made small, preferably zero, when $\Delta H(0)N_0(0)$ is different from zero.*

This objective should be understood as follows. Usually the plant transmission $T_0(s)$ is determined by finding a compromise between the requirements upon the mean square tracking error and the mean square input. Once $T_0(s)$ has been chosen, the transfer function $N_0(s)$ from the reference variable to the plant input is fixed. The given $T_0(s)$ and $N_0(s)$ can be realized in many different ways, for example, by first choosing the transfer function $G(s)$ in the feedback link and then adjusting the transfer function $P(s)$ in the link from the reference variable so that the desired $T_0(s)$ is achieved. Now Design Objective 2.8 states that this realization should be chosen so that

$$S_0(j\omega) = \frac{1}{1 + H_0(j\omega)G(j\omega)} \qquad \textbf{2-185}$$

is small over the frequency band of $|\Delta H(j\omega)N_0(j\omega)|^2 \Sigma_r(\omega)$. The latter function is known when some idea about $\Delta H(j\omega)$ is available and $T_0(j\omega)$ has been decided upon. We note that making the sensitivity function $S_0(j\omega)$ small is a requirement that is also necessary to reduce the effect of disturbances in the control system, as we found in Section 2.7. As noted in Section 2.7, $S_0(0)$ can be made zero by introducing integrating action (Problem 2.3).

We conclude this section with an interpretation of the function $S_0(s)$. From **2-179** and **2-171** it follows that

$$\frac{\Delta T(s)}{T_0(s)} = S_1(s) \frac{\Delta H(s)}{H_0(s)}. \qquad \textbf{2-186}$$

Thus $S_1(s)$ relates the relative change in the plant transfer function $H(s)$ to the resulting relative change in the control system transmission $T(s)$. When the changes in the plant transfer function are restricted in magnitude, we can approximate $S_1(j\omega) \simeq S_0(j\omega)$. This interpretation of the function $S_0(s)$ is a classical concept due to Bode (see, e.g., Horowitz, 1963). $S_0(s)$ is called the *sensitivity function* of the closed-loop system, since it gives information about the sensitivity of the control system transmission to changes in the plant transfer function.

Example 2.12. *The effect of parameter variations on the position servo*

Let us analyze the sensitivity to parameter changes in Design I of the position servo (Example 2.4, Section 2.3). The sensitivity function for this design is given by

$$S_0(s) = \frac{s(s + \alpha)}{s^2 + \alpha s + \kappa \lambda}. \qquad \textbf{2-187}$$

Plots of $|S(j\omega)|$ for various values of the gain λ have been given in Fig. 2.26. It is seen that for $\lambda = 15$ V/rad, which is the most favorable value of the gain, protection against the effect of parameter variations is achieved up to about 3 rad/s. To be more specific, let us assume that the parameter variations are caused by variations in the moment of inertia J. Since the plant parameters α and κ are given by (Example 2.4)

$$\alpha = \frac{B}{J}, \quad \kappa = \frac{k}{J}, \qquad \text{2-188}$$

it is easily found that for small variations ΔJ in J we can write

$$\frac{\Delta H(s)}{H(s)} \simeq -\frac{s}{s+\alpha}\frac{\Delta J}{J}, \qquad \text{2-189}$$

where

$$H(s) = \frac{\kappa}{s(s+\alpha)} \qquad \text{2-190}$$

is the plant transfer function. We note the following.

1. For zero frequency we have

$$\frac{\Delta H(0)}{H(0)} = 0, \qquad \text{2-191}$$

no matter what value ΔJ has. Since $T(0) = 1$, and consequently $\Delta T(0) = 0$, this means that the response to changes in the set point of the tracking system is always correct, independent of the inertial load of the servo.

2. We see from **2-189** that as a function of ω the effect of a variation in the moment of inertia upon the plant transfer function increases up to the break frequency $\alpha = 4.6$ rad/s and stays constant from there onward. From the behavior of the sensitivity function, it follows that for low frequencies (up to about 3 rad/s) the effect of a variation in the moment of inertia upon the transmission is attenuated and that especially for low frequencies a great reduction results.

To illustrate the control system sensitivity, in Fig. 2.30 the response of the closed-loop system to a step in the reference variable is given for the cases

$$\frac{\Delta J}{J} = 0, -0.3, \text{ and } +0.3. \qquad \text{2-192}$$

Taking into account that a step does not have a particularly small frequency band, the control system compensates the parameter variation quite satisfactorily.

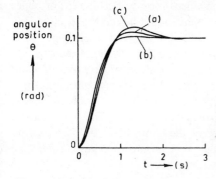

Fig. 2.30. The effect of parameter variations on the response of the position servo, Design I, to a step of 0.1 rad in the reference variable: (a) Nominal inertial load; (b) inertial load 1.3 of nominal; (c) inertial load 0.7 of nominal.

2.10* THE OPEN-LOOP STEADY-STATE EQUIVALENT CONTROL SCHEME

The potential advantages of closed-loop control may be very clearly brought to light by comparing closed-loop control systems to their so-called open-loop steady-state equivalents. This section is devoted to a discussion of such open-loop equivalent control systems, where we limit ourselves to the time-invariant case.

Consider a time-invariant closed-loop control system and denote the transfer matrix from the reference variable r to the plant input u by $N(s)$. Then we can always construct an open-loop control system (see Fig. 2.31) that has the same transfer matrix $N(s)$ from the reference variable r to the plant input u. As a result, the transmission of both the closed-loop system and the newly constructed open-loop control system is given by

$$T(s) = K(s)N(s), \qquad \text{2-193}$$

where $K(s)$ is the transfer matrix of the plant from the plant input u to the controlled variable z. For reasons explained below, we call the open-loop system *steady-state equivalent* to the given closed-loop system.

In most respects the open-loop steady-state equivalent proves to be inferior to the closed-loop control system. Often, however, it is illuminating to

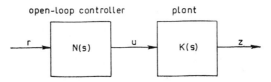

Fig. 2.31. The open-loop steady-state equivalent control system.

study the open-loop equivalent of a given closed-loop system since it provides a reference situation with a performance that should be improved upon. We successively compare closed-loop control systems and their open-loop equivalents according to the following aspects of control system performance: *stability*; *steady-state tracking properties*; *transient behavior*; *effect of plant disturbances*; *effect of observation noise*; *sensitivity to plant variations*.

We first consider *stability*. We immediately see that the characteristic values of the equivalent open-loop control system consist of the characteristic values of the plant, together with those of the controller (compare Section 1.5.4). This means, among other things, that *an unstable plant cannot be stabilized by an open-loop controller*. Since stability is a basic design objective, there is little point in considering open-loop equivalents when the plant is not asymptotically stable.

Let us assume that the plant and the open-loop equivalent are asymptotically stable. We now consider the *steady-state tracking properties* of both control systems. Since the systems have equal transmissions and equal transfer matrices from the reference variable to the plant input, their steady-state mean square tracking errors and mean square input are also equal. This explains the name steady-state equivalent. This also means that *from the point of view of tracking performance there is no need to resort to closed-loop control*.

We proceed to the *transient properties*. Since among the characteristic values of the open-loop equivalent control system the characteristic values of the plant appear unchanged, obviously *no improvement in the transient properties can be obtained by open-loop control*, in contrast to closed-loop control. By transient properties we mean the response of the control system to nonzero initial conditions of the plant.

Next we consider the *effect of disturbances*. As in Section 2.7, we assume that the disturbance variable can be written as the sum of a constant and a variable part. Since in the multivariable case we can write for the contribution of the disturbance variable to the controlled variable in the closed-loop system

$$\mathbf{Z}(s) = [I + H(s)G(s)]^{-1}D(sI - A)^{-1}\mathbf{V}_p(s), \qquad \textbf{2-194}$$

it follows that the contribution of the disturbance variable to the mean square tracking error of the *closed-loop* system can be expressed as

$C_{e\infty}$ (with disturbance) $- C_{e\infty}$ (without disturbance)

$$= \operatorname{tr} \left\{ S^T(0)V_0 S(0)W_e + \int_{-\infty}^{\infty} S(j\omega)\Sigma_{v0}(\omega)S^T(-j\omega)W_e \, df \right\}, \qquad \textbf{2-195}$$

2.10 Open-Loop Steady-State Equivalence

where we have used the results of Sections 1.10.3 and 1.10.4, and where

$$S(s) = [I + H(s)G(s)]^{-1},$$

$$\Sigma_{v0}(\omega) = D(j\omega I - A)^{-1}\Sigma_{vp}(\omega)(-j\omega I - A^T)^{-1}D^T, \qquad \textbf{2-196}$$

$$V_0 = D(-A)^{-1}E\{v_{p0}v_{p0}^T\}(-A^T)^{-1}D^T.$$

In analogy with the single-input single-output case, $S(s)$ is called the *sensitivity matrix* of the system. The matrix A is assumed to be nonsingular.

Let us now consider the equivalent open-loop system. Here the contribution of the disturbance to the controlled variable is given by

$$\mathbf{Z}(s) = D(sI - A)^{-1}\mathbf{V}_p(s). \qquad \textbf{2-197}$$

Assuming that the open-loop equivalent control system is asymptotically stable, it is easily seen that the increase in the steady-state mean square tracking error due to the disturbance in the *open-loop* system can be expressed as

$C_{e\infty}$ (with disturbance) $- C_{e\infty}$ (without disturbance)

$$= \mathrm{tr}\left\{W_e V_0 + \int_{-\infty}^{\infty}\Sigma_{v0}(\omega)W_e\,df\right\}. \qquad \textbf{2-198}$$

We see from **2-198** that the increase in the mean square tracking error is completely independent of the controller, hence is not affected by the open-loop control system design. Clearly, *in an open-loop control system disturbance reduction is impossible.*

Since the power spectral density matrix $\Sigma_{v0}(\omega)$ may be ill-known, it is of some interest to establish whether or not there exists a condition that guarantees that in a closed-loop control system the disturbance is reduced as compared to the open-loop equivalent irrespective of Σ_{v0}. Let us rewrite the increase **2-195** in the mean square tracking error of a closed-loop system as follows:

$C_{e\infty}$ (with disturbance) $- C_{e\infty}$ (without disturbance)

$$= \mathrm{tr}\left\{S^T(0)W_e S(0)V_0 + \int_{-\infty}^{\infty}S^T(-j\omega)W_e S(j\omega)\Sigma_{v0}(\omega)\,df\right\}, \qquad \textbf{2-199}$$

where $S(s)$ is the sensitivity matrix of the system. A comparison with **2-198** leads to the following statement.

Theorem 2.1. *Consider a time-invariant asymptotically stable closed-loop control system where the controlled variable is also the observed variable and where the plant is asymptotically stable. Then the increase in the steady-state mean square tracking error due to the plant disturbance is less than or* at least equal to that for the open-loop steady-state equivalent, regardless of the properties of the plant disturbance, if and only if

$$S^T(-j\omega)W_e S(j\omega) \leq W_e \quad \text{for all real } \omega. \quad \textbf{2-200}$$

The proof of this theorem follows from the fact that, given any two nonnegative-definite Hermitian matrices M_1 and M_2, then $M_1 \geq M_2$ implies and is implied by tr $(M_1 N) \geq$ tr $(M_2 N)$ for any nonnegative-definite Hermitian matrix N.

The condition **2-200** is especially convenient for single-input single-output systems, where $S(s)$ is a scalar function so that **2-200** reduces to

$$|S(j\omega)| \leq 1 \quad \text{for all real } \omega. \quad \textbf{2-201}$$

Usually, it is simpler to verify this condition in terms of the return difference function

$$J(s) = \frac{1}{S(s)} = 1 + H(s)G(s). \quad \textbf{2-202}$$

With this we can rewrite **2-201** as

$$|J(j\omega)| \geq 1 \quad \text{for all real } \omega. \quad \textbf{2-203}$$

Also, for multiinput multioutput systems it is often more convenient to verify **2-200** in terms of the return difference matrix

$$J(s) = S^{-1}(s) = I + H(s)G(s). \quad \textbf{2-204}$$

In this connection the following result is useful.

Theorem 2.2. *Let* $J(s) = S^{-1}(s)$. *Then the three following statements are equivalent*:

(a) $S^T(-j\omega)W_e S(j\omega) \leq W_e$,

(b) $J^T(-j\omega)W_e J(j\omega) \geq W_e$, **2-205**

(c) $J(j\omega)W_e^{-1}J^T(-j\omega) \geq W_e^{-1}$.

The proof is left as an exercise.

Thus we have seen that open-loop systems are inferior to closed-loop control systems from the point of view of disturbance reduction. In all fairness it should be pointed out, however, that in open-loop control systems the plant disturbance causes no increase in the mean square input.

The next item of consideration is the *effect of observation noise*. Obviously, in open-loop control systems observation noise does not affect either the mean square tracking error or the mean square input, since there is no feedback link that introduces the observation noise into the system. In this respect the open-loop equivalent is superior to the closed-loop system.

2.10 Open-Loop Steady-State Equivalence

Our final point of consideration is the *sensitivity to plant variations*. Let us first consider the single-input single-output case, and let us derive the mean square tracking error attributable to a plant variation for an open-loop control system. Since an open-loop control system has a unity sensitivity function, it follows from **2-184** that under the assumptions of Section 2.9 the mean square tracking error resulting from a plant variation is given by

$$C_{e\infty} \text{ (open-loop)} \simeq |\Delta H(0)N_0(0)|^2 R_0 + \int_{-\infty}^{\infty} |\Delta H(j\omega)N_0(j\omega)|^2 \Sigma_r(\omega) \, df. \quad \textbf{2-206}$$

Granting that $N_0(s)$ is decided upon from considerations involving the nominal mean square tracking error and input, we conclude from this expression that the sensitivity to a plant transfer function variation of an open-loop control system is not influenced by the control system design. Apparently, *protection against plant variations cannot be achieved through open-loop control*.

For the closed-loop case, the mean square tracking error attributable to plant variations is given by **2-184**:

$$C_{e\infty} \text{ (closed-loop)} \simeq |S_0(0)|^2 |\Delta H(0)N_0(0)|^2 R_0$$
$$+ \int_{-\infty}^{\infty} |S_0(j\omega)|^2 |\Delta H(j\omega)N_0(j\omega)|^2 \Sigma_r(\omega) \, df. \quad \textbf{2-207}$$

A comparison of **2-206** and **2-207** shows that the closed-loop system is always less sensitive to plant variations than the equivalent open-loop system, no matter what the nature of the plant variations and the properties of the reference variable are, if the sensitivity function satisfies the inequality

$$|S_0(j\omega)| \leq 1 \quad \text{for all } \omega. \quad \textbf{2-208}$$

Thus we see that the condition that guarantees that the closed-loop system is less sensitive than the open-loop system to disturbances also makes the system less sensitive to plant variations.

In the case of disturbance attenuation, the condition **2-208** generalizes to

$$S_0^T(-j\omega)W_e S_0(j\omega) \leq W_e, \quad \text{for all } \omega, \quad \textbf{2-209}$$

for the multivariable case. It can be proved (Cruz and Perkins, 1964; Kreindler, 1968a) that the condition **2-209** *guarantees that the increase in the steady-state mean square tracking error due to (small) plant variations in a closed-loop system is always less than or equal to that for the open-loop steady-state equivalent, regardless of the nature of the plant variation and the properties of the reference variable.*

We conclude this section with Table 2.2, which summarizes the points of agreement and difference between closed-loop control schemes and their open-loop steady-state equivalents.

Table 2.2 Comparison of Closed-Loop and Open-Loop Designs

Feature	Closed-loop design	Open-loop steady-state equivalent
Stability	Unstable plant can be stabilized	Unstable plant cannot be stabilized
Steady-state mean square tracking error and input attributable to reference variable	Identical performance if the plant is asymptotically stable.	
Transient behavior	Great improvement in response to initial conditions is possible	No improvement in response to initial conditions is possible
Effect of disturbances	Effect on mean square tracking error can be greatly reduced; mean square input is increased	Full effect on mean square tracking error; mean square input is not affected
Effect of observation noise	Both mean square tracking error and mean square input are increased	No effect on mean square tracking error or mean square input
Effect of plant variations	Effect on mean square tracking error can be greatly reduced	Full effect on mean square tracking error

2.11 CONCLUSIONS

In this chapter we have given a description of control problems and of the various aspects of the performance of a control system. It has been shown that closed-loop control schemes can give very attractive performances. Various rules have been developed which can be applied when designing a control system.

Very little advice has been offered, however, on the question how to select the precise form of the controller. This problem is considered in the following chapters. We formulate the problem of finding a suitable compromise for the requirement of a small mean square tracking error without an overly large mean square input as a mathematical optimization problem. This optimization problem will be developed and solved in stages in Chapters 3–5. Its solution enables us to determine, explicitly and quantitatively, suitable control schemes.

2.12 PROBLEMS

2.1. *The control of the angular velocity of a motor*

Consider a dc motor described by the differential equation

$$J\frac{dc(t)}{dt} + Bc(t) = m(t), \qquad \textbf{2-210}$$

where $c(t)$ is the angular velocity of the motor, $m(t)$ the torque applied to the shaft of the motor, J the moment of inertia, and B the friction coefficient. Suppose that

$$m(t) = ku(t), \qquad \textbf{2-211}$$

where $u(t)$ is the electric voltage applied to the motor and k the torque coefficient. Inserting **2-211** into **2-210**, we write the system differential equation as

$$\frac{dc(t)}{dt} + \alpha c(t) = \kappa u(t). \qquad \textbf{2-212}$$

The following numerical values are used:

$$\alpha = 0.5 \text{ s}^{-1}, \qquad \kappa = 150 \text{ rad}/(\text{V s}^2), \qquad J = 0.01 \text{ kg m}^2. \qquad \textbf{2-213}$$

It is assumed that the angular velocity is both the observed and the controlled variable. We study the simple proportional control scheme where the input voltage is given by

$$u(t) = -\lambda c(t) + \rho r(t). \qquad \textbf{2-214}$$

Here $r(t)$ is the reference variable and λ and ρ are gains to be determined. The system is to be made into a tracking system.

(a) Determine the values of the feedback gain λ for which the closed-loop system is asymptotically stable.

(b) For each value of the feedback gain λ, determine the gain ρ such that the tracking system exhibits a zero steady-state error response to a step in the reference variable. In the remainder of the problem, the gain ρ is always chosen so that this condition is satisfied.

(c) Suppose that the reference variable is exponentially correlated noise with an rms value of 30 rad/s and a break frequency of 1 rad/s. Determine the feedback gain such that the rms input voltage to the dc motor is 2 V. What is the rms tracking error for this gain? Sketch a Bode plot of the transmission of the control system for this gain. What is the 10% cutoff frequency? Compare this to the 10% cutoff frequency of the reference variable and comment on the magnitude of the rms tracking error as compared to the rms value of the reference variable. What is the 10% settling time of the response of the system to a step in the reference variable?

(d) Suppose that the system is disturbed by a stochastically varying torque on the shaft of the dc motor, which can be described as exponentially correlated noise with an rms value of 0.1732 N m and a break frequency of 1 rad/s. Compute the increases in the steady-state mean square tracking error and mean square input attributable to the disturbance for the values of λ and ρ selected under (c). Does the disturbance significantly affect the performance of the system?

(e) Suppose that the measurement of the angular velocity is afflicted by additive measurement noise which can be represented as exponentially correlated noise with an rms value of 0.1 rad/s and a break frequency of 100 rad/s. Does the measurement noise seriously impede the performance of the system?

(f) Suppose that the dc motor exhibits variations in the form of changes in the moment of inertia J, attributable to load variations. Consider the off-nominal values 0.005 kg m² and 0.02 kg m² for the moment of inertia. How do these extreme variations affect the response of the system to steps in the reference variable when the gains λ and ρ are chosen as selected under (c)?

2.2. *A decoupled control system design for the stirred tank*

Consider the stirred tank control problem as described in Examples 2.2 (Section 2.2.2) and 2.8 (Section 2.5.3). The state differential equation of the plant is given by

$$\dot{x}(t) = \begin{pmatrix} -0.01 & 0 \\ 0 & -0.02 \end{pmatrix} x(t) + \begin{pmatrix} 1 & 1 \\ -0.25 & 0.75 \end{pmatrix} u(t), \qquad \text{2-215}$$

and the controlled variable by

$$z(t) = \begin{pmatrix} 0.01 & 0 \\ 0 & 1 \end{pmatrix} x(t). \qquad \text{2-216}$$

(a) Show that the plant can be completely decoupled by choosing

$$u(t) = Qu'(t), \qquad \text{2-217}$$

where Q is a suitable 2 × 2 matrix and where $u'(t) = \text{col}\,[\mu_1'(t), \mu_2'(t)]$ is a new input to the plant.

(b) Using (a), design a closed-loop control system, analogous to that designed in Example 2.8, which is completely decoupled, where $T(0) = I$, and where each link has a 10% cutoff frequency of 0.01 rad/s.

2.3. *Integrating action*

Consider a time-invariant single-input single-output plant where the controlled variable is also the observed variable, that is, $C = D$, and which has a nonsingular A-matrix. For the suppression of constant disturbances, the sensitivity function $S(j\omega)$ should be made small, preferably zero, at $\omega = 0$. $S(s)$ is given by

$$S(s) = \frac{1}{1 + H(s)G(s)}, \qquad \text{2-218}$$

where $H(s)$ is the plant transfer function and $G(s)$ the controller transfer function (see Fig. 2.25). Suppose that it is possible to find a rational function $Q(s)$ such that the controller with transfer function

$$G(s) = \frac{1}{s} Q(s) \qquad \text{2-219}$$

makes the closed-loop system asymptotically stable. We say that this controller introduces *integrating action*. Show that for this control system $S(0) = 0$, provided $H(0)Q(0)$ is nonzero. Consequently, controllers with integrating action can completely suppress constant disturbances.

2.4*. *Constant disturbances in plants with a singular A-matrix*

Consider the effect of constant disturbances in a control system satisfying the assumptions 1 through 5 of Section 2.7, but where the matrix A of the plant is singular, that is, the plant contains integration.

(a) Show that the contribution of the constant part of the disturbance to the steady-state mean square tracking error can be expressed as

$$\lim_{s \to 0} E\{v_{p0}^T(-sI - A^T)^{-1} D^T S(-s) S(s) D(sI - A)^{-1} v_{p0}\}. \qquad \text{2-220}$$

We distinguish between the two cases (b) and (c).

(b) Assume that the disturbances enter the system in such a way that

$$\lim_{s \to 0} D(sI - A)^{-1} v_{p0} \qquad \text{2-221}$$

is always finite. This means that constant disturbances always result in finite, constant equivalent errors at the controlled variable despite the integrating nature of the plant. Show that in this case
 (i) Design Objective 2.5 applies without modification, and
 (ii) $S(0) = 0$, provided

$$\lim_{s \to 0} sH(s)G(s) \qquad \text{2-222}$$

is nonzero.

Here $H(s)$ is the plant transfer function and $G(s)$ the transfer function in the feedback link (see Fig. 2.25). This result shows that in a plant with integration where constant disturbances always result in finite, constant equivalent errors at the controlled variable, constant disturbances are completely suppressed (provided **2-222** is satisfied, which implies that neither the plant nor the controller transfer function has a zero at the origin).

(c) We now consider the case where **2-221** is not finite. Suppose that

$$\lim_{s \to 0} s^k D(sI - A)^{-1} v_{p0} \qquad \text{2-223}$$

is finite, where k is the least positive integer for which this is true. Show that in this case

$$\lim_{s \to 0} \frac{S(s)}{s^k} \qquad \text{2-224}$$

should be made small, preferably zero, to achieve a small constant error at the controlled variable. Show that **2-224** can be made equal to zero by letting

$$G(s) = \frac{1}{s^{k-m_0+1}} Q(s), \qquad \text{2-225}$$

where $Q(s)$ is a rational function of s such that $Q(0) \neq 0$ and $Q(0) \neq \infty$, and where m_0 is the least integer m such that

$$\lim_{s \to 0} s^m H(s) \qquad \text{2-226}$$

is finite.

3 OPTIMAL LINEAR STATE FEEDBACK CONTROL SYSTEMS

3.1 INTRODUCTION

In Chapter 2 we gave an exposition of the problems of linear control theory. In this chapter we begin to build a theory that can be used to solve the problems outlined in Chapter 2. The main restriction of this chapter is that we assume that the complete state $x(t)$ of the plant can be accurately measured at all times and is available for feedback. Although this is an unrealistic assumption for many practical control systems, the theory of this chapter will prove to be an important foundation for the more general case where we do not assume that $x(t)$ is completely accessible.

Much attention of this chapter is focused upon regulator problems, that is, problems where the goal is to maintain the state of the system at a desired value. We shall see that linear control theory provides powerful tools for solving such problems. Both the deterministic and the stochastic versions of the optimal linear regulator problem are studied in detail. Important extensions of the regulator problem—the nonzero set point regulator and the optimal linear tracking problem—also receive considerable attention.

Other topics dealt with are the numerical solution of Riccati equations, asymptotic properties of optimal control laws, and the sensitivity of linear optimal state feedback systems.

3.2 STABILITY IMPROVEMENT OF LINEAR SYSTEMS BY STATE FEEDBACK

3.2.1 Linear State Feedback Control

In Chapter 2 we saw that an important aspect of feedback system design is the stability of the control system. Whatever we want to achieve with the control system, its stability must be assured. Sometimes the main goal of a feedback design is actually to stabilize a system if it is initially unstable, or to improve its stability if transient phenomena do not die out sufficiently fast.

194 Optimal Linear State Feedback Control Systems

The purpose of this section is to investigate how the stability properties of linear systems can be improved by state feedback.

Consider the linear time-varying system with state differential equation

$$\dot{x}(t) = A(t)x(t) + B(t)u(t). \qquad 3\text{-}1$$

If we suppose that the complete state can be accurately measured at all times, it is possible to implement a *linear control law* of the form

$$u(t) = -F(t)x(t) + u'(t), \qquad 3\text{-}2$$

where $F(t)$ is a time-varying *feedback gain matrix* and $u'(t)$ a new input. If this control law is connected to the system **3-1**, the closed-loop system is described by the state differential equation

$$\dot{x}(t) = [A(t) - B(t)F(t)]x(t) + B(t)u'(t). \qquad 3\text{-}3$$

The stability of this system depends of course on the behavior of $A(t)$ and $B(t)$ but also on that of the gain matrix $F(t)$. It is convenient to introduce the following terminology.

Definition 3.1. *The linear control law*

$$u(t) = -F(t)x(t) + u'(t) \qquad 3\text{-}4$$

is called an **asymptotically stable control law** *for the system*

$$\dot{x}(t) = A(t)x(t) + B(t)u(t) \qquad 3\text{-}5$$

if the closed-loop system

$$\dot{x}(t) = [A(t) - B(t)F(t)]x(t) + B(t)u'(t) \qquad 3\text{-}6$$

is asymptotically stable.

If the system **3-5** is *time-invariant*, and we choose a constant matrix F, the stability of the control law **3-4** is determined by the characteristic values of the matrix $A - BF$. In the next section we find that under a mildly restrictive condition (namely, the system must be completely controllable), all closed-loop characteristic values can be arbitrarily located in the complex plane by choosing F suitably (with the restriction of course that complex poles occur in complex conjugate pairs). If all the closed-loop poles are placed in the left-half plane, the system is of course asymptotically stable.

We also see in the next section that for single-input systems, that is, systems with a scalar input u, usually a unique gain matrix F is found for a given set of closed-loop poles. Melsa (1970) lists a FORTRAN computer program to determine this matrix. In the multiinput case, however, a given set of poles can usually be achieved with many different choices of F.

Example 3.1. *Stabilization of the inverted pendulum*

The state differential equation of the inverted pendulum positioning system of Example 1.1 (Section 1.2.3) is given by

$$\dot{x}(t) = \begin{pmatrix} 0 & 1 & 0 & 0 \\ 0 & -\dfrac{F}{M} & 0 & 0 \\ 0 & 0 & 0 & 1 \\ -\dfrac{g}{L'} & 0 & \dfrac{g}{L'} & 0 \end{pmatrix} x(t) + \begin{pmatrix} 0 \\ \dfrac{1}{M} \\ 0 \\ 0 \end{pmatrix} \mu(t). \qquad 3\text{-}7$$

Let us consider the time-invariant control law

$$\mu(t) = -(\phi_1, \phi_2, \phi_3, \phi_4)x(t). \qquad 3\text{-}8$$

It follows that for the system **3-7** and control law **3-8** we have

$$A - BF = \begin{pmatrix} 0 & 1 & 0 & 0 \\ -\dfrac{\phi_1}{M} & -\dfrac{F+\phi_2}{M} & -\dfrac{\phi_3}{M} & -\dfrac{\phi_4}{M} \\ 0 & 0 & 0 & 1 \\ -\dfrac{g}{L'} & 0 & \dfrac{g}{L'} & 0 \end{pmatrix}. \qquad 3\text{-}9$$

The characteristic polynomial of this matrix is

$$s^4 + s^3 \frac{F+\phi_2}{M} + s^2 \left(\frac{\phi_1}{M} - \frac{g}{L'}\right) - s\frac{F+\phi_2+\phi_4}{M}\frac{g}{L'} - \frac{\phi_1+\phi_3}{M}\frac{g}{L'}. \qquad 3\text{-}10$$

Now suppose that we wish to assign all closed-loop poles to the location $-\alpha$. Then the closed-loop characteristic polynomial should be given by

$$(s+\alpha)^4 = s^4 + 4\alpha s^3 + 6\alpha^2 s^2 + 4\alpha^3 s + \alpha^4. \qquad 3\text{-}11$$

Equating the coefficients of **3-10** and **3-11**, we find the following equations in ϕ_1, ϕ_2, ϕ_3, and ϕ_4:

$$\frac{F+\phi_2}{M} = 4\alpha,$$

$$\frac{\phi_1}{M} - \frac{g}{L'} = 6\alpha^2,$$

$$-\frac{F+\phi_2+\phi_4}{M}\frac{g}{L'} = 4\alpha^3,$$

$$-\frac{\phi_1+\phi_3}{M}\frac{g}{L'} = \alpha^4.$$

3-12

With the numerical values of Example 1.1 and with $\alpha = 3\ s^{-1}$, we find from these linear equations the following control law:

$$\mu(t) = -(65.65,\ 11.00,\ -72.60,\ -21.27)x(t). \qquad \textbf{3-13}$$

Example 3.2. *Stirred tank*

The stirred tank of Example 1.2 (Section 1.2.3) is an example of a multi-input system. With the numerical values of Example 1.2, the linearized state differential equation of the system is

$$\dot{x}(t) = \begin{pmatrix} -0.01 & 0 \\ 0 & -0.02 \end{pmatrix} x(t) + \begin{pmatrix} 1 & 1 \\ -0.25 & 0.75 \end{pmatrix} u(t). \qquad \textbf{3-14}$$

Let us consider the time-invariant control law

$$u(t) = -\begin{pmatrix} \phi_{11} & \phi_{12} \\ \phi_{21} & \phi_{22} \end{pmatrix} x(t). \qquad \textbf{3-15}$$

It follows from **3-14** and **3-15** that the closed-loop characteristic polynomial is given by

$$\det(sI - A + BF) = s^2 + s(0.03 + \phi_{11} - 0.25\phi_{12} + \phi_{21} + 0.75\phi_{22})$$
$$+ (0.0002 + 0.02\phi_{11} - 0.0025\phi_{12} + 0.02\phi_{21} + 0.0075\phi_{22} + \phi_{11}\phi_{22} - \phi_{12}\phi_{21}). \qquad \textbf{3-16}$$

We can see at a glance that a given closed-loop characteristic polynomial can be achieved for many different values of the gain factors ϕ_{ij}. For example, the three following feedback gain matrices

$$F_a = \begin{pmatrix} 1.1 & 3.7 \\ 0 & 0 \end{pmatrix}, \quad F_b = \begin{pmatrix} 0 & 0 \\ 1.1 & -1.2333 \end{pmatrix} \quad \text{and} \quad F_c = \begin{pmatrix} 0.1 & 0 \\ 0 & 0.1 \end{pmatrix} \qquad \textbf{3-17}$$

all yield the closed-loop characteristic polynomial $s^2 + 0.2050s + 0.01295$, so that the closed-loop characteristic values are $-0.1025 \pm j0.04944$. We note that in the control law corresponding to the first gain matrix the second component of the input is not used, the second feedback matrix leaves the first component untouched, while in the third control law both inputs control the system.

In Fig. 3.1 are sketched the responses of the three corresponding closed-loop systems to the initial conditions

$$\xi_1(0) = 0\ \text{m}^3, \qquad \xi_2(0) = 0.1\ \text{kmol/m}^3. \qquad \textbf{3-18}$$

Note that even though the closed-loop poles are the same the differences in the three responses are very marked.

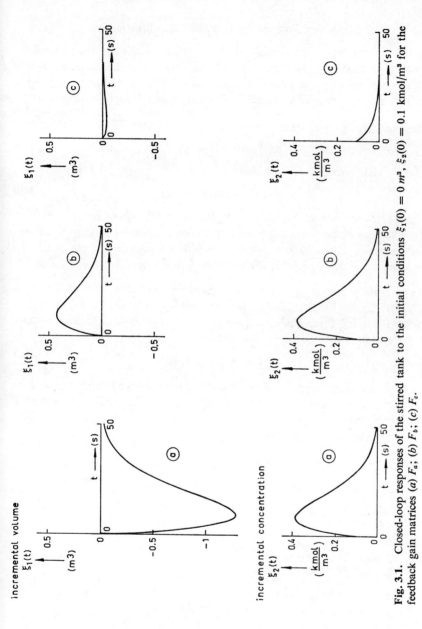

Fig. 3.1. Closed-loop responses of the stirred tank to the initial conditions $\xi_1(0) = 0\ m^3$, $\xi_2(0) = 0.1$ kmol/m³ for the feedback gain matrices (a) F_a; (b) F_b; (c) F_c.

3.2.2* Conditions for Pole Assignment and Stabilization

In this section we state precisely (1) under what conditions the closed-loop poles of a time-invariant linear system can be arbitrarily assigned to any location in the complex plane by linear state feedback, and (2) under what conditions the system can be stabilized. First, we have the following result.

Theorem 3.1. *Consider the linear time-invariant system*

$$\dot{x}(t) = Ax(t) + Bu(t) \qquad \text{3-19}$$

with the time-invariant control law

$$u(t) = -Fx(t) + u'(t). \qquad \text{3-20}$$

Then the closed-loop characteristic values, that is, the characteristic values of $A - BF$, *can be arbitrarily located in the complex plane (with the restriction that complex characteristic values occur in complex conjugate pairs) by choosing F suitably if and only if the system* **3-19** *is completely controllable.*

A complete proof of this theorem is given by Wonham (1967a), Davison (1968b), Chen (1968b), and Heymann (1968). Wolovich (1968) considers the time-varying case. We restrict our proof to single-input systems. Suppose that the system with the state differential equation

$$\dot{x}(t) = Ax(t) + b\mu(t), \qquad \text{3-21}$$

where $\mu(t)$ is a scalar input, is completely controllable. Then we know from Section 1.9 that there exists a state transformation $x'(t) = T^{-1}x(t)$, where T is a nonsingular transformation matrix, which transforms the system **3-19** into its phase-variable canonical form:

$$\dot{x}'(t) = \begin{pmatrix} 0 & 1 & 0 & \cdots\cdots & 0 \\ 0 & 0 & 1 & 0 & \cdots\cdots & 0 \\ \cdots\cdots\cdots\cdots\cdots\cdots\cdots\cdots\cdots \\ 0 & \cdots\cdots\cdots & 0 & 1 \\ -\alpha_0 & -\alpha_1 & \cdots\cdots & -\alpha_{n-1} \end{pmatrix} x'(t) + \begin{pmatrix} 0 \\ 0 \\ \cdots \\ 0 \\ 1 \end{pmatrix} \mu(t). \qquad \text{3-22}$$

Here the numbers α_i, $i = 0, 1, \cdots, n-1$ are the coefficients of the characteristic polynomial of the system **3-21**, that is, $\det(sI - A) = s^n + \alpha_{n-1}s^{n-1} \cdots + \alpha_1 s + \alpha_0$. Let us write **3-22** more compactly as

$$\dot{x}'(t) = A'x'(t) + b'\mu(t). \qquad \text{3-23}$$

Consider now the linear control law

$$\mu(t) = -f'x'(t) + \mu'(t), \qquad \text{3-24}$$

3.2 Stability Improvement by State Feedback

where f' is the row vector

$$f' = (\phi_1, \phi_2, \cdots, \phi_n). \qquad 3\text{-}25$$

If this control law is connected to the system, the closed-loop system is described by the state differential equation

$$\dot{x}'(t) = (A' - b'f')x'(t) + b'\mu'(t). \qquad 3\text{-}26$$

It is easily seen that the matrix $A' - b'f$ is given by

$$A' - b'f' = \begin{pmatrix} 0 & 1 & 0 \cdots \cdots \cdots \cdots 0 \\ 0 & 0 & 1 \; 0 \cdots \cdots \cdots 0 \\ \cdots \cdots \cdots \cdots \cdots \cdots \cdots \cdots \cdots \cdots \\ 0 \cdots \cdots \cdots \cdots \cdots \cdots \cdots 0 & 1 \\ -\alpha_0 - \phi_1 & -\alpha_1 - \phi_2 \cdots \cdots \cdots \cdots -\alpha_{n-1} - \phi_n \end{pmatrix}. \qquad 3\text{-}27$$

This clearly shows that the characteristic polynomial of the matrix $A' - b'f'$ has the coefficients $(\alpha_i - \phi_{i+1})$, $i = 0, 1, \cdots, n - 1$. Since the ϕ_i, $i = 1, 2, \cdots, n$, are arbitrarily chosen real numbers, the coefficients of the closed-loop characteristic polynomial can be given any desired values, which means that the closed-loop poles can be assigned to arbitrary locations in the complex plane (provided complex poles occur in complex conjugate pairs).

Once the feedback law in terms of the transformed state variable has been chosen, it can immediately be expressed in terms of the original state variable $x(t)$ as follows:

$$\mu(t) = -f'x'(t) + \mu'(t) = -f'T^{-1}x(t) + \mu'(t) = -fx(t) + \mu'(t). \qquad 3\text{-}28$$

This proves that if **3-19** is completely controllable, the closed-loop characteristic values may be arbitrarily assigned. For the proof of the converse of this statement, see the end of the proof of Theorem 3.2. Since the proof for multiinput systems is somewhat more involved we omit it. As we have seen in Example 3.2, for multiinput systems there usually are many solutions for the feedback gain matrix F for a given set of closed-loop characteristic values.

Through Theorem 3.1 it is always possible to stabilize a completely controllable system by state feedback, or to improve its stability, by assigning the closed-loop poles to locations in the left-half complex plane. The theorem gives no guidance, however, as to where in the left-half complex plane the closed-loop poles should be located. Even more uncertainty occurs in the multiinput case where the same closed-loop pole configuration can be achieved by various control laws. This uncertainty is removed by optimal linear regulator theory, which is discussed in the remainder of this chapter.

200 Optimal Linear State Feedback Control Systems

Theorem 3.1 implies that it is always possible to stabilize a completely controllable linear system. Suppose, however, that we are confronted with a time-invariant system that is not completely controllable. From the discussion of stabilizability in Section 1.6.4, it can be shown that stabilizability, as the name expresses, is precisely the condition that allows us to stabilize a not completely controllable time-invariant system by a time-invariant linear control law (Wonham, 1967a):

Theorem 3.2. *Consider the linear time-invariant system*

$$\dot{x}(t) = Ax(t) + Bu(t) \qquad 3\text{-}29$$

with the time-invariant control law

$$u(t) = -Fx(t) + u'(t). \qquad 3\text{-}30$$

Then it is possible to find a constant matrix F such that the closed-loop system is asymptotically stable if and only if the system **3-29** *is stabilizable.*

The proof of this theorem is quite simple. From Theorem 1.26 (Section 1.6.3), we know that the system can be transformed into the controllability canonical form

$$\dot{x}'(t) = \begin{pmatrix} A'_{11} & A'_{12} \\ 0 & A'_{22} \end{pmatrix} x'(t) + \begin{pmatrix} B'_1 \\ 0 \end{pmatrix} u(t), \qquad 3\text{-}31$$

where the pair $\{A'_{11}, B'_1\}$ is completely controllable. Consider the linear control law

$$u(t) = -(F'_1, \; F'_2) x'(t) + u'(t). \qquad 3\text{-}32$$

For the closed-loop system we find

$$\begin{aligned}
\dot{x}'(t) &= \begin{pmatrix} A'_{11} & A'_{12} \\ 0 & A'_{22} \end{pmatrix} x'(t) - \begin{pmatrix} B'_1 \\ 0 \end{pmatrix} (F'_1, \; F'_2) x'(t) + \begin{pmatrix} B'_1 \\ 0 \end{pmatrix} u'(t) \\
&= \begin{pmatrix} A'_{11} - B'_1 F'_1 & A'_{12} - B'_1 F'_2 \\ 0 & A'_{22} \end{pmatrix} x'(t) + \begin{pmatrix} B'_1 \\ 0 \end{pmatrix} u'(t).
\end{aligned} \qquad 3\text{-}33$$

The characteristic values of the compound matrix in this expression are the characteristic values of $A'_{11} - B'_1 F'_1$ together with those of A'_{22}. Now if the system **3-29** is stabilizable, A'_{22} is asymptotically stable, and since the pair $\{A'_{11}, B'_1\}$ is completely controllable, it is always possible to find an F'_1 such that $A'_{11} - B'_1 F'_1$ is stable. This proves that if **3-29** is stabilizable it is always possible to find a feedback law that stabilizes the system. Conversely, if one can find a feedback law that stabilizes the system, A'_{22} must be asymptotically stable, hence the system is stabilizable. This proves the other direction of the theorem.

The proof of the theorem shows that, if the system is stabilizable but not completely controllable, only some of the closed-loop poles can be arbitrarily located since the characteristic values of A'_{22} are not affected by the control law. This proves one direction of Theorem 3.1.

3.3 THE DETERMINISTIC LINEAR OPTIMAL REGULATOR PROBLEM

3.3.1 Introduction

In Section 3.2 we saw that under a certain condition (complete controllability) a time-invariant linear system can always be stabilized by a linear feedback law. In fact, more can be done. Because the closed-loop poles can be located anywhere in the complex plane, the system can be stabilized; but, moreover, by choosing the closed-loop poles far to the left in the complex plane, the convergence to the zero state can be made arbitrarily fast. To make the system move fast, however, large input amplitudes are required. In any practical problem the input amplitudes must be bounded; this imposes a limit on the distance over which the closed-loop poles can be moved to the left. These considerations lead quite naturally to the formulation of an optimization problem, where we take into account both the speed of convergence of the state to zero and the magnitude of the input amplitudes.

To introduce this optimization problem, we temporarily divert our attention from the question of the pole locations, to return to it in Section 3.8.

Consider the linear time-varying system with state differential equation

$$\dot{x}(t) = A(t)x(t) + B(t)u(t), \qquad \text{3-34}$$

and let us study the problem of bringing this system from an arbitrary initial state to the zero state as quickly as possible (in Section 3.7 we consider the case where the desired state is not the zero state). There are many criteria that express how fast an initial state is reduced to the zero state; a very useful one is the quadratic integral criterion

$$\int_{t_0}^{t_1} x^T(t) R_1(t) x(t) \, dt. \qquad \text{3-35}$$

Here $R_1(t)$ is a nonnegative-definite symmetric matrix. The quantity $x^T(t)R_1(t)x(t)$ is a measure of the extent to which the state at time t deviates from the zero state; the weighting matrix $R_1(t)$ determines how much weight is attached to each of the components of the state. The integral 3-35 is a criterion for the cumulative deviation of $x(t)$ from the zero state during the interval $[t_0, t_1]$.

As we saw in Chapter 2, in many control problems it is possible to identify a controlled variable $z(t)$. In the linear models we employ, we usually have

$$z(t) = D(t)x(t). \qquad \textbf{3-36}$$

If the actual problem is to reduce the controlled variable $z(t)$ to zero as fast as possible, the criterion **3-35** can be modified to

$$\int_{t_0}^{t_1} z^T(t)R_3(t)z(t)\,dt, \qquad \textbf{3-37}$$

where $R_3(t)$ is a positive-definite symmetric weighting matrix. It is easily seen that **3-37** is equivalent to **3-35**, since with **3-36** we can write

$$\int_{t_0}^{t_1} z^T(t)R_3(t)z(t)\,dt = \int_{t_0}^{t_1} x^T(t)R_1(t)x(t)\,dt, \qquad \textbf{3-38}$$

where

$$R_1(t) = D^T(t)R_3(t)D(t). \qquad \textbf{3-39}$$

If we now attempt to find an optimal input to the system by minimizing the quantity **3-35** or **3-37**, we generally run into the difficulty that indefinitely large input amplitudes result. To prevent this we include the input in the criterion; we thus consider

$$\int_{t_0}^{t_1} [z^T(t)R_3(t)z(t) + u^T(t)R_2(t)u(t)]\,dt, \qquad \textbf{3-40}$$

where $R_2(t)$ is a positive-definite symmetric weighting matrix. The inclusion of the second term in the criterion reduces the input amplitudes if we attempt to make the total value of **3-40** as small as possible. The relative importance of the two terms in the criterion is determined by the matrices R_3 and R_2.

If it is very important that the terminal state $x(t_1)$ is as close as possible to the zero state, it is sometimes useful to extend **3-40** with a third term as follows

$$\int_{t_0}^{t_1} [z^T(t)R_3(t)z(t) + u^T(t)R_2(t)u(t)]\,dt + x^T(t_1)P_1 x(t_1), \qquad \textbf{3-41}$$

where P_1 is a nonnegative-definite symmetric matrix.

We are now in a position to introduce the deterministic linear optimal regulator problem:

Definition 3.2. *Consider the linear time-varying system*

$$\dot{x}(t) = A(t)x(t) + B(t)u(t), \qquad \textbf{3-42}$$

where

$$x(t_0) = x_0, \qquad \textbf{3-43}$$

with the controlled variable

$$z(t) = D(t)x(t). \qquad \textbf{3-44}$$

3.3 The Deterministic Linear Optimal Regulator

Consider also the criterion

$$\int_{t_0}^{t_1} [z^T(t)R_3(t)z(t) + u^T(t)R_2(t)u(t)]\, dt + x^T(t_1)P_1 x(t_1), \qquad 3\text{-}45$$

*where P_1 is a nonnegative-definite symmetric matrix and $R_3(t)$ and $R_2(t)$ are positive-definite symmetric matrices for $t_0 \leq t \leq t_1$. Then the problem of determining an input $u^0(t)$, $t_0 \leq t \leq t_1$, for which the criterion is minimal is called the **deterministic linear optimal regulator problem**.*

Throughout this chapter, and indeed throughout this book, it is understood that $A(t)$ is a continuous function of t and that $B(t)$, $D(t)$, $R_3(t)$, and $R_2(t)$ are piecewise continuous functions of t, and that all these matrix functions are bounded.

A special case of the regulator problem is the time-invariant regulator problem:

Definition 3.3. *If all matrices occurring in the formulation of the deterministic linear optimal regulator problem are constant, we refer to it as the **time-invariant deterministic linear optimal regulator problem**.*

We continue this section with a further discussion of the formulation of the regulator problem. First, we note that in the regulator problem, as it stands in Definition 3.2, we consider only the *transient* situation where an arbitrary initial state must be reduced to the zero state. The problem formulation does not include disturbances or a reference variable that should be tracked; these more complicated situations are discussed in Section 3.6.

A difficulty of considerable interest is how to choose the weighting matrices R_3, R_2, and P_1 in the criterion **3-45**. This must be done in the following manner. Usually it is possible to define three quantities, the *integrated square regulating error*, the *integrated square input*, and the *weighted square terminal error*. The integrated square regulating error is given by

$$\int_{t_0}^{t_1} z^T(t)W_e(t)z(t)\, dt, \qquad 3\text{-}46$$

where $W_e(t)$, $t_0 \leq t \leq t_1$, is a weighting matrix such that $z^T(t)W_e(t)z(t)$ is properly dimensioned and has physical significance. We discussed the selection of such weighting matrices in Chapter 2. Furthermore, the integrated square input is given by

$$\int_{t_0}^{t_1} u^T(t)W_u(t)u(t)\, dt, \qquad 3\text{-}47$$

where the weighting matrix $W_u(t)$, $t_0 \leq t \leq t_1$, is similarly selected. Finally, the weighted square terminal error is given by

$$x^T(t_1)W_t x(t_1), \qquad 3\text{-}48$$

where also W_t is a suitable weighting matrix. We now consider various problems, such as:

1. Minimize the integrated square regulating error with the integrated square input and the weighted square terminal error constrained to certain maximal values.
2. Minimize the weighted square terminal error with the integrated square input and the integrated square regulating error constrained to certain maximal values.
3. Minimize the integrated square input with the integrated square regulating error and the weighted square terminal error constrained to certain maximal values.

All these versions of the problem can be studied by considering the minimization of the criterion

$$\rho_1 \int_{t_0}^{t_1} z^T(t)W_e(t)z(t)\,dt + \rho_2 \int_{t_0}^{t_1} u^T(t)W_u(t)u(t)\,dt + \rho_3 x^T(t_1)W_t x(t_1), \quad \textbf{3-49}$$

where the constants ρ_1, ρ_2, and ρ_3 are suitably chosen. The expression **3-45** is exactly of this form. Let us, for example, consider the important case where the terminal error is unimportant and where we wish to minimize the integrated square regulating error with the integrated square input constrained to a certain maximal value. Since the terminal error is of no concern, we set $\rho_3 = 0$. Since we are minimizing the integrated square regulating error, we take $\rho_1 = 1$. We thus consider the minimization of the quantity

$$\int_{t_0}^{t_1} [z^T(t)W_e(t)z(t) + \rho_2 u^T(t)W_u(t)u(t)]\,dt. \quad \textbf{3-50}$$

The scalar ρ_2 now plays the role of a Lagrange multiplier. To determine the appropriate value of ρ_2, we solve the problem for many different values of ρ_2. This provides us with a graph as indicated in Fig. 3.2, where the integrated square regulating error is plotted versus the integrated square input with ρ_2 as a parameter. As ρ_2 decreases, the integrated square regulating error decreases but the integrated square input increases. From this plot we can determine the value of ρ_2 that gives a sufficiently small regulating error without excessively large inputs.

From the same plot we can solve the problem where we must minimize the integrated square input with a constrained integrated square regulating error. Other versions of the problem formulation can be solved in a similar manner. We thus see that the regulator problem, as formulated in Definition 3.2, is quite versatile and can be adapted to various purposes.

3.3 The Deterministic Linear Optimal Regulator

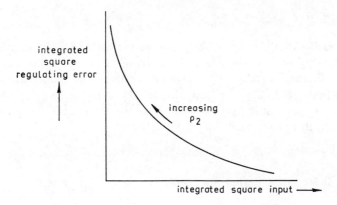

Fig. 3.2. Integrated square regulating error versus integrated square input, with $\rho_1 = 1$ and $\rho_3 = 0$.

We see in later sections that the solution of the regulator problem can be given in the form of a linear control law which has several useful properties. This makes the study of the regulator problem an interesting and practical proposition.

Example 3.3. *Angular velocity stabilization problem*

As a first example, we consider an angular velocity stabilization problem. The plant consists of a dc motor the shaft of which has the angular velocity $\xi(t)$ and which is driven by the input voltage $\mu(t)$. The system is described by the scalar state differential equation

$$\dot{\xi}(t) = -\alpha\xi(t) + \kappa\mu(t), \qquad \textbf{3-51}$$

where α and κ are given constants. We consider the problem of stabilizing the angular velocity $\xi(t)$ at a desired value ω_0. In the formulation of the general regulator problem we have chosen the origin of state space as the equilibrium point. Since in the present problem the desired equilibrium position is $\xi(t) \equiv \omega_0$, we shift the origin. Let μ_0 be the constant input voltage to which ω_0 corresponds as the steady-state angular velocity. Then μ_0 and ω_0 are related by

$$0 = -\alpha\omega_0 + \kappa\mu_0. \qquad \textbf{3-52}$$

Introduce now the new state variable

$$\xi'(t) = \xi(t) - \omega_0. \qquad \textbf{3-53}$$

Then with the aid of **3-52**, it follows from **3-51** that $\xi'(t)$ satisfies the state differential equation

$$\dot{\xi}'(t) = -\alpha\xi'(t) + \kappa\mu'(t), \qquad \textbf{3-54}$$

where

$$\mu'(t) = \mu(t) - \mu_0. \quad \text{3-55}$$

This shows that the problem of bringing the system **3-51** from an arbitrary initial state $\xi(t_0) = \omega_1$ to the state $\xi = \omega_0$ is equivalent to bringing the system **3-51** from the initial state $\xi(t_0) = \omega_1 - \omega_0$ to the equilibrium state $\xi = 0$. Thus, without restricting the generality of the example, we consider the problem of regulating the system **3-51** about the zero state. The controlled variable ζ in this problem obviously is the state ξ:

$$\zeta(t) = \xi(t). \quad \text{3-56}$$

As the optimization criterion, we choose

$$\int_{t_0}^{t_1} [\zeta^2(t) + \rho\mu^2(t)]\, dt + \pi_1 \xi^2(t_1), \quad \text{3-57}$$

with $\rho > 0$, $\pi_1 \geq 0$. This criterion ensures that the deviations of $\xi(t)$ from zero are restricted [or, equivalently, that $\xi(t)$ stays close to ω_0], that $\mu(t)$ does not assume too large values [or, equivalently, $\mu(t)$ does not deviate too much from μ_0], and that the terminal state $\xi(t_1)$ will be close to zero [or, equivalently, that $\xi(t_1)$ will be close to ω_0]. The values of ρ and π_1 must be determined by trial and error. For α and κ we use the following numerical values:

$$\begin{aligned} \alpha &= 0.5 \text{ s}^{-1}, \\ \kappa &= 150 \text{ rad/(V s}^2). \end{aligned} \quad \text{3-58}$$

Example 3.4. *Position control*

In Example 2.4 (Section 2.3), we discussed position control by a dc motor. The system is described by the state differential equation

$$\dot{x}(t) = \begin{pmatrix} 0 & 1 \\ 0 & -\alpha \end{pmatrix} x(t) + \begin{pmatrix} 0 \\ \kappa \end{pmatrix} \mu(t), \quad \text{3-59}$$

where $x(t)$ has as components the angular position $\xi_1(t)$ and the angular velocity $\xi_2(t)$ and where the input variable $\mu(t)$ is the input voltage to the dc amplifier that drives the motor. We suppose that it is desired to bring the angular position to a constant value ξ_{10}. As in the preceding example, we make a shift in the origin of the state space to obtain a standard regulator problem. Let us define the new state variable $x'(t)$ with components

$$\begin{aligned} \xi_1'(t) &= \xi_1(t) - \xi_{10}, \\ \xi_2'(t) &= \xi_2(t). \end{aligned} \quad \text{3-60}$$

3.3 The Deterministic Linear Optimal Regulator

A simple substitution shows that $x'(t)$ satisfies the state differential equation

$$\dot{x}'(t) = \begin{pmatrix} 0 & 1 \\ 0 & -\alpha \end{pmatrix} x'(t) + \begin{pmatrix} 0 \\ \kappa \end{pmatrix} \mu(t). \qquad 3\text{-}61$$

Note that in contrast to the preceding example we need not define a new input variable. This results from the fact that the angular position can be maintained at any constant value with a zero input. Since the system **3-61** is identical to **3-59**, we omit the primes and consider the problem of regulating **3-59** about the zero state.

For the controlled variable we choose the angular position:

$$\zeta(t) = \xi_1(t) = (1, 0)x(t). \qquad 3\text{-}62$$

An appropriate optimization criterion is

$$\int_{t_0}^{t_1} [\zeta^2(t) + \rho \mu^2(t)] \, dt. \qquad 3\text{-}63$$

The positive scalar weighting coefficient ρ determines the relative importance of each term of the integrand. The following numerical values are used for α and κ:

$$\begin{aligned} \alpha &= 4.6 \text{ s}^{-1}, \\ \kappa &= 0.787 \text{ rad}/(\text{V s}^2). \end{aligned} \qquad 3\text{-}64$$

3.3.2 Solution of the Regulator Problem

In this section we solve the deterministic optimal regulator problem using elementary methods of the calculus of variations. It is convenient to rewrite the criterion **3-45** in the form

$$\int_{t_0}^{t_1} [x^T(t)R_1(t)x(t) + u^T(t)R_2(t)u(t)] \, dt + x^T(t_1)P_1 x(t_1), \qquad 3\text{-}65$$

where $R_1(t)$ is the nonnegative-definite symmetric matrix

$$R_1(t) = D^T(t)R_3(t)D(t). \qquad 3\text{-}66$$

Suppose that the input that minimizes this criterion exists and let it be denoted by $u^0(t)$, $t_0 \leq t \leq t_1$. Consider now the input

$$u(t) = u^0(t) + \varepsilon \tilde{u}(t), \qquad t_0 \leq t \leq t_1, \qquad 3\text{-}67$$

where $\tilde{u}(t)$ is an arbitrary function of time and ε is an arbitrary number. We shall check how this change in the input affects the criterion **3-65**. Owing to the change in the input, the state will change, say from $x^0(t)$ (the optimal behavior) to

$$x(t) = x^0(t) + \varepsilon \tilde{x}(t), \qquad t_0 \leq t \leq t_1. \qquad 3\text{-}68$$

208 Optimal Linear State Feedback Control Systems

This defines $\tilde{x}(t)$, which we now determine. The solution $x(t)$ as given by 3-68 must satisfy the state differential equation 3-42 with $u(t)$ chosen according to 3-67. This yields

$$\dot{x}^0(t) + \varepsilon \dot{\tilde{x}}(t) = A(t)x^0(t) + \varepsilon A(t)\tilde{x}(t) + B(t)u^0(t) + \varepsilon B(t)\tilde{u}(t). \quad \text{3-69}$$

Since the optimal solution must also satisfy the state differential equation, we have

$$\dot{x}^0(t) = A(t)x^0(t) + B(t)u^0(t). \quad \text{3-70}$$

Subtraction of 3-69 and 3-70 and cancellation of ε yields

$$\dot{\tilde{x}}(t) = A(t)\tilde{x}(t) + B(t)\tilde{u}(t). \quad \text{3-71}$$

Since the initial state does not change if the input changes from $u^0(t)$ to $u^0(t) + \varepsilon \tilde{u}(t)$, $t_0 \le t \le t_1$, we have $\tilde{x}(t_0) = 0$, and the solution of 3-71 using **1-61** can be written as

$$\tilde{x}(t) = \int_{t_0}^{t} \Phi(t, \tau) B(\tau) \tilde{u}(\tau) \, d\tau, \quad \text{3-72}$$

where $\Phi(t, t_0)$ is the transition matrix of the system 3-71. We note that $\tilde{x}(t)$ does not depend upon ε. We now consider the criterion 3-65. With 3-67 and 3-68 we can write

$$\int_{t_0}^{t_1} [x^T(t)R_1(t)x(t) + u^T(t)R_2(t)u(t)] \, dt + x^T(t_1)P_1 x(t_1)$$

$$= \int_{t_0}^{t_1} [x^{0T}(t)R_1(t)x^0(t) + u^{0T}(t)R_2(t)u^0(t)] \, dt + x^{0T}(t_1)P_1 x^0(t_1)$$

$$+ 2\varepsilon \left\{ \int_{t_0}^{t_1} [\tilde{x}^T(t)R_1(t)x^0(t) + \tilde{u}^T(t)R_2(t)u^0(t)] \, dt + \tilde{x}^T(t_1)P_1 x^0(t_1) \right\}$$

$$+ \varepsilon^2 \left\{ \int_{t_0}^{t_1} [\tilde{x}^T(t)R_1(t)\tilde{x}(t) + \tilde{u}^T(t)R_2(t)\tilde{u}(t)] \, dt + \tilde{x}^T(t_1)P_1 \tilde{x}(t_1) \right\}. \quad \text{3-73}$$

Since $u^0(t)$ is the optimal input, changing the input from $u^0(t)$ to the input 3-67 can only increase the value of the criterion. This implies that, as a function of ε, 3-73 must have a minimum at $\varepsilon = 0$. Since 3-73 is a quadratic expression in ε, it can assume a minimum for $\varepsilon = 0$ only if its first derivative with respect to ε is zero at $\varepsilon = 0$. Thus we must have

$$\int_{t_0}^{t_1} [\tilde{x}^T(t)R_1(t)x^0(t) + \tilde{u}^T(t)R_2(t)u^0(t)] \, dt + \tilde{x}^T(t_1)P_1 x^0(t_1) = 0. \quad \text{3-74}$$

Substitution of 3-72 into 3-74 yields after an interchange of the order of

3.3 The Deterministic Linear Optimal Regulator

integration and a change of variables

$$\int_{t_0}^{t_1} \tilde{u}^T(t)\left\{B^T(t)\int_t^{t_1}\Phi^T(\tau,t)R_1(\tau)x^0(\tau)\,d\tau + R_2(t)u^0(t)\right.$$
$$\left. + B^T(t)\Phi^T(t_1,t)P_1 x^0(t_1)\right\}dt = 0. \quad \text{3-75}$$

Let us now abbreviate,

$$p(t) = \int_t^{t_1}\Phi^T(\tau,t)R_1(\tau)x^0(\tau)\,d\tau + \Phi^T(t_1,t)P_1 x^0(t_1). \quad \text{3-76}$$

With this abbreviation 3-75 can be written more compactly as

$$\int_{t_0}^{t_0}\tilde{u}^T(t)\{B^T(t)p(t) + R_2(t)u^0(t)\}\,dt = 0. \quad \text{3-77}$$

This can be true for every $\tilde{u}(t)$, $t_0 \leq t \leq t_1$, only if

$$B^T(t)p(t) + R_2(t)u^0(t) = 0, \quad t_0 \leq t \leq t_1. \quad \text{3-78}$$

By the assumption that $R_2(t)$ is nonsingular for $t_0 \leq t \leq t_1$, we can write

$$u^0(t) = -R_2^{-1}(t)B^T(t)p(t), \quad t_0 \leq t \leq t_1. \quad \text{3-79}$$

If $p(t)$ were known, this relation would give us the optimal input at time t.

We convert the relation 3-76 for $p(t)$ into a differential equation. First, we see by setting $t = t_1$ that

$$p(t_1) = P_1 x^0(t_1). \quad \text{3-80}$$

By differentiating 3-76 with respect to t, we find

$$\dot{p}(t) = -R_1(t)x^0(t) - A^T(t)p(t), \quad \text{3-81}$$

where we have employed the relationship [Theorem 1.2(d), Section 1.3.1]

$$\frac{d}{dt}\Phi^T(t_0,t) = -A^T(t)\Phi^T(t_0,t). \quad \text{3-82}$$

We are now in a position to state the *variational equations*. Substitution of 3-79 into the state differential equation yields

$$\dot{x}^0(t) = A(t)x^0(t) - B(t)R_2^{-1}(t)B^T(t)p(t). \quad \text{3-83}$$

Together with 3-81 this forms a set of $2n$ simultaneous linear differential equations in the n components of $x^0(t)$ and the n components of $p(t)$. We term $p(t)$ the *adjoint variable*. The $2n$ boundary conditions for the differential equations are

$$x^0(t_0) = x_0 \quad \text{3-84}$$

and

$$p(t_1) = P_1 x'(t_1). \quad \text{3-85}$$

210   Optimal Linear State Feedback Control Systems

We see that the boundary conditions hold at opposite ends of the interval $[t_0, t_1]$, which means that we are faced with a two-point boundary value problem. To solve this boundary value problem, let us write the simultaneous differential equations 3-83 and 3-81 in the form

$$\begin{pmatrix} \dot{x}^0(t) \\ \dot{p}(t) \end{pmatrix} = \begin{pmatrix} A(t) & -B(t)R_2^{-1}(t)B^T(t) \\ -R_1(t) & -A^T(t) \end{pmatrix} \begin{pmatrix} x^0(t) \\ p(t) \end{pmatrix}. \quad \text{3-86}$$

Consider this the state differential equation of an $2n$-dimensional linear system with the transition matrix $\Theta(t, t_0)$. We partition this transition matrix corresponding to 3-86 as

$$\Theta(t, t_0) = \begin{pmatrix} \Theta_{11}(t, t_0) & \Theta_{12}(t, t_0) \\ \Theta_{21}(t, t_0) & \Theta_{22}(t, t_0) \end{pmatrix}. \quad \text{3-87}$$

With this partitioning we can express the state at an intermediate time t in terms of the state and adjoint variable at the terminal time t_1 as follows:

$$x^0(t) = \Theta_{11}(t, t_1)x^0(t_1) + \Theta_{12}(t, t_1)p(t_1). \quad \text{3-88}$$

With the terminal condition 3-85, it follows

$$x^0(t) = [\Theta_{11}(t, t_1) + \Theta_{12}(t, t_1)P_1]x^0(t_1). \quad \text{3-89}$$

Similarly, we can write for the adjoint variable

$$p(t) = \Theta_{21}(t, t_1)x^0(t_1) + \Theta_{22}(t, t_1)p(t_1)$$
$$= [\Theta_{21}(t, t_1) + \Theta_{22}(t, t_1)P_1]x^0(t_1). \quad \text{3-90}$$

Elimination of $x^0(t_1)$ from 3-89 and 3-90 yields

$$p(t) = [\Theta_{21}(t, t_1) + \Theta_{22}(t, t_1)P_1][\Theta_{11}(t, t_1) + \Theta_{12}(t, t_1)P_1]^{-1}x^0(t). \quad \text{3-91}$$

The expression 3-91 shows that there exists a linear relation between $p(t)$ and $x^0(t)$ as follows

$$p(t) = P(t)x^0(t), \quad \text{3-92}$$

where

$$P(t) = [\Theta_{21}(t, t_1) + \Theta_{22}(t, t_1)P_1][\Theta_{11}(t, t_1) + \Theta_{12}(t, t_1)P_1]^{-1}. \quad \text{3-93}$$

With 3-79 we obtain for the optimal input to the system

$$u^0(t) = -F(t)x^0(t), \quad \text{3-94}$$

where

$$F(t) = R_2^{-1}(t)B^T(t)P(t). \quad \text{3-95}$$

This is the solution of the regulator problem, which has been derived under the assumption that an optimal solution exists. We summarize our findings as follows.

3.3 The Deterministic Linear Optimal Regulator

Theorem 3.3. *Consider the deterministic linear optimal regulator problem. Then the optimal input can be generated through a linear control law of the form*

$$u^0(t) = -F(t)x^0(t), \quad \text{3-96}$$

where

$$F(t) = R_2^{-1}(t)B^T(t)P(t). \quad \text{3-97}$$

The matrix $P(t)$ is given by

$$P(t) = [\Theta_{21}(t, t_1) + \Theta_{22}(t, t_1)P_1][\Theta_{11}(t, t_1) + \Theta_{12}(t, t_1)P_1]^{-1}, \quad \text{3-98}$$

where $\Theta_{11}(t, t_0)$, $\Theta_{12}(t, t_0)$, $\Theta_{21}(t, t_0)$, and $\Theta_{22}(t, t_0)$ are obtained by partitioning the transition matrix $\Theta(t, t_0)$ of the state differential equation

$$\begin{pmatrix} \dot{x}(t) \\ \dot{p}(t) \end{pmatrix} = \begin{pmatrix} A(t) & -B(t)R_2^{-1}(t)B^T(t) \\ -R_1(t) & -A^T(t) \end{pmatrix} \begin{pmatrix} x(t) \\ p(t) \end{pmatrix}, \quad \text{3-99}$$

where

$$R_1(t) = D^T(t)R_3(t)D(t). \quad \text{3-100}$$

This theorem gives us the solution of the regulator problem in the form of a *linear control law*. The control law automatically generates the optimal input for *any* initial state. A block diagram interpretation is given in Fig. 3.3 which very clearly illustrates the closed-loop nature of the solution.

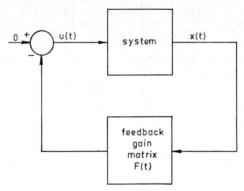

Fig. 3.3. The feedback structure of the optimal linear regulator.

The formulation of the regulator problem as given in Definition 3.2 of course does not impose this closed-loop form of the solution. We can just as easily derive an open-loop representation of the solution. At time t_0 the expression **3-89** reduces to

$$x_0 = [\Theta_{11}(t_0, t_1) + \Theta_{12}(t_0, t_1)P_1]x^0(t_1). \quad \text{3-101}$$

Solving **3-101** for $x^0(t_1)$ and substituting the result into **3-90**, we obtain
$$p(t) = [\Theta_{21}(t, t_1) + \Theta_{22}(t, t_1)P_1][\Theta_{11}(t_0, t_1) + \Theta_{12}(t_0, t_1)P_1]^{-1}x_0. \quad \textbf{3-102}$$
This gives us from **3-79**
$$u^0(t) = -R_2^{-1}(t)B^T(t)[\Theta_{21}(t, t_1) + \Theta_{22}(t, t_1)P_1][\Theta_{11}(t_0, t_1) + \Theta_{12}(t_0, t_1)P_1]^{-1}x_0,$$
$$t_0 \leq t \leq t_1. \quad \textbf{3-103}$$

For a given x_0 this yields the prescribed behavior of the input. The corresponding behavior of the state follows by substituting $x(t_1)$ as obtained from **3-101** into **3-89**:
$$x^0(t) = [\Theta_{11}(t, t_1) + \Theta_{12}(t, t_1)P_1][\Theta_{11}(t_0, t_1) + \Theta_{12}(t_0, t_1)P_1]^{-1}x_0. \quad \textbf{3-104}$$

In view of what we learned in Chapter 2 about the many advantages of closed-loop control, for practical implementation we prefer of course the closed-loop form of the solution **3-96** to the open-loop form **3-103**. In Section 3.6, where we deal with the stochastic regulator problem, it is seen that state feedback is not only preferable but in fact imperative.

Example 3.5. *Angular velocity stabilization*

The angular velocity stabilization problem of Example 3.3 (Section 3.3.1) is the simplest possible nontrivial application of the theory of this section. The combined state and adjoint variable equations **3-99** are now given by

$$\begin{pmatrix} \dot{\xi}(t) \\ \dot{\pi}(t) \end{pmatrix} = \begin{pmatrix} -\alpha & -\dfrac{\kappa^2}{\rho} \\ -1 & \alpha \end{pmatrix} \begin{pmatrix} \xi(t) \\ \pi(t) \end{pmatrix}. \quad \textbf{3-105}$$

The transition matrix corresponding to this system of differential equations can be found to be

$$\Theta(t, t_0) = \begin{pmatrix} \dfrac{\gamma - \alpha}{2\gamma} e^{\gamma(t-t_0)} + \dfrac{\gamma + \alpha}{2\gamma} e^{-\gamma(t-t_0)} & -\dfrac{\kappa^2}{2\rho\gamma}[e^{\gamma(t-t_0)} - e^{-\gamma(t-t_0)}] \\ -\dfrac{1}{2\gamma}[e^{\gamma(t-t_0)} - e^{-\gamma(t-t_0)}] & \dfrac{\gamma + \alpha}{2\gamma} e^{\gamma(t-t_0)} + \dfrac{\gamma - \alpha}{2\gamma} e^{-\gamma(t-t_0)} \end{pmatrix},$$
$$\textbf{3-106}$$

where
$$\gamma = \sqrt{\alpha^2 + \dfrac{\kappa^2}{\rho}}. \quad \textbf{3-107}$$

To simplify the notation we write the transition matrix as
$$\Theta(t, t_0) = \begin{pmatrix} \theta_{11}(t, t_0) & \theta_{12}(t, t_0) \\ \theta_{21}(t, t_0) & \theta_{22}(t, t_0) \end{pmatrix}. \quad \textbf{3-108}$$

3.3 The Deterministic Linear Optimal Regulator

It follows from **3-103** and **3-104** that in open-loop form the optimal input and state are given by

$$\mu^0(t) = -\frac{\kappa}{\rho} \frac{\theta_{21}(t, t_1) + \theta_{22}(t, t_1)\pi_1}{\theta_{11}(t_0, t_1) + \theta_{12}(t_0, t_1)\pi_1} \xi_0, \qquad \text{3-109}$$

$$\xi^0(t) = \frac{\theta_{11}(t, t_1) + \theta_{12}(t, t_1)\pi_1}{\theta_{11}(t_0, t_1) + \theta_{12}(t_0, t_1)\pi_1} \xi_0. \qquad \text{3-110}$$

Figure 3.4 shows the optimal trajectories and the behavior of the optimal input for different values of the weighting factor ρ. The following numerical values have been used:

$$\alpha = 0.5 \text{ s}^{-1},$$
$$\kappa = 150 \text{ rad}/(\text{V s}^2), \qquad \text{3-111}$$
$$t_0 = 0 \text{ s}, \qquad t_1 = 1 \text{ s}.$$

The weighting coefficient π_1 has in this case been set to zero. The figure clearly shows that as ρ decreases the input amplitude grows, whereas the settling time becomes smaller.

Figure 3.5 depicts the influence of the weighting coefficient π_1; the factor ρ is kept constant. It is seen that as π_1 increases the terminal state tends to be closer to the zero state at the expense of a slightly larger input amplitude toward the end of the interval.

Suppose now that it is known that the deviations in the initial state are usually not larger than ± 100 rad/s and that the input amplitudes should be limited to ± 3 V. Then we see from the figures that a suitable choice for ρ is about 1000. The value of π_1 affects the behavior only near the terminal time.

Let us now consider the feedback form of the solution. It follows from Theorem 3.3 that the optimal trajectories of Figs. 3.4 and 3.5 can be generated by the control law

$$\mu^0(t) = -F(t)\xi(t), \qquad \text{3-112}$$

where the time-varying scalar gain $F(t)$ is given by

$$F(t) = \frac{\kappa}{\rho} \frac{\theta_{21}(t, t_1) + \theta_{22}(t, t_1)\pi_1}{\theta_{11}(t, t_1) + \theta_{12}(t, t_1)\pi_1}. \qquad \text{3-113}$$

Figure 3.6 shows the behavior of the gain $F(t)$ corresponding to the various numerical values used in Figs. 3.4 and 3.5. Figure 3.6 exhibits quite clearly that in most cases the gain factor $F(t)$ is constant during almost the whole interval $[t_0, t_1]$. Only near the end do deviations occur. We also see that $\pi_1 = 0.19$ gives a constant gain factor over the entire interval. Such a gain factor would be very desirable from a practical point of view since the implementation of a time-varying gain is complicated and costly. Comparison

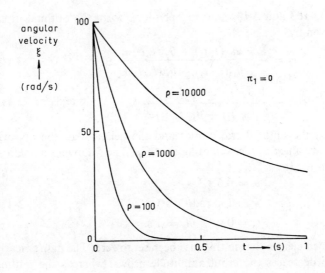
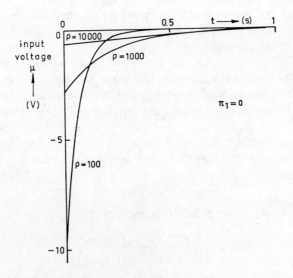

Fig. 3.4. The behavior of state and input for the angular velocity stabilization problem for different values of ρ.

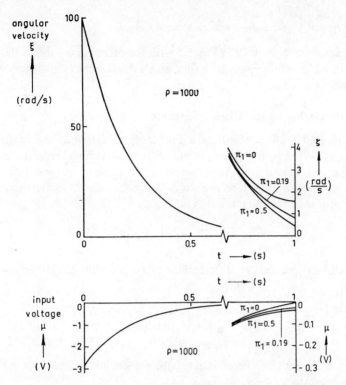

Fig. 3.5. The behavior of state and input for the angular velocity stabilization problem for different values of π_1. Note the changes in the vertical scales near the end of the interval

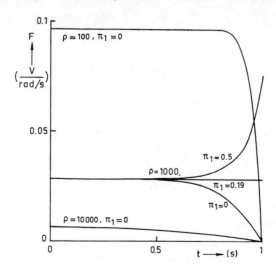

Fig. 3.6. The behavior of the optimal feedback gain factor for the angular velocity stabilization problem for various values of ρ and π_1.

of the curves for $\pi_1 = 0.19$ in Fig. 3.5 with the other curves shows that there is little point in letting F vary with time unless the terminal state is very heavily weighted.

3.3.3 Derivation of the Riccati Equation

We proceed with establishing a few more facts about the matrix $P(t)$ as given by **3-98**. In our further analysis, $P(t)$ plays a crucial role. It is possible to derive a differential equation for $P(t)$. To achieve this we differentiate $P(t)$ as given by **3-98** with respect to t. Using the rule for differentiating the inverse of a time-dependent matrix $M(t)$,

$$\frac{d}{dt} M^{-1}(t) = -M^{-1}(t)\dot{M}(t)M^{-1}(t), \qquad \textbf{3-114}$$

which can be proved by differentiating the identity $M(t)M^{-1}(t) = I$, we obtain

$$\begin{aligned}\dot{P}(t) = &[\dot{\Theta}_{21}(t, t_1) + \dot{\Theta}_{22}(t, t_1)P_1][\Theta_{11}(t, t_1) + \Theta_{12}(t, t_1)P_1]^{-1} \\ &- [\Theta_{21}(t, t_1) + \Theta_{22}(t, t_1)P_1][\Theta_{11}(t, t_1) + \Theta_{12}(t, t_1)P_1]^{-1} \\ &\cdot [\dot{\Theta}_{11}(t, t_1) + \dot{\Theta}_{12}(t, t_1)P_1][\Theta_{11}(t, t_1) + \Theta_{12}(t, t_1)P_1]^{-1},\end{aligned} \quad \textbf{3-115}$$

where a dot denotes differentiation with respect to t. Since $\Theta(t, t_0)$ is the transition matrix of **3-99**, we have

$$\begin{aligned}\dot{\Theta}_{11}(t, t_1) &= A(t)\Theta_{11}(t, t_1) - B(t)R_2^{-1}(t)B^T(t)\Theta_{21}(t, t_1), \\ \dot{\Theta}_{12}(t, t_1) &= A(t)\Theta_{12}(t, t_1) - B(t)R_2^{-1}(t)B^T(t)\Theta_{22}(t, t_1), \\ \dot{\Theta}_{21}(t, t_1) &= -R_1(t)\Theta_{11}(t, t_1) - A^T(t)\Theta_{21}(t, t_1), \\ \dot{\Theta}_{22}(t, t_1) &= -R_1(t)\Theta_{12}(t, t_1) - A^T(t)\Theta_{22}(t, t_1).\end{aligned} \quad \textbf{3-116}$$

Substituting all this into **3-115**, we find after rearrangement the following differential equation for $P(t)$:

$$-\dot{P}(t) = R_1(t) - P(t)B(t)R_2^{-1}(t)B^T(t)P(t) + P(t)A(t) + A^T(t)P(t). \quad \textbf{3-117}$$

The boundary condition for this differential equation is found by setting $t = t_1$ in **3-98**. It follows that

$$P(t_1) = P_1. \qquad \textbf{3-118}$$

The matrix differential equation thus derived resembles the well-known scalar differential equation

$$\frac{dy}{dx} + \alpha(x)y + \beta(x)y^2 = \gamma(x), \qquad \textbf{3-119}$$

where x is the independent and y the dependent variable, and $\alpha(x)$, $\beta(x)$,

3.3 The Deterministic Linear Optimal Regulator

and $\gamma(x)$ are known functions of x. This equation is known as the Riccati equation (Davis, 1962). Consequently, we refer to **3-117** as a *matrix Riccati equation* (Kalman, 1960).

We note that since the matrix P_1 that occurs in the terminal condition for $P(t)$ is symmetric, and since the matrix differential equation for $P(t)$ is also symmetric, the solution $P(t)$ must be symmetric for all $t_0 \leq t \leq t_1$. This symmetry will often be used, especially when computing P.

We now find an interpretation for the matrix $P(t)$. The optimal closed-loop system is described by the state differential equation

$$\dot{x}(t) = [A(t) - B(t)F(t)]x(t). \qquad \textbf{3-120}$$

Let us consider the optimization criterion **3-65** computed over the interval $[t, t_1]$. We write

$$\int_t^{t_1} [x^T(\tau)R_1(\tau)x(\tau) + u^T(\tau)R_2(\tau)u(\tau)] \, d\tau + x^T(t_1)P_1x(t_1)$$

$$= \int_t^{t_1} x^T(\tau)[R_1(\tau) + F^T(\tau)R_2(\tau)F(\tau)]x(\tau) \, d\tau + x^T(t_1)P_1x(t_1), \qquad \textbf{3-121}$$

since

$$u(\tau) = -F(\tau)x(\tau). \qquad \textbf{3-122}$$

From the results of Section 1.11.5 (Theorem 1.54), we know that **3-121** can be written as

$$x^T(t)\tilde{P}(t)x(t), \qquad \textbf{3-123}$$

where $\tilde{P}(t)$ is the solution of the matrix differential equation

$$-\dot{\tilde{P}}(t) = R_1(t) + F^T(t)R_2(t)F(t)$$
$$+ \tilde{P}(t)[A(t) - B(t)F(t)] + [A(t) - B(t)F(t)]^T\tilde{P}(t), \qquad \textbf{3-124}$$

with

$$\tilde{P}(t_1) = P_1.$$

Substituting $F(t) = R_2^{-1}(t)B^T(t)P(t)$ into **3-124** yields

$$-\dot{\tilde{P}}(t) = R_1(t) + P(t)B(t)R_2^{-1}(t)B^T(t)P(t) + \tilde{P}(t)A(t)$$
$$- \tilde{P}(t)B(t)R_2^{-1}(t)B^T(t)P(t) + A^T(t)\tilde{P}(t)$$
$$- P(t)B(t)R_2^{-1}(t)B^T(t)\tilde{P}(t). \qquad \textbf{3-125}$$

We claim that the solution of this matrix differential equation is precisely

$$\tilde{P}(t) = P(t). \qquad \textbf{3-126}$$

This is easily seen since substitution of $P(t)$ for $\tilde{P}(t)$ reduces the differential equation **3-125** to

$$-\dot{P}(t) = R_1(t) - P(t)B(t)R_2^{-1}(t)B^T(t)P(t) + P(t)A(t) + A^T(t)P(t). \quad \textbf{3-127}$$

This is the matrix Riccati equation **3-117** which is indeed satisfied by $P(t)$; also, the terminal condition is correct. This derivation also shows that $P(t)$ must be nonnegative-definite since **3-121** is a nonnegative expression because R_1, R_2, and P_1 are nonnegative-definite.

We summarize our conclusions as follows.

Theorem 3.4. *The optimal input for the deterministic optimal linear regulator is generated by the linear control law*

$$u^0(t) = -F^0(t)x^0(t), \quad \textbf{3-128}$$

where

$$F^0(t) = R_2^{-1}(t)B^T(t)P(t). \quad \textbf{3-129}$$

Here the symmetric nonnegative-definite matrix $P(t)$ satisfies the matrix Riccati equation

$$-\dot{P}(t) = R_1(t) - P(t)B(t)R_2^{-1}(t)B^T(t)P(t) + P(t)A(t) + A^T(t)P(t), \quad \textbf{3-130}$$

with the terminal condition

$$P(t_1) = P_1, \quad \textbf{3-131}$$

and where

$$R_1(t) = D^T(t)R_3(t)D(t).$$

For the optimal solution we have

$$\int_t^{t_1} [x^{0T}(\tau)R_1(\tau)x^0(\tau) + u^{0T}(\tau)R_2(\tau)u^0(\tau)]\,d\tau + x^{0T}(t_1)P_1 x^0(t_1)$$

$$= x^{0T}(t)P(t)x^0(t), \quad t \leq t_1. \quad \textbf{3-132}$$

We see that the matrix $P(t)$ not only gives us the optimal feedback law but also allows us to evaluate the value of the criterion for any given initial state and initial time.

From the derivation of this section, we extract the following result (Wonham, 1968a), which will be useful when we consider the stochastic linear optimal regulator problem and the optimal observer problem.

Lemma 3.1. *Consider the matrix differential equation*

$$-\dot{\tilde{P}}(t) = R_1(t) + F^T(t)R_2(t)F(t) + \tilde{P}(t)[A(t) - B(t)F(t)]$$
$$+ [A(t) - B(t)F(t)]^T \tilde{P}(t), \quad \textbf{3-133}$$

3.3 The Deterministic Linear Optimal Regulator

with the terminal condition

$$\tilde{P}(t_1) = P_1, \qquad \text{3-134}$$

where $R_1(t)$, $R_2(t)$, $A(t)$ and $B(t)$ are given time-varying matrices of appropriate dimensions, with $R_1(t)$ nonnegative-definite and $R_2(t)$ positive-definite for $t_0 \leq t \leq t_1$, and P_1 nonnegative-definite. Let $F(t)$ be an arbitrary continuous matrix function for $t_0 \leq t \leq t_1$. Then for $t_0 \leq t \leq t_1$

$$\tilde{P}(t) \geq P(t), \qquad \text{3-135}$$

where $P(t)$ is the solution of the matrix Riccati equation

$$-\dot{P}(t) = R_1(t) - P(t)B(t)R_2^{-1}(t)B^T(t)P(t) + P(t)A(t) + A^T(t)P(t), \qquad \text{3-136}$$

$$P(t_1) = P_1. \qquad \text{3-137}$$

The inequality **3-135** converts into an equality if

$$F(\tau) = R_2^{-1}(\tau)B^T(\tau)P(\tau) \quad \text{for } t \leq \tau \leq t_1. \qquad \text{3-138}$$

The lemma asserts that $\tilde{P}(t)$ is "minimized" in the sense stated in **3-135** by choosing F as indicated in **3-138**. The proof is simple. The quantity

$$x^T(t)\tilde{P}(t)x(t) \qquad \text{3-139}$$

is the value of the criterion **3-121** if the system is controlled with the arbitrary linear control law

$$u(\tau) = -F(\tau)x(\tau), \quad t \leq \tau \leq t_1. \qquad \text{3-140}$$

The optimal control law, which happens to be linear and is therefore also the best linear control law, yields $x^T(t)P(t)x(t)$ for the criterion (Theorem 3.4), so that

$$x^T(t)\tilde{P}(t)x(t) \geq x^T(t)P(t)x(t) \quad \text{for all } x(t). \qquad \text{3-141}$$

This proves **3-135**.

We conclude this section with a remark about the existence of the solution of the regulator problem. It can be proved that under the conditions formulated in Definition 3.2 the deterministic linear optimal regulator problem always has a unique solution. The existence of the solution of the regulator problem also guarantees (1) the existence of the inverse matrix in **3-98**, and (2) the fact that the matrix Riccati equation **3-130** with the terminal condition **3-131** has the unique solution **3-98**. Some references on the existence of the solutions of the regulator problem and Riccati equations are Kalman (1960), Athans and Falb (1966), Kalman and Englar (1966), Wonham (1968a), Bucy (1967a, b), Moore and Anderson (1968), Bucy and Joseph (1968), and Schumitzky (1968).

220   Optimal Linear State Feedback Control Systems

Example 3.6. *Angular velocity stabilization*

Let us continue Example 3.5. $P(t)$ is in this case a scalar function and satisfies the scalar Riccati equation

$$-\dot{P}(t) = 1 - \frac{\kappa^2}{\rho} P^2(t) - 2\alpha P(t), \qquad \text{3-142}$$

with the terminal condition

$$P(t_1) = \pi_1. \qquad \text{3-143}$$

In this scalar situation the Riccati equation **3-142** can be solved directly. In view of the results obtained in Example 3.5, however, we prefer to use **3-98**, and we write

$$P(t) = \frac{\theta_{21}(t, t_1) + \theta_{22}(t, t_1)\pi_1}{\theta_{11}(t, t_1) + \theta_{12}(t, t_1)\pi_1}, \qquad t \leq t_1, \qquad \text{3-144}$$

with the θ_{ij} defined as in Example 3.5. Figure 3.7 shows the behavior of $P(t)$ for some of the cases previously considered. We note that $P(t)$, just as the gain factor $F(t)$, has the property that it is constant during almost the entire interval except near the end. (This is not surprising since $P(t)$ and $F(t)$ differ by a constant factor.)

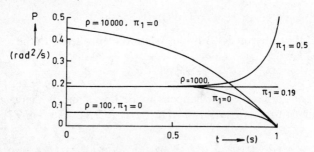

Fig. 3.7. The behavior of $P(t)$ for the angular velocity stabilization problem for various values of ρ and π_1.

3.4 STEADY-STATE SOLUTION OF THE DETERMINISTIC LINEAR OPTIMAL REGULATOR PROBLEM

3.4.1 Introduction and Summary of Main Results

In the preceding section we considered the problem of minimizing the criterion

$$\int_{t_0}^{t_1} [z^T(t)R_3(t)z(t) + u^T(t)R_2(t)u(t)]\, dt + x^T(t_1)P_1 x(t_1) \qquad \text{3-145}$$

3.4 Steady-State Solution of the Regulator Problem

for the system
$$\dot{x}(t) = A(t)x(t) + B(t)u(t),$$
$$z(t) = D(t)x(t),$$
3-146

where the terminal time t_1 is finite. From a practical point of view, it is often natural to consider very long control periods $[t_0, t_1]$. In this section we therefore extensively study the asymptotic behavior of the solution of the deterministic regulator problem as $t_1 \to \infty$.

The main results of this section can be summarized as follows.

1. *As the terminal time t_1 approaches infinity, the solution $P(t)$ of the matrix Riccati equation*

$$-\dot{P}(t) = D^T(t)R_3(t)D(t) - P(t)B(t)R_2^{-1}(t)B^T(t)P(t) + A^T(t)P(t) + P(t)A(t),$$
3-147

with the terminal condition
$$P(t_1) = P_1,$$
3-148

*generally approaches a **steady-state solution** $\bar{P}(t)$ that is independent of P_1.*

The conditions under which this result holds are precisely stated in Section 3.4.2. We shall also see that in the time-invariant case, that is, when the matrices A, B, D, R_3, and R_2 are constant, the steady-state solution $\bar{P}$, not surprisingly, is also constant and is a solution of the *algebraic Riccati equation*

$$0 = D^T R_3 D - \bar{P}BR_2^{-1}B^T\bar{P} + A^T\bar{P} + \bar{P}A.$$
3-149

It is easily recognized that $\bar{P}$ is nonnegative-definite. We prove that in general (the precise conditions are given) the steady-state solution $\bar{P}$ is the only solution of the algebraic Riccati equation that is nonnegative-definite, so that it can be uniquely determined.

Corresponding to the steady-state solution of the Riccati equation, we obtain of course the *steady-state control law*

$$u(t) = -\bar{F}(t)x(t),$$
3-150

where
$$\bar{F}(t) = R_2^{-1}(t)B^T(t)\bar{P}(t).$$
3-151

It will be proved that this steady-state control law minimizes the criterion 3-145 with t_1 replaced with ∞. Of great importance is the following:

2. *The steady-state control law is in general asymptotically stable.*

Again, precise conditions will be given. Intuitively, it is not difficult to understand this fact. Since

$$\int_{t_0}^{\infty} [z^T(t)R_3(t)z(t) + u^T(t)R_2(t)u(t)]\, dt$$
3-152

exists for the steady-state control law, it follows that in the closed-loop system $u(t) \to 0$ and $z(t) \to 0$ as $t \to \infty$. In general, this can be true only if $x(t) \to 0$, which means that the closed-loop system is asymptotically stable.

Fact 2 is very important since we now have the means to devise linear feedback systems that are asymptotically stable and at the same time possess optimal transient properties in the sense that any nonzero initial state is reduced to the zero state in an optimal fashion. For time-invariant systems this is a welcome addition to the theory of stabilization outlined in Section 3.2. There we saw that any time-invariant system in general can be stabilized by a linear feedback law, and that the closed-loop poles can be arbitrarily assigned. The solution of the regulator problem gives us a prescription to assign these poles in a rational manner. We return to the question of the optimal closed-loop pole distribution in Section 3.8.

Example 3.7. *Angular velocity stabilization*

For the angular velocity stabilization problem of Examples 3.3, 3.5, and 3.6, the solution of the Riccati equation is given by **3-144**. It is easily found with the aid of **3-106** that as $t_1 \to \infty$,

$$P(t) \to \bar{P} = \frac{\rho}{\kappa^2}\left(-\alpha + \sqrt{\alpha^2 + \frac{\kappa^2}{\rho}}\right). \qquad \textbf{3-153}$$

$\bar{P}$ can also be found by solving the algebraic equation **3-149** which in this case reduces to

$$0 = 1 - \frac{\kappa^2}{\rho}\bar{P}^2 - 2\alpha\bar{P}. \qquad \textbf{3-154}$$

This equation has the solutions

$$\frac{\rho}{\kappa^2}\left(-\alpha \pm \sqrt{\alpha^2 + \frac{\kappa^2}{\rho}}\right). \qquad \textbf{3-155}$$

Since $\bar{P}$ must be nonnegative, it follows immediately that **3-153** is the correct solution.

The corresponding steady-state gain is given by

$$\bar{F} = \frac{1}{\kappa}\left(-\alpha + \sqrt{\alpha^2 + \frac{\kappa^2}{\rho}}\right). \qquad \textbf{3-156}$$

By substituting

$$\mu(t) = -\bar{F}\xi(t) \qquad \textbf{3-157}$$

into the system state differential equation, it follows that the closed-loop system is described by the state differential equation

$$\dot{\xi}(t) = -\sqrt{\alpha^2 + \frac{\kappa^2}{\rho}}\,\xi(t). \qquad \textbf{3-158}$$

3.4 Steady-State Solution of the Regulator Problem

Obviously, this system is asymptotically stable.

Example 3.8. *Position control*

As a more complicated example, we consider the position control problem of Example 3.4 (Section 3.3.1). The steady-state solution $\bar{P}$ of the Riccati equation **3-147** must now satisfy the equation

$$0 = \begin{pmatrix} 1 \\ 0 \end{pmatrix}(1, 0) - \bar{P}\begin{pmatrix} 0 \\ \kappa \end{pmatrix}\frac{1}{\rho}(0, \kappa)\bar{P} + \begin{pmatrix} 0 & 0 \\ 1 & -\alpha \end{pmatrix}\bar{P} + \bar{P}\begin{pmatrix} 0 & 1 \\ 0 & -\alpha \end{pmatrix}. \quad \textbf{3-159}$$

Let $\bar{P}_{ij}$, $i, j = 1, 2$, denote the elements of $\bar{P}$. Then using the fact that $\bar{P}_{12} = \bar{P}_{21}$, the following algebraic equations are obtained from **3-159**

$$0 = 1 - \frac{\kappa^2}{\rho}\bar{P}_{12}^2,$$

$$0 = -\frac{\kappa^2}{\rho}\bar{P}_{12}\bar{P}_{22} + \bar{P}_{11} - \alpha\bar{P}_{12}, \quad \textbf{3-160}$$

$$0 = -\frac{\kappa^2}{\rho}\bar{P}_{22}^2 + 2\bar{P}_{12} - 2\alpha\bar{P}_{22}.$$

These equations have several solutions, but it is easy to verify that the only nonnegative-definite solution is given by

$$\bar{P}_{11} = \frac{\sqrt{\rho}}{\kappa}\sqrt{\alpha^2 + \frac{2\kappa}{\sqrt{\rho}}},$$

$$\bar{P}_{12} = \bar{P}_{21} = \frac{\sqrt{\rho}}{\kappa}, \quad \textbf{3-161}$$

$$\bar{P}_{22} = \frac{\rho}{\kappa^2}\left(-\alpha + \sqrt{\alpha^2 + \frac{2\kappa}{\sqrt{\rho}}}\right).$$

The corresponding steady-state feedback gain matrix can be found to be

$$\bar{F} = \left(\frac{1}{\sqrt{\rho}}, \frac{1}{\kappa}\left(-\alpha + \sqrt{\alpha^2 + \frac{2\kappa}{\sqrt{\rho}}}\right)\right). \quad \textbf{3-162}$$

Thus the input is given by

$$\mu(t) = -\bar{F}x(t). \quad \textbf{3-163}$$

It is easily found that the optimal closed-loop system is described by the state differential equation

$$\dot{x}(t) = \begin{pmatrix} 0 & 1 \\ -\frac{\kappa}{\sqrt{\rho}} & -\sqrt{\alpha^2 + \frac{2\kappa}{\sqrt{\rho}}} \end{pmatrix}x(t). \quad \textbf{3-164}$$

224   Optimal Linear State Feedback Control Systems

The closed-loop characteristic polynomial can be computed to be

$$s^2 + s\sqrt{\alpha^2 + \frac{2\kappa}{\sqrt{\rho}}} + \frac{\kappa}{\sqrt{\rho}}. \qquad \textbf{3-165}$$

The closed-loop characteristic values are

$$\frac{1}{2}\left(-\sqrt{\alpha^2 + \frac{2\kappa}{\sqrt{\rho}}} \pm \sqrt{\alpha^2 - \frac{2\kappa}{\sqrt{\rho}}}\right). \qquad \textbf{3-166}$$

Figure 3.8 gives the loci of the closed-loop characteristic values as ρ varies. It is interesting to see that as ρ decreases the closed-loop poles go to infinity along two straight lines that make an angle of $\pi/4$ with the negative real axis. Asymptotically, the closed-loop poles are given by

$$\frac{\kappa^{1/2}}{\rho^{1/4}} \tfrac{1}{2}\sqrt{2}(-1 \pm j) \qquad \text{as } \rho \to 0. \qquad \textbf{3-167}$$

Figure 3.9 shows the response of the steady-state optimal closed-loop system

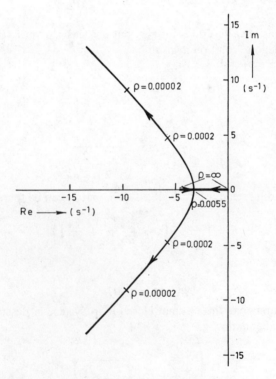

Fig. 3.8. Loci of the closed-loop roots of the position control system as a function of ρ.

3.4 Steady-State Solution of the Regulator Problem

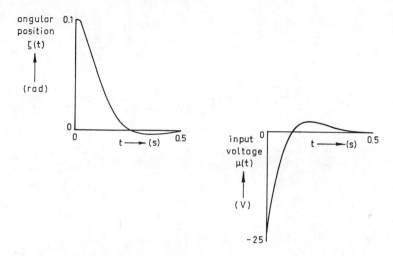

Fig. 3.9. Response of the optimal position control system to the initial state $\xi_1(0) = 0.1$ rad, $\xi_2(0) = 0$ rad/s.

corresponding to the following numerical values:

$$\kappa = 0.787 \text{ rad}/(\text{V s}^2),$$
$$\alpha = 4.6 \text{ s}^{-1}, \qquad \qquad \textbf{3-168}$$
$$\rho = 0.00002 \text{ rad}^2/\text{V}^2.$$

The corresponding gain matrix is

$$\bar{F} = (223.6, 18.69), \qquad \qquad \textbf{3-169}$$

while the closed-loop poles can be computed to be $-9.658 \pm j9.094$. We observe that the present design is equivalent to the position and velocity feedback design of Example 2.4 (Section 2.3). The gain matrix **3-169** is optimal from the point of view of transient response. It is interesting to note that the present design method results in a second-order system with relative damping of nearly $\frac{1}{2}\sqrt{2}$, which is exactly what we found in Example 2.7 (Section 2.5.2) to be the most favorable design.

To conclude the discussion we remark that it follows from Example 3.4 that if $x(t)$ is actually the deviation of the state from a certain equilibrium state x_0 which is not the zero state, $x(t)$ in the control law **3-163** should be replaced with $x'(t)$, where

$$x'(t) = \begin{pmatrix} \xi_1(t) - \xi_{10} \\ \xi_2(t) \end{pmatrix}. \qquad \qquad \textbf{3-170}$$

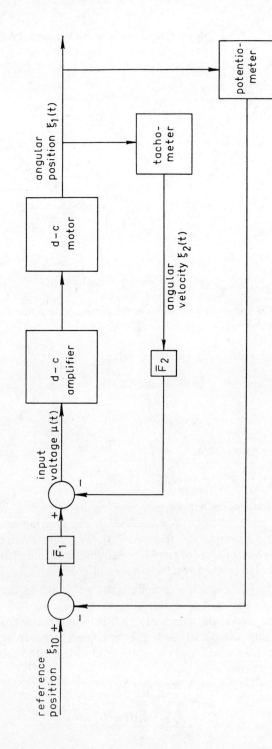

Fig. 3.10. Block diagram of the optimal position control system.

3.4 Steady-State Solution of the Regulator Problem

Here ξ_{10} is the desired angular position. This results in the control law

$$\mu(t) = -\bar{F}_1[\xi_1(t) - \xi_{10}] - \bar{F}_2\xi_2(t), \qquad \text{3-171}$$

where $\bar{F} = (\bar{F}_1, \bar{F}_2)$. The block diagram corresponding to this control law is given in Fig. 3.10.

Example 3.9. *Stirred tank*

As another example, we consider the stirred tank of Example 1.2 (Section 1.2.3). Suppose that it is desired to stabilize the outgoing flow $F(t)$ and the outgoing concentration $c(t)$. We therefore choose as the controlled variable

$$z(t) = y(t) = \begin{pmatrix} 0.01 & 0 \\ 0 & 1 \end{pmatrix} x(t), \qquad \text{3-172}$$

where we use the numerical values of Example 1.2. To determine the weighting matrix R_3, we follow the same argument as in Example 2.8 (Section 2.5.3). The nominal value of the outgoing flow is 0.02 m³/s. A 10% change corresponds to 0.002 m³/s. The nominal value of the outgoing concentration is 1.25 kmol/m³. Here a 10% change corresponds to about 0.1 kmol/m³. Suppose that we choose R_3 diagonal with diagonal elements σ_1 and σ_2. Then

$$z^T(t)R_3z(t) = \sigma_1\zeta_1^2(t) + \sigma_2\zeta_2^2(t), \qquad \text{3-173}$$

where $z(t) = \text{col }(\zeta_1(t), \zeta_2(t))$. Then if a 10% change in the outgoing flow is to make about the same contribution to the criterion as a 10% change in the outgoing concentration, we must have

$$\sigma_1(0.002)^2 \simeq \sigma_2(0.1)^2, \qquad \text{3-174}$$

or

$$\frac{\sigma_1}{\sigma_2} \simeq 2500. \qquad \text{3-175}$$

Let us therefore select

$$\sigma_1 = 50, \qquad \sigma_2 = \tfrac{1}{50}, \qquad \text{3-176}$$

or

$$R_3 = \begin{pmatrix} 50 & 0 \\ 0 & 0.02 \end{pmatrix}. \qquad \text{3-177}$$

To choose R_2 we follow a similar approach. A 10% change in the feed F_1 corresponds to 0.0015 m³/s, while a 10% change in the feed F_2 corresponds to 0.0005 m³/s. Let us choose $R_2 = \text{diag }(\rho_1, \rho_2)$. Then the 10% changes in F_1 and F_2 contribute an amount of

$$\rho_1(0.0015)^2 + \rho_2(0.0005)^2 \qquad \text{3-178}$$

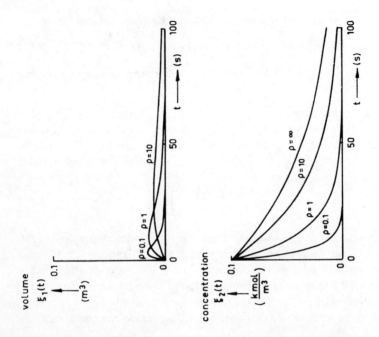
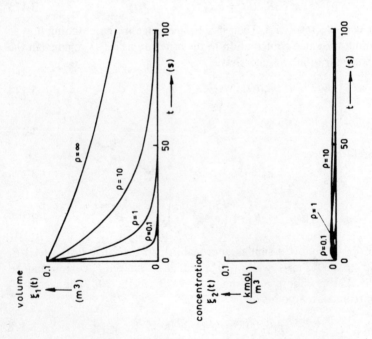

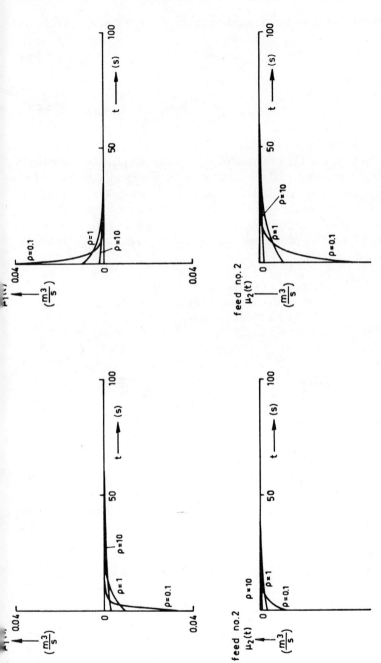

Fig. 3.11. Closed-loop responses of the regulated stirred tank for various values of the weighting factor ρ. Left column: Responses of incremental volume, concentration, feed no. 1, and feed no. 2 to the initial state $x(0) = \text{col}(0.1, 0)$. Right column: Responses of incremental volume, concentration, feed no. 1, and feed no. 2 to the initial state $x(0) = \text{col}(0, 0.1)$.

230 Optimal Linear State Feedback Control Systems

to the criterion. Both terms contribute equally if

$$\frac{\rho_1}{\rho_2} = \frac{1}{9}. \qquad \text{3-179}$$

We therefore select

$$R_2 = \rho \begin{pmatrix} \frac{1}{3} & 0 \\ 0 & 3 \end{pmatrix}, \qquad \text{3-180}$$

where ρ is a scalar constant to be determined.

Figure 3.11 depicts the behavior of the optimal steady-state closed-loop system for $\rho = \infty$, 10, 1, and 0.1. The case $\rho = \infty$ corresponds to the open-loop system (no control at all). We see that as ρ decreases a faster and faster response is obtained at the cost of larger and larger input amplitudes. Table 3.1 gives the closed-loop characteristic values as a function of ρ. We see that in all cases a system is obtained with closed-loop poles that are well inside the left-half complex plane.

Table 3.1   Locations of the Steady-State Optimal Closed-Loop Poles as a Function of ρ for the Regulated Stirred Tank

ρ	Optimal closed-loop poles (s^{-1})	
∞	-0.01	-0.02
10	-0.02952,	-0.04523
1	-0.07517,	-0.1379
0.1	-0.2310,	-0.4345

We do not list here the gain matrices $\bar{F}$ found for each value of ρ, but it turns out that they are not diagonal, as opposed to what we considered in Example 2.8. The feedback schemes obtained in the present example are optimal in the sense that they are the best compromises between the requirement of maximal speed of response and the limitations on the input amplitudes.

Finally, we observe from the plots of Fig. 3.11 that the closed-loop system shows relatively little interaction, that is, the response to an initial disturbance in the concentration hardly affects the tank volume, and vice versa.

3.4.2*   Steady-State Properties of Optimal Regulators

In this subsection and the next we give precise results concerning the steady-state properties of optimal regulators. This section is devoted to the general,

3.4 Steady-State Solution of the Regulator Problem

time-varying case; in the next section the time-invariant case is investigated in much more detail. Most of the results in the present section are due to Kalman (1960). We more or less follow his exposition.

We first state the following result.

Theorem 3.5. *Consider the matrix Riccati equation*

$$-\dot{P}(t) = D^T(t)R_3(t)D(t) - P(t)B(t)R_2^{-1}(t)B^T(t)P(t) + A^T(t)P(t) + P(t)A(t). \quad \text{3-181}$$

Suppose that $A(t)$ is continuous and bounded, that $B(t)$, $D(t)$, $R_3(t)$, and $R_2(t)$ are piecewise continuous and bounded on $[t_0, \infty)$, and furthermore that

$$R_3(t) \geq \alpha I, \qquad R_2(t) \geq \beta I, \qquad \text{for all } t, \quad \text{3-182}$$

where α and β are positive constants.

(i) *Then if the system*

$$\dot{x}(t) = A(t)x(t) + B(t)u(t),$$
$$z(t) = D(t)x(t), \quad \text{3-183}$$

is either

    (a) *completely controllable, or*
    (b) *exponentially stable,*

the solution $P(t)$ of the Riccati equation **3-181** *with the terminal condition $P(t_1) = 0$ converges to a nonnegative-definite matrix function $\bar{P}(t)$ as $t_1 \to \infty$. $\bar{P}(t)$ is a solution of the Riccati equation* **3-181**.

(ii) *Moreover, if the system* **3-183** *is either*

    (c) *both uniformly completely controllable and uniformly completely reconstructible, or*
    (d) *exponentially stable,*

the solution $P(t)$ of the Riccati equation **3-181** *with the terminal condition $P(t_1) = P_1$ converges to $\bar{P}(t)$ as $t_1 \to \infty$ for any $P_1 \geq 0$.*

The proof of the first part of this theorem is not very difficult. From Theorem 3.4 (Section 3.3.3), we know that for finite t_1

$$x^T(t)P(t)x(t) = \min_{\substack{u(\tau), \\ t \leq \tau \leq t_1}} \left\{ \int_t^{t_1} [z^T(\tau)R_3(\tau)z(\tau) + u^T(\tau)R_2(\tau)u(\tau)] \, d\tau \right\}. \quad \text{3-184}$$

Of course this expression is a function of the terminal time t_1. We first establish that as a function of t_1 this expression has an upper bound. If the system is completely controllable [assumption (a)], there exists an input that transfers the state $x(t)$ to the zero state at some time t_1'. For this input we can compute the criterion

$$\int_t^{t_1'} [z^T(\tau)R_3(\tau)z(\tau) + u^T(\tau)R_2(\tau)u(\tau)] \, d\tau. \quad \text{3-185}$$

232 Optimal Linear State Feedback Control Systems

This number is an upper bound for **3-184**, since obviously we can take $u(t) = 0$ for $t \geq t_1'$.

If the system is exponentially stable (Section 1.4.1), $x(t)$ converges exponentially to zero if we let $u(t) \equiv 0$. Then

$$\int_t^{t_1}[z^T(\tau)R_3(\tau)z(\tau) + u^T(\tau)R_2(\tau)u(\tau)]\,d\tau = \int_t^{t_1} z^T(\tau)R_3(\tau)z(\tau)\,d\tau \quad \textbf{3-186}$$

converges to a finite number as $t_1 \to \infty$, since $D(t)$ and $R_3(t)$ are assumed to be bounded. This number is an upper bound for **3-184**.

Thus we have shown that as a function of t_1 the expression **3-184** has an upper bound under either assumption (a) or (b). Furthermore, it is reasonably obvious that as a function of t_1 this expression is monotonically nondecreasing. Suppose that this were not true. Then there must exist a t_1' and t_1'' with $t_1'' > t_1'$ such that for $t_1 = t_1''$ the criterion is smaller than for $t_1 = t_1'$. Now apply the input that is optimal for t_1'' over the interval $[t_0, t_1']$. Since the integrand of the criterion is nonnegative, the criterion for this smaller interval must give a value that is less than or equal to the criterion for the larger interval $[t_0, t_1'']$. This is a contradiction, hence **3-184** must be a monotonically nondecreasing function of t_1.

Since as a function of t_1 the expression **3-184** is bounded from above and monotonically nondecreasing, it must have a limit as $t_1 \to \infty$. Since $x(t)$ is arbitrary, each of the elements of $P(t)$ has a limit, hence $P(t)$ has a limit that we denote as $\bar{P}(t)$. That $\bar{P}(t)$ is nonnegative-definite and symmetric is obvious. That $\bar{P}(t)$ is a solution of the matrix Riccati equation follows by the continuity of the solutions of the Riccati equation with respect to initial conditions. Following Kalman (1960), let $\Pi(t; P_1, t_1)$ denote the solution of the matrix Riccati equation with the terminal condition $P_1(t_1) = P_1$. Then

$$\bar{P}(t) = \lim_{t_2 \to \infty} \Pi(t; 0, t_2) = \lim_{t_2 \to \infty} \Pi[t; \Pi(t_1; 0, t_2), t_1]$$

$$= \Pi[t; \lim_{t_2 \to \infty} \Pi(t_1; 0, t_2), t_1]$$

$$= \Pi[t, \bar{P}(t_1), t_1], \quad \textbf{3-187}$$

which shows that $\bar{P}_1(t)$ is indeed a solution of the Riccati equation. The proof of the remainder of Theorem 3.5 will be deferred for a moment.

We refer to $\bar{P}(t)$ as the *steady-state* solution of the Riccati equation. To this steady-state solution corresponds the *steady-state optimal control law*

$$u(t) = -\bar{F}(t)x(t), \quad \textbf{3-188}$$

where

$$F(t) = R_2^{-1}(t)B^T(t)\bar{P}(t). \quad \textbf{3-189}$$

3.4 Steady-State Solution of the Regulator Problem

Concerning the stability of the steady-state control law, we have the following result.

Theorem 3.6. *Consider the deterministic linear optimal regulator problem and suppose that the assumptions of Theorem 3.5 concerning A, B, D, R_3 and R_2 are satisfied. Then if the system*

$$\dot{x}(t) = A(t)x(t) + B(t)u(t),$$
$$z(t) = D(t)x(t), \quad \text{3-190}$$

is either

(a) *both uniformly completely controllable and uniformly completely reconstructible, or*

(b) *exponentially stable,*

the following facts hold:

(i) *The steady-state optimal control law*

$$u(t) = -R_2^{-1}(t)B^T(t)\bar{P}(t)x(t) \quad \text{3-191}$$

is exponentially stable.

(ii) *The steady-state control law* **3-191** *minimizes*

$$\lim_{t_1 \to \infty} \left\{ \int_{t_0}^{t_1} [z^T(t)R_3(t)z(t) + u^T(t)R_2(t)u(t)] \, dt + x^T(t_1)P_1x(t_1) \right\} \quad \text{3-192}$$

for all $P_1 \geq 0$. The minimal value of the criterion **3-192**, *which is achieved by the steady-state control law, is given by*

$$x^T(t_0)\bar{P}(t_0)x(t_0). \quad \text{3-193}$$

A rigorous proof of these results is given by Kalman (1960). We only make the theorem plausible. If condition (a) or (b) of Theorem 3.6 is satisfied, also condition (a) or (b) of Theorem 3.5 holds. It follows that the solution of the Riccati equation **3-181** with $P(t_1) = 0$ converges to $\bar{P}(t)$ as $t_1 \to \infty$. For the corresponding steady-state control law, we have

$$\int_{t_0}^{\infty} [z^T(t)R_3(t)z(t) + u^T(t)R_2(t)u(t)] \, dt = x^T(t_0)\bar{P}(t_0)x(t_0). \quad \text{3-194}$$

Since the integral converges and $R_3(t)$ and $R_2(t)$ satisfy the conditions **3-182**, both $z(t)$ and $u(t)$ must converge to zero as $t \to \infty$. Suppose now that the closed-loop system is not asymptotically stable. Then there exists an initial state such that $x(t)$ does not approach zero while $z(t) \to 0$ and $u(t) \to 0$. This is clearly in conflict with the complete reconstructibility of the system if (a) holds, or with the assumption of exponential stability of the system if (b) holds. Hence the closed-loop system must be asymptotically stable. That it moreover is exponentially stable follows from the uniformity properties.

This settles part (i) of the theorem. Part (ii) can be shown as follows. Suppose that there exists another control law that yields a smaller value for

234 Optimal Linear State Feedback Control Systems

3-192. Because the criterion 3-192 yields a finite value when the steady-state optimal control law is used, this other control law must also yield a finite value. Then, by the same argument as for the steady-state control law, this other control law must be asymptotically stable. This means that for this control law

$$\lim_{t_1 \to \infty} \left\{ \int_{t_0}^{t_1} [z^T(t)R_3(t)z(t) + u^T(t)R_2(t)u(t)] \, dt + x^T(t_1)P_1 x(t_1) \right\}$$

$$= \int_{t_0}^{\infty} [z^T(t)R_3(t)z(t) + u^T(t)R_2(t)u(t)] \, dt. \quad \text{3-195}$$

But since the right-hand side of this expression is minimized by the steady-state control law, there cannot be another control law that yields a smaller value for the left-hand side. This proves part (ii) of Theorem 3.6. This moreover proves the second part of Theorem 3.5, since under assumptions (c) or (d) of this theorem the steady-state feedback law minimizes the criterion 3-192 for all $P_1 \geq 0$, which implies that the Riccati equation converges to $\bar{P}(t)$ for all $P_1 \geq 0$.

We illustrate the results of this section as follows.

Example 3.10. *Reel-winding mechanism*

As an example of a simple time-varying system, consider the reel-winding mechanism of Fig. 3.12. A dc motor drives a reel on which a wire is being

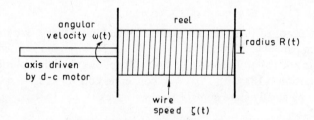

Fig. 3.12. Schematic representation of a reel-winding mechanism.

wound. The speed at which the wire runs on to the reel is to be kept constant. Because of the increasing diameter of the reel, the moment of inertia increases; moreover, to keep the wire speed constant, the angular velocity must decrease. Let $\omega(t)$ be the angular velocity of the reel, $J(t)$ the moment of inertia of reel and motor armature, and $\mu(t)$ the input voltage to the power amplifier that drives the dc motor. Then we have

$$\frac{d}{dt}[J(t)\omega(t)] = \kappa\mu(t) - \phi\omega(t), \quad \text{3-196}$$

3.4 Steady-State Solution of the Regulator Problem

where κ is a constant which expresses the proportionality of the torque of the motor and the input voltage, and where ϕ is a friction coefficient. Furthermore, let $R(t)$ denote the radius of the reel; then the speed $\zeta(t)$ at which the wire is wound is given by

$$\zeta(t) = R(t)\omega(t). \tag{3-197}$$

Let us introduce the state variable

$$\xi(t) = J(t)\omega(t). \tag{3-198}$$

The system is then described by the equations

$$\dot{\xi}(t) = -\frac{\phi}{J(t)}\xi(t) + \kappa\mu(t),$$

$$\zeta(t) = \frac{R(t)}{J(t)}\xi(t). \tag{3-199}$$

We assume that the reel speed is so controlled that the wire speed is kept constant at the value ζ_0. The time dependence of J and R can then be established as follows. Suppose that during a short time dt the radius increases from R to $R + dR$. The increase in the volume of wire wound upon the reel is proportional to $R\,dR$. The volume is also proportional to dt, since the wire is wound with a supposedly constant speed. Thus we have

$$R\,dR = c\,dt, \tag{3-200}$$

where c is a constant. This yields after integration

$$R(t) = \sqrt{R^2(0) + ht}, \tag{3-201}$$

where h is another constant. However, if the radius increases from R to $R + dR$, the moment of inertia increases with an amount that is proportional to $R\,dR\,R^2 = R^3\,dR$. Thus we have

$$dJ = c'R^3\,dR, \tag{3-202}$$

where c' is a constant. This yields after integration

$$J(t) = J(0) + h'[R^4(t) - R^4(0)], \tag{3-203}$$

where h' is another constant.

Let us now consider the problem of regulating the system such that the wire speed is kept at the constant value ζ_0. The nominal solution $\zeta_0(t)$, $\mu_0(t)$ that corresponds to this situation can be found as follows. If $\zeta_0(t) \equiv \zeta_0$, we have

$$\xi_0(t) = \frac{J(t)}{R(t)}\zeta_0. \tag{3-204}$$

236 Optimal Linear State Feedback Control Systems

The nominal input is found from the state differential equation:

$$\mu_0(t) = \frac{1}{\kappa}\left[\dot{\xi}_0(t) + \frac{\phi}{J(t)}\xi_0(t)\right] = \frac{1}{\kappa}\left[\frac{d}{dt}\!\left(\frac{J(t)}{R(t)}\right) + \frac{\phi}{R(t)}\right]\zeta_0. \quad \text{3-205}$$

Let us now define the shifted state, input, and controlled variables:

$$\begin{aligned}\xi'(t) &= \xi(t) - \xi_0(t),\\ \mu'(t) &= \mu(t) - \mu_0(t),\\ \zeta'(t) &= \zeta(t) - \zeta_0(t).\end{aligned} \quad \text{3-206}$$

These variables satisfy the equations

$$\dot{\xi}'(t) = -\frac{\phi}{J(t)}\xi'(t) + \kappa\mu'(t),$$

$$\zeta'(t) = \frac{R(t)}{J(t)}\xi'(t). \quad \text{3-207}$$

Let us choose the criterion

$$\int_{t_0}^{t_1}[\zeta'^2(t) + \rho\mu'^2(t)]\,dt. \quad \text{3-208}$$

Then the Riccati equation takes the form

$$-\dot{P}(t) = \frac{R^2(t)}{J^2(t)} - P^2(t)\frac{\kappa^2}{\rho} - 2\frac{\phi}{J(t)}P(t), \quad \text{3-209}$$

with the terminal condition

$$P(t_1) = 0. \quad \text{3-210}$$

$P(t)$ is in this case a scalar function. The scalar feedback gain factor is given by

$$F(t) = \frac{\kappa}{\rho}P(t). \quad \text{3-211}$$

We choose the following numerical values:

$$\begin{aligned}J(t) &= 0.02 + 66.67[R^4(t) - R^4(0)] \text{ kg m}^2,\\ R(t) &= \sqrt{0.01 + 0.0005t} \text{ m},\\ \phi &= 0.01 \text{ kg m}^2/\text{s},\\ \kappa &= 0.1 \text{ kg m}^2 \text{ rad}/(\text{V s}^2),\\ \rho &= 0.06 \text{ m}^2/(\text{V}^2\text{s}^2).\end{aligned} \quad \text{3-212}$$

Figure 3.13 shows the behavior of the optimal gain factor $F(t)$ for the terminal

3.4 Steady-State Solution of the Regulator Problem

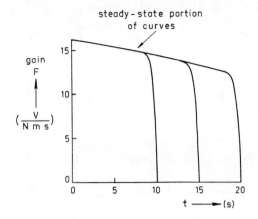

Fig. 3.13. Behavior of the optimal gain factor for the reel-winding problem for various values of the terminal time t_1.

times $t_1 = 10$, 15, and 20 s. We note that for each value of t_1 the gain exhibits an identical steady-state behavior; only near the terminal time do deviations occur. It is clearly shown that the steady-state gain is time-varying. It is not convenient to implement such a time-varying gain. In the present case a practically adequate performance might probably just as well be obtained through a time-invariant feedback gain.

3.4.3* Steady-State Properties of the Time-Invariant Optimal Regulator

In this section we study the steady-state properties of the time-invariant optimal linear regulator. We are able to state sufficient and necessary conditions under which the Riccati equation has a steady-state solution and under which the steady-state optimal closed-loop system is stable. Most of these facts have been given by Wonham (1968a), Lukes (1968), and Mårtensson (1971).

Our results can be summarized as follows.

Theorem 3.7. *Consider the time-invariant regulator problem for the system*

$$\dot{x}(t) = Ax(t) + Bu(t),$$
$$z(t) = Dx(t), \qquad \textbf{3-213}$$

and the criterion

$$\int_{t_0}^{t_1} [z^T(t)R_3 z(t) + u^T(t)R_2 u(t)]\, dt + x^T(t_1)P_1 x(t_1), \qquad \textbf{3-214}$$

with $R_3 > 0$, $R_2 > 0$, $P_1 \geq 0$. The associated Riccati equation is given by

$$-\dot{P}(t) = D^T R_3 D - P(t) B R_2^{-1} B^T P(t) + A^T P(t) + P(t) A, \quad \text{3-215}$$

with the terminal condition

$$P(t_1) = P_1. \quad \text{3-216}$$

(a) *Assume that $P_1 = 0$. Then as $t_1 \to \infty$ the solution of the Riccati equation approaches a constant steady-state value $\bar{P}$ if and only if the system possesses no poles that are at the same time unstable, uncontrollable, and reconstructible.*
(b) *If the system 3-213 is both stabilizable and detectable, the solution of the Riccati equation 3-215 approaches the unique value $\bar{P}$ as $t_1 \to \infty$ for every $P_1 \geq 0$.*
(c) *If $\bar{P}$ exists, it is a nonnegative-definite symmetric solution of the algebraic Riccati equation*

$$0 = D^T R_3 D - \bar{P} B R_2^{-1} B^T \bar{P} + A^T \bar{P} + \bar{P} A. \quad \text{3-217}$$

If the system 3-213 is stabilizable and detectable, $\bar{P}$ is the unique nonnegative-definite symmetric solution of the algebraic Riccati equation 3-217.
(d) *If $\bar{P}$ exists, it is strictly positive-definite if and only if the system 3-213 is completely reconstructible.*
(e) *If $\bar{P}$ exists, the steady-state control law*

$$u(t) = -\bar{F} x(t), \quad \text{3-218}$$

where

$$\bar{F} = R_2^{-1} B^T \bar{P}, \quad \text{3-219}$$

is asymptotically stable if and only if the system 3-213 is stabilizable and detectable.
(f) *If the system 3-213 is stabilizable and detectable, the steady-state control law minimizes*

$$\lim_{t_1 \to \infty} \left\{ \int_{t_0}^{t_1} [z^T(t) R_3 z(t) + u^T(t) R_2 u(t)] \, dt + x^T(t_1) P_1 x(t_1) \right\} \quad \text{3-220}$$

for all $P_1 \geq 0$. For the steady-state control law, the criterion 3-220 takes the value

$$x^T(t_0) \bar{P} x(t_0). \quad \text{3-221}$$

We first prove part (a) of this theorem. Suppose that the system is not completely reconstructible. Then it can be transformed into reconstructibility canonical form as follows.

$$\dot{x}(t) = \begin{pmatrix} A_{11} & 0 \\ A_{21} & A_{22} \end{pmatrix} x(t) + \begin{pmatrix} B_1 \\ B_2 \end{pmatrix} u(t), \quad \text{3-222}$$

$$z(t) = (D_1, \quad 0) x(t),$$

3.4 Steady-State Solution of the Regulator Problem

where the pair $\{A_{11}, D_1\}$ is completely reconstructible. Partitioning the solution $P(t)$ of the Riccati equation **3-215** according to the partitioning in **3-222** as

$$P(t) = \begin{pmatrix} P_{11}(t) & P_{12}(t) \\ P_{12}^T(t) & P_{22}(t) \end{pmatrix}, \qquad 3\text{-}223$$

it is easily found that the Riccati equation **3-215** reduces to the following three matrix equations

$$-\dot{P}_{11}(t) = D_1^T R_3 D_1 - [P_{11}(t)B_1 + P_{12}(t)B_2]R_2^{-1}$$
$$\cdot [B_1^T P_{11}(t) + B_2^T P_{12}^T(t)] + A_{11}^T P_{11}(t)$$
$$+ A_{21}^T P_{12}^T(t) + P_{11}(t)A_{11} + P_{12}(t)A_{21}, \qquad 3\text{-}224$$

$$-\dot{P}_{12}(t) = -[P_{11}(t)B_1 + P_{12}(t)B_2]R_2^{-1}[B_1^T P_{12}(t) + B_2^T P_{22}(t)]$$
$$+ A_{11}^T P_{12}(t) + A_{21}^T P_{22}(t) + P_{12}(t)A_{22}, \qquad 3\text{-}225$$

$$-\dot{P}_{22}(t) = -[P_{12}^T(t)B_1 + P_{22}(t)B_2]R_2^{-1}[B_1^T P_{12}(t) + B_2^T P_{22}(t)]$$
$$+ A_{22}^T P_{22}(t) + P_{22}(t)A_{22}. \qquad 3\text{-}226$$

It is easily seen that with the terminal conditions $P_{11}(t_1) = 0$, $P_{12}(t_1) = 0$, and $P_{22}(t_1) = 0$ Eqs. **3-225** and **3-226** are satisfied by

$$P_{12}(t) = 0, \qquad P_{22}(t) = 0, \qquad t \leq t_1. \qquad 3\text{-}227$$

With these identities **3-224** reduces to

$$-\dot{P}_{11}(t) = D_1^T R_3 D_1 - P_{11}(t)B_1 R_2^{-1} B_1^T P_{11}(t) + A_{11}^T P_{11}(t) + P_{11}(t)A_{11},$$
$$P_{11}(t_1) = 0. \qquad 3\text{-}228$$

It follows from this that the unreconstructible poles of the system, that is, the characteristic values of A_{22}, do not affect the convergence of $P_{11}(t)$ as $t_1 \to \infty$, hence that the convergence of $P(t)$ is also not affected by the unreconstructible poles. To investigate the convergence of $P(t)$, we can therefore as well assume for the time being that the system **3-213** is completely reconstructible.

Let us now transform the system **3-213** into controllability canonical form and thus represent it as follows:

$$\dot{x}(t) = \begin{pmatrix} A_{11} & A_{12} \\ 0 & A_{22} \end{pmatrix} x(t) + \begin{pmatrix} B_1 \\ 0 \end{pmatrix} u(t), \qquad 3\text{-}229$$

$$z(t) = (D_1, \quad D_2)x(t),$$

where the pair $\{A_{11}, B_1\}$ is completely controllable. Suppose now that the system is not stabilizable so that A_{22} is not asymptotically stable. Then obviously there exist initial states of the form col $(0, x_{20})$ such that $x(t) \to \infty$ no matter how $u(t)$ is chosen. By the assumed complete reconstructibility, for such initial states

$$\int_{t_0}^{t_1} [z^T(t)R_3 z(t) + u^T(t)R_2 u(t)]\, dt \qquad \text{3-230}$$

will never converge to a finite number as $t_1 \to \infty$. This proves that $P(t)$ also will not converge to a finite value as $t_1 \to \infty$ if the system **3-213** is not stabilizable. However, if **3-213** is stabilizable, we can always find a feedback law that makes the closed-loop system stable. For this feedback law **3-230** converges to a finite number as $t_1 \to \infty$; this number is an upper bound for the minimal value of the criterion. As in Section 3.4.2, we can argue that the minimal value of **3-230** is a monotonically nondecreasing function of t_1. This proves that the minimal value of **3-230** has a limit as $t_1 \to \infty$, hence that $P(t)$ as solved from **3-215** with $P(t_1) = 0$ has a limit $\bar{P}$ as $t_1 \to \infty$. This terminates the proof of part (a) of the theorem.

We defer the proof of parts (b) and (c) for a moment. Part (d) is easily recognized to be valid. Suppose that the system is not completely reconstructible. Then, as we have seen in the beginning of the proof of (a), when the system is represented in reconstructibility canonical form, and $P_1 = 0$, $P(t)$ can be represented in the form

$$\begin{pmatrix} P_{11}(t) & 0 \\ 0 & 0 \end{pmatrix}, \qquad \text{3-231}$$

which very clearly shows that $\bar{P}$, if it exists, is singular. This proves that if $\bar{P}$ is strictly positive-definite the system must be completely reconstructible. To prove the converse assume that the system is completely reconstructible and that $\bar{P}$ is singular. Then there exists a nonzero initial state such that

$$\int_{t_0}^{\infty} [z^T(t)R_3 z(t) + u^T(t)R_2 u(t)]\, dt = 0. \qquad \text{3-232}$$

Since $R_3 > 0$ and $R_2 > 0$, this implies that

$$u(t) = 0 \quad \text{and} \quad z(t) = 0 \quad \text{for } t \geq t_0. \qquad \text{3-233}$$

But this would mean that there is a nonzero initial state that causes a zero input response of $z(t)$ that is zero for all t. This is in contradiction to the assumption of complete reconstructibility, and therefore the assumption that $\bar{P}$ is singular is false. This terminates the proof of part (d).

We now consider the proof of part (e). We assume that $\bar{P}$ exists. This means that the system has no unstable, uncontrollable poles that are reconstructible.

3.4 Steady-State Solution of the Regulator Problem

We saw in the proof of (a) that in the reconstructibility canonical representation of the system $\bar{P}$ is given in the form

$$\begin{pmatrix} \bar{P}_{11} & 0 \\ 0 & 0 \end{pmatrix}. \qquad \text{3-234}$$

This shows that the steady-state feedback gain matrix is of the form

$$\bar{F} = R_2^{-1}(B_1^T, \ B_2^T)\begin{pmatrix} \bar{P}_{11} & 0 \\ 0 & 0 \end{pmatrix} = (R_2^{-1}B_1^T\bar{P}_{11}, \ 0). \qquad \text{3-235}$$

This in turn means that the steady-state feedback gain matrix leaves the unreconstructible part of the system completely untouched, which implies that if the steady-state control law is to make the closed-loop system asymptotically stable, the unreconstructible part of the system must be asymptotically stable, that is, the open-loop system must be detectable. Moreover, if the closed-loop system is to be asymptotically stable, the open-loop system must be stabilizable, otherwise no control law, hence not the steady-state control law either, can make the closed-loop system stable. Thus we see that stabilizability and detectability are necessary conditions for the steady-state control law to be asymptotically stable.

Stabilizability and detectability are also sufficient to guarantee asymptotic stability. We have already seen that the steady-state control law does not affect and is not affected by the unreconstructible part of the system; therefore, if the system is detectable, we may as well omit the unreconstructible part and assume that the system is completely reconstructible. Let us represent the system in controllability canonical form as in 3-229. Partitioning the matrix $P(t)$ according to the partitioning of 3-229, we write:

$$P(t) = \begin{pmatrix} P_{11}^T(t) & P_{12}(t) \\ P_{12}^T(t) & P_{22}(t) \end{pmatrix}. \qquad \text{3-236}$$

It is not difficult to find from the Riccati equation 3-215 that $P_{11}(t)$ is the solution of

$$-\dot{P}_{11}(t) = D_1^T R_3 D_1 - P_{11}(t)B_1 R_2^{-1} B_1^T P_{11}(t) + A_{11}^T P_{11}(t) + P_{11}(t)A_{11},$$

$$P_{11}(t_1) = 0. \qquad \text{3-237}$$

We see that this is the usual Riccati-type equation. Now since the pair $\{A_{11}, B_1\}$ is completely controllable, we know from Theorem 3.5 that $P_{11}(t)$ has an asymptotic solution $\bar{P}_{11}$ as $t_1 \to \infty$ such that $A_{11} - B_1 \bar{F}_1$, where $\bar{F}_1 = R_2^{-1} B_1^T \bar{P}_{11}$, is asymptotically stable. The control law for the whole system **3-229** is given by

$$F = (F_1, F_2) = R_2^{-1}(B_1^T, \ 0)\begin{pmatrix} \bar{P}_{11} & \bar{P}_{12} \\ \bar{P}_{12}^T & \bar{P}_{22} \end{pmatrix} = (R_2^{-1}B_1^T\bar{P}_{11}, \ R_2^{-1}B_1^T\bar{P}_{12}). \quad \textbf{3-238}$$

With this control law the closed-loop system is described by

$$\dot{x}(t) = \begin{pmatrix} A_{11} - B_1\bar{F}_1 & A_{12} - B_1\bar{F}_2 \\ 0 & A_{22} \end{pmatrix} x(t). \quad \textbf{3-239}$$

Clearly, if the open-loop system is stabilizable, the closed-loop system is asymptotically stable since both $A_{11} - B_1\bar{F}_1$ and A_{22} are asymptotically stable. This proves that detectability and stabilizability are sufficient conditions to guarantee that the closed-loop steady-state control law will be asymptotically stable. This terminates the proof of (e).

Consider now part (f) of the theorem. Obviously, the steady-state control law minimizes

$$\int_{t_0}^{\infty} [z^T(t)R_3 z(t) + u^T(t)R_2 u(t)] \, dt, \quad \textbf{3-240}$$

and the minimal value of this criterion is given by $x^T(t_0)\bar{P}x(t_0)$. Let us now consider the criterion

$$\lim_{t_1 \to \infty} \left\{ \int_{t_0}^{t_1} [z^T(t)R_3 z(t) + u^T(t)R_2 u(t)] \, dt + x^T(t_1)P_1 x(t_1) \right\}, \quad \textbf{3-241}$$

with $P_1 \geq 0$. If the system is stabilizable and detectable, for the steady-state control law the criterion **3-241** is equal to

$$\int_{t_0}^{\infty} [\bar{z}^T(t)R_3 \bar{z}(t) + \bar{u}^T(t)R_2 \bar{u}(t)] \, dt = x^T(t_0)\bar{P}x(t_0), \quad \textbf{3-242}$$

where $\bar{z}$ and $\bar{u}$ are the controlled variable and input generated by the steady-state control law. We claim that the steady-state control law not only minimizes **3-240**, but also **3-241**. Suppose that there exists another control law that gives a smaller value of **3-241**, so that for this control law

$$\int_{t_0}^{\infty} [z^T(t)R_3 z(t) + u^T(t)R_2 u(t)] \, dt + \lim_{t_1 \to \infty} x^T(t_1)P_1 x(t_1) < x^T(t_0)\bar{P}x(t_0). \quad \textbf{3-243}$$

Because $P_1 \geq 0$ this would imply that for this feedback law

$$\int_{t_0}^{\infty} [z^T(t)R_3 z(t) + u^T(t)R_2 u(t)] \, dt < x_0^T(t_0)\bar{P}x(t_0). \quad \textbf{3-244}$$

But since we know that the left-hand side of this expression is minimized by the steady-state control law, and no value of the criterion less than

$x^T(t_0)\bar{P}x(t_0)$ can be achieved, this is a contradiction, which means that 3-241 is also minimized by the steady-state control law. This terminates the proof of part (f).

We now return to part (b) of the theorem. The fact stated in (b) immediately follows from (f). Consider now part (c). In general, the algebraic Riccati equation has many solutions (see Problem 3.8). If $\bar{P}$ exists, it is a nonnegative-definite solution of the algebraic Riccati equation because $\bar{P}$ must be a solution of the Riccati differential equation 3-215. Suppose that the system 3-213 is stabilizable and detectable, and let P' be any nonnegative-definite solution of the algebraic Riccati equation. Consider the Riccati differential equation 3-215 with the terminal condition $P_1 = P'$. Obviously, the solution of the Riccati equation is $P(t) = P'$, $t \leq t$. Then the steady-state solution $\bar{P}$ must also be given by P'. This proves that any nonnegative-definite solution P' of the algebraic Riccati equation is the steady-state solution $\bar{P}$, hence that the steady-state value $\bar{P}$ is the unique nonnegative-definite solution of the algebraic Riccati equation. This terminates the proof of (c), and also the proof of the whole theorem.

Comments. We conclude this section with the following comments. Parts (b) and (c) state that stabilizability and detectability are sufficient conditions for the Riccati equation to converge to a unique $\bar{P}$ for all $P_1 \geq 0$ and for the algebraic Riccati equation to have a unique nonnegative-definite solution. That these conditions are not necessary can be seen from simple examples.

Furthermore, it may very well happen that although $\bar{P}$ does not exist,

$$\bar{F} = \lim_{t_1 \to \infty} R_2^{-1} B^T P(t) \qquad \text{3-245}$$

does exist.

It is not difficult to conclude that the steady-state control law $u(t) = -\bar{F}x(t)$, if it exists, changes only the locations of those open-loop poles that are both controllable and reconstructible. Therefore an unfavorable situation may arise when a system possesses uncontrollable or unreconstructible poles, in particular if these poles are unstable. Unfortunately, it is usually impossible to change the structure of the system so as to make uncontrollable poles controllable. If a system possesses unreconstructible poles with undesirable locations, it is often possible, however, to redefine the controlled variable such that the system no longer has unreconstructible poles.

3.4.4* Solution of the Time-Invariant Regulator Problem by Diagonalization

In this section we further investigate the steady-state solution of the time-invariant regulator problem. This first of all provides us with a method for

244 Optimal Linear State Feedback Control Systems

computing the steady-state solution $\bar{P}$ of the Riccati equation, and moreover puts us into a position to derive information about the closed-loop regulator poles and the closed-loop behavior of the regulator. Throughout the section we assume that the open-loop system is both stabilizable and detectable.

In Section 3.3.2 we saw that the regulator problem can be solved by considering the linear differential equation

$$\begin{pmatrix} \dot{x}(t) \\ \dot{p}(t) \end{pmatrix} = Z \begin{pmatrix} x(t) \\ p(t) \end{pmatrix}, \qquad \text{3-246}$$

where Z is the constant matrix

$$Z = \begin{pmatrix} A & -BR_2^{-1}B^T \\ -R_1 & -A^T \end{pmatrix}. \qquad \text{3-247}$$

Here $R_1 = D^T R_3 D$. Correspondingly, we have the boundary conditions

$$x(t_0) = x_0, \qquad \text{3-248a}$$
$$p(t_1) = P_1 x(t_1). \qquad \text{3-248b}$$

From Sections 3.3.2 and 3.3.3 (Eq. **3-92**), we know that $p(t)$ and $x(t)$ are related by

$$p(t) = P(t)x(t), \qquad \text{3-249}$$

where $P(t)$ is the solution of the matrix Riccati equation with the terminal condition $P(t_1) = P_1$. Suppose now that we choose

$$P_1 = \bar{P}, \qquad \text{3-250}$$

where $\bar{P}$ is the steady-state solution of the Riccati equation. Then the Riccati equation obviously has the solution

$$P(t) = \bar{P}, \qquad t_0 \leq t \leq t_1. \qquad \text{3-251}$$

This shows that the steady-state solution can be obtained by replacing the *terminal* condition **3-248b** with the *initial* condition

$$p(t_0) = \bar{P}x(t_0). \qquad \text{3-252}$$

Solving the differential equation **3-246** with the initial conditions **3-248a** and **3-252** gives us the steady-state behavior of the state and adjoint variable.

We study the solution of this initial value problem by diagonalization of the matrix Z. It can be shown by elementary determinant manipulations that

$$\det(-sI - Z) = \det(sI - Z). \qquad \text{3-253}$$

Consequently, $\det(sI - Z)$ is a polynomial in s^2 which shows that, if λ is a

3.4 Steady-State Solution of the Regulator Problem

characteristic value of Z, $-\lambda$ is also a characteristic value. Let us for simplicity assume that the characteristic values of Z are all distinct (for the more general case, see Problem 3.9). This allows us to diagonalize Z as follows:

$$Z = W \begin{pmatrix} \Lambda & 0 \\ 0 & -\Lambda \end{pmatrix} W^{-1}. \qquad \text{3-254}$$

Here Λ is a diagonal matrix which is constructed as follows. If a characteristic value λ of Z has a strictly positive real part, it is a diagonal element of Λ; $-\lambda$ is automatically placed in $-\Lambda$. If λ has zero real part, one of the pair λ, $-\lambda$ is arbitrarily assigned to Λ and the other to $-\Lambda$. The matrix W is composed of the characteristic vectors of Z; the ith column vector of W is the characteristic vector of Z corresponding to the characteristic value in the ith diagonal position of diag $(\Lambda, -\Lambda)$.

Let us now consider the differential equation

$$\begin{pmatrix} \dot{z}_1(t) \\ \dot{z}_2(t) \end{pmatrix} = \begin{pmatrix} \Lambda & 0 \\ 0 & -\Lambda \end{pmatrix} \begin{pmatrix} z_1(t) \\ z_2(t) \end{pmatrix}, \qquad \text{3-255}$$

where

$$\begin{pmatrix} z_1(t) \\ z_2(t) \end{pmatrix} = W^{-1} \begin{pmatrix} x(t) \\ p(t) \end{pmatrix}. \qquad \text{3-256}$$

We partition W^{-1} as follows:

$$W^{-1} = V = \begin{pmatrix} V_{11} & V_{12} \\ V_{21} & V_{22} \end{pmatrix}. \qquad \text{3-257}$$

Then we can write

$$\begin{aligned} z_1(t) &= V_{11} x(t) + V_{12} p(t) \\ &= (V_{11} + V_{12} \bar{P}) x(t). \end{aligned} \qquad \text{3-258}$$

We know that the steady-state solution is stable, that is, $x(t) \to 0$ as $t \to \infty$. This also implies that $z_1(t) \to 0$ as $t \to \infty$. From **3-255**, however, we see that

$$z_1(t) = e^{\Lambda(t-t_0)} z_1(t_0). \qquad \text{3-259}$$

Since the characteristic values of Λ all have zero or positive real parts, $z_1(t)$ can converge to zero only if $z_1(t_0) = 0$. According to **3-258**, this can be the case for all x_0 if and only if $\bar{P}$ satisfies the relation

$$V_{11} + V_{12} \bar{P} = 0. \qquad \text{3-260}$$

If V_{12} is nonsingular, we can solve for $\bar{P}$ as follows:

$$\bar{P} = -V_{12}^{-1} V_{11}. \qquad \text{3-261}$$

In any case $\bar{P}$ must satisfy **3-260**. Let us suppose that **3-260** does not have a

246 Optimal Linear State Feedback Control Systems

unique nonnegative-definite solution for $\bar{P}$ and let P' be any nonnegative-definite solution. Consider now the differential equation 3-246 with the terminal condition

$$p(t_1) = P'x(t_1). \qquad 3\text{-}262$$

We can write the solution in the form

$$\begin{pmatrix} x(t) \\ p(t) \end{pmatrix} = \begin{pmatrix} W_{11} & W_{12} \\ W_{21} & W_{22} \end{pmatrix} \begin{pmatrix} e^{\Lambda(t-t_1)} & 0 \\ 0 & e^{-\Lambda(t-t_1)} \end{pmatrix} \begin{pmatrix} V_{11} & V_{12} \\ V_{21} & V_{22} \end{pmatrix} \begin{pmatrix} x(t_1) \\ p(t_1) \end{pmatrix}, \qquad 3\text{-}263$$

where W has also been partitioned. Substitution of 3-262 gives

$$\begin{pmatrix} x(t) \\ p(t) \end{pmatrix} = \begin{pmatrix} W_{11} & W_{12} \\ W_{21} & W_{22} \end{pmatrix} \begin{pmatrix} e^{\Lambda(t-t_1)} & 0 \\ 0 & e^{-\Lambda(t-t_1)} \end{pmatrix} \begin{pmatrix} (V_{11} + V_{12}P')x(t_1) \\ (V_{21} + V_{22}P')x(t_1) \end{pmatrix}. \qquad 3\text{-}264$$

By using the fact that P' is a solution of 3-260, this can be further worked out; we obtain

$$x(t) = W_{12}e^{-\Lambda(t-t_1)}(V_{21} + V_{22}P')x(t_1), \qquad 3\text{-}265\text{a}$$

$$p(t) = W_{22}e^{-\Lambda(t-t_1)}(V_{21} + V_{22}P')x(t_1). \qquad 3\text{-}265\text{b}$$

For $t = t_0$ the first of these equations reduces to

$$x_0 = W_{12}e^{-\Lambda(t_0-t_1)}(V_{21} + V_{22}P')x(t_1). \qquad 3\text{-}266$$

Since the two-point boundary value problem must have a solution for all x_0, the matrix that relates x_0 and $x(t_1)$ must be nonsingular (otherwise this equation would not have a solution if x_0 is not in the range of this matrix). In fact, since any $t \leq t_1$ can be considered as the initial time for the interval $[t, t_1]$, the matrix

$$W_{12}e^{-\Lambda(t-t_1)}(V_{21} + V_{22}P') \qquad 3\text{-}267$$

must be nonsingular for all $t \leq t_1$. Solving 3-265a for $x(t_1)$ and substituting this into 3-265b yields

$$p(t) = W_{22}e^{-\Lambda(t-t_1)}(V_{21} + V_{22}P')(V_{21} + V_{22}P')^{-1}e^{\Lambda(t-t_1)}W_{12}^{-1}x(t), \qquad 3\text{-}268$$

or (O'Donnell, 1966)

$$p(t) = W_{22}W_{12}^{-1}x(t). \qquad 3\text{-}269$$

Apparently, solving the two-point boundary value problem with the terminal condition $P(t_1) = P'$ yields a solution of the form

$$p(t) = \hat{P}x(t), \qquad 3\text{-}270$$

where $\hat{P}$ is constant. Since this solution is independent of the terminal time t_1, $\hat{P}$ is also the steady-state solution $\bar{P}$ of the Riccati equation as $t_1 \to \infty$. Since, as we know from Theorem 3.7, this steady-state solution is unique, we cannot but conclude that

$$\bar{P} = W_{22} W_{12}^{-1}. \qquad 3\text{-}271$$

This argument shows that W_{12} is nonsingular and that $\bar{P}$ can be represented in the form 3-271. Since the partitioned blocks of V and W have a special relationship, it can also be shown that V_{12} is nonsingular, hence also that 3-261 is a valid expression (Problem 3.12).

In addition to these results, we can obtain the following interesting conclusion. By solving 3-266 for $x(t_1)$ and substituting the result into 3-265a, we find

$$x(t) = W_{12} e^{-\Lambda(t-t_0)} W_{12}^{-1} x_0. \qquad 3\text{-}272$$

This shows very explicitly that the characteristic values of the steady-state closed-loop system are precisely the diagonal elements of $-\Lambda$ (O'Donnell, 1966). Since the closed-loop system is known to be asymptotically stable, it follows that the diagonal elements of $-\Lambda$ have strictly negative real parts. Since these characteristic values are obtained from the characteristic values of Z, this means that Z cannot have any characteristic values with zero real parts, and that the steady-state closed-loop characteristic values are precisely those characteristic values of Z that have negative real parts (Letov, 1960).

We summarize these conclusions as follows.

Theorem 3.8. *Consider the time-invariant deterministic linear optimal regulator problem and suppose that the pair $\{A, B\}$ is stabilizable and the pair $\{A, D\}$ detectable. Define the $2n \times 2n$ matrix*

$$Z = \begin{pmatrix} A & -BR_2^{-1}B^T \\ -D^T R_3 D & -A^T \end{pmatrix}, \qquad 3\text{-}273$$

and assume that Z has $2n$ distinct characteristic values. Then
(a) If λ is a characteristic value of Z, $-\lambda$ also is a characteristic value. Z has no characteristic values with zero real parts.
(b) The characteristic values of the steady-state closed-loop optimal regulator are those characteristic values of Z that have negative real parts.
(c) If Z is diagonalized in the form

$$Z = W \begin{pmatrix} \Lambda & 0 \\ 0 & -\Lambda \end{pmatrix} W^{-1}, \qquad 3\text{-}274$$

where the diagonal matrix Λ has as diagonal elements the characteristic values of Z with positive real parts, the steady-state solution of the Riccati equation 3-215 can be written as

$$\bar{P} = W_{22} W_{12}^{-1} = -V_{12}^{-1} V_{11}, \qquad 3\text{-}275$$

where the W_{ij} and V_{ij}, $i, j = 1, 2$, are obtained by partitioning W and $V = W^{-1}$, respectively. The inverse matrix in both expressions exists.

(d) *The response of the steady-state closed-loop optimal regulator can be written as*

$$x(t) = W_{12} e^{-\Lambda(t-t_0)} W_{12}^{-1} x_0. \qquad 3\text{-}276$$

The diagonalization approach discussed in this section is further pursued in Problems 3.8 through 3.12.

3.5 NUMERICAL SOLUTION OF THE RICCATI EQUATION

3.5.1 Direct Integration

In this section we discuss various methods for the numerical solution of the Riccati equation, which is of fundamental importance for the regulator problem and, as we see in Chapter 4, also for state reconstruction problems. The matrix Riccati equation is given by

$$-\dot{P}(t) = R_1(t) - P(t)B(t)R_2^{-1}(t)B^T(t)P(t) + A^T(t)P(t) + P(t)A(t), \qquad 3\text{-}277$$

with the terminal condition

$$P(t_1) = P_1.$$

A direct approach results from considering **3-277** a set of n^2 simultaneous nonlinear first-order differential equations (assuming that $P(t)$ is an $n \times n$ matrix) and using any standard numerical technique to integrate these equations backward from t_1. The most elementary method is Euler's method, where we write

$$P(t - \Delta t) \simeq P(t) - \dot{P}(t)\,\Delta t, \qquad 3\text{-}278$$

and compute $P(t)$ for $t = t_1 - \Delta t, t_1 - 2\Delta t, \cdots$. If the solution converges to a constant value, such as usually occurs in the time-invariant case, some stopping rule is needed. A disadvantage of this approach is that for sufficient accuracy usually a quite small value of Δt is required, which results in a large number of steps. Also, the symmetry of $P(t)$ tends to be destroyed because of numerical errors. This can be remedied by symmetrizing after each step, that is, replacing $P(t)$ with $\frac{1}{2}[P(t) + P^T(t)]$. Alternatively, the symmetry of $P(t)$ can be exploited by reducing **3-277** to a set of $\frac{1}{2}n(n + 1)$ simultaneous first-order differential equations, which results in an appreciable saving of computer time. A further discussion of the method of direct integration may be found in Bucy and Joseph (1968).

The method of direct integration is applicable to both the time-varying and the time-invariant case. If only steady-state solutions for time-invariant problems are required, the methods presented in Sections 3.5.3 and 3.5.4 are more effective.

We finally point out the following. In order to realize a time-varying control law, the entire behavior of $F(t)$ for $t_0 \leq t \leq t_1$ must be stored. It seems attractive to circumvent this as follows. By off-line integration $P(t_0)$ can be computed. Then the Riccati equation 3-277 is integrated on-line, with the correct initial value $P(t_0)$, and the feedback gain matrix is obtained, on-line, from $F(t) = R_2^{-1}(t)B^T(t)P(t)$. This method usually leads to unsatisfactory results, however, since in the forward direction the Riccati equation 3-277 is unstable, which causes computational inaccuracies that increase with t (Kalman, 1960).

3.5.2 The Kalman–Englar Method

When a complete solution is required of the *time-invariant* Riccati equation, a convenient approach (Kalman and Englar, 1966) is based upon the following expression, which derives from 3-98:

$$P(t_{i+1}) = [\Theta_{21}(t_{i+1}, t_i) + \Theta_{22}(t_{i+1}, t_i)P(t_i)][\Theta_{11}(t_{i+1}, t_i) + \Theta_{12}(t_{i+1}, t_i)P(t_i)]^{-1},$$
3-279

where
$$t_{i+1} = t_i - \Delta t. \qquad 3\text{-}280$$

The matrices $\Theta_{ij}(t, t_0)$ are obtained by partitioning the transition matrix $\Theta(t, t_0)$ of the system

$$\begin{pmatrix} \dot{x}(t) \\ \dot{p}(t) \end{pmatrix} = Z \begin{pmatrix} x(t) \\ p(t) \end{pmatrix}, \qquad 3\text{-}281$$

where
$$Z = \begin{pmatrix} A & -BR_2^{-1}B^T \\ -D^T R_3 D & -A^T \end{pmatrix}. \qquad 3\text{-}282$$

We can compute $\Theta(t_{i+1}, t_i)$ once and for all as

$$\Theta(t_{i+1}, t_i) = e^{-Z\Delta t}, \qquad 3\text{-}283$$

which can be evaluated according to the power series method of Section 1.3.2. The solution of the Riccati equation is then found by repeated application of **3-279**. It is advantageous to symmetrize after each step.

Numerical difficulties occur when Δt is chosen too large. Vaughan (1969) discusses these difficulties in some detail. They manifest themselves in near-singularity of the matrix to be inverted in **3-279**. It has been shown by Vaughan that a very small Δt is required when the real parts of the characteristic values of Z have a large spread. For most problems there exists a Δt small enough to obtain accurate results. Long computing times may result, however, especially when the main interest is in the steady-state solution.

3.5.3* Solution by Diagonalization

In order to obtain the steady-state solution of the time-invariant Riccati equation, the results derived in Section 3.4.4 by diagonalizing the $2n \times 2n$ matrix Z are useful. Here the asymptotic solution is expressed as

$$\bar{P} = W_{22}W_{12}^{-1}, \qquad \text{3-284}$$

where W_{22} and W_{12} are obtained by partitioning a matrix W as follows:

$$W = \begin{pmatrix} W_{11} & W_{12} \\ W_{21} & W_{22} \end{pmatrix}. \qquad \text{3-285}$$

The matrix W consists of the characteristic vectors of the matrix Z so arranged that the first n columns of W correspond to the characteristic values of Z with positive real parts, and the last n columns of W to the characteristic values of Z with negative real parts.

Generally, some or all of the characteristic vectors of Z may be complex so that W_{22} and W_{12} may be complex matrices. Complex arithmetic can be avoided as follows. Since if e is a characteristic vector of Z corresponding to a characteristic value λ with negative real part, its complex conjugate $\bar{e}$ is also a characteristic vector corresponding to a characteristic value $\bar{\lambda}$ with a negative real part, the last n columns of W will contain besides real column vectors only complex conjugate pairs of column vectors. Then it is always possible to perform a nonsingular linear transformation

$$\begin{pmatrix} W'_{12} \\ W'_{22} \end{pmatrix} = \begin{pmatrix} W_{12} \\ W_{22} \end{pmatrix} U, \qquad \text{3-286}$$

such that every pair of complex conjugate column vectors e and $\bar{e}$ in col (W_{12}, W_{22}) is replaced with two real vectors Re (e) and Im (e) in col (W'_{12}, W'_{22}). Then

$$W'_{22}W'^{-1}_{12} = (W_{22}U)(W_{12}U)^{-1} = W_{22}W_{12}^{-1}, \qquad \text{3-287}$$

which shows that W'_{22} and W'_{12} can be used to compute $\bar{P}$ instead of W_{22} and W_{12}.

Let us summarize this method of obtaining $\bar{P}$:

(a) Form the matrix Z and use any standard numerical technique to compute those characteristic vectors that correspond to characteristic values with negative real parts.

(b) Form from these n characteristic vectors a $2n \times n$ matrix

$$\begin{pmatrix} W'_{12} \\ W'_{22} \end{pmatrix}, \qquad \text{3-288}$$

where W'_{21} and W'_{22} are $n \times n$ submatrices, as follows. If e is a real characteristic vector, let e be one of the columns of **3-288**. If e and $\bar{e}$ form a complex conjugate pair, let Re (e) be one column of **3-288** and Im (e) another.

(c) Compute $\bar{P}$ as

$$\bar{P} = W'_{22}W'^{-1}_{12}. \qquad \textbf{3-289}$$

The efficiency of this method depends upon the efficiency of the subprogram that computes the characteristic vectors of Z. Van Ness (1969) has suggested a characteristic vector algorithm that is especially suitable for problems of this type. The algorithm as outlined above has been successfully applied for solving high-order Riccati equations (Freestedt, Webber, and Bass, 1968; Blackburn and Bidwell, 1968; Hendricks and Haynes, 1968). Fath (1969) presents a useful modification of the method.

The diagonalization approach can also be employed to obtain not only the asymptotic solution of the Riccati equation but the complete behavior of $P(t)$ by the formulas of Problem 3.11.

A different method for computing the asymptotic solution $\bar{P}$ is to use the identity (see Problem 3.10)

$$\phi(Z)\begin{pmatrix} I \\ \bar{P} \end{pmatrix} = 0, \qquad \textbf{3-290}$$

where $\phi(s)$ is obtained by factoring

$$\det(sI - Z) = \phi(s)\phi(-s), \qquad \textbf{3-291}$$

such that the roots of $\phi(s)$ are precisely the characteristic values of Z with negative real parts. Clearly, $\phi(s)$ is the characteristic polynomial of the steady-state closed-loop optimal system. Here $\det(sI - Z)$ can be obtained by the Leverrier algorithm of Section 1.5.1, or by any standard technique for obtaining characteristic values of matrices. Both favorable (Freestedt, Webber, and Bass, 1968) and unfavorable (Blackburn and Bidwell, 1968; Hendricks and Haynes, 1968) experiences with this method have been reported.

3.5.4* Solution by the Newton–Raphson Method

In this subsection a method is discussed for computing the steady-state solution of the time-invariant Riccati equation, which is quite different from the previous methods. It is based upon repeated solution of a linear matrix equation of the type

$$0 = A^T P + PA + R, \qquad \textbf{3-292}$$

which has been discussed in Section 1.11.3.

The steady-state solution $\bar{P}$ of the Riccati equation must satisfy the algebraic Riccati equation
$$0 = R_1 - PSP + A^T P + PA, \qquad \text{3-293}$$
where
$$S = BR_2^{-1}B^T. \qquad \text{3-294}$$
Consider the matrix function
$$F(P) = R_1 - PSP + A^T P + PA. \qquad \text{3-295}$$
The problem is to find the nonnegative-definite symmetric matrix $\bar{P}$ that satisfies
$$F(\bar{P}) = 0. \qquad \text{3-296}$$

We derive an iterative procedure. Suppose that at the k-th stage a solution P_k has been obtained, which is not much different from the desired solution $\bar{P}$, and let us write
$$\bar{P} = P_k + \tilde{P}. \qquad \text{3-297}$$
If $\tilde{P}$ is small we can approximate $F(\bar{P})$ by omitting quadratic terms in $\tilde{P}$ and we obtain
$$F(\bar{P}) \simeq R_1 - P_k S P_k - P_k S \tilde{P} - \tilde{P} S P_k - A^T(P_k + \tilde{P}) + (P_k + \tilde{P})A. \qquad \text{3-298}$$
The basic idea of the Newton–Raphson method is to estimate $\tilde{P}$ by setting the right-hand side of **3-298** equal to zero. If the estimate of $\tilde{P}$ so obtained is denoted as $\tilde{P}_k$ and we let
$$P_{k+1} = P_k + \tilde{P}_k, \qquad \text{3-299}$$
then we find by setting the right-hand side of **3-298** equal to zero:
$$0 = R_1 + P_k S P_k + P_{k+1} A_k + A_k^T P_{k+1}, \qquad \text{3-300}$$
where
$$A_k = A - S P_k. \qquad \text{3-301}$$
Equation **3-300** is of the type **3-292**, for which efficient methods of solution exist (see Section 1.11.3). We have thus obtained the following algorithm.

(a) Choose a suitable P_0 and set the iteration index k equal to 0.
(b) Solve P_{k+1} from **3-300**.
(c) If convergence is obtained, stop; otherwise, increase k by one and return to (b).

Kleinman (1968) and McClamroch (1969) have shown that if the algebraic Riccati equation has a unique nonnegative-definite solution, P_k and P_{k+1} satisfy
$$P_{k+1} \leq P_k, \qquad k = 0, 1, 2, \ldots, \qquad \text{3-302}$$

and

$$\lim_{k \to \infty} P_k = \bar{P}, \quad \text{3-303}$$

provided P_0 is so chosen that

$$A_0 = A - SP_0 \quad \text{3-304}$$

is asymptotically stable. This means that the convergence of the scheme is assured if the initial estimate is suitably chosen. If the initial estimate is incorrectly selected, however, convergence to a different solution of the algebraic Riccati equation may occur, or no convergence at all may result. If A is asymptotically stable, a safe choice is $P_0 = 0$. If A is not asymptotically stable, the initial choice may present difficulties. Wonham and Cashman (1968), Man and Smith (1969), and Kleinman (1970b) give methods for selecting P_0 when A is not asymptotically stable.

The main problem with this approach is **3-292**, which must be solved many times over. Although it is linear, the numerical effort may still be rather formidable, since the number of linear equations that must be solved at each iteration increases rapidly with the dimension of the problem (for $n = 15$ this number is 120). In Section 1.11.3 several numerical approaches to solving **3-292** are referenced. In the literature favorable experiences using the Newton–Raphson method to solve Riccati equations has been reported with up to 15-dimensional problems (Blackburn, 1968; Kleinman, 1968, 1970a).

3.6 STOCHASTIC LINEAR OPTIMAL REGULATOR AND TRACKING PROBLEMS

3.6.1 Regulator Problems with Disturbances—The Stochastic Regulator Problem

In the preceding sections we discussed the deterministic linear optimal regulator problem. The solution of this problem allows us to tackle purely transient problems where a linear system has a disturbed initial state, and it is required to return the system to the zero state as quickly as possible while limiting the input amplitude. There exist practical problems that can be formulated in this manner, but much more common are problems where there are disturbances that act uninterruptedly upon the system, and that tend to drive the state away from the zero state. The problem is then to design a feedback configuration through which initial offsets are reduced as quickly as possible, but which also counteracts the effects of disturbances as much as possible in the steady-state situation. The solution of this problem will bring us into a position to synthesize the controllers that have been asked for in

254   Optimal Linear State Feedback Control Systems

Chapter 2. For the time being we maintain the assumption that the complete state of the system can be accurately observed at each instant of time.

The effect of the disturbances can be accounted for by suitably extending the system description. We consider systems described by

$$\dot{x}(t) = A(t)x(t) + B(t)u(t) + v(t),$$
$$z(t) = D(t)x(t),$$
  3-305

where $u(t)$ is the input variable, $z(t)$ is the controlled variable, and $v(t)$ represents disturbances that act upon the system. We mathematically represent the disturbances as a stochastic process, which we model as the output of a linear system driven by white noise. Thus we assume that $v(t)$ is given by

$$v(t) = D_d(t)x_d(t).$$
  3-306

Here $x_d(t)$ is the solution of

$$\dot{x}_d(t) = A_d(t)x_d(t) + w(t),$$
  3-307

where $w(t)$ is white noise. We furthermore assume that both $x(t_0)$ and $x_d(t_0)$ are stochastic variables.

We combine the description of the system and the disturbances by defining an augmented state vector $\tilde{x}(t) = \text{col}\,[x(t), x_d(t)]$, which from **3-305**, **3-306**, and **3-307** can be seen to satisfy

$$\dot{\tilde{x}}(t) = \begin{pmatrix} A(t) & D_d(t) \\ 0 & A_d(t) \end{pmatrix} \tilde{x}(t) + \begin{pmatrix} B(t) \\ 0 \end{pmatrix} u(t) + \begin{pmatrix} 0 \\ w(t) \end{pmatrix}.$$
  3-308

In terms of the augmented state, the controlled variable is given by

$$z(t) = (D(t), 0)\tilde{x}(t).$$
  3-309

We note in passing that **3-308** represents a system that is not completely controllable (from u).

We now turn our attention to the optimization criterion. In the deterministic regulator problem, we considered the quadratic integral criterion

$$\int_{t_0}^{t_1} [z^T(t)R_3(t)z(t) + u^T(t)R_2(t)u(t)]\,dt + x^T(t_1)P_1 x(t_1).$$
  3-310

For a given input $u(t)$, $t_0 \leq t \leq t_1$, and a given realization of the disturbances $v(t)$, $t_0 \leq t \leq t_1$, this criterion is a measure for the deviations $z(t)$ and $u(t)$ from zero. A priori, however, this criterion cannot be evaluated because of the stochastic nature of the disturbances. We therefore average over all possible realizations of the disturbances and consider the criterion

$$E\left\{\int_{t_0}^{t_1}[z^T(t)R_3(t)z(t) + u^T(t)R_2(t)u(t)]\,dt + x^T(t_1)P_1 x(t_1)\right\}.$$
  3-311

3.6 Stochastic Regulator and Tracking Problems

In terms of the augmented state $\tilde{x}(t) = \text{col}\,[x(t), x_d(t)]$, this criterion can be expressed as

$$E\left\{\int_{t_0}^{t_1}[z^T(t)R_3(t)z(t) + u^T(t)R_2(t)u(t)]\,dt + \tilde{x}^T(t_1)\tilde{P}_1\tilde{x}(t_1)\right\}, \quad \text{3-312}$$

where

$$\tilde{P}_1 = \begin{pmatrix} P_1 & 0 \\ 0 & 0 \end{pmatrix}. \quad \text{3-313}$$

It is obvious that the problem of minimizing 3-312 for the system 3-308 is nothing but a special case of the general problem of minimizing

$$E\left\{\int_{t_0}^{t_1}[z^T(t)R_3(t)z(t) + u^T(t)R_2(t)u(t)]\,dt + x^T(t_1)P_1 x(t_1)\right\} \quad \text{3-314}$$

for the system

$$\dot{x}(t) = A(t)x(t) + B(t)u(t) + w(t), \quad \text{3-315}$$

where $w(t)$ is white noise and where $x(t_0)$ is a stochastic variable. We refer to this problem as the stochastic linear optimal regulator problem:

Definition 3.4. *Consider the system described by the state differential equation*

$$\dot{x}(t) = A(t)x(t) + B(t)u(t) + w(t) \quad \text{3-316}$$

with initial state

$$x(t_0) = x_0 \quad \text{3-317}$$

and controlled variable

$$z(t) = D(t)x(t). \quad \text{3-318}$$

In 3-316 $w(t)$ is white noise with intensity $V(t)$. The initial state x_0 is a stochastic variable, independent of the white noise w, with

$$E\{x_0 x_0^T\} = Q_0. \quad \text{3-319}$$

Consider the criterion

$$E\left\{\int_{t_0}^{t_1}[z^T(t)R_3(t)z(t) + u^T(t)R_2(t)u(t)]\,dt + x^T(t_1)P_1 x(t_1)\right\}, \quad \text{3-320}$$

where $R_3(t)$ and $R_2(t)$ are positive-definite symmetric matrices for $t_0 \leq t \leq t_1$ and P_1 is nonnegative-definite symmetric. Then the problem of determining for each t, $t_0 \leq t \leq t_1$, the input $u(t)$ as a function of all information from the past such that the criterion is minimized is called the **stochastic linear optimal regulator problem**. *If all matrices in the problem formulation are constant, we refer to it as the* **time-invariant stochastic linear optimal regulator problem**.

The solution of this problem is discussed in Section 3.6.3.

Example 3.11. *Stirred tank*

In Example 1.37 (Section 1.11.4), we considered an extension of the model of the stirred tank where disturbances in the form of fluctuations in the concentrations of the feeds are incorporated. The extended system model is given by

$$\dot{x}(t) = \begin{pmatrix} -\dfrac{1}{2\theta} & 0 & 0 & 0 \\ 0 & -\dfrac{1}{\theta} & \dfrac{F_{10}}{V_0} & \dfrac{F_{20}}{V_0} \\ 0 & 0 & -\dfrac{1}{\theta_1} & 0 \\ 0 & 0 & 0 & -\dfrac{1}{\theta_2} \end{pmatrix} x(t) + \begin{pmatrix} 1 & 1 \\ \dfrac{c_{10}-c_0}{V_0} & \dfrac{c_{20}-c_0}{V_0} \\ 0 & 0 \\ 0 & 0 \end{pmatrix} u(t)$$

$$+ \begin{pmatrix} 0 & 0 \\ 0 & 0 \\ 1 & 0 \\ 0 & 1 \end{pmatrix} w(t), \quad \text{3-321}$$

where $w(t)$ is white noise with intensity

$$V = \begin{pmatrix} \dfrac{2\sigma_1^2}{\theta_1} & 0 \\ 0 & \dfrac{2\sigma_2^2}{\theta_2} \end{pmatrix}. \quad \text{3-322}$$

Here the components of the state are, respectively, the incremental volume of fluid, the incremental concentration in the tank, the incremental concentration of the feed F_1, and the incremental concentration of the feed F_2. Let us consider as previously the incremental outgoing flow and the incremental outgoing concentration as the components of the controlled variable. Thus we have

$$z(t) = \begin{pmatrix} \dfrac{1}{2\theta} & 0 & 0 & 0 \\ 0 & 1 & 0 & 0 \end{pmatrix} x(t). \quad \text{3-323}$$

The stochastic optimal regulator problem now consists in determining the input $u(t)$ such that a criterion of the form

$$E\left\{ \int_{t_0}^{t_1} [z^T(t)R_3 z(t) + u^T(t)R_2 u(t)]\, dt + x^T(t_1)P_1 x(t_1) \right\} \quad \text{3-324}$$

is minimized. We select the weighting matrices R_3 and R_2 in exactly the same manner as in Example 3.9 (Section 3.4.1), while we choose P_1 to be the zero matrix.

3.6.2 Stochastic Tracking Problems

We have introduced the stochastic optimal regulator problem by considering regulator problems with disturbances. Stochastic regulator problems also arise when we formulate *stochastic optimal tracking problems*. Consider the linear system

$$\dot{x}(t) = A(t)x(t) + B(t)u(t), \qquad \text{3-325}$$

with the controlled variable

$$z(t) = D(t)x(t). \qquad \text{3-326}$$

Suppose we wish the controlled variable to follow as closely as possible a *reference variable* $z_r(t)$ which we model as the output of a linear differential system driven by white noise:

$$z_r(t) = D_r(t)x_r(t), \qquad \text{3-327}$$

with

$$\dot{x}_r(t) = A_r(t)x_r(t) + w(t). \qquad \text{3-328}$$

Here $w(t)$ is white noise with given intensity $V(t)$. The system equations and the reference model equations can be combined by defining the augmented state $\tilde{x}(t) = \text{col}\,[x(t), x_r(t)]$, which satisfies

$$\dot{\tilde{x}}(t) = \begin{pmatrix} A(t) & 0 \\ 0 & A_r(t) \end{pmatrix} \tilde{x}(t) + \begin{pmatrix} B(t) \\ 0 \end{pmatrix} u(t) + \begin{pmatrix} 0 \\ w(t) \end{pmatrix}. \qquad \text{3-329}$$

In passing, we note that this system (just as that of **3-308**) is not completely controllable from u.

To obtain an *optimal* tracking system, we consider the criterion

$$E\left\{\int_{t_0}^{t_1} \{[z(t) - z_r(t)]^T R_3(t)[z(t) - z_r(t)] + u^T(t)R_2(t)u(t)\}\,dt\right\}, \qquad \text{3-330}$$

where $R_3(t)$ and $R_2(t)$ are suitable weighting matrices. This criterion expresses that the controlled variable should be close to the reference variable, while the input amplitudes should be restricted. In fact, for $R_3(t) = W_e(t)$ and $R_2(t) = \rho W_u(t)$, the criterion reduces to

$$\int_{t_0}^{t_1} [C_e(t) + \rho C_u(t)]\,dt, \qquad \text{3-331}$$

where $C_e(t)$ and $C_u(t)$ denote the mean square tracking error and the mean

258 Optimal Linear State Feedback Control Systems

square input, respectively, as defined in Chapter 2 (Section 2.3):

$$C_e(t) = E\{e^T(t)W_e(t)e(t)\},$$
$$C_u(t) = E\{u^T(t)W_u(t)u(t)\}.$$
3-332

Here $e(t)$ is the tracking error

$$e(t) = z(t) - z_r(t).$$
3-333

The weighting coefficient ρ must be adjusted so as to obtain the smallest possible mean square tracking error for a given value of the mean square input.

The criterion 3-330 can be expressed in terms of the augmented state $x(t)$ as follows:

$$E\left\{\int_{t_0}^{t_1} [\tilde{z}^T(t)R_3(t)\tilde{z}(t) + u^T(t)R_2(t)u(t)]\,dt\right\},$$
3-334

where

$$\tilde{z}(t) = (D(t), -D_r(t))\tilde{x}(t).$$
3-335

Obviously, the problem of minimizing the criterion 3-334 for the system 3-329 is a special case of the stochastic linear optimal regulator problem of Definition 3.4.

Without going into detail we point out that tracking problems with disturbances also can be converted into stochastic regulator problems by the state augmentation technique.

In conclusion, we note that the approach of this subsection is entirely in line with the approach of Chapter 2, where we represented reference variables as having a variable part and a constant part. In the present section we have set the constant part equal to zero; in Section 3.7.1 we deal with nonzero constant references.

Example 3.12. *Angular velocity tracking system*

Consider the angular velocity control system of Example 3.3 (Section 3.3.1). Suppose we wish that the angular velocity, which is the controlled variable $\zeta(t)$, follows as accurately as possible a reference variable $\zeta_r(t)$, which may be described as exponentially correlated noise with time constant θ and rms value σ. Then we can model the reference process as (see Example 1.36, Section 1.11.4)

$$\zeta_r(t) = \xi_r(t),$$
3-336

where $\xi_r(t)$ is the solution of

$$\dot{\xi}_r(t) = -\frac{1}{\theta}\xi_r(t) + w(t).$$
3-337

The white noise $w(t)$ has intensity $2\sigma^2/\theta$. Since the system state differential equation is

$$\dot{\xi}(t) = -\alpha\xi(t) + \kappa\mu(t),$$
3-338

3.6 Stochastic Regulator and Tracking Problems

the augmented state differential equation is given by

$$\dot{\tilde{x}}(t) = \begin{pmatrix} -\alpha & 0 \\ 0 & -\dfrac{1}{\theta} \end{pmatrix} \tilde{x}(t) + \begin{pmatrix} \kappa \\ 0 \end{pmatrix} \mu(t) + \begin{pmatrix} 0 \\ 1 \end{pmatrix} w(t), \qquad 3\text{-}339$$

with $\tilde{x}(t) = \text{col}\,[\xi(t), \xi_r(t)]$. For the optimization criterion we choose

$$E\left\{ \int_{t_0}^{t_1} [[\zeta(t) - \zeta_r(t)]^2 + \rho \mu^2(t)]\, dt \right\}, \qquad 3\text{-}340$$

where ρ is a suitable weighting factor. This criterion can be rewritten as

$$E\left\{ \int_{t_0}^{t_1} [\tilde{\zeta}^2(t) + \rho \mu^2(t)]\, dt \right\}, \qquad 3\text{-}341$$

where

$$\tilde{\zeta}(t) = (1, -1)\tilde{x}(t). \qquad 3\text{-}342$$

The problem of minimizing **3-341** for the system described by **3-339** and **3-342** constitutes a stochastic optimal regulator problem.

3.6.3 Solution of the Stochastic Linear Optimal Regulator Problem

In Section 3.6.1 we formulated the stochastic linear optimal regulator problem. This problem (Definition 3.4) exhibits an essential difference from the deterministic regulator problem because the white noise makes it impossible to predict exactly how the system is going to behave. Because of this, the best policy is obviously not to determine the input $u(t)$ over the control period $[t_0, t_1]$ *a priori*, but to reconsider the situation at each intermediate instant t on the basis of all available information.

At the instant t the further behavior of the system is entirely determined by the present state $x(t)$, the input $u(\tau)$ for $\tau \geq t$, and the white noise $w(\tau)$ for $\tau \geq t$. All the information from the past that is relevant for the future is contained in the state $x(t)$. Therefore we consider control laws of the form

$$u(t) = g[x(t), t], \qquad 3\text{-}343$$

which prescribe an input corresponding to each possible value of the state at time t.

The use of such control laws presupposes that each component of the state can be accurately measured at all times. As we have pointed out before, this is an unrealistic assumption. This is even more so in the stochastic case where the state in general includes components that describe the disturbances or the reference variable; it is very unlikely that these components can be easily measured. We postpone the solution of this difficulty until after

260 Optimal Linear State Feedback Control Systems

Chapter 4, however, where the reconstruction of the state from incomplete and inaccurate measurements is discussed.

In preceding sections we have obtained the solution of the deterministic regulator problem in the feedback form 3-343. For the stochastic version of the problem, we have the surprising result that the presence of the white noise term $w(t)$ in the system equation 3-316 does not alter the solution except to increase the minimal value of the criterion. We first state this fact and then discuss its proof:

Theorem 3.9. *The optimal **linear** solution of the stochastic linear optimal regulator problem is to choose the input according to the linear control law*

$$u(t) = -F^0(t)x(t), \qquad \text{3-344}$$

where

$$F^0(t) = R_2^{-1}(t)B^T(t)P(t). \qquad \text{3-345}$$

Here $P(t)$ is the solution of the matrix Riccati equation

$$-\dot{P}(t) = R_1(t) - P(t)B(t)R_2^{-1}(t)B^T(t)P(t) + A^T(t)P(t) + P(t)A(t) \qquad \text{3-346}$$

with the terminal condition

$$P(t_1) = P_1. \qquad \text{3-347}$$

Here we abbreviate as usual

$$R_1(t) = D^T(t)R_3(t)D(t). \qquad \text{3-348}$$

The minimal value of the criterion is given by

$$\operatorname{tr}\left[P(t_0)Q_0 + \int_{t_0}^{t_1} P(t)V(t)\,dt\right]. \qquad \text{3-349}$$

It is observed that this theorem gives only the best *linear* solution of the stochastic regulator problem. Since we limit ourselves to linear systems, this is quite satisfactory. It can be proved, however, that the linear feedback law is optimal (without qualification) when the white noise $w(t)$ is Gaussian (Kushner, 1967, 1971; Åström, 1970).

To prove the theorem let us suppose that the system is controlled through the linear control law

$$u(t) = -F(t)x(t). \qquad \text{3-350}$$

Then the closed-loop system is described by the differential equation

$$\dot{x}(t) = [A(t) - B(t)F(t)]x(t) + w(t) \qquad \text{3-351}$$

and we can write for the criterion 3-320

$$E\left\{\int_{t_0}^{t_1} x^T(t)[R_1(t) + F^T(t)R_2(t)F(t)]x(t)\,dt + x^T(t_1)P_1 x(t_1)\right\}. \qquad \text{3-352}$$

3.6 Stochastic Regulator and Tracking Problems

We know from Theorem 1.54 (Section 1.11.5) that the criterion can be expressed as

$$\text{tr}\left[\tilde{P}(t_0)Q_0 + \int_{t_0}^{t_1} \tilde{P}(t)V(t)\,dt\right], \qquad \text{3-353}$$

where $\tilde{P}(t)$ is the solution of the matrix differential equation

$$-\dot{\tilde{P}}(t) = [A(t) - B(t)F(t)]^T \tilde{P}(t)$$
$$+ \tilde{P}(t)[A(t) - B(t)F(t)] + R_1(t) + F^T(t)R_2(t)F(t), \qquad \text{3-354}$$

with the terminal condition

$$\tilde{P}(t_1) = P_1. \qquad \text{3-355}$$

Now Lemma 3.1 (Section 3.3.3) states that $\tilde{P}(t)$ satisfies the inequality

$$\tilde{P}(t) \geq P(t) \qquad \text{3-356}$$

for all $t_0 \leq t \leq t_1$, where $P(t)$ is the solution of the Riccati equation **3-346** with the terminal condition **3-347**. The inequality **3-356** converts into an equality if F is chosen as

$$F^0(\tau) = R_2^{-1}(\tau)B^T(\tau)P(\tau), \qquad t \leq \tau \leq t_1. \qquad \text{3-357}$$

The inequality **3-356** implies that

$$\text{tr}\,[\tilde{P}(t)\Gamma] \geq \text{tr}\,[P(t)\Gamma] \qquad \text{3-358}$$

for any nonnegative-definite matrix Γ. This shows very clearly that **3-353** is minimized by choosing F according to **3-357**. For this choice of F, the criterion **3-353** is given by **3-349**. This terminates the proof that the control law **3-345** is the optimal linear control law.

Theorem 3.9 puts us into a position to solve various types of problems. In Sections 3.6.1 and 3.6.2, we showed that the stochastic linear optimal regulator problem may originate from regulator problems for disturbed systems, or from optimal tracking problems. In both cases the problem has a special structure. We now briefly discuss the properties of the solutions that result from these special structures.

In the case of a regulator with disturbances, the system state differential and output equations take the partitioned form **3-308**, **3-309**. Suppose that we partition the solution $P(t)$ of the Riccati equation **3-346** according to the partitioning $\tilde{x}(t) = \text{col}\,[x(t), x_d(t)]$ as

$$P(t) = \begin{pmatrix} P_{11}(t) & P_{12}(t) \\ P_{12}^T(t) & P_{22}(t) \end{pmatrix}. \qquad \text{3-359}$$

If, accordingly, the optimal feedback gain matrix is partitioned as

$$F^0(t) = (F_1(t), F_2(t)), \qquad \text{3-360}$$

it is not difficult to see that
$$F_1(t) = R_2^{-1}(t)B^T(t)P_{11}(t), \qquad \text{3-361}$$
$$F_2(t) = R_2^{-1}(t)B^T(t)P_{12}(t).$$

Furthermore, it can be found by partitioning the Riccati equation that P_{11}, P_{12}, and P_{22} are the solutions of the matrix differential equations

$$\begin{cases} -\dot{P}_{11}(t) = D^T(t)R_3(t)D(t) - P_{11}(t)B(t)R_2^{-1}(t)B^T(t)P_{11}(t) \\ \qquad\qquad + A^T(t)P_{11}(t) + P_{11}(t)A(t), \\ P_{11}(t_1) = P_1, \end{cases} \quad \text{3-362}$$

$$\begin{cases} -\dot{P}_{12}(t) = P_{11}(t)D_d(t) + [A(t) - B(t)F_1(t)]^T P_{12}(t) + P_{12}(t)A_d(t), \\ P_{12}(t_1) = 0, \end{cases} \quad \text{3-363}$$

$$\begin{cases} -\dot{P}_{22}(t) = -P_{12}^T(t)B(t)R_2^{-1}(t)B^T(t)P_{12}(t) + D_d^T(t)P_{12}(t) + P_{12}^T(t)D_d(t) \\ \qquad\qquad + A_d^T(t)P_{22}(t) + P_{22}(t)A_d(t), \\ P_{22}(t_1) = 0. \end{cases} \quad \text{3-364}$$

We observe that P_{11}, and therefore also F_1, is completely independent of the properties of the disturbances, and is in fact obtained by solving the deterministic regulator problem with the disturbances omitted. Once P_{11} and F_1 have been found, **3-363** can be solved to determine P_{12} and from this F_2. The control system structure is given in Fig. 3.14. Apparently, the *feedback link*,

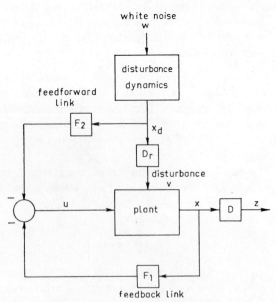

Fig. 3.14. Structure of the optimal state feedback regulator with disturbances.

3.6 Stochastic Regulator and Tracking Problems

that is, *the link from the state x to the input u is completely independent of the properties of the disturbances*. The *feedforward link*, that is, the link from the state of the disturbances x_d to the input u, is of course dependent upon the properties of the disturbances.

A similar conclusion can be reached for optimal tracking problems. Here it turns out that with the structures 3-329 and 3-335 of the state differential and output equations the feedback gain matrix can be partitioned as

$$F^0(t) = (F_1(t), -F_2(t)) \quad \text{3-365}$$

(note the minus sign that has been introduced), where

$$F_1(t) = R_2^{-1}(t)B^T(t)P_{11}(t),$$
$$F_2(t) = -R_2^{-1}(t)B^T(t)P_{12}(t). \quad \text{3-366}$$

Here the matrices P_{11}, P_{12}, and P_{22} are obtained by partitioning the matrix P according to the partitioning $\tilde{x}(t) = \text{col}\,[x(t), x_r(t)]$; they satisfy the matrix differential equations

$$\begin{cases} -\dot{P}_{11}(t) = D^T(t)R_3(t)D(t) - P_{11}(t)B(t)R_2^{-1}(t)B^T(t)P_{11}(t) \\ \qquad\qquad\qquad + A^T(t)P_{11}(t) + P_{11}(t)A(t), \\ P(t_1) = 0, \end{cases} \quad \text{3-367}$$

$$\begin{cases} -\dot{P}_{12}(t) = -D^T(t)R_3(t)D_r(t) + [A(t) - B(t)F_1(t)]^T P_{12}(t) \\ \qquad\qquad\qquad + P_{12}(t)A_r(t), \\ P_{12}(t_1) = 0, \end{cases} \quad \text{3-368}$$

$$\begin{cases} -\dot{P}_{22}(t) = D_r^T(t)R_3(t)D_r(t) - P_{12}^T(t)B(t)R_2^{-1}(t)B^T(t)P_{12}(t) \\ \qquad\qquad\qquad + A_r^T(t)P_{22}(t) + P_{22}(t)A_r(t), \\ P_{22}(t_1) = 0. \end{cases} \quad \text{3-369}$$

We conclude that for the optimal tracking system as well *the feedback link is independent of the properties of the reference variable*, while the *feedforward link* is of course influenced by the properties of the reference variable. A schematic representation of the optimal tracking system is given in Fig. 3.15.

Let us now return to the general stochastic optimal regulator problem. In practice we are usually confronted with control periods that are very long, which means that we are interested in the case where $t_1 \to \infty$. In the deterministic regulator problem, we saw that normally the Riccati equation 3-346 has a steady-state solution $\bar{P}(t)$ as $t_1 \to \infty$, and that the corresponding steady-state control law $\bar{F}(t)$ is optimal for half-infinite control periods. It is not difficult to conjecture (Kushner, 1971) that the steady-state control law

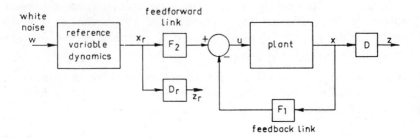

Fig. 3.15. Structure of the optimal state feedback tracking system.

is optimal for the stochastic regulator in the sense that it minimizes

$$\lim_{t_1 \to \infty} \frac{1}{t_1 - t_0} E\left\{\int_{t_0}^{t_1} [z^T(t)R_3(t)z(t) + u^T(t)R_2(t)u(t)] \, dt\right\}, \qquad 3\text{-}370$$

if this expression exists for the steady-state control law, with respect to all other control laws for which **3-370** exists. For the steady-state optimal control law, the criterion **3-370** is given by

$$\lim_{t_1 \to \infty} \frac{1}{t_1 - t_0} \int_{t_0}^{t_1} \operatorname{tr} [\bar{P}(t)V(t)] \, dt, \qquad 3\text{-}371$$

if it exists (compare **3-349**). Moreover, it is recognized that for a time-invariant stochastic regulator problem and an asymptotically stable time-invariant control law the expression **3-370** is equal to

$$\lim_{t \to \infty} E\{z^T(t)R_3 z(t) + u^T(t)R_2 u(t)\}. \qquad 3\text{-}372$$

From this it immediately follows that the steady-state optimal control law minimizes **3-372** with respect to all other time-invariant control laws. We see from **3-371** that the minimal value of **3-372** is given by

$$\operatorname{tr} (\bar{P}V). \qquad 3\text{-}373$$

We observe that if $R_3 = W_e$ and $R_2 = \rho W_u$, where W_e and W_u are the weighting matrices in the mean square tracking error and the mean square input (as introduced in Section 2.5.1), the expression **3-372** is precisely

$$C_{e\infty} + \rho C_{u\infty}. \qquad 3\text{-}374$$

Here $C_{e\infty}$ is the steady-state mean square tracking error and $C_{u\infty}$ the steady-state mean square input. To compute $C_{e\infty}$ and $C_{u\infty}$ *separately*, as usually is required, it is necessary to set up the complete closed-loop system equations and derive from these the differential equation for the variance matrix of the state. From this variance matrix all mean square quantities of interest can be obtained.

Example 3.13. *Stirred tank regulator*

In Example 3.11 we described a stochastic regulator problem arising from the stirred tank problem. Let us, in addition to the numerical values of Example 1.2 (Section 1.2.3), assume the following values:

$$\theta_1 = 40 \text{ s},$$
$$\theta_2 = 50 \text{ s},$$
$$\sigma_1 = 0.1 \text{ kmol/m}^3,$$
$$\sigma_2 = 0.2 \text{ kmol/m}^3.$$

3-375

Just as in Example 3.9 (Section 3.4.1), we choose the weighting matrices R_3 and R_2 as follows.

$$R_3 = \begin{pmatrix} 50 & 0 \\ 0 & 0.02 \end{pmatrix}, \quad R_2 = \rho \begin{pmatrix} \frac{1}{3} & 0 \\ 0 & 3 \end{pmatrix},$$

3-376

where ρ is to be selected. The optimal control law has been computed for $\rho = 10$, 1, and 0.1, as in Example 3.9, but the results are not listed here. It turns out, of course, that the feedback gains from the plant state variables are not affected by the inclusion of the disturbances in the system model. This means that the closed-loop poles are precisely those listed in Table 3.1.

In order to evaluate the detailed performance of the system, the steady-state variance matrix

$$\bar{Q} = \lim_{t \to \infty} E\{x(t)x^T(t)\}$$

3-377

has been computed from the matrix equation

$$0 = (A - B\bar{F})\bar{Q} + \bar{Q}(A - B\bar{F})^T + V.$$

3-378

The steady-state variance matrix of the input can be found as follows:

$$\lim_{t \to \infty} E\{u(t)u^T(t)\} = \lim_{t \to \infty} E\{\bar{F}x(t)x^T(t)\bar{F}^T\} = \bar{F}\bar{Q}\bar{F}^T.$$

3-379

From these variance matrices the rms values of the components of the controlled variable and the input variable are easily obtained. Table 3.2 lists the results. The table shows very clearly that as ρ decreases the fluctuations in the outgoing concentration become more and more reduced. The fluctuations in the outgoing flow caused by the control mechanism also eventually decrease with ρ. All this happens of course at the expense of an increase in the fluctuations in the incoming feeds. Practical considerations must decide which value of ρ is most suitable.

Table 3.2 Rms Values for Stirred-Tank Regulator

	Steady-state rms values of			
	Incremental outgoing flow (m³/s)	Incremental concentration (kmol/m³)	Incremental feed	
ρ			No. 1 (m³/s)	No. 2 (m³/s)
∞	0	0.06124	0	0
10	0.0001038	0.03347	0.0008957	0.0006980
1	0.00003303	0.008238	0.001567	0.001487
0.1	0.000004967	0.001127	0.001769	0.001754

Example 3.14. *Angular velocity tracking system*

Let us consider the angular velocity tracking problem as outlined in Example 3.12. To solve this problem we exploit the special structure of the tracking problem. It follows from **3-365** that the optimal tracking law is given by

$$\mu(t) = -F_1(t)\xi(t) + F_2(t)\xi_r(t). \quad \textbf{3-380}$$

The feedback gain $F_1(t)$ is independent of the properties of the reference variable and in fact has already been computed in previous examples where we considered the angular velocity regulation problem. From Example 3.7 (Section 3.4.1), it follows that the steady-state value of the feedback gain is given by

$$\bar{F}_1 = \frac{1}{\kappa}\left(-\alpha + \sqrt{\alpha^2 + \frac{\kappa^2}{\rho}}\right), \quad \textbf{3-381}$$

while the steady-state value of P_{11} is

$$\bar{P}_{11} = \frac{\rho}{\kappa^2}\left(-\alpha + \sqrt{\alpha^2 + \frac{\kappa^2}{\rho}}\right). \quad \textbf{3-382}$$

By using **3-368**, it follows that the steady-state value of P_{12} can be solved from

$$0 = -1 - \sqrt{\alpha^2 + \frac{\kappa^2}{\rho}}\,\bar{P}_{12} - \frac{1}{\theta}\bar{P}_{12}. \quad \textbf{3-383}$$

Solution yields

$$\bar{P}_{12} = -\frac{1}{\frac{1}{\theta} + \sqrt{\alpha^2 + \frac{\kappa^2}{\rho}}}, \quad \textbf{3-384}$$

3.6 Stochastic Regulator and Tracking Problems

so that

$$\bar{F}_2 = \frac{\dfrac{\kappa}{\rho}}{\dfrac{1}{\theta} + \sqrt{\alpha^2 + \dfrac{\kappa^2}{\rho}}}.\qquad\text{3-385}$$

Finally, solution of **3-369** for $\bar{P}_{22}$ gives

$$\bar{P}_{22} = \frac{\theta}{2} \cdot \frac{\alpha^2 + \dfrac{1}{\theta^2} + \dfrac{2}{\theta}\sqrt{\alpha^2 + \dfrac{\kappa^2}{\rho}}}{\left(\dfrac{1}{\theta} + \sqrt{\alpha^2 + \dfrac{\kappa^2}{\rho}}\right)^2}.\qquad\text{3-386}$$

Let us choose the following numerical values:

$$\begin{aligned}\alpha &= 0.5 \text{ s}^{-1},\\ \kappa &= 150 \text{ rad}/(\text{V s}^2),\\ \theta &= 1 \text{ s},\\ \sigma &= 30 \text{ rad/s},\\ \rho &= 1000 \text{ rad}^2/(\text{V}^2 \text{ s}^2).\end{aligned}\qquad\text{3-387}$$

This yields the following numerical results:

$$\bar{F}_1 = 0.02846, \qquad \bar{F}_2 = 0.02600,\qquad\text{3-388}$$

$$\bar{P} = \begin{pmatrix} 0.1897 & -0.1733 \\ -0.1733 & 0.1621 \end{pmatrix}.\qquad\text{3-389}$$

From **3-373** it follows that

$$\lim_{t \to \infty} [E\{\tilde{\zeta}^2(t)\} + \rho E\{\mu^2(t)\}] = \text{tr}\,(\bar{P}V),\qquad\text{3-390}$$

where $\tilde{\zeta}(t) = \xi(t) - \xi_r(t)$. Since in the present problem

$$V = \begin{pmatrix} 0 & 0 \\ 0 & \dfrac{2\sigma^2}{\theta} \end{pmatrix} = \begin{pmatrix} 0 & 0 \\ 0 & 1800 \end{pmatrix},\qquad\text{3-391}$$

we find that

$$\lim_{t \to \infty} [E\{\tilde{\zeta}^2(t)\} + \rho E\{\mu^2(t)\}] = 291.8 \text{ rad}^2/\text{s}^2.\qquad\text{3-392}$$

We can use **3-392** to obtain rough estimates of the rms tracking error and rms input voltage as follows. First, we have from **3-392**

$$\lim_{t \to \infty} E\{\tilde{\zeta}^2(t)\} < 291.8 \text{ rad}^2/\text{s}^2.\qquad\text{3-393}$$

It follows that

$$\text{steady-state rms tracking error} < 17.08 \text{ rad/s.} \quad \textbf{3-394}$$

Similarly, it follows from **3-392**

$$\lim_{t \to \infty} E\{\mu^2(t)\} < \frac{291.8}{\rho} = 0.2918 \text{ V}^2. \quad \textbf{3-395}$$

We conclude that

$$\text{steady-state rms input voltage} < 0.5402 \text{ V.} \quad \textbf{3-396}$$

Exact values for the rms tracking error and rms input voltage can be found by computing the steady-state variance matrix of the state $\tilde{x}(t)$ of the closed-loop augmented system. This system is described by the equation

$$\dot{\tilde{x}}(t) = \begin{pmatrix} -\alpha & 0 \\ 0 & -\dfrac{1}{\theta} \end{pmatrix} \tilde{x}(t) + \begin{pmatrix} \kappa \\ 0 \end{pmatrix}(-\bar{F}_1, \bar{F}_2)\tilde{x}(t) + \begin{pmatrix} 0 \\ 1 \end{pmatrix} w(t), \quad \textbf{3-397}$$

or

$$\dot{\tilde{x}}(t) = \begin{pmatrix} -\alpha - \kappa \bar{F}_1 & \kappa \bar{F}_2 \\ 0 & -\dfrac{1}{\theta} \end{pmatrix} \tilde{x}(t) + \begin{pmatrix} 0 \\ 1 \end{pmatrix} w(t). \quad \textbf{3-398}$$

As a result, the steady-state variance matrix $\bar{Q}$ of $\tilde{x}(t)$, is the solution of the matrix equation

$$0 = \begin{pmatrix} -\alpha - \kappa \bar{F}_1 & \kappa \bar{F}_2 \\ 0 & -\dfrac{1}{\theta} \end{pmatrix} \bar{Q} + \bar{Q} \begin{pmatrix} -\alpha - \kappa \bar{F}_1 & 0 \\ \kappa \bar{F}_2 & -\dfrac{1}{\theta} \end{pmatrix} + \begin{pmatrix} 0 & 0 \\ 0 & \dfrac{2\sigma^2}{\theta} \end{pmatrix}.$$

$$\textbf{3-399}$$

Numerical solution yields

$$\bar{Q} = \begin{pmatrix} 497.5 & 608.4 \\ 608.4 & 900.0 \end{pmatrix}. \quad \textbf{3-400}$$

The steady-state mean square tracking error can be expressed as

$$\lim_{t \to \infty} E\{[\xi(t) - \xi_r(t)]^2\} = \bar{Q}_{11} - 2\bar{Q}_{12} + \bar{Q}_{22}$$

$$= 180.7 \text{ rad}^2/\text{s}^2, \quad \textbf{3-401}$$

where the $\bar{Q}_{ij}$ are the entries of $\bar{Q}$. Similarly, the mean square input is given by

$$\lim_{t \to \infty} E\{[-\bar{F}_1\xi(t) + \bar{F}_2\xi_r(t)]^2\} = \bar{F}_1^2\bar{Q}_{11} - 2\bar{F}_1\bar{F}_2\bar{Q}_{12} + \bar{F}_2^2\bar{Q}_{22}$$

$$= 0.1110 \text{ V}^2. \quad \textbf{3-402}$$

3.6 Stochastic Regulator and Tracking Problems

In Table 3.3 the estimated and actual rms values are compared. Also given are the open-loop rms values, that is, the rms values without any control at all. It is seen that the estimated rms tracking error and input voltage are a little on the large side, but that they give a very good indication of the orders of magnitude. We moreover see that the control is not very good since the rms tracking error of 13.44 rad/s is not small as compared to the rms value of the

Table 3.3 Numerical Results for the Angular Velocity Tracking System

	Steady-state rms tracking error (rad/s)	Steady-state rms input voltage (V)
Open-loop	30	0
Estimated closed-loop	<17.08	<0.5402
Actual closed-loop	13.44	0.3333

reference variable of 30 rad/s. Since the rms input is quite small, however, there seems to be room for considerable improvement. This can be achieved by choosing the weighting coefficient ρ much smaller (see Problem 3.5).

Let us check the reference variable and closed-loop system bandwidths for the present example. The reference variable break frequency is $1/\theta = 1$ rad/s. Substituting the control law into the system equation, we find for the closed-loop system equation

$$\dot{\xi}(t) = -\sqrt{\alpha^2 + \frac{\kappa^2}{\rho}}\,\xi(t) + \bar{F}_2\kappa\xi_r(t). \qquad \textbf{3-403}$$

This is a first-order system with break frequency

$$\sqrt{\alpha^2 + \frac{\kappa^2}{\rho}} = 4.769 \text{ rad/s.} \qquad \textbf{3-404}$$

Since the power spectral density of the reference variable, which is exponentially correlated noise, decreases relatively slowly with increasing frequency, the difference in break frequencies of the reference variable and the closed-loop system is not large enough to obtain a sufficiently small tracking error.

3.7 REGULATORS AND TRACKING SYSTEMS WITH NONZERO SET POINTS AND CONSTANT DISTURBANCES

3.7.1 Nonzero Set Points

In our discussion of regulator and tracking problems, we have assumed up to this point that the zero state is always the desired equilibrium state of the system. In practice, it is nearly always true, however, that the desired equilibrium state, which we call the *set point* of the state, is a constant point in state space, different from the origin. This kind of discrepancy can be removed by shifting the origin of the state space to this point, and this is what we have always done in our examples. This section, however, is devoted to the case where the set point may be variable; that is, we assume that the set point is constant over long periods of time but that from time to time it is shifted. This is a common situation in practice.

We limit our discussion to the time-invariant case. Consider the linear time-invariant system with state differential equation

$$\dot{x}(t) = Ax(t) + Bu(t), \qquad \text{3-405}$$

where the controlled variable is given by

$$z(t) = Dx(t). \qquad \text{3-406}$$

Let us suppose that the set point of the controlled variable is given by z_0. Then in order to maintain the system at this set point, a constant input u_0 must be found (diCaprio and Wang, 1969) that holds the state at a point x_0 such that

$$z_0 = Dx_0. \qquad \text{3-407}$$

It follows from the state differential equation that x_0 and u_0 must be related by

$$0 = Ax_0 + Bu_0. \qquad \text{3-408}$$

Whether or not the system can be maintained at the given set point depends on whether **3-407** and **3-408** can be solved for u_0 for the given value of z_0. We return to this question, but let us suppose for the moment that a solution exists. Then we define the *shifted input*, the *shifted state*, and the *shifted controlled variable*, respectively, as

$$\begin{aligned} u'(t) &= u(t) - u_0, \\ x'(t) &= x(t) - x_0, \\ z'(t) &= z(t) - z_0. \end{aligned} \qquad \text{3-409}$$

3.7 Nonzero Set Points and Constant Disturbances

It is not difficult to find, by solving these equations for u, x, and z, substituting the result into the state differential equation **3-405** and the output equation **3-406**, and using **3-407** and **3-408**, that the shifted variables satisfy the equations

$$\dot{x}'(t) = Ax(t) + Bu'(t),$$
$$z'(t) = Dx'(t).$$
$$\text{3-410}$$

Suppose now that at a given time the set point is suddenly shifted from one value to another. Then in terms of the shifted system equations **3-410**, the system suddenly acquires a nonzero initial state. In order to let the system achieve the new set point in an orderly fashion we propose to effect the transition such that an optimization criterion of the form

$$\int_{t_0}^{t_1} [z'^T(t)R_3 z'(t) + u'^T(t)R_2 u'(t)]\, dt + x'^T(t_1)P_1 x'(t_1) \quad \text{3-411}$$

is minimized. Let us assume that this shifted regulator problem possesses a steady-state solution in the form of the time-invariant asymptotically stable steady-state control law

$$u'(t) = -\bar{F}x'(t). \quad \text{3-412}$$

Application of this control law ensures that, in terms of the original system variables, the system is transferred to the new set point as quickly as possible without excessively large transient input amplitudes.

Let us see what form the control law takes in terms of the original system variables. We write from **3-412** and **3-409**:

$$u(t) = -\bar{F}x(t) + u_0 + \bar{F}x_0. \quad \text{3-413}$$

This shows that the control law is of the form

$$u(t) = -\bar{F}x(t) + u_0', \quad \text{3-414}$$

where the constant vector u_0' is to be determined such that in the steady-state situation the controlled variable $z(t)$ assumes the given value z_0. We now study the question under what conditions u_0' can be found.

Substitution of **3-414** into the system state differential equation yields

$$\dot{x}(t) = (A - B\bar{F})x(t) + Bu_0'. \quad \text{3-415}$$

Since the closed-loop system is asymptotically stable, as $t \to \infty$ the state reaches a steady-state values x_0 that satisfies

$$0 = \bar{A}x_0 + Bu_0'. \quad \text{3-416}$$

Here we have abbreviated

$$\bar{A} = A - B\bar{F}. \quad \text{3-417}$$

Since the closed-loop system is asymptotically stable, $\bar{A}$ has all of its characteristic values in the left-half complex plane and is therefore nonsingular; consequently, we can solve **3-416** for x_0:

$$x_0 = (-\bar{A})^{-1} B u_0'. \qquad \textbf{3-418}$$

If the set point z_0 of the controlled variable is to be achieved, we must therefore have

$$z_0 = D(-\bar{A})^{-1} B u_0'. \qquad \textbf{3-419}$$

When considering the problem of solving this equation for u_0' for a given value of z_0, three cases must be distinguished:

(a) *The dimension of z is greater than that of u:* Then **3-419** has a solution for special values of z_0 only; in general, no solution exists. In this case we attempt to control the variable $z(t)$ with an input $u(t)$ of smaller dimension; since we have too few degrees of freedom, it is not surprising that no solution can generally be found.

(b) *The dimensions of u and z are the same*, that is, a sufficient number of degrees of freedom is available to control the system. In this case **3-419** can be solved for u_0' provided $D(-\bar{A})^{-1}B$ is nonsingular; assuming this to be the case (we shall return to this), we find

$$u_0' = [D(-\bar{A})^{-1}B]^{-1} z_0, \qquad \textbf{3-420}$$

which yields for the optimal input to the tracking system

$$u(t) = -\bar{F}x(t) + [D(-\bar{A})^{-1}B]^{-1} z_0. \qquad \textbf{3-421}$$

(c) *The dimension of z is less than that of u:* In this case there are too many degrees of freedom and **3-419** has many solutions. We can choose one of these solutions, but it is more advisable to reformulate the tracking problem by adding components to the controlled variable.

On the basis of these considerations, we henceforth assume that

$$\dim(z) = \dim(u), \qquad \textbf{3-422}$$

so that case (b) applies. We see that

$$D(-\bar{A})^{-1}B = H_c(0), \qquad \textbf{3-423}$$

where

$$H_c(s) = D(sI - \bar{A})^{-1}B. \qquad \textbf{3-424}$$

We call $H_c(s)$ the *closed-loop transfer matrix*, since it is the transfer matrix from $u'(t)$ to $z(t)$ for the system

$$\begin{aligned}\dot{x}(t) &= Ax(t) + Bu(t), \\ z(t) &= Dx(t), \\ u(t) &= -\bar{F}x(t) + u'(t).\end{aligned} \qquad \textbf{3-425}$$

3.7 Nonzero Set Points and Constant Disturbances

In terms of $H_c(0)$ the optimal control law **3-421** can be written as

$$u(t) = -\bar{F}x(t) + H_c^{-1}(0)z_0. \qquad \textbf{3-426}$$

As we have seen, this control law has the property that after a step change in the set point z_0 the system is transferred to the new set point as quickly as possible without excessively large transient input amplitudes. Moreover, this control law of course makes the system return to the set point from any initial state in an optimal manner. We call **3-426** the *nonzero set point optimal control law*. It has the property that it statically decouples the control system, that is, the transmission $T(s)$ of the control system (the transfer matrix from the set point z_0 to the controlled variable z) has the property that $T(0) = I$.

We now study the question under what conditions $H_c(0)$ has an inverse. It will be proved that this property can be directly ascertained from the open-loop system equations

$$\begin{aligned}\dot{x}(t) &= Ax(t) + Bu(t), \\ z(t) &= Dx(t).\end{aligned} \qquad \textbf{3-427}$$

Consider the following string of equalities

$$\begin{aligned}\det[H_c(s)] &= \det[D(sI - A + B\bar{F})^{-1}B] \\ &= \det[D(sI - A)^{-1}\{I + B\bar{F}(sI - A)^{-1}\}^{-1}B] \\ &= \det[D(sI - A)^{-1}\{I - B\bar{F}[I + (sI - A)^{-1}B\bar{F}]^{-1}(sI - A)^{-1}\}B] \\ &= \det[D(sI - A)^{-1}B]\det[I - \bar{F}(sI - A + B\bar{F})^{-1}B] \\ &= \det[D(sI - A)^{-1}B]\det[I - (sI - A + B\bar{F})^{-1}B\bar{F}] \\ &= \det[D(sI - A)^{-1}B]\det[(sI - A + B\bar{F})^{-1}]\det(sI - A) \\ &= \frac{\det[D(sI - A)^{-1}B]\det(sI - A)}{\det(sI - A + B\bar{F})} \\ &= \frac{\psi(s)}{\phi_c(s)}. \qquad \textbf{3-428}\end{aligned}$$

Here we have used Lemma 1.1 (Section 1.5.3) twice. The polynomial $\psi(s)$ is defined by

$$\det[H(s)] = \frac{\psi(s)}{\phi(s)}, \qquad \textbf{3-429}$$

where $H(s)$ is the open-loop transfer matrix

$$H(s) = D(sI - A)^{-1}B, \qquad \textbf{3-430}$$

and $\phi(s)$ the open-loop characteristic polynomial

$$\phi(s) = \det(sI - A). \qquad \text{3-431}$$

Finally, $\phi_c(s)$ is the closed-loop characteristic polynomial

$$\phi_c(s) = \det(sI - A + B\bar{F}). \qquad \text{3-432}$$

We see from **3-428** that the zeroes of the closed-loop transfer matrix are the same as those of the open-loop transfer matrix. We also see that

$$\det[D(-\bar{A})^{-1}B] = \det[H_c(0)] = \frac{\psi(0)}{\phi_c(0)} \qquad \text{3-433}$$

is zero if and only if $\psi(0) = 0$. Thus the condition $\psi(0) \neq 0$ guarantees that $D(-\bar{A})^{-1}B$ is nonsingular, hence that the nonzero set point control law exists. These results can be summarized as follows.

Theorem 3.10. *Consider the time-invariant system*

$$\begin{aligned}\dot{x}(t) &= Ax(t) + Bu(t), \\ z(t) &= Dx(t),\end{aligned} \qquad \text{3-434}$$

where z and u have the same dimensions. Consider any asymptotically stable time-invariant control law

$$u(t) = -Fx(t) + u'(t). \qquad \text{3-435}$$

Let $H(s)$ be the open-loop transfer matrix

$$H(s) = D(sI - A)^{-1}B, \qquad \text{3-436}$$

and $H_c(s)$ the closed-loop transfer matrix

$$H_c(s) = D(sI - A + BF)^{-1}B. \qquad \text{3-437}$$

Then $H_c(0)$ is nonsingular and the controlled variable $z(t)$ can under steady-state conditions be maintained at any constant value z_0 by choosing

$$u'(t) = H_c^{-1}(0)z_0 \qquad \text{3-438}$$

if and only if $H(s)$ has a nonzero numerator polynomial that has no zeroes at the origin.

It is noted that the theorem is stated for any asymptotically stable control law and not only for the steady-state optimal control law.

The discussion of this section has been confined to deterministic regulators. Of course stochastic regulators (including tracking problems) can also have nonzero set points. The theory of this section applies to stochastic regulators

3.7 Nonzero Set Points and Constant Disturbances

without modification; the nonzero set point optimal control law for the stochastic regulator is also given by

$$u(t) = -\bar{F}x(t) + H_c^{-1}(0)z_0. \qquad 3\text{-}439$$

Example 3.15. *Position control system*

Let us consider the position control system of Example 3.4 (Section 3.3.1). In Example 3.8 (Section 3.4.1), we found the optimal steady-state control law. It is not difficult to find from the results of Example 3.8 that the closed-loop transfer function is given by

$$H_c(s) = \frac{\kappa}{s^2 + s\sqrt{\alpha^2 + \dfrac{2\kappa}{\sqrt{\rho}}} + \dfrac{\kappa}{\sqrt{\rho}}}. \qquad 3\text{-}440$$

If follows from **3-435** and **3-438** that the nonzero set point optimal control law is given by

$$\mu(t) = -\bar{F}x(t) + \frac{1}{H_c(0)}\zeta_0$$

$$= -\frac{1}{\sqrt{\rho}}\xi_1(t) - \frac{1}{\kappa}\left(-\alpha + \sqrt{\alpha^2 + \frac{2\kappa}{\sqrt{\rho}}}\right)\xi_2(t) + \frac{1}{\sqrt{\rho}}\zeta_0, \qquad 3\text{-}441$$

where ζ_0 is the set point for the angular position. This is precisely the control law **3-171** that we found in Example 3.8 from elementary considerations.

Example 3.16. *Stirred tank*

As an example of a multivariable system, we consider the stirred-tank regulator problem of Example 3.9 (Section 3.4.1). For $\rho = 1$ (where ρ is defined as in Example 3.9), the regulator problem yields the steady-state feedback gain matrix

$$\bar{F} = \begin{pmatrix} 0.1009 & -0.09708 \\ 0.01681 & 0.05475 \end{pmatrix}. \qquad 3\text{-}442$$

It is easily found that the corresponding closed-loop transfer matrix is given by

$$H_c(s) = D(sI - A + B\bar{F})^{-1}B$$

$$= \frac{1}{s^2 + 0.2131s + 0.01037}$$

$$\cdot \begin{pmatrix} 0.01s + 0.0007475 & 0.01s + 0.001171 \\ -0.25s - 0.01931 & 0.75s + 0.1084 \end{pmatrix}. \qquad 3\text{-}443$$

276   Optimal Linear State Feedback Control Systems

From this the nonzero set point optimal control law can be found to be

$$u(t) = -\bar{F}x(t) + \begin{pmatrix} 10.84 & -0.1171 \\ 1.931 & 0.07475 \end{pmatrix} z_0. \qquad \textbf{3-444}$$

Figure 3.16 gives the response of the closed-loop system to step changes in the components of the set point z_0. Here the set point of the outgoing flow is

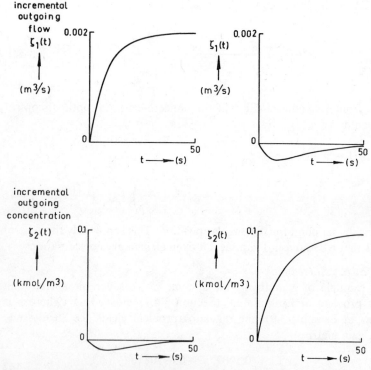

Fig. 3.16. The responses of the stirred tank as a nonzero set point regulating system. Left column: Responses of the incremental outgoing flow and concentration to a step of $0.002 \, m^3/s$ in the set point of the flow. Right column: Responses of the incremental outgoing flow and concentration to a step of $0.1 \, kmol/m^3$ in the set point of the concentration.

changed by $0.002 \, m^3/s$, which amounts to 10% of the nominal value, while the set point of the outgoing concentration is changed by $0.1 \, kmol/m^3$, which is 8% of the nominal value. We note that the control system exhibits a certain amount of dynamic *coupling* or *interaction*, that is, a change in the set point of one of the components of the controlled variable transiently affects the other component. The effect is small, however.

3.7.2* Constant Disturbances

In this subsection we discuss a method for counteracting the effect of constant disturbances in time-invariant regulator systems. As we saw in Chapter 2, in regulators and tracking systems where high precision is required, it is important to eliminate the effect of constant disturbances completely. This can be done by the application of integrating action. We introduce integrating action in the context of state feedback control by first extending the usual regulator problem, and then consider the effect of constant disturbances in the corresponding modified closed-loop control system configuration.

Consider the time-invariant system with state differential equation

$$\dot{x}(t) = Ax(t) + Bu(t), \qquad \text{3-445}$$

with $x(t_0)$ given and with the controlled variable

$$z(t) = Dx(t). \qquad \text{3-446}$$

We add to the system variables the "integral state" $q(t)$ (Newell and Fisher, 1971; Shih, 1970; Porter, 1971), defined by

$$\dot{q}(t) = z(t), \qquad \text{3-447}$$

with $q(t_0)$ given. One can now consider the problem of minimizing a criterion of the form

$$\int_{t_0}^{\infty} [z^T(t)R_3 z(t) + q^T(t)R_3' q(t) + u^T(t)R_2 u(t)] \, dt, \qquad \text{3-448}$$

where R_3, R_3', and R_2 are suitably chosen weighting matrices. The first term of the integrand forces the controlled variable to zero, while the second term forces the integral state, that is, the total area under the response of the controlled variable, to go to zero. The third term serves, as usual, to restrict the input amplitudes.

Let us assume that by minimizing an expression of the form **3-448**, or by any other method, a time-invariant control law

$$u(t) = -F_1 x(t) - F_2 q(t) \qquad \text{3-449}$$

is determined that stabilizes the augmented system described by **3-445**, **3-446**, and **3-447**. (We defer for a moment the question under which conditions such an asymptotically stable control law exists.) Suppose now that a constant disturbance occurs in the system, so that we must replace the state differential equation **3-445** with

$$\dot{x}(t) = Ax(t) + Bu(t) + v_0, \qquad \text{3-450}$$

where v_0 is a constant vector. Since the presence of the constant disturbance does not affect the asymptotic stability of the system, we have

$$\lim_{t \to \infty} \dot{q}(t) = 0, \qquad 3\text{-}451$$

or, from **3-447**,

$$\lim_{t \to \infty} z(t) = 0. \qquad 3\text{-}452$$

This means that *the control system with the asymptotically stable control law* **3-449** *has the property that the effect of constant disturbances on the controlled variable eventually vanishes*. Since this is achieved by the introduction of the integral state q, this control scheme is a form of integral control. Figure 3.17 depicts the integral control scheme.

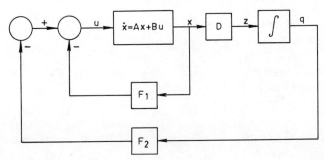

Fig. 3.17. State feedback integral control.

Let us now consider the mechanism that effects the suppression of the constant disturbance. The purpose of the multivariable integration of **3-447** is to generate a constant contribution u_0 to the input that counteracts the effect of the constant disturbance on the controlled variable. Thus let us consider the response of the system **3-450** to the input

$$u(t) = -F_1 x(t) + u_0. \qquad 3\text{-}453$$

Substitution of this expression into the state differential equation **3-450** yields

$$\dot{x}(t) = (A - BF_1)x(t) + Bu_0 + v_0. \qquad 3\text{-}454$$

In equilibrium conditions the state assumes a constant value x_0 that must satisfy the relation

$$0 = \bar{A}x_0 + Bu_0 + v_0, \qquad 3\text{-}455$$

where

$$\bar{A} = A - BF_1. \qquad 3\text{-}456$$

Solution for x_0 yields

$$x_0 = (-\bar{A})^{-1} Bu_0 + (-\bar{A})^{-1} v_0, \qquad 3\text{-}457$$

provided $\bar{A}$ is nonsingular. The corresponding equilibrium value z_0 of the controlled variable is given by

$$z_0 = Dx_0 = D(-\bar{A})^{-1}Bu_0 + D(-\bar{A})^{-1}v_0. \qquad \textbf{3-458}$$

When we now consider the question whether or not a value of u_0 exists that makes $z_0 = 0$, we obviously obtain the same conditions as in Section 3.7.1, broken down to the three following cases.

(a) *The dimension of z is greater than that of u:* In this case the equation

$$0 = D(-\bar{A})^{-1}Bu_0 + D(-\bar{A})^{-1}v_0 \qquad \textbf{3-459}$$

represents more equations than variables, which means that in general no solution exists. The number of degrees of freedom is too small, and the steady-state error in z cannot be eliminated.

(b) *The dimension of z equals that of u:* In this case a solution exists if and only if

$$D(-\bar{A})^{-1}B = H_c(0) \qquad \textbf{3-460}$$

is nonsingular, where

$$H_c(s) = D(sI - \bar{A})^{-1}B \qquad \textbf{3-461}$$

is the closed-loop transfer matrix. As we saw in Theorem 3.10, $H_c(0)$ is nonsingular if and only if the open-loop transfer matrix $H(s) = D(sI - A)^{-1}B$ has no zeroes at the origin.

(c) *The dimension of z is less than that of u:* In this case there are too many degrees of freedom and the dimension of z can be increased by adding components to the controlled variable.

On the basis of these considerations, we from now on restrict ourselves to the case where dim (z) = dim (u). Then the present analysis shows that a *necessary* condition for the successful operation of the integral scheme under consideration is that the open-loop transfer matrix $H(s) = D(sI - A)^{-1}B$ have no zeroes at the origin. In fact, it can be shown, by a slight extension of the argument of Power and Porter (1970) involving the controllability canonical form of the system **3-445**, that necessary and sufficient conditions for the existence of an asymptotically stable control law of the form **3-449** are that

(i) the system **3-445** is stabilizable; and
(ii) the open-loop transfer matrix $H(s) = D(sI - A)^{-1}B$ has no zeroes at the origin.

Power and Porter (1970) and Davison and Smith (1971) prove that necessary and sufficient conditions for arbitrary placement of the closed-loop system poles are that the system **3-445** be completely controllable and that the open-loop transfer matrix have no zeroes at the origin. Davison and Smith (1971) state the latter condition in an alternative form.

280 Optimal Linear State Feedback Control Systems

In the literature alternative approaches to determining integral control schemes can be found (see, e.g., Anderson and Moore, 1971, Chapter 10; Johnson, 1971b).

Example 3.17. *Integral control of the positioning system*

Let us consider the positioning system of previous examples and assume that a constant disturbance can enter into the system in the form of a constant torque τ_0 on the shaft of the motor. We thus modify the state differential equation 3-59 to

$$\dot{x}(t) = \begin{pmatrix} 0 & 1 \\ 0 & -\alpha \end{pmatrix} x(t) + \begin{pmatrix} 0 \\ \kappa \end{pmatrix} \mu(t) + \begin{pmatrix} 0 \\ \gamma \end{pmatrix} \tau_0, \qquad 3\text{-}462$$

where $\gamma = 1/J$, with J the moment of inertia of all the rotating parts. As before, the controlled variable is given by

$$\zeta(t) = (1, 0)x(t). \qquad 3\text{-}463$$

We add to the system the scalar integral state $q(t)$, defined by

$$\dot{q}(t) = \zeta(t). \qquad 3\text{-}464$$

From Example 3.15 we know that the open-loop transfer function has no zeroes at the origin; moreover, the system is completely controllable so that we expect no difficulties in finding an integral control system. Let us consider the optimization criterion

$$\int_{t_0}^{\infty} [\zeta^2(t) + \lambda q^2(t) + \rho \mu^2(t)] \, dt. \qquad 3\text{-}465$$

As in previous examples, we choose

$$\rho = 0.00002 \text{ rad}^2/\text{V}^2. \qquad 3\text{-}466$$

Inspection of Fig. 3.9 shows that in the absence of integral control $q(t)$ will reach a steady-state value of roughly 0.01 rad s for the given initial condition. Choosing

$$\lambda = 10 \text{ s}^{-2} \qquad 3\text{-}467$$

can therefore be expected to affect the control scheme significantly.

Numerical solution of the corresponding regulator problem with the numerical values of Example 3.4 (Section 3.3.1) and $\gamma = 0.1 \text{ kg}^{-1} \text{ m}^{-2}$ yields the steady-state control law

$$\mu(t) = -F_1 x(t) - F_2 q(t), \qquad 3\text{-}468$$

with

$$F_1 = (299.8, 22.37),$$
$$F_2 = 707.1. \qquad 3\text{-}469$$

The corresponding closed-loop characteristic values are $-9.519 \pm j9.222$ s^{-1} and -3.168 s^{-1}. Upon comparison with the purely proportional scheme of Example 3.8 (Section 3.4,1), we note that the proportional part of the feedback, represented by F_1, has hardly changed (compare **3-169**), and that the corresponding closed-loop poles, which are $-9.658 \pm j9.094$ s^{-1} in Example 3.8 also have moved very little. Figure 3.18 gives the response of the integral

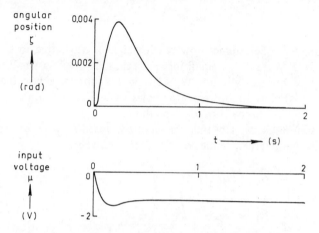

Fig. 3.18. Response of the integral position control system to a constant torque of 10 N m on the shaft of the motor.

control system from zero initial conditions to a constant torque τ_0 of 10 N m on the shaft of the motor. The maximum deviation of the angular displacement caused by this constant torque is about 0.004 rad.

3.8* ASYMPTOTIC PROPERTIES OF TIME-INVARIANT OPTIMAL CONTROL LAWS

3.8.1* Asymptotic Behavior of the Optimal Closed-Loop Poles

In Section 3.2 we saw that the stability of time-invariant linear state feedback control systems can be achieved or improved by assigning the closed-loop poles to suitable locations in the left-half complex plane. We were not able to determine which pole patterns are most desirable, however. In Sections 3.3 and 3.4, the theory of linear optimal state feedback control systems was developed. For time-invariant optimal systems, a question of obvious interest concerns the closed-loop pole patterns that result. This section is devoted to a study of these patterns. This will supply valuable information about the response that can be expected from optimal regulators.

282   Optimal Linear State Feedback Control Systems

Suppose that in the time-invariant regulator problem we let

$$R_2 = \rho N, \qquad \text{3-470}$$

where N is a positive-definite symmetric matrix and ρ a positive scalar. With this choice of R_2, the optimization criterion is given by

$$\int_{t_0}^{\infty} [z^T(t) R_3 z(t) + \rho u^T(t) N u(t)] \, dt. \qquad \text{3-471}$$

The parameter ρ determines how much weight is attributed to the input; a large value of ρ results in small input amplitudes, while a small value of ρ permits large input amplitudes. We study in this subsection how the locations of the optimal closed-loop regulator poles vary as a function of ρ. For this investigation we employ root locus methods.

In Section 3.4.4 we saw that the optimal closed-loop poles are the left-half plane characteristic values of the matrix Z, where

$$Z = \begin{pmatrix} A & -BR_2^{-1}B^T \\ -R_1 & -A^T \end{pmatrix} = \begin{pmatrix} A & -\dfrac{1}{\rho} BN^{-1}B^T \\ -D^T R_3 D & -A^T \end{pmatrix}. \qquad \text{3-472}$$

Using Lemma 1.2 (Section 1.5.4) and Lemma 1.1 (Section 1.5.3), we expand $\det(sI - Z)$ as follows:

$$\begin{aligned}
\det(sI - Z) &= \det \begin{pmatrix} sI - A & \dfrac{1}{\rho} BN^{-1}B^T \\ D^T R_3 D & sI + A^T \end{pmatrix} \\
&= \det(sI - A) \\
&\quad \cdot \det\left[(sI + A^T) - D^T R_3 D (sI - A)^{-1} \dfrac{1}{\rho} BN^{-1}B^T\right] \\
&= \det(sI - A) \det(sI + A^T) \\
&\quad \cdot \det\left[I - D^T R_3 D (sI - A)^{-1} \dfrac{1}{\rho} BN^{-1}B^T (sI + A^T)^{-1}\right] \\
&= \det(sI - A)(-1)^n \det(-sI - A) \\
&\quad \cdot \det\left[I + \dfrac{1}{\rho} N^{-1}B^T(-sI - A^T)^{-1} D^T R_3 D (sI - A)^{-1} B\right] \\
&= (-1)^n \phi(s)\phi(-s) \det\left[I + \dfrac{1}{\rho} N^{-1}H^T(-s) R_3 H(s)\right], \qquad \text{3-473}
\end{aligned}$$

where n is the dimension of the state x, and

$$\phi(s) = \det(sI - A),$$
$$H(s) = D(sI - A)^{-1}B. \qquad \textbf{3-474}$$

For simplicity, we first study the case where both the input u and the controlled variable z are scalars, while

$$R_3 = 1, \quad N = 1. \qquad \textbf{3-475}$$

We return to the multiinput multioutput case at the end of this section. It follows from **3-473** that in the single-input single-output case the closed-loop poles are the left-half plane zeroes of

$$(-1)^n \phi(s)\phi(-s)\left[1 + \frac{1}{\rho} H(-s)H(s)\right], \qquad \textbf{3-476}$$

where $H(s)$ is now a scalar transfer function. Let us represent $H(s)$ in the form

$$H(s) = \frac{\psi(s)}{\phi(s)}, \qquad \textbf{3-477}$$

where $\psi(s)$ is the numerator polynomial of $H(s)$. It follows that the closed-loop poles are the left-half plane roots of

$$\phi(s)\phi(-s) + \frac{1}{\rho} \psi(s)\psi(-s) = 0. \qquad \textbf{3-478}$$

We can apply two techniques in determining the loci of the closed-loop poles. The first method is to recognize that **3-478** is a function of s^2, to substitute $s^2 = s'$, and to find the root loci in the s'-plane. The closed-loop poles are then obtained as the left-half plane square roots of the roots in the s'-plane. This is the *root-square locus* method (Chang, 1961).

For our purposes it is more convenient to trace the loci in the s-plane. Let us write

$$\psi(s) = \alpha \prod_{i=1}^{p} (s - \nu_i),$$
$$\phi(s) = \prod_{i=1}^{n} (s - \pi_i), \qquad \textbf{3-479}$$

where the ν_i, $i = 1, 2, \cdots, p$, are the zeroes of $H(s)$, and the π_i, $i = 1, 2, \cdots, n$, the poles of $H(s)$. To bring **3-478** in standard form, we rewrite it with **3-479** as

$$\prod_{i=1}^{n} (s - \pi_i)(s + \pi_i) + (-1)^{n-p} \frac{\alpha^2}{\rho} \prod_{i=1}^{p} (s - \nu_i)(s + \nu_i) = 0. \qquad \textbf{3-480}$$

284 Optimal Linear State Feedback Control Systems

Applying the rules of Section 1.5.5, we conclude the following.

(a) As $\rho \to 0$, of the $2n$ roots of **3-480** a total number of $2p$ asymptotically approach the p zeroes ν_i, $i = 1, 2, \cdots, p$, and their negatives $-\nu_i$, $i = 1, 2, \cdots, p$.

(b) As $\rho \to 0$, the other $2(n - p)$ roots of **3-480** asymptotically approach straight lines which intersect in the origin and make angles with the positive real axis of

$$\frac{k\pi}{n - p}, \quad k = 0, 1, 2, \cdots, 2n - 2p - 1, \quad n - p \text{ odd,} \qquad \text{3-481}$$

$$\frac{(k + \tfrac{1}{2})\pi}{n - p}, \quad k = 0, 1, 2, \cdots, 2n - 2p - 1, \quad n - p \text{ even.}$$

(c) As $\rho \to 0$, the $2(n - p)$ faraway roots of **3-480** are asymptotically at a distance

$$\left(\frac{\alpha^2}{\rho}\right)^{1/[2(n-p)]} \qquad \text{3-482}$$

from the origin.

(d) As $\rho \to \infty$, the $2n$ roots of **3-480** approach the n poles π_i, $i = 1, 2, \cdots, n$, and their negatives $-\pi_i$, $i = 1, 2, \cdots, n$.

Since the optimal closed-loop poles are the left-half plane roots of **3-480** we easily conclude the following (Kalman, 1964).

Theorem 3.11. *Consider the steady-state solution of the single-input single-output regulator problem with $R_3 = 1$ and $R_2 = \rho$. Assume that the open-loop system is stabilizable and detectable and let its transfer function be given by*

$$H(s) = \frac{\alpha \prod_{i=1}^{p} (s - \nu_i)}{\prod_{i=1}^{n} (s - \pi_i)}, \quad \alpha \neq 0, \qquad \text{3-483}$$

where the π_i, $i = 1, 2, \cdots, n$, are the characteristic values of the system. Then we have the following.

(a) *As $\rho \downarrow 0$, p of the n optimal closed-loop characteristic values asymptotically approach the numbers $\hat{\nu}_i$, $i = 1, 2, \cdots, p$, where*

$$\hat{\nu}_i = \begin{cases} \nu_i & \text{if } \operatorname{Re}(\nu_i) \leq 0, \\ -\nu_i & \text{if } \operatorname{Re}(\nu_i) > 0. \end{cases} \qquad \text{3-484}$$

(b) *As $\rho \downarrow 0$, the remaining $n - p$ optimal closed-loop characteristic values asymptotically approach straight lines which intersect in the origin and make*

3.8 Asymptotic Properties

angles with the negative real axis of

$$\pm l\frac{\pi}{n-p}, \quad l = 0, 1, \cdots, \frac{n-p-1}{2}, \quad n-p \text{ odd},$$ 3-485

$$\pm \frac{(l+\tfrac{1}{2})\pi}{n-p}, \quad l = 0, 1, \cdots, \frac{n-p}{2} - 1, \quad n-p \text{ even}.$$

These faraway closed-loop characteristic values are asymptotically at a distance

$$\omega_0 = \left(\frac{\alpha^2}{\rho}\right)^{1/[2(n-p)]}$$ 3-486

from the origin.
(c) *As* $\rho \to \infty$, *the* n *closed-loop characteristic values approach the numbers* $\hat{\pi}_i, i = 1, 2, \cdots, n$, *where*

$$\hat{\pi}_i = \begin{cases} \pi_i & \text{if } \operatorname{Re}(\pi_i) \leq 0, \\ -\pi_i & \text{if } \operatorname{Re}(\pi_i) > 0. \end{cases}$$ 3-487

The configuration of poles indicated by (b) is known as a *Butterworth configuration* of order $n-p$ with radius ω_0 (Weinberg, 1962). In Fig. 3.19

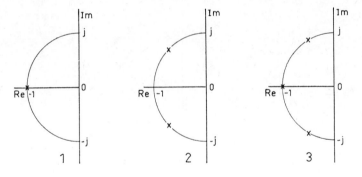

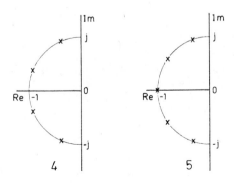

Fig. 3.19. Butterworth pole configurations of orders one through five and unit radii.

286 Optimal Linear State Feedback Control Systems

some low-order Butterworth configurations are indicated. In the next section we investigate what responses correspond to such configurations.

Figure 3.20 gives an example of the behavior of the closed-loop poles for a fictitious open-loop pole-zero configuration. Crosses mark the open-loop poles, circles the open-loop zeroes. Since the excess of poles over zeroes is two, a second-order Butterworth configuration results as $\rho \downarrow 0$. The remaining

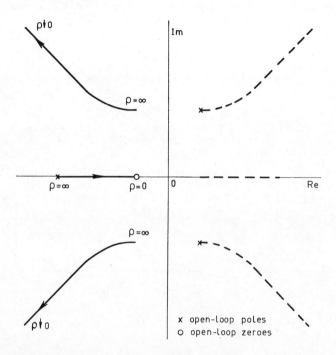

Fig. 3.20. Root loci of the characteristic values of the matrix Z (dashed and solid lines) and of the closed-loop poles (solid lines only) for a single-input single-output system with a fictitious open-loop pole-zero configuration.

closed-loop pole approaches the open-loop zero as $\rho \downarrow 0$. For $\rho \to \infty$ the closed-loop poles approach the single left-half plane open-loop pole and the mirror images of the two right-half plane open-loop poles.

We now return to the multiinput case. Here we must investigate the roots of

$$\phi(s)\phi(-s) \det \left[I + \frac{1}{\rho} N^{-1} H^T(-s) R_3 H(s) \right] = 0. \qquad \textbf{3-488}$$

The problem of determining the root loci for this expression is not as simple

3.8 Asymptotic Properties

as in the single-input case. Evaluation of the determinant leads to an expression of the form

$$\sum_{i=0}^{n} \alpha_i(1/\rho) s^{2i} = 0, \qquad \textbf{3-489}$$

where the functions $\alpha_i(1/\rho)$, $i = 0, 1, 2, \cdots, n$ are polynomials in $1/\rho$. Rosenau (1968) has given rules that are helpful in obtaining root loci for such an expression. We are only interested in the asymptotic behavior of the roots as $\rho \to 0$ and $\rho \to \infty$. The roots of **3-488** are also the roots of

$$\phi(s)\phi(-s) \det [\rho I + N^{-1} H^T(-s) R_3 H(s)] = 0. \qquad \textbf{3-490}$$

As $\rho \to 0$ some of the roots go to infinity; those that stay finite approach the zeroes of

$$\phi(s)\phi(-s) \det [N^{-1} H^T(-s) R_3 H(s)], \qquad \textbf{3-491}$$

provided this expression is not identically zero. Let us suppose that $H(s)$ is a square transfer matrix (in Section 3.7 we saw that this is a natural assumption). Then we know from Section 1.5.3 that

$$\det [H(s)] = \frac{\psi(s)}{\phi(s)}, \qquad \textbf{3-492}$$

where $\psi(s)$ is a polynomial at most of degree $n - k$, with n the dimension of the system and k the dimension of u and z. As a result, we can write for **3-491**

$$\frac{\det (R_3)}{\det (N)} \psi(-s)\psi(s). \qquad \textbf{3-493}$$

Thus it follows that as $\rho \downarrow 0$ those roots of **3-490** that stay finite approach the zeroes of the transfer matrix $H(s)$ and their negatives. This means that those optimal closed-loop poles of the regulator that stay finite approach those zeroes of $H(s)$ that have negative real parts and the negatives of the zeroes that have nonnegative real parts.

It turns out (Rosenau, 1968) that as $\rho \downarrow 0$ the far-off closed-loop regulator poles, that is, those poles that go to infinity, generally do not form a single Butterworth configuration, such as in the single-input case, but that they group into *several* Butterworth configurations of different orders and different radii (see Examples 3.19 and 3.21). A rough estimate of the distance of the faraway poles from the origin can be obtained as follows. Let $\phi_c(s)$ denote the closed-loop characteristic polynomial. Then we have

$$\phi_c(s)\phi_c(-s) = \phi(s)\phi(-s) \det \left[I + \frac{1}{\rho} N^{-1} H^T(-s) R_3 H(s) \right]. \qquad \textbf{3-494}$$

288   Optimal Linear State Feedback Control Systems

For small ρ we can approximate the right-hand side of this expression by

$$\phi(s)\phi(-s)\det\left[\frac{1}{\rho}N^{-1}H^T(-s)R_3H(s)\right] = \frac{\det(R_3)}{\rho^k \det(N)}\psi(s)\psi(-s) \quad \text{3-495}$$

where k is the dimension of the input u. Let us write

$$\psi(s) = \alpha\prod_{i=1}^{p}(s - \nu_i). \quad \text{3-496}$$

Then the leading term in **3-491** is given by

$$\alpha^2 \frac{\det(R_3)}{\rho^k \det(N)}(-1)^p s^{2p}. \quad \text{3-497}$$

This shows that the polynomial $\phi_c(s)\phi_c(-s)$ contains the following terms

$$\phi_c(s)\phi_c(-s) = (-1)^n s^{2n} + \cdots + \alpha^2\frac{\det(R_3)}{\rho^k \det(N)}(-1)^p s^{2p} + \cdots. \quad \text{3-498}$$

The terms given are the term with the highest power of s and the term with the highest power of $1/\rho$. An approximation of the faraway roots of this polynomial (for small ρ) is obtained from

$$(-1)^n s^{2n} + \alpha^2\frac{\det(R_3)}{\rho^k \det(N)}(-1)^p s^{2p} = 0. \quad \text{3-499}$$

It follows that the closed-loop poles are approximated by the left-half plane solutions of

$$(-1)^{(n-p-1)/[2(n-p)]}\left(\alpha^2\frac{\det(R_3)}{\rho^k \det(N)}\right)^{1/[2(n-p)]}. \quad \text{3-500}$$

This first approximation indicates a Butterworth configuration of order $n - p$. We use this expression to estimate the distance of the faraway poles to the origin; this (crude) estimate is given by

$$\left(\alpha^2 \frac{\det(R_3)}{\rho^k \det(N)}\right)^{1/[2(n-p)]}. \quad \text{3-501}$$

We consider finally the behavior of the closed-loop poles for $\rho \to \infty$. In this case we see from **3-494** that the characteristic values of the matrix Z approach the roots of $\phi(s)\phi(-s)$. This means that the closed-loop poles approach the numbers $\hat{\pi}_i$, $i = 1, 2, \cdots, n$, as given by **3-487**.

We summarize our results for the multiinput case as follows.

Theorem 3.12. *Consider the steady-state solution of the multiinput time-invariant regulator problem. Assume that the open-loop system is stabilizable and detectable, that the input u and the controlled variable z have the same*

3.8 Asymptotic Properties

dimension k, and that the state x has dimension n. Let $H(s)$ be the $k \times k$ open-loop transfer matrix

$$H(s) = D(sI - A)^{-1}B. \qquad \text{3-502}$$

Suppose that $\phi(s)$ is the open-loop characteristic polynomial and write

$$\det[H(s)] = \frac{\psi(s)}{\phi(s)} = \frac{\alpha \prod_{i=1}^{p}(s - \nu_i)}{\prod_{i=1}^{n}(s - \pi_i)}. \qquad \text{3-503}$$

Assume that $\alpha \neq 0$ and take $R_2 = \rho N$ with $N > 0$, $\rho > 0$.

(a) Then as $\rho \to 0$, p of the optimal closed-loop regulator poles approach the values $\hat{\nu}_i$, $i = 1, 2, \cdots, p$, where

$$\hat{\nu}_i = \begin{cases} \nu_i & \text{if } \operatorname{Re}(\nu_i) \leq 0 \\ -\nu_i & \text{if } \operatorname{Re}(\nu_i) > 0. \end{cases} \qquad \text{3-504}$$

The remaining closed-loop poles go to infinity and group into several Butterworth configurations of different orders and different radii. A rough estimate of the distance of the faraway closed-loop poles to the origin is

$$\left(\alpha^2 \frac{\det(R_3)}{\rho^k \det(N)}\right)^{1/[2(n-p)]}. \qquad \text{3-505}$$

(b) As $\rho \to \infty$, the n closed-loop regulator poles approach the numbers $\hat{\pi}_i$, $i = 1, 2, \cdots, n$, where

$$\hat{\pi}_i = \begin{cases} \pi_i & \text{if } \operatorname{Re}(\pi_i) \leq 0 \\ -\pi_i & \text{if } \operatorname{Re}(\pi_i) > 0. \end{cases} \qquad \text{3-506}$$

We conclude this section with the following comments. When ρ is very small, large input amplitudes are permitted. As a result, the system can move fast, which is reflected in a great distance of the faraway poles from the origin. Apparently, Butterworth pole patterns give good responses. Some of the closed-loop poles, however, do not move away but shift to the locations of open-loop zeroes. As is confirmed later in this section, in systems with left-half plane zeroes only these nearby poles are "canceled" by the open-loop zeroes, which means that their effect in the controlled variable response is not noticeable.

The case $\rho = \infty$ corresponds to a very heavy constraint on the input amplitudes. It is interesting to note that the "cheapest" stabilizing control law ("cheap" in terms of input amplitude) is a control law that relocates the unstable system poles to their mirror images in the left-half plane.

Problem 3.14 gives some information concerning the asymptotic behavior of the closed-loop poles for systems for which $\dim(u) \neq \dim(z)$.

Example 3.18. *Position control system*

In Example 3.8 (Section 3.4.1), we studied the locations of the closed-loop poles of the optimal position control system as a function of the parameter ρ. As we have seen, the closed-loop poles approach a Butterworth configuration of order two. This is in agreement with the results of this section. Since the open-loop transfer function

$$H(s) = \frac{\kappa}{s(s + \alpha)} \qquad \text{3-507}$$

has no zeroes, both closed-loop poles go to infinity as $\rho \downarrow 0$.

Example 3.19. *Stirred tank*

As an example of a multiinput multioutput system consider the stirred tank regulator problem of Example 3.9 (Section 3.4.1). From Example 1.15 (Section 1.5.3), we know that the open-loop transfer matrix is given by

$$H(s) = \begin{pmatrix} \dfrac{0.01}{s + 0.01} & \dfrac{0.01}{s + 0.01} \\ \dfrac{-0.25}{s + 0.02} & \dfrac{0.75}{s + 0.02} \end{pmatrix}. \qquad \text{3-508}$$

For this transfer matrix we have

$$\det [H(s)] = \frac{0.01}{(s + 0.01)(s + 0.02)}. \qquad \text{3-509}$$

Apparently, the transfer matrix has no zeroes; all closed-loop poles are therefore expected to go to ∞ as $\rho \downarrow 0$. With the numerical values of Example 3.9 for R_3 and N, we find for the characteristic polynomial of the matrix Z

$$s^4 + s^2\left(-0.5 \times 10^{-3} - \frac{0.02416}{\rho}\right)$$
$$+ \left(0.4 \times 10^{-7} + \frac{0.7416 \times 10^{-5}}{\rho} + \frac{10^{-4}}{\rho^2}\right). \qquad \text{3-510}$$

Figure 3.21 gives the behavior of the two closed-loop poles as ρ varies. Apparently, each pole traces a first-order Butterworth pattern. The asymptotic behavior of the roots for $\rho \downarrow 0$ can be found by solving the equation

$$s^4 - \frac{0.02416}{\rho} s^2 + \frac{10^{-4}}{\rho^2} = 0, \qquad \text{3-511}$$

which yields for the asymptotic closed-loop pole locations

$$-\frac{0.1373}{\sqrt{\rho}} \quad \text{and} \quad -\frac{0.07280}{\sqrt{\rho}}. \qquad \text{3-512}$$

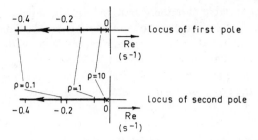

Fig. 3.21. Loci of the closed-loop roots for the stirred tank regulator. The locus on top originates from -0.02, the one below from -0.01.

The estimate **3-505** yields for the distance of the faraway poles to the origin

$$\frac{0.1}{\sqrt{\rho}}. \qquad \text{3-513}$$

We see that this is precisely the geometric average of the values **3-512**.

Example 3.20. *Pitch control of an airplane*

As an example of a more complicated system, we consider the longitudinal motions of an airplane (see Fig. 3.22). These motions are characterized by the velocity u along the x-axis of the airplane, the velocity w along the z-axis of the airplane, the pitch θ, and the pitch rate $q = \dot{\theta}$. The x- and z-axes are rigidly connected to the airplane. The x-axis is chosen to coincide with the horizontal axis when the airplane performs a horizontal stationary flight.

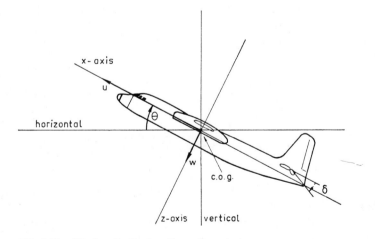

Fig. 3.22. The longitudinal motions of an airplane.

292   Optimal Linear State Feedback Control Systems

The control variables for these motions are the engine thrust T and the elevator deflection δ. The equations of motion can be linearized around a nominal solution which consists of horizontal flight with constant speed. It can be shown (Blakelock, 1965) that the linearized longitudinal equations of motion are independent of the lateral motions of the plane.

We choose the components of the state as follows:

$$\begin{aligned}
\xi_1(t) &= u(t), &&\text{incremental speed along } x\text{-axis,} \\
\xi_2(t) &= w(t), &&\text{speed along } z\text{-axis,} \\
\xi_3(t) &= \theta(t), &&\text{pitch,} \\
\xi_4(t) &= q(t), &&\text{pitch rate.}
\end{aligned}$$
3-514

The input variable, this time denoted by c, we define as

$$c(t) = \begin{pmatrix} T(t) \\ \delta(t) \end{pmatrix} \quad \begin{matrix} \text{incremental engine thrust,} \\ \text{elevator deflection.} \end{matrix}$$
3-515

With these definitions the state differential equations can be found from the inertial and aerodynamical laws governing the motion of the airplane (Blakelock, 1965). For a particular medium-weight transport aircraft under cruising conditions, the following linearized state differential equation results:

$$\dot{x}(t) = \begin{pmatrix} -0.01580 & 0.02633 & -9.810 & 0 \\ -0.1571 & -1.030 & 0 & 120.5 \\ 0 & 0 & 0 & 1 \\ 0.0005274 & -0.01652 & 0 & -1.466 \end{pmatrix} x(t)$$

$$+ \begin{pmatrix} 0.0006056 & 0 \\ 0 & -9.496 \\ 0 & 0 \\ 0 & -5.565 \end{pmatrix} c(t).$$
3-516

Here the following physical units are employed: u and w in m/s, θ in rad, q in rad/s, T in N, and δ in rad.

In this example we assume that the thrust is constant, so that the elevator deflection $\delta(t)$ is the only control variable. With this the system is described by the state differential equation

$$\dot{x}(t) = \begin{pmatrix} -0.01580 & 0.02633 & -9.810 & 0 \\ -0.1571 & -1.030 & 0 & 120.5 \\ 0 & 0 & 0 & 1 \\ 0.0005274 & -0.01652 & 0 & -1.466 \end{pmatrix} x(t)$$
$$+ \begin{pmatrix} 0 \\ -9.496 \\ 0 \\ -5.565 \end{pmatrix} \delta(t). \qquad \text{3-517}$$

As the controlled variable we choose the pitch $\theta(t)$:

$$\theta(t) = (0, 0, 1, 0)x(t). \qquad \text{3-518}$$

It can be found that the transfer function from the elevator deflection $\delta(t)$ to the pitch $\theta(t)$ is given by

$$\frac{-5.565s^2 - 5.663s - 0.1112}{s^4 + 2.512s^3 - 3.544s^2 + 0.06487s + 0.03079}. \qquad \text{3-519}$$

The poles of the transfer function are

$$\begin{aligned} -0.006123 \pm j0.09353, \\ -1.250 \pm j1.394, \end{aligned} \qquad \text{3-520}$$

while the zeroes are given by

$$-0.02004 \quad \text{and} \quad -0.9976. \qquad \text{3-521}$$

The loci of the closed-loop poles can be found by machine computation. They are given in Fig. 3.23. As expected, the faraway poles group into a Butterworth pattern of order two and the nearby closed-loop poles approach the open-loop zeroes. The system is further discussed in Example 3.22.

Example 3.21. *The control of the longitudinal motions of an airplane*

In Example 3.20 we considered the control of the pitch of an airplane through the elevator deflection. In the present example we extend the system by controlling, in addition to the pitch, the speed along the x-axis. As an additional control variable, we use the incremental engine thrust $T(t)$. Thus we choose for the input variable

$$c(t) = \begin{pmatrix} T(t) \\ \delta(t) \end{pmatrix} \quad \begin{array}{l} \text{incremental engine thrust,} \\ \text{elevator deflection,} \end{array} \qquad \text{3-522}$$

294 Optimal Linear State Feedback Control Systems

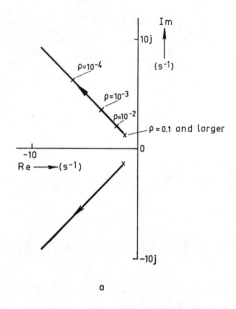

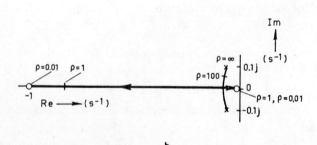

Fig. 3.23. Loci of the closed-loop poles of the pitch stabilization system. (*a*) Faraway poles; (*b*) nearby poles.

and for the controlled variable

$$z(t) = \begin{pmatrix} u(t) \\ \theta(t) \end{pmatrix} \quad \begin{array}{l} \text{incremental speed along the } x\text{-axis,} \\ \text{pitch.} \end{array} \qquad \textbf{3-523}$$

From the system state differential equation **3-516**, it can be computed that the system transfer matrix has the numerator polynomial

$$\psi(s) = -0.003370(s + 1.002), \qquad \textbf{3-524}$$

which results in a single open-loop zero at -1.002. The open-loop poles are at $-0.006123 \pm j0.09353$ and $-1.250 \pm j1.394$.

Before analyzing the problem any further, we must establish the weighting matrices R_3 and N. For both we adopt a diagonal form and to determine their values we proceed in essentially the same manner as in Example 3.9 (Section 3.4.1) for the stirred tank. Suppose that $R_3 = \text{diag}(\sigma_1, \sigma_2)$. Then

$$z^T(t)R_3 z(t) = \sigma_1 u^2(t) + \sigma_2 \theta^2(t). \tag{3-525}$$

Now let us assume that a deviation of 10 m/s in the speed along the x-axis is considered to be about as bad as a deviation of 0.2 rad (12°) in the pitch. We therefore select σ_1 and σ_2 such that

$$\sigma_1 (10)^2 = \sigma_2 (0.2)^2, \tag{3-526}$$

or

$$\frac{\sigma_1}{\sigma_2} = 0.0004. \tag{3-527}$$

Thus we choose

$$R_3 = \begin{pmatrix} 0.02 & 0 \\ 0 & 50 \end{pmatrix}, \tag{3-528}$$

where for convenience we have let $\det(R_3) = 1$. Similarly, suppose that $N = \text{diag}(\rho_1, \rho_2)$ so that

$$c^T(t)Nc(t) = \rho_1 T^2(t) + \rho_2 \delta^2(t). \tag{3-529}$$

To determine ρ_1 and ρ_2, we assume that a deviation of 500 N in the engine thrust is about as acceptable as a deviation of 0.2 rad (12°) in the elevator deflection. This leads us to select

$$\rho_1 (500)^2 = \rho_2 (0.2)^2, \tag{3-530}$$

which results in the following choice of N:

$$N = \begin{pmatrix} 0.0004 & 0 \\ 0 & 2500 \end{pmatrix}. \tag{3-531}$$

With these values of R_3 and N, the relation 3-505 gives us the following estimate for the distance of the far-off poles:

$$\omega_0 = \left(\alpha^2 \frac{\det(R_3)}{\rho^k \det(N)} \right)^{1/[2(n-p)]} = \frac{0.15}{\rho^{1/3}}. \tag{3-532}$$

The closed-loop pole locations must be found by machine computation. Table 3.4 lists the closed-loop poles for various values of ρ and also gives the estimated radius ω_0. We note first that one of the closed-loop poles approaches the open-loop zero at -1.002. Furthermore, we see that ω_0 is

296   Optimal Linear State Feedback Control Systems

only a very crude estimate for the distance of the faraway poles from the origin.

The complete closed-loop loci are sketched in Fig. 3.24. It is noted that the appearance of these loci is quite different from those for single-input systems. Two of the faraway poles assume a second-order Butterworth configuration, while the third traces a first-order Butterworth pattern. The system is further discussed in Example 3.24.

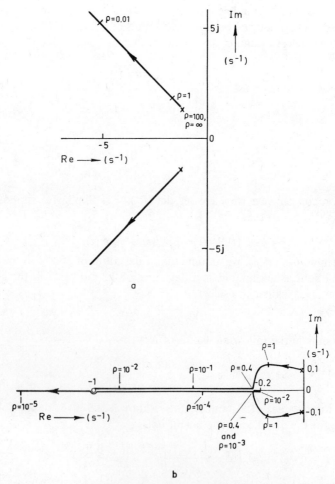

Fig. 3.24. Loci of the closed-loop poles for the longitudinal motion control system. (*a*) Faraway poles; (*b*) nearby pole and one faraway pole. For clarity the coinciding portions of the loci on the real axis are represented as distinct lines; in reality they coincide with the real axis.

3.8 Asymptotic Properties

Table 3.4 Closed-Loop Poles for the Longitudinal Motion Stability Augmentation System

ρ	Closed-loop poles (s^{-1})		ω_0 (s^{-1})
∞	$-0.006123 \pm j0.09353$	$-1.250 \pm j1.394$	0
1	$-0.1734 \pm j0.1184$	$-1.263 \pm j1.415$	0.15
10^{-1}	$-0.5252 \quad -0.2166$	$-1.376 \pm j1.564$	0.32
10^{-2}	$-0.8877 \quad -0.2062$	$-1.986 \pm j2.179$	0.70
10^{-3}	$-0.9745 \quad -0.2431$	$-3.484 \pm j3.609$	1.5
10^{-4}	$-0.9814 \quad -0.4806$	$-6.241 \pm j6.312$	3.2
10^{-5}	$-1.020 \quad -1.344$	$-11.14 \pm j11.18$	7.0
10^{-6}	$-1.003 \quad -4.283$	$-19.83 \pm j19.83$	15
10^{-8}	$-1.002 \quad -42.82$	$-62.73 \pm j62.73$	70

3.8.2* Asymptotic Properties of the Single-Input Single-Output Nonzero Set Point Regulator

In this section we discuss the single-input single-output nonzero set point optimal regulator in the light of the results of Section 3.8.1. Consider the single-input system

$$\dot{x}(t) = Ax(t) + b\mu(t) \qquad \text{3-533}$$

with the scalar controlled variable

$$\zeta(t) = dx(t). \qquad \text{3-534}$$

Here b is a column vector and d a row vector. From Section 3.7 we know that the nonzero set point optimal control law is given by

$$\mu(t) = -\bar{f}x(t) + \frac{1}{H_c(0)}\zeta_0, \qquad \text{3-535}$$

where $\bar{f}$ is the row vector

$$\bar{f} = \frac{1}{\rho}b^T\bar{P}, \qquad \text{3-536}$$

with $\bar{P}$ the solution of the appropriate Riccati equation. Furthermore, $H_c(s)$ is the closed-loop transfer function

$$H_c(s) = d(sI - A + b\bar{f})^{-1}b, \qquad \text{3-537}$$

and ζ_0 is the set point for the controlled variable.

In order to study the response of the regulator to a step change in the set point, let us replace ζ_0 with a time-dependent variable $\zeta_0(t)$. The interconnection of the open-loop system and the nonzero set point optimal

298 Optimal Linear State Feedback Control Systems

control law is then described by

$$\dot{x}(t) = (A - b\bar{f})x(t) + b\frac{1}{H_c(0)}\zeta_0(t), \quad \text{3-538}$$

$$\zeta(t) = dx(t).$$

Laplace transformation yields for the transfer function $T(s)$ from the variable set point $\zeta_0(t)$ to the controlled variable $\zeta(t)$:

$$T(s) = d(sI - A + b\bar{f})^{-1}b\frac{1}{H_c(0)}. \quad \text{3-539}$$

Let us consider the closed-loop transfer function $d(sI - A + b\bar{f})^{-1}b$. Obviously,

$$d(sI - A + b\bar{f})^{-1}b = \frac{\psi_c(s)}{\phi_c(s)}, \quad \text{3-540}$$

where $\phi_c(s) = \det(sI - A + b\bar{f})$ is the closed-loop characteristic polynomial and $\psi_c(s)$ is another polynomial. Now we saw in Section 3.7 (Eq. 3-428) that the numerator of the determinant of a square transfer matrix $D(sI - A + BF)^{-1}B$ is independent of the feedback gain matrix F and is equal to the numerator polynomial of the open-loop transfer matrix $D(sI - A)^{-1}B$. Since in the single-input single-output case the determinant of the transfer function reduces to the transfer function itself, we can immediately conclude that $\psi_c(s)$ equals $\psi(s)$, which is defined from

$$H(s) = \frac{\psi(s)}{\phi(s)}. \quad \text{3-541}$$

Here $H(s) = d(sI - A)^{-1}b$ is the open-loop transfer function and $\phi(s) = \det(sI - A)$ the open-loop characteristic polynomial.

As a result of these considerations, we conclude that

$$T(s) = \frac{\psi(s)}{\phi_c(s)}\frac{\phi_c(0)}{\psi(0)}. \quad \text{3-542}$$

Let us write

$$\psi(s) = \alpha \prod_{i=1}^{p}(s - \nu_i), \quad \text{3-543}$$

where the ν_i, $i = 1, 2, \cdots, p$, are the zeroes of $H(s)$. Then it follows from Theorem 3.11 that as $\rho \downarrow 0$ we can write for the closed-loop characteristic polynomial

$$\phi_c(s) \simeq \prod_{i=1}^{p}(s - \hat{\nu}_i)\prod_{i=1}^{n-p}(s - \eta_i\omega_0), \quad \text{3-544}$$

where the $\hat{\nu}_i$, $i = 1, 2, \cdots, p$, are defined by **3-484**, the η_i, $i = 1, 2, \cdots,$

3.8 Asymptotic Properties

$n - p$, form a Butterworth configuration of order $n - p$ and radius 1, and where

$$\omega_0 = \left(\frac{\alpha^2}{\rho}\right)^{1/[2(n-p)]} \qquad \text{3-545}$$

Substitution of **3-544** into **3-542** yields the following approximation for $T(s)$:

$$T(s) \simeq \frac{1}{\prod_{i=1}^{n-p}\left(-\dfrac{s}{\eta_i \omega_0} + 1\right)} \prod_{i=1}^{p} \left(\frac{-\dfrac{s}{\nu_i} + 1}{-\dfrac{s}{\hat{\nu}_i} + 1}\right). \qquad \text{3-546}$$

This we rewrite as

$$T(s) \simeq \frac{1}{\chi_{n-p}(s/\omega_0)} \prod_{i=1}^{p} \left(\frac{-\dfrac{s}{\nu_i} + 1}{-\dfrac{s}{\hat{\nu}_i} + 1}\right), \qquad \text{3-547}$$

where $\chi_{n-p}(s)$ is a *Butterworth polynomial* of order $n - p$, that is, $\chi_{n-p}(s)$ is defined by

$$\chi_{n-p}(s) = \prod_{i=1}^{n-p}\left(-\frac{s}{\eta_i} + 1\right). \qquad \text{3-548}$$

Table 3.5 lists some low-order Butterworth polynomials (Weinberg, 1962).

Table 3.5 Butterworth Polynomials of Orders One through Five

$\chi_1(s) = s + 1$
$\chi_2(s) = s^2 + 1.414s + 1$
$\chi_3(s) = s^3 + 2s^2 + 2s + 1$
$\chi_4(s) = s^4 + 2.613s^3 + 3.414s^2 + 2.613s + 1$
$\chi_5(s) = s^5 + 3.236s^4 + 5.236s^3 + 5.236s^2 + 3.236s + 1$

The expression **3-547** shows that, if the open-loop transfer function has zeroes in the left-half plane only, the control system transfer function $T(s)$ approaches

$$\frac{1}{\chi_{n-p}(s/\omega_0)} \qquad \text{3-549}$$

as $\rho \downarrow 0$. We call this a *Butterworth transfer function* of order $n - p$ and break frequency ω_0. In Figs. 3.25 and 3.26, plots are given of the step responses and Bode diagrams of systems with Butterworth transfer functions

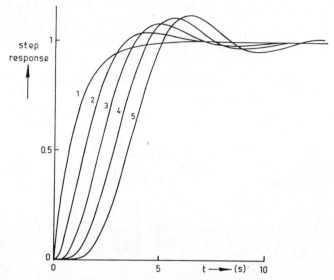

Fig. 3.25. Step responses of systems with Butterworth transfer functions of orders one through five with break frequencies 1 rad/s.

of various orders. The plots of Fig. 3.25 give an indication of the type of response obtained to steps in the set point. This response is asymptotically independent of the open-loop system poles and zeroes (provided the latter are in the left-half complex plane). We also see that by choosing ρ small enough the break frequency ω_0 can be made arbitrarily high, and correspondingly the settling time of the step response can be made arbitrarily small. An extremely fast response is of course obtained at the expense of large input amplitudes.

This analysis shows that the response of the controlled variable to changes in the set point is dominated by the far-off poles $\eta_i \omega_0$, $i = 1, 2, \cdots, n - p$. The nearby poles, which nearly coincide with the open-loop zeroes, have little effect on the response of the controlled variable because they nearly cancel against the zeroes. As we see in the next section, the far-off poles dominate not only the response of the controlled variable to changes in the set point but also the response to arbitrary initial conditions. As can easily be seen, and as illustrated in the examples, the nearby poles *do* show up in the *input*. The settling time of the tracking error is therefore determined by the faraway poles, but that of the input by the nearby poles.

The situation is less favorable for systems with *right-half plane zeroes*. Here the transmission $T(s)$ contains extra factors of the form

$$\frac{s + \hat{\nu}_i}{s - \hat{\nu}_i}$$

3-550

3.8 Asymptotic Properties 301

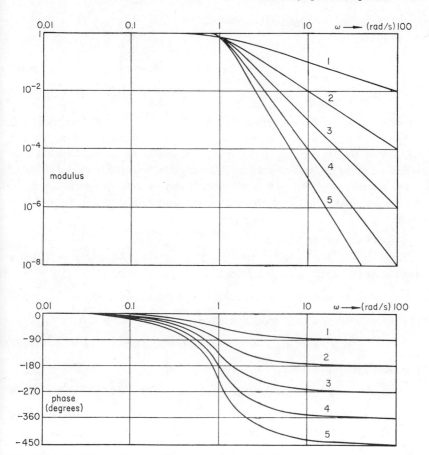

Fig. 3.26. Modulus and phase of Butterworth transfer functions of orders one through five with break frequencies 1 rad/s.

and the tracking error response is dominated by the nearby pole at $\hat{v}_i$. This points to an inherent limitation in the speed of response of systems with right-half plane zeroes. In the next subsection we further pursue this topic. First, however, we summarize the results of this section:

Theorem 3.13. *Consider the nonzero set point optimal control law* **3-535** *for the time-invariant, single-input single-output, stabilizable and detectable system*

$$\dot{x}(t) = Ax(t) + b\mu(t),$$
$$\zeta(t) = dx(t),$$
3-551

where $R_3 = 1$ and $R_2 = \rho$. Then as $\rho \downarrow 0$ the control system transmission $T(s)$ (i.e., the closed-loop transfer function from the variable set point $\zeta_0(t)$

to the controlled variable $\zeta(t)$) approaches

$$T(s) \to \frac{1}{\chi_{n-p}(s/\omega_0)} \prod_{i=1}^{p} \left(\frac{-\dfrac{s}{v_i} + 1}{-\dfrac{s}{\hat{v}_i} + 1} \right), \qquad \text{3-552}$$

where $\chi_{n-p}(s)$ is a Butterworth polynomial of order $n - p$ and radius 1, n is the order of the system, p is the number of zeroes of the open-loop transfer function of the system, ω_0 is the asymptotic radius of the Butterworth configuration of the faraway closed-loop poles as given by **3-486**, v_i, $i = 1, 2, \cdots, p$, are the zeroes of the open-loop transfer function, and $\hat{v}_i$, $i = 1, 2, \cdots, p$, are the open-loop transfer function zeroes mirrored into the left-half complex plane.

Example 3.22. *Pitch control*

Consider the pitch control problem of Example 3.20. For $\rho = 0.01$ the steady-state feedback gain matrix can be computed to be

$$\bar{f} = (-0.0001174, 0.002813, -10.00, -1.619). \qquad \text{3-556}$$

The corresponding closed-loop characteristic polynomial is given by

$$\phi_c(s) = s^4 + 11.49s^3 + 66.43s^2 + 56.84s + 1.112. \qquad \text{3-557}$$

The closed-loop poles are

$$-0.02004, \; -0.9953, \; \text{and} \; -0.5239 \pm j5.323. \qquad \text{3-558}$$

We see that the first two poles are very close to the open-loop zeroes at -0.2004 and -0.9976. The closed-loop transfer function is given by

$$H_c(s) = \frac{\psi(s)}{\phi_c(s)} = \frac{-5.565s^2 - 5.663s - 0.1112}{s^4 + 11.49s^3 + 66.43s^2 + 56.84s + 1.112}, \qquad \text{3-559}$$

so that $H_c(0) = -0.1000$. As a result, the nonzero set point control law is given by

$$\delta(t) = -\bar{f}x(t) - 10.00\theta_0(t), \qquad \text{3-560}$$

where $\theta_0(t)$ is the set point of the pitch.

Figure 3.27 depicts the response of the system to a step of 0.1 rad in the set point $\theta_0(t)$. It is seen that the pitch θ quickly settles at the desired value; its response is completely determined by the second-order Butterworth configuration at $-5.239 \pm j5.323$. The pole at -0.9953 (corresponding to a time constant of about 1 s) shows up most clearly in the response of the speed along the z-axis w and can also be identified in the behavior of the elevator deflection δ. The very slow motion with a time constant of 50 s, which

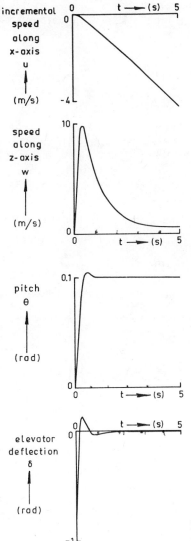

Fig. 3.27. Response of the pitch control system to a step of 0.1 rad in the pitch angle set point.

corresponds to the pole at -0.02004, is represented in the response of the speed along the x-axis u, the speed along the z-axis w, and also in the elevator deflection δ, although this is not visible in the plot. It takes about 2 min for u and w to settle at the steady-state values -49.16 and 7.754 m/s.

Note that this control law yields an initial elevator deflection of -1 rad which, practically speaking, is far too large.

Example 3.23. *System with a right-half plane zero*

As a second example consider the single-input system with state differential equation

$$\dot{x}(t) = \begin{pmatrix} 0 & 1 \\ 0 & -2 \end{pmatrix} x(t) + \begin{pmatrix} 0 \\ 1 \end{pmatrix} \mu(t). \qquad 3\text{-}561$$

Let us choose for the controlled variable

$$\zeta(t) = (1, \ -1)x(t). \qquad 3\text{-}562$$

This system has the open-loop transfer function

$$H(s) = \frac{-s+1}{s(s+2)}, \qquad 3\text{-}563$$

and therefore has a zero in the right-half plane. Consider for this system the criterion

$$\int_{t_0}^{\infty} [\zeta^2(t) + \rho\mu^2(t)] \, dt. \qquad 3\text{-}564$$

It can be found that the corresponding Riccati equation has the steady-state solution

$$\bar{P} = \begin{pmatrix} 1 + \sqrt{1 + 4\rho + 2\sqrt{\rho}} & \sqrt{\rho} \\ \sqrt{\rho} & \rho\left(-2 + \sqrt{4 + \dfrac{1}{\rho} + \dfrac{2}{\sqrt{\rho}}}\right) \end{pmatrix}. \qquad 3\text{-}565$$

The corresponding steady-state feedback gain vector is

$$\bar{f} = \left(\frac{1}{\sqrt{\rho}}, \ -2 + \sqrt{4 + \frac{1}{\rho} + \frac{2}{\sqrt{\rho}}}\right). \qquad 3\text{-}566$$

The closed-loop poles can be found to be

$$\frac{1}{2}\left(-\sqrt{4 + \frac{1}{\rho} + \frac{2}{\sqrt{\rho}}} \pm \sqrt{4 + \frac{1}{\rho} - \frac{2}{\sqrt{\rho}}}\right). \qquad 3\text{-}567$$

Figure 3.28 gives a sketch of the loci of the closed-loop poles. As expected, one of the closed-loop poles approaches the mirror image of the right-half plane zero, while the other pole goes to $-\infty$ along the real axis.

For $\rho = 0.04$ the closed-loop characteristic polynomial is given by

$$s^2 + 6.245s + 5, \qquad 3\text{-}568$$

and the closed-loop poles are located at -0.943 and -5.302. The closed-loop

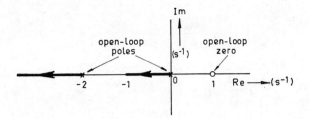

Fig. 3.28. Loci of the closed-loop poles for a system with a right-half plane zero.

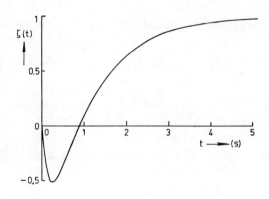

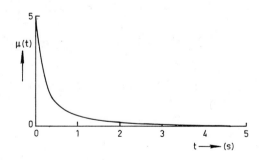

Fig. 3.29. Response of a closed-loop system with a right-half plane zero to a unit step in the set point.

transfer function is

$$H_c(s) = \frac{\psi(s)}{\phi_c(s)} = \frac{-s+1}{s^2 + 6.245s + 5}, \qquad \text{3-569}$$

so that $H_c(0) = 0.2$. The steady-state feedback gain vector is

$$\bar{f} = (5, \quad 4.245). \qquad \text{3-570}$$

As a result, the nonzero set point control law is

$$\mu(t) = -(5, \quad 4.245)x(t) + 5\zeta_0(t). \qquad \text{3-571}$$

Figure 3.29 gives the response of the closed-loop system to a step in the set point $\zeta_0(t)$. We see that in this case the response is dominated by the closed-loop pole at -0.943. It is impossible to obtain a response that is faster and at the same time has a smaller integrated square tracking error.

3.8.3* The Maximally Achievable Accuracy of Regulators and Tracking Systems

In this section we study the steady-state solution of the Riccati equation as ρ approaches zero in

$$R_2 = \rho N. \qquad \text{3-572}$$

The reason for our interest in this asymptotic solution is that it will give us insight into the maximally achievable accuracy of regulator and tracking systems when no limitations are imposed upon the input amplitudes.

This section is organized as follows. First, the main results are stated in the form of a theorem. The proof of this theorem (Kwakernaak and Sivan, 1972), which is long and technical, is omitted. The remainder of the section is devoted to a discussion of the results and to examples.

We first state the main results:

Theorem 3.14. *Consider the time-invariant stabilizable and detectable linear system*

$$\begin{aligned}\dot{x}(t) &= Ax(t) + Bu(t), \\ z(t) &= Dx(t),\end{aligned} \qquad \text{3-573}$$

where B and D are assumed to have full rank. Consider also the criterion

$$\int_{t_0}^{\infty} [z^T(t)R_3 z(t) + u^T(t)R_2 u(t)]\, dt \qquad \text{3-574}$$

where $R_3 > 0$, $R_2 > 0$. Let

$$R_2 = \rho N, \qquad \text{3-575}$$

with $N > 0$ and ρ a positive scalar, and let $\bar{P}_\rho$ be the steady-state solution of the Riccati equation

$$-\dot{P}_\rho(t) = D^T R_3 D - P_\rho(t) B R_2^{-1} B^T P_\rho(t) + A^T P_\rho(t) + P_\rho(t) A,$$
$$P_\rho(t_1) = 0. \qquad 3\text{-}576$$

Then the following facts hold.
(a) The limit

$$\lim_{\rho \downarrow 0} \bar{P}_\rho = P_0 \qquad 3\text{-}577$$

exists.
(b) Let $z_\rho(t)$, $t \geq t_0$, denote the response of the controlled variable for the regulator that is steady-state optimal for $R_2 = \rho N$. Then

$$\lim_{\rho \downarrow 0} \int_{t_0}^{\infty} z_\rho^T(t) R_3 z_\rho(t) \, dt = x^T(t_0) P_0 x(t_0). \qquad 3\text{-}578$$

(c) If dim $(z) >$ dim (u), then $P_0 \neq 0$.
(d) If dim $(z) =$ dim (u) and the numerator polynomial $\psi(s)$ of the open-loop transfer matrix $H(s) = D(sI - A)^{-1}B$ is nonzero, $P_0 = 0$ if and only if $\psi(s)$ has zeroes with nonpositive real parts only.
(e) If dim $(z) <$ dim (u), then a sufficient condition for P_0 to be 0 is that there exists a rectangular matrix M such that the numerator polynomial $\psi(s)$ of the square transfer matrix $D(sI - A)^{-1}BM$ is nonzero and has zeroes with nonpositive real parts only.

A discussion of the significance of the various parts of the theorem now follows. Item (a) states that, as we let the weighting coefficient of the input ρ decrease, the criterion

$$\int_{t_0}^{\infty} [z_\rho^T(t) R_3 z_\rho(t) + \rho u_\rho^T(t) N u_\rho(t)] \, dt = x^T(t_0) \bar{P}_\rho x(t_0) \qquad 3\text{-}579$$

approaches a limit $x^T(t_0) P_0 x(t_0)$. If we identify R_3 with W_e and N with W_u, the expression **3-579** can be rewritten as

$$\int_{t_0}^{\infty} C_{e,\rho}(t) \, dt + \rho \int_{t_0}^{\infty} C_{u,\rho}(t) \, dt, \qquad 3\text{-}580$$

where $C_{e,\rho}(t) = z_\rho^T(t) W_e z_\rho(t)$ is the weighted square regulating error and $C_{u,\rho}(t) = u_\rho^T(t) W_u u_\rho(t)$ the weighted square input. It follows from item (b) of the theorem that as $\rho \downarrow 0$, of the two terms in **3-580** the first term, that is, the integrated square regulating error, fully accounts for the two terms together so that in the limit the integrated square regulating error is given by

$$\lim_{\rho \downarrow 0} \int_{t_0}^{\infty} C_{e,\rho}(t) \, dt = x^T(t_0) P_0 x(t_0). \qquad 3\text{-}581$$

308   Optimal Linear State Feedback Control Systems

If the weighting coefficient ρ is zero, no costs are spared in the sense that no limitations are imposed upon the input amplitudes. Clearly, under this condition the greatest accuracy in regulation is achieved in the sense that the integrated square regulation error is the least that can ever be obtained.

Parts (c), (d), and (e) of the theorem are concerned with the conditions under which $P_0 = 0$, which means that ultimately perfect regulation is approached since

$$\lim_{\rho \downarrow 0} \int_{t_0}^{\infty} C_{e,\rho}(t)\, dt = 0. \qquad \text{3-582}$$

Part (c) of the theorem states that, if the dimension of the controlled variable is greater than that of the input, perfect regulation is impossible. This is very reasonable, since in this case the number of degrees of freedom to control the system is too small. In order to determine the maximal accuracy that can be achieved, P_0 must be computed. Some remarks on how this can be done are given in Section 4.4.4.

In part (d) the case is considered where the number of degrees of freedom is sufficient, that is, the input and the controlled variable have the same dimensions. Here the maximally achievable accuracy is dependent upon the properties of the open-loop system transfer matrix $H(s)$. Perfect regulation is possible only when the numerator polynomial $\psi(s)$ of the transfer matrix has no right-half plane zeroes (assuming that $\psi(s)$ is not identical to zero). This can be made intuitively plausible as follows. Suppose that at time 0 the system is in the initial state x_0. Then in terms of Laplace transforms the response of the controlled variable can be expressed as

$$\mathbf{Z}(s) = H(s)\mathbf{U}(s) + D(sI - A)^{-1}x_0, \qquad \text{3-583}$$

where $\mathbf{Z}(s)$ and $\mathbf{U}(s)$ are the Laplace transforms of z and u, respectively. $\mathbf{Z}(s)$ can be made identical to zero by choosing

$$\mathbf{U}(s) = -H^{-1}(s)D(sI - A)^{-1}x_0. \qquad \text{3-584}$$

The input $u(t)$ in general contains delta functions and derivatives of delta functions at time 0. These delta functions instantaneously transfer the system from the state x_0 at time 0 to a state $x(0^+)$ that has the property that $z(0^+) = Dx(0^+) = 0$ *and* that $z(t)$ can be maintained at 0 for $t > 0$ (Sivan, 1965). Note that in general the state $x(t)$ undergoes a delta function and derivative of delta function type of trajectory at time 0 but that $z(t)$ moves from $z(0) = Dx_0$ to 0 directly, without infinite excursions, as can be seen by inserting **3-584** into **3-583**.

The expression **3-584** leads to a stable behavior of the input only if the inverse transfer matrix $H^{-1}(s)$ is stable, that is if the numerator polynomial $\psi(s)$ of $H(s)$ has no right-half plane zeroes. The reason that the input **3-584**

3.8 Asymptotic Properties

cannot be used in the case that $H^{-1}(s)$ has unstable poles is that although the input **3-584** drives the controlled variable $z(t)$ to zero and maintains $z(t)$ at zero, the input itself grows indefinitely (Levy and Sivan, 1966). By our problem formulation such inputs are ruled out, so that in this case **3-584** is not the limiting input as $\rho \downarrow 0$ and, in fact, costless regulation cannot be achieved.

Finally, in part (e) of the theorem, we see that if dim (z) < dim (u), then $P_0 = 0$ if the situation can be reduced to that of part (d) by replacing the input u with an input u' of the form

$$u'(t) = Mu(t). \qquad \textbf{3-585}$$

The existence of such a matrix M is not a necessary condition for P_0 to be zero, however.

Theorem 3.14 extends some of the results of Section 3.8.2. There we found that for single-input single-output systems without zeroes in the right-half complex plane the response of the controlled variable to steps in the set point is asymptotically completely determined by the faraway closed-loop poles and not by the nearby poles. The reason is that the nearby poles are canceled by the zeroes of the system. Theorem 3.14 leads to more general conclusions. It states that for multiinput multioutput systems without zeroes in the right-half complex plane the integrated square regulating error goes to zero asymptotically. This means that for small values of ρ the closed-loop response of the controlled variable to any initial condition of the system is very fast, which means that this response is determined by the faraway closed-loop poles only. Consequently, also in this case the effect of the nearby poles is canceled by the zeroes. The slow motion corresponding to the nearby poles of course shows up in the response of the input variable, so that in general the input can be expected to have a much longer settling time than the controlled variable. For illustrations we refer to the examples.

It follows from the theory that optimal regulator systems can have "hidden modes" which do not appear in the controlled variable but which do appear in the state and the input. These modes may impair the operation of the control system. Often this phenomenon can be remedied by redefining or extending the controlled variable so that the requirements upon the system are more faithfully reflected.

It also follows from the theory that systems with right-half plane zeroes are fundamentally deficient in their capability to regulate since the mirror images of the right-half plane zeroes appear as nearby closed-loop poles which are not canceled by zeroes. If these right-half plane zeroes are far away from the origin, however, their detrimental effect may be limited.

It should be mentioned that ultimate accuracy can of course never be achieved since this would involve infinite feedback gains and infinite input amplitudes. The results of this section, however, give an idea of the ideal performance of which the system is capable. In practice, this limit may not nearly be approximated because of the constraints on the input amplitudes.

So far the discussion has been confined to the deterministic regulator problem. Let us now briefly consider the stochastic regulator problem, which includes tracking problems. As we saw in Section 3.6, we have for the stochastic regulator problem

$$C_{e\infty,\rho} + \rho C_{u\infty,\rho} = \text{tr}\,(\bar{P}V), \qquad 3\text{-}586$$

where $C_{e\infty}$ and $C_{u\infty}$ indicate the steady-state mean square regulation error and the steady-state mean square input, respectively. It immediately follows that

$$\lim_{\rho \downarrow 0} (C_{e\infty,\rho} + \rho C_{u\infty,\rho}) = \text{tr}\,(P_0 V). \qquad 3\text{-}587$$

It is not difficult to argue [analogously to the proof of part (b) of Theorem 3.14] that of the two terms in **3-587** the first term fully accounts for the left-hand side so that

$$\lim_{\rho \downarrow 0} C_{e\infty,\rho} = \text{tr}\,(P_0 V). \qquad 3\text{-}588$$

This means that perfect stochastic regulation ($P_0 = 0$) can be achieved under the same conditions for which perfect deterministic regulation is possible. It furthermore is easily verified that, for the regulator with nonwhite disturbances (Section 3.6.1) and for the stochastic tracking problem (Section 3.6.2), perfect regulation or tracking, respectively, is achieved if and only if in both cases the *plant* transfer matrix $H(s) = D(sI - A)^{-1}B$ satisfies the conditions outlined in Theorem 3.14. This shows that it is the plant alone that determines the maximally achievable accuracy and not the properties of the disturbances or the reference variable.

In conclusion, we note that Theorem 3.14 gives no results for the case in which the numerator polynomial $\psi(s)$ is identical to zero. This case rarely seems to occur, however.

Example 3.24. *Control of the longitudinal motions of an airplane*

As an example of a multiinput system, we consider the regulation of the longitudinal motions of an airplane as described in Example 3.21. For $\rho = 10^{-6}$ we found in Example 3.21 that the closed-loop poles are -1.003, -4.283, and $-19.83 \pm j19.83$. The first of these closed-loop poles practically coincides with the open-loop zero at -1.002.

Figure 3.30 shows the response of the closed-loop system to an initial deviation in the speed along the x-axis u, and to an initial deviation in the pitch θ. It is seen that the response of the speed along the x-axis is determined

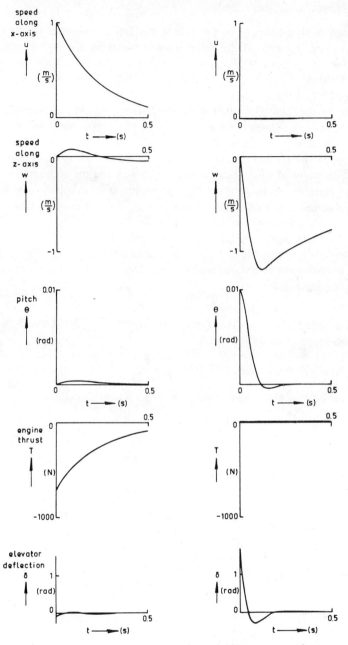

Fig. 3.30. Closed-loop responses of a longitudinal stability augmentation system for an airplane. Left column: Responses to the initial state $u(0) = 1$ m/s, while all other components of the initial state are zero. Right column: Response to the initial state $\theta(0) = 0.01$ rad, while all other components of the initial state are zero.

mainly by a time constant of about 0.24 s which corresponds to the pole at -4.283. The response of the pitch is determined by the Butterworth configuration at $-19.83 \pm j19.83$. The slow motion with a time constant of about 1 s that corresponds to the pole at -1.003 only affects the response of the speed along the z-axis w.

We note that the controlled system exhibits very little *interaction* in the sense that the restoration of the speed along the x-axis does not result in an appreciable deviation of the pitch, and conversely.

Finally, it should be remarked that the value $\rho = 10^{-6}$ is not suitable from a practical point of view. It causes far too large a change in the engine thrust and the elevator angle. In addition, the engine is unable to follow the fast thrust changes that this control law requires. Further investigation should take into account the dynamics of the engine.

The example confirms, however, that since the plant has no right-half plane zeroes an arbitrarily fast response can be obtained, and that the nearby pole that corresponds to the open-loop zero does not affect the response of the controlled variable.

Example 3.25. *A system with a right-half plane zero*

In Example 3.23 we saw that the system described by **3-561** and **3-562** with the open-loop transfer function

$$H(s) = \frac{-s + 1}{s(s + 2)} \qquad 3\text{-}589$$

has the following steady-state solution of the Riccati equation

$$\bar{P} = \begin{pmatrix} 1 + \sqrt{1 + 4\rho + 2\sqrt{\rho}} & \sqrt{\rho} \\ \sqrt{\rho} & \rho\left(-2 + \sqrt{4 + \frac{1}{\rho} + \frac{2}{\sqrt{\rho}}}\right) \end{pmatrix}. \qquad 3\text{-}590$$

As ρ approaches zero, $\bar{P}$ approaches P_0, where

$$P_0 = \begin{pmatrix} 2 & 0 \\ 0 & 0 \end{pmatrix}. \qquad 3\text{-}591$$

As we saw in Example 3.23, in the limit $\rho \downarrow 0$ the response is dominated by the closed-loop pole at -1.

3.9* SENSITIVITY OF LINEAR STATE FEEDBACK CONTROL SYSTEMS

In Chapter 2 we saw that a very important property of a feedback system is its ability to suppress disturbances and to compensate for parameter changes.

3.9 Sensitivity

In this section we investigate to what extent optimal regulators and tracking systems possess these properties. When we limit ourselves to time-invariant problems and consider only the steady-state case, where the terminal time is at infinity, the optimal regulator and tracking systems we have derived have the structure of Fig. 3.31. The optimal control law can generally be represented

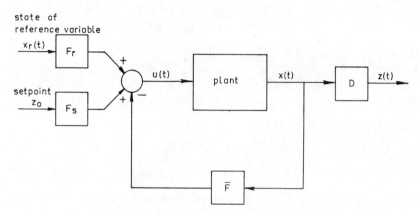

Fig. 3.31. The structure of a time-invariant linear state feedback control system.

in the form
$$u(t) = -\bar{F}x(t) + F_r x_r(t) + F_s z_0, \qquad \text{3-592}$$
where $x_r(t)$ is the state of the reference variable, z_0 the set point, and $\bar{F}$, F_r, and F_s are constant matrices. The matrix $\bar{F}$ is given by
$$\bar{F} = R_2^{-1} B^T \bar{P}, \qquad \text{3-593}$$
where $\bar{P}$ is the nonnegative-definite solution of the algebraic Riccati equation
$$0 = D^T R_3 D - \bar{P} B R_2^{-1} B^T \bar{P} + A^T \bar{P} + \bar{P} A. \qquad \text{3-594}$$

In Chapter 2 (Section 2.10) we saw that the ability of the closed-loop system to suppress disturbances or to compensate for parameter changes as compared to an equivalent open-loop configuration is determined by the behavior of the return difference matrix $J(s)$. Let us derive $J(s)$ in the present case. The transfer matrix of the plant is given by $(sI - A)^{-1}B$, while that of the feedback link is simply $\bar{F}$. Thus the return difference matrix is
$$J(s) = I + (sI - A)^{-1} B \bar{F}. \qquad \text{3-595}$$

Note that we consider the complete state $x(t)$ as the controlled variable (see Section 2.10).

We now derive an expression for $J(s)$ starting from the algebraic Riccati equation **3-594**. Addition and substraction of an extra term $s\bar{P}$ yields after rearrangement

$$0 = D^T R_3 D - \bar{P} B R_2^{-1} B^T \bar{P} - (-sI - A^T)\bar{P} - \bar{P}(sI - A). \quad \text{3-596}$$

Premultiplication by $B^T(-sI - A^T)^{-1}$ and postmultiplication by $(sI - A)^{-1}B$ gives

$$0 = B^T(-sI - A^T)^{-1}(-\bar{P} B R_2^{-1} B^T \bar{P} + D^T R_3 D^T)(sI - A)^{-1}B$$
$$- B^T \bar{P}(sI - A)^{-1}B - B^T(-sI - A^T)^{-1}\bar{P} B. \quad \text{3-597}$$

This can be rearranged as follows:

$$[I + B^T(-sI - A^T)^{-1}\bar{P} B R_2^{-1}] R_2 [I + R_2^{-1} B^T \bar{P}(sI - A)^{-1}B]$$
$$= R_2 + B^T(-sI - A^T)^{-1} D^T R_3 D(sI - A)^{-1}B. \quad \text{3-598}$$

After substitution of $R_2^{-1} B^T \bar{P} = \bar{F}$, this can be rewritten as

$$[I + B^T(-sI - A^T)^{-1} \bar{F}^T] R_2 [I + \bar{F}(sI - A)^{-1}B]$$
$$= R_2 + H^T(-s) R_3 H(s), \quad \text{3-599}$$

where $H(s) = D(sI - A)^{-1}B$. Premultiplication of both sides of **3-599** by $\bar{F}^T$ and postmultiplication by $\bar{F}$ yields after a simple manipulation

$$[I + \bar{F}^T B^T(-sI - A^T)^{-1}] \bar{F}^T R_2 \bar{F} [I + (sI - A)^{-1} B \bar{F}]$$
$$= \bar{F}^T R_2 \bar{F} + \bar{F}^T H^T(-s) R_3 H(s) \bar{F}, \quad \text{3-600}$$

or

$$J^T(-s) \bar{F}^T R_2 \bar{F} J(s) = \bar{F}^T R_2 \bar{F} + \bar{F}^T H^T(-s) R_3 H(s) \bar{F}. \quad \text{3-601}$$

If we now substitute $s = j\omega$, we see that the second term on the right-hand side of this expression is nonnegative-definite Hermitian; this means that we can write

$$J^T(-j\omega) W J(j\omega) \geq W \quad \text{for all real } \omega, \quad \text{3-602}$$

where

$$W = \bar{F}^T R_2 \bar{F}. \quad \text{3-603}$$

We know from Section 2.10 that a condition of the form **3-602** guarantees disturbance suppression and compensation of parameter changes as compared to the equivalent open-loop system *for all frequencies*. This is a useful result. We know already from Section 3.6 that the optimal regulator gives *optimal* protection against *white* noise disturbances entering at the input side of the plant. The present result shows, however, that protection against disturbances is not restricted to this special type of disturbances only. By the same token, compensation of parameter changes is achieved.

Thus we have obtained the following result (Kreindler, 1968b; Anderson and Moore, 1971).

3.9 Sensitivity

Theorem 3.15. *Consider the system configuration of Fig. 3.31, where the "plant" is the detectable and stabilizable time-invariant system*

$$\dot{x}(t) = Ax(t) + Bu(t). \qquad \text{3-604}$$

Let the feedback gain matrix be given by

$$\bar{F} = R_2^{-1} B^T \bar{P}, \qquad \text{3-605}$$

where $\bar{P}$ is the nonnegative-definite solution of the algebraic Riccati equation

$$0 = D^T R_3 D - \bar{P} B R_2^{-1} B^T \bar{P} + A^T \bar{P} + \bar{P} A. \qquad \text{3-606}$$

Then the return difference

$$J(s) = I + (sI - A)^{-1} B \bar{F} \qquad \text{3-607}$$

satisfies the inequality

$$J^T(-j\omega) W J(j\omega) \geq W \quad \text{for all real } \omega, \qquad \text{3-608}$$

where

$$W = \bar{F}^T R_2 \bar{F}. \qquad \text{3-609}$$

For an extension of this result to time-varying systems, we refer the reader to Kreindler (1969).

It is clear that with the configuration of Fig. 3.31 improved protection is achieved only against disturbances and parameter variations *inside* the feedback loop. In particular, variations in D fully affect the controlled variable $z(t)$. It frequently happens, however, that D does not exhibit variations. This is especially the case if the controlled variable is composed of components of the state vector, which means that $z(t)$ is actually inside the loop (see Fig. 3.32).

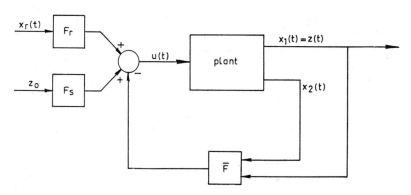

Fig. 3.32. Example of a situation in which the controlled variable is inside the feedback loop.

Theorem 3.15 has the shortcoming that the weighting matrix $\bar{F}^T R_2 \bar{F}$ is known only *after* the control law has been computed; this makes it difficult to choose the design parameters R_3 and R_2 such as to achieve a given weighting matrix. We shall now see that under certain conditions it is possible to determine an asymptotic expression for W. In Section 3.8.3 it was found that if dim (z) = dim (u), and the open-loop transfer matrix $H(s) = D(sI - A)^{-1}B$ does not have any right-half plane zeroes, the solution $\bar{P}$ of the algebraic Riccati equation approaches the zero matrix as the weighting matrix R_2 approaches the zero matrix. A glance at the algebraic Riccati equation 3-594 shows that this implies that

$$\bar{P}BR_2^{-1}B^T\bar{P} \to D^T R_3 D \qquad \text{3-610}$$

as $R_2 \to 0$, or, since $R_2^{-1}B^T\bar{P} = \bar{F}$, that

$$\bar{F}^T R_2 \bar{F} \to D^T R_3 D \qquad \text{3-611}$$

as $R_2 \to 0$. This proves that the weighting matrix W in the sensitivity criterion 3-608 approaches $D^T R_3 D$ as $R_2 \to 0$.

We have considered the entire state $x(t)$ as the feedback variable. This means that the weighted square tracking error is

$$x^T(t)Wx(t). \qquad \text{3-612}$$

From the results we have just obtained, it follows that as $R_2 \to 0$ this can be replaced with

$$x^T(t)D^T R_3 D x(t) = z^T(t) R_3 z(t). \qquad \text{3-613}$$

This means (see Section 2.10) that in the limit $R_2 \to 0$ the controlled variable receives all the protection against disturbances and parameter variations, and that the components of the controlled variable are weighted by R_3. This is a useful result because it is the controlled variable we are most interested in.

The property derived does *not* hold, however, for plants with zeroes in the right-half plane, or with too few inputs, because here $\bar{P}$ does not approach the zero matrix.

We summarize our conclusions:

Theorem 3.16. *Consider the weighting matrix*

$$W = \bar{F}^T R_2 \bar{F}, \qquad \text{3-614}$$

where

$$\bar{F} = R_2^{-1} B^T \bar{P}, \qquad \text{3-615}$$

with $\bar{P}$ the nonnegative-definite symmetric solution of

$$0 = D^T R_3 D - \bar{P}BR_2^{-1}B^T\bar{P} + A^T\bar{P} + \bar{P}A. \qquad \text{3-616}$$

If the conditions are satisfied (Theorem 3.14) under which $\bar{P} \to 0$ *as* $R_2 \to 0$, *then*

$$W \to D^T R_3 D \qquad \text{3-617}$$

as $R_2 \to 0$.

The results of this section indicate in a general way that state feedback systems offer protection against disturbances and parameter variations. Since sensitivity matrices are not very convenient to work with, indications as to what to do for specific parameter variations are not easily found. The following general conclusions are valid, however.

1. As the weighting matrix R_2 is decreased the protection against disturbances and parameter variations improves, since the feedback gains increase. For plants with zeroes in the left-half complex plane only, the break frequency up to which protection is obtained is determined by the faraway closed-loop poles, which move away from the origin as R_2 decreases.

2. For plants with zeroes in the left-half plane only, most of the protection extends to the controlled variable. The weight attributed to the various components of the controlled variable is determined by the weighting matrix R_3.

3. For plants with zeroes in the right-half plane, the break frequency up to which protection is obtained is limited by those nearby closed-loop poles that are not canceled by zeroes.

Example 3.26. *Position control system*

As an illustration of the theory of this section, let us perform a brief sensitivity analysis of the position control system of Example 3.8 (Section 3.4.1). With the numerical values given, it is easily found that the weighting matrix in the sensitivity criterion is given by

$$W = \bar{F}^T R_2 \bar{F} = \begin{pmatrix} 1 & 0.08364 \\ 0.08364 & 0.006994 \end{pmatrix}. \qquad \text{3-618}$$

This is quite close to the limiting value

$$D^T R_3 D = \begin{pmatrix} 1 & 0 \\ 0 & 0 \end{pmatrix}. \qquad \text{3-619}$$

To study the sensitivity of the closed-loop system to parameter variations, in Fig. 3.33 the response of the closed-loop system is depicted for nominal and off-nominal conditions. Here the off-nominal conditions are caused by a change in the inertia of the load driven by the position control system. The curves *a* correspond to the nominal case, while in the case of curves *b* and *c* the combined inertia of load and armature of the motor is $\frac{2}{3}$ of nominal

318 Optimal Linear State Feedback Control Systems

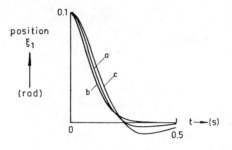

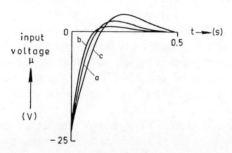

Fig. 3.33. The effect of parameter variations on the response of the position control system: (*a*) Nominal load; (*b*) inertial load $\frac{2}{3}$ of nominal; (*c*) inertial load $\frac{3}{2}$ of nominal.

and $\frac{3}{2}$ of nominal, respectively. A change in the total moment of inertia by a certain factor corresponds to division of the constants α and κ by the same factor. Thus $\frac{2}{3}$ of the nominal moment of inertia yields 6.9 and 1.18 for α and κ, respectively, while $\frac{3}{2}$ of the nominal moment of inertia results in the values 3.07 and 0.525 for α and κ, respectively. Figure 3.33 vividly illustrates the limited effect of relatively large parameter variations.

3.10 CONCLUSIONS

This chapter has dealt with state feedback control systems where all the components of the state can be accurately measured at all times. We have discussed quite extensively how linear state feedback control systems can be designed that are optimal in the sense of a quadratic integral criterion. Such systems possess many useful properties. They can be made to exhibit a satisfactory transient response to nonzero initial conditions, to an external reference variable, and to a change in the set point. Moreover, they have excellent stability characteristics and are insensitive to disturbances and parameter variations.

All these properties can be achieved in the desired measure by appropriately choosing the controlled variable of the system and properly adjusting the weighting matrices R_3 and R_2. The results of Sections 3.8 and 3.9, which concern the asymptotic properties and the sensitivity properties of steady-state control laws, give considerable insight into the influence of the weighting matrices.

A major objection to the theory of this section, however, is that very often it is either too costly or impossible to measure all components of the state. To overcome this difficulty, we study in Chapter 4 the problem of reconstructing the state of the system from incomplete and inaccurate measurements. Following this in Chapter 5 it is shown how the theory of linear state feedback control can be integrated with the theory of state reconstruction to provide a general theory of optimal linear feedback control.

3.11 PROBLEMS

3.1. *Stabilization of the position control system*

Consider the position control system of Example 3.4 (Section 3.3.1). Determine the set of all linear control laws that stabilize the position control system.

3.2. *Position control of a frictionless dc motor*

A simplification of the regulator problem of Example 3.4 (Section 3.3.1) occurs when we neglect the friction in the motor; the state differential equation then takes the form

$$\dot{x}(t) = \begin{pmatrix} 0 & 1 \\ 0 & 0 \end{pmatrix} x(t) + \begin{pmatrix} 0 \\ \kappa \end{pmatrix} \mu(t), \qquad \text{3-620}$$

where $x(t) = \text{col}\,[\xi_1(t), \xi_2(t)]$. Take as the controlled variable

$$\zeta(t) = (1,\ 0)x(t), \qquad \text{3-621}$$

and consider the criterion

$$\int_{t_0}^{t_1} [\zeta^2(t) + \rho\mu^2(t)]\,dt. \qquad \text{3-622}$$

(a) Determine the steady-state solution $\bar{P}$ of the Riccati equation.

(b) Determine the steady-state control law.

(c) Compute the closed-loop poles. Sketch the loci of the closed-loop poles as ρ varies.

(d) Use the numerical values $\kappa = 150 \text{ rad}/(\text{V s}^2)$ and $\rho = 2.25 \text{ rad}^2/\text{V}^2$ and determine by computation or simulation the response of the closed-loop system to the initial condition $\xi_1(0) = 0.1 \text{ rad}$, $\xi_2(0) = 0 \text{ rad/s}$.

3.3. *Regulation of an amplidyne*
Consider the amplidyne of Problem 1.2.

(a) Suppose that the output voltage is to be kept at a constant value e_{20}. Denote the nominal input voltage as e_{00} and represent the system in terms of a shifted state variable with zero as nominal value.

(b) Choose as the controlled variable

$$\zeta(t) = e_2(t) - e_{20}, \qquad \text{3-623}$$

and consider the criterion

$$\int_{t_0}^{t_1} [\zeta^2(t) + \rho \mu'^2(t)] \, dt \qquad \text{3-624}$$

where

$$\mu'(t) = e_0(t) - e_{00}. \qquad \text{3-625}$$

Find the steady-state solution of the resulting regulator problem for the following numerical values:

$$\begin{aligned}
\frac{R_1}{L_1} &= 10 \text{ s}^{-1}, & \frac{R_2}{L_2} &= 1 \text{ s}^{-1}, \\
R_1 &= 5 \, \Omega, & R_2 &= 10 \, \Omega, \\
k_1 &= 20 \text{ V/A}, & k_2 &= 50 \text{ V/A}, \\
\rho &= 0.025.
\end{aligned} \qquad \text{3-626}$$

(c) Compute the closed-loop poles.

(d) Compute or simulate the response of the closed-loop system to the initial conditions $x(0) = \text{col}(1, 0)$ and $x(0) = \text{col}(0, 1)$.

3.4. *Stochastic position control system*
Consider the position control problem of Example 3.4 (Section 3.3.1) but assume that in addition to the input a stochastically varying torque operates upon the system so that the state differential equation **3-59** must be extended as follows:

$$\dot{x}'(t) = \begin{pmatrix} 0 & 1 \\ 0 & -\alpha \end{pmatrix} x'(t) + \begin{pmatrix} 0 \\ \kappa \end{pmatrix} \mu(t) + \begin{pmatrix} 0 \\ \nu(t) \end{pmatrix}. \qquad \text{3-627}$$

Here $\nu(t)$ represents the effect of the disturbing torque. We model $\nu(t)$ as exponentially correlated noise:

$$\dot{\nu}(t) = -\frac{1}{\theta} \nu(t) + \omega(t), \qquad \text{3-628}$$

where $\omega(t)$ is white noise with intensity $2\sigma^2/\theta$.

(a) Consider the controlled variable

$$\zeta(t) = (1, \; 0) x'(t) \qquad \text{3-629}$$

and the criterion

$$E\left\{\int_{t_0}^{t_1}[\zeta^2(t) + \rho\mu^2(t)]\,dt\right\}.\qquad 3\text{-}630$$

Find the steady-state solution of the corresponding stochastic regulator problem.

(b) Use the numerical values

$$\kappa = 0.787 \text{ rad}/(\text{V s}^2),$$
$$\alpha = 4.6 \text{ s}^{-1},\qquad 3\text{-}631$$
$$\sigma = 5 \text{ rad/s}^2,$$
$$\theta = 1 \text{ s}.$$

Compute the steady-state rms values of the controlled variable $\zeta(t)$ and the input $\mu(t)$ for $\rho = 0.2 \times 10^{-4}$ rad^2/V^2.

3.5. *Angular velocity tracking system*

Consider the angular velocity tracking problem of Examples 3.12 (Section 3.6.2) and 3.14 (Section 3.6.3). In Example 3.14 we found that the value of ρ that was chosen ($\rho = 1000$) leaves considerable room for improvement.

(a) Vary ρ and select that value of ρ that results in a steady-state rms input voltage of 3 V.

(b) Compute the corresponding steady-state rms tracking error.

(c) Compute the corresponding break frequency of the closed-loop system and compare this to the break frequency of the reference variable.

3.6. *Nonzero set point regulator for an amplidyne*

Consider Problem 3.3 where a regulator has been derived for an amplidyne.

(a) Using the results of this problem, find the nonzero set point regulator.

(b) Simulate or calculate the response of the regulator to a step in the output voltage set point of 10 V.

3.7. *Extension of the regulator problem*

Consider the linear time-varying system

$$\dot{x}(t) = A(t)x(t) + B(t)u(t) \qquad 3\text{-}632$$

with the generalized quadratic criterion

$$\int_{t_0}^{t_1}[x^T(t)R_1(t)x(t) + 2x^T(t)R_{12}(t)u(t) + u^T(t)R_2(t)u(t)]\,dt + x^T(t_1)P_1x(t_1),$$

$$3\text{-}633$$

where $R_1(t)$, $R_{12}(t)$, and $R_2(t)$ are matrices of appropriate dimensions.

322 Optimal Linear State Feedback Control Systems

(a) Show that the problem of minimizing **3-633** for the system **3-632** can be reformulated as minimizing the criterion

$$\int_{t_0}^{t_1} [x^T(t)R_1'(t)x(t) + u'^T(t)R_2(t)u'(t)]\, dt + x^T(t_1)P_1 x(t_1) \qquad \text{3-634}$$

for the system
$$\dot{x}(t) = A'(t)x(t) + B(t)u'(t), \qquad \text{3-635}$$
where
$$\begin{aligned} R_1'(t) &= R_1(t) - R_{12}(t)R_2^{-1}(t)R_{12}^T(t), \\ u'(t) &= u(t) + R_2^{-1}(t)R_{12}^T(t)x(t), \\ A'(t) &= A(t) - B(t)R_2^{-1}(t)R_{12}^T(t) \end{aligned} \qquad \text{3-636}$$

(Kalman, 1964; Anderson, 1966a; Anderson and Moore, 1971).

(b) Show that **3-633** is minimized for the system **3-632** by letting
$$u(t) = -F^0(t)x(t), \qquad \text{3-637}$$
where
$$F^0(t) = R_2^{-1}(t)[B^T(t)P(t) + R_{12}^T(t)], \qquad \text{3-638}$$
with $P(t)$ the solution of the matrix Riccati equation

$$\begin{aligned} -\dot{P}(t) = {}& [A(t) - B(t)R_2^{-1}(t)R_{12}^T(t)]^T P(t) \\ & + P(t)[A(t) - B(t)R_2^{-1}(t)R_{12}^T(t)] \\ & + R_1(t) - R_{12}(t)R_2^{-1}(t)R_{12}^T(t) \\ & - P(t)B(t)R_2^{-1}(t)B^T(t)P(t), \qquad t \leq t_1, \end{aligned} \qquad \text{3-639}$$
$$P(t_1) = P_1.$$

(c) For arbitrary $F(t)$, $t \leq t_1$, let $\tilde{P}(t)$ be the solution of the matrix differential equation

$$\begin{aligned} -\dot{\tilde{P}}(t) = {}& [A(t) - B(t)F(t)]^T \tilde{P}(t) + \tilde{P}(t)[A(t) - B(t)F(t)] \\ & + R_1(t) - R_{12}(t)F(t) - F^T(t)R_{12}^T(t) \\ & + F^T(t)R_2(t)F(t), \qquad t \leq t_1, \end{aligned} \qquad \text{3-640}$$
$$P(t_1) = P_1.$$

Show that by choosing $F(t)$ equal to $F^0(t)$, $\tilde{P}(t)$ is minimized in the sense that $\tilde{P}(t) \geq P(t)$, $t \leq t_1$, where $P(t)$ is the solution of **3-639**. *Remark:* The proof of (c) follows from (b). One can also prove that **3-637** is the best *linear* control law by rearranging **3-640** and applying Lemma 3.1 (Section 3.3.3) to it.

3.8*. *Solutions of the algebraic Riccati equation* (O'Donnell, 1966; Anderson, 1966b; Potter, 1964)

Consider the algebraic Riccati equation

$$0 = R_1 - \bar{P}BR_2^{-1}B^T\bar{P} + \bar{P}A + A^T\bar{P}. \qquad \text{3-641}$$

Let Z be the matrix

$$Z = \begin{pmatrix} A & -BR_2^{-1}B^T \\ -R_1 & -A^T \end{pmatrix}. \qquad \text{3-642}$$

Z can always be represented as

$$Z = WJW^{-1}, \qquad \text{3-643}$$

where J is the Jordan canonical form of Z. It is always possible to arrange the columns of W such that J can be partitioned as

$$J = \begin{pmatrix} J_{11} & 0 \\ J_{21} & J_{22} \end{pmatrix}. \qquad \text{3-644}$$

Here J_{11}, J_{21} and J_{22} are $n \times n$ blocks. Partition W accordingly as

$$W = \begin{pmatrix} W_{11} & W_{12} \\ W_{21} & W_{22} \end{pmatrix}. \qquad \text{3-645}$$

(a) Consider the equality

$$ZW = WJ, \qquad \text{3-646}$$

and show by considering the 12- and 22-blocks of this equality that if W_{12} is nonsingular $\bar{P} = W_{22}W_{12}^{-1}$ is a solution of the algebraic Riccati equation. Note that in this manner many solutions can be obtained by permuting the order of the characteristic values in J.

(b) Show also that the characteristic values of the matrix $A - BR_2^{-1}B^T W_{22}W_{12}^{-1}$ are precisely the characteristic values of J_{22} and that the (generalized) characteristic vectors of this matrix are the columns of W_{12}. *Hint*: Evaluate the 12-block of the identity **3-646**.

3.9*. *Steady-state solution of the Riccati equation by diagonalization*

Consider the $2n \times 2n$ matrix Z as given by **3-247** and suppose that it cannot be diagonalized. Then Z can be represented as

$$Z = WJW^{-1}, \qquad \text{3-647}$$

where J is the Jordan canonical form of Z, and W is composed of the characteristic vectors and generalized characteristic vectors of Z. It is always possible to arrange the columns of W such that J can be partitioned as follows

$$J = \begin{pmatrix} J_{11} & 0 \\ J_{21} & J_{22} \end{pmatrix}, \qquad \text{3-648}$$

where the $n \times n$ matrix J_{11} has as diagonal elements those characteristic values of Z that have positive real parts and half of those that have zero real parts. Partition W and $V = W^{-1}$ accordingly as

$$W = \begin{pmatrix} W_{11} & W_{12} \\ W_{21} & W_{22} \end{pmatrix}, \quad V = \begin{pmatrix} V_{11} & V_{12} \\ V_{21} & V_{22} \end{pmatrix}. \quad \text{3-649}$$

Assume that $\{A, B\}$ is stabilizable and $\{A, D\}$ detectable. Follow the argument of Section 3.4.4 closely and show that for the present case the following conclusions hold.

(a) The steady-state solution $\bar{P}$ of the Riccati equation

$$-\dot{P}(t) = R_1 - P(t)BR_2^{-1}B^TP(t) + A^TP(t) + P(t)A \quad \text{3-650}$$

satisfies

$$V_{11} + V_{12}\bar{P} = 0. \quad \text{3-651}$$

(b) W_{12} is nonsingular and

$$\bar{P} = W_{22}W_{12}^{-1}. \quad \text{3-652}$$

(c) The steady-state optimal behavior of the state is given by

$$x(t) = W_{12}e^{J_{22}(t-t_0)}W_{12}^{-1}x(t_0). \quad \text{3-653}$$

Hence Z has no characteristic values with zero real parts, and the steady-state closed-loop poles consist of those characterstic values of Z that have negative real parts. *Hint:* Show that

$$e^{Jt} = \begin{pmatrix} e^{J_{11}t} & 0 \\ X(t) & e^{J_{22}t} \end{pmatrix}, \quad \text{3-654}$$

where the precise form of $X(t)$ is unimportant.

3.10*. *Bass' relation for $\bar{P}$* (Bass, 1967)

Consider the algebraic Riccati equation

$$0 = R_1 - \bar{P}BR_2^{-1}B^T\bar{P} + A^T\bar{P} + \bar{P}A \quad \text{3-655}$$

and suppose that the conditions are satisfied under which it has a unique nonnegative-definite symmetric solution. Let the matrix Z be given by

$$Z = \begin{pmatrix} A & -BR_2^{-1}B^T \\ -R_1 & -A^T \end{pmatrix}. \quad \text{3-656}$$

It follows from Theorem 3.8 (Section 3.4.4) that Z has no characteristic values with zero real parts. Factor the characteristic polynomial of Z as follows

$$\det(sI - Z) = \phi(s)\phi(-s) \quad \text{3-657}$$

such that the roots of $\phi(s)$ have strictly negative real parts. Show that $\bar{P}$

satisfies the relation:

$$\phi(Z)\begin{pmatrix} I \\ P \end{pmatrix} = 0. \qquad 3\text{-}658$$

Hint: Write $\phi(Z) = \phi(WJW^{-1}) = W\phi(J)W^{-1} = W\phi(J)V$ where $V = W^{-1}$ and $J = \text{diag}(\Lambda, -\Lambda)$ in the notation of Section 3.4.4.

3.11*. *Negative exponential solution of the Riccati equation* (Vaughan, 1969)

Using the notation of Section 3.4.4, show that the solution of the time-invariant Riccati equation

$$-\dot{P}(t) = R_1 - P(t)BR_2^{-1}B^T P(t) + A^T P(t) + P(t)A,$$
$$P(t_1) = P_1, \qquad 3\text{-}659$$

can be expressed as follows:

$$P(t) = [W_{22} + W_{21}G(t_1 - t)][W_{12} + W_{11}G(t_1 - t)]^{-1}, \qquad 3\text{-}660$$

where

$$G(t) = e^{-\Lambda t} S e^{-\Lambda t}, \qquad 3\text{-}661$$

with

$$S = (V_{11} + V_{12}P_1)(V_{21} + V_{22}P_1)^{-1}. \qquad 3\text{-}662$$

Show with the aid of Problem 3.12 that S can also be written in terms of W as

$$S = -(W_{22} - P_1 W_{12})^{-1}(W_{21} - P_1 W_{11}). \qquad 3\text{-}663$$

3.12*. *The relation between W and V*

Consider the matrix Z as defined in Section 3.4.4.

(a) Show that if $e = \text{col}(e', e'')$, where e' and e'' both are n-dimensional vectors, is a right characteristic vector of Z corresponding to the characteristic value λ, that is, $Ze = \lambda e$, then $(e''^T, -e'^T)$ is a left characteristic vector of Z corresponding to the characteristic value $-\lambda$, that is,

$$(e''^T, -e'^T)Z = -\lambda(e''^T, -e'^T). \qquad 3\text{-}664$$

(b) Assume for simplicity that all characteristic values λ_i, $i = 1, 2, \cdots, 2n$, of Z are distinct and let the corresponding characteristic vectors be given by e_i, $i = 1, 2, \cdots, 2n$. Scale the e_i such that if the characteristic vector $e = \text{col}(e', e'')$ corresponds to a characteristic value λ, and $f = \text{col}(f', f'')$ corresponds to $-\lambda$, then

$$f''^T e' - f'^T e'' = 1. \qquad 3\text{-}665$$

Show that if W is a matrix of which the columns are e_i, $i = 1, 2, \cdots, 2n$, and we partition

$$W = \begin{pmatrix} W_{11} & W_{12} \\ W_{21} & W_{22} \end{pmatrix}, \qquad 3\text{-}666$$

then (O'Donnell, 1966; Walter, 1970)

$$W^{-1} = V = \begin{pmatrix} W_{22}^T & -W_{12}^T \\ -W_{21}^T & W_{11}^T \end{pmatrix}. \qquad 3\text{-}667$$

Hint: Remember that left and right characteristic vectors for different characteristic values are orthogonal.

3.13*. *Frequency domain solution of regulator problems*

For single-input time-invariant systems in phase-variable canonical form, the regulator problem can be conveniently solved in the frequency domain. Let

$$\dot{x}(t) = Ax(t) + b\mu(t) \qquad 3\text{-}668$$

be given in phase-variable canonical form and consider the problem of minimizing

$$\int_{t_0}^{\infty} [\zeta^2(t) + \rho\mu^2(t)] \, dt, \qquad 3\text{-}669$$

where

$$\zeta(t) = dx(t). \qquad 3\text{-}670$$

(a) Show that the closed-loop characteristic polynomial can be found by factorization of the polynomial

$$1 + \frac{1}{\rho} H(s)H(-s), \qquad 3\text{-}671$$

where $H(s)$ is the open-loop transfer function $H(s) = d(sI - A)^{-1}b$.

(b) For a given closed-loop characteristic polynomial, show how the corresponding control law

$$\mu(t) = -\bar{f}x(t) \qquad 3\text{-}672$$

can be found. *Hint:* Compare Section 3.2.

3.14*. *The minimum number of faraway closed-loop poles*

Consider the problem of minimizing

$$\int_{t_0}^{\infty} [x^T(t)R_1 x(t) + \rho u^T(t) N u(t)] \, dt, \qquad 3\text{-}673$$

where $R_1 \geq 0$, $N > 0$, and $\rho > 0$, for the system

$$\dot{x}(t) = Ax(t) + Bu(t). \qquad 3\text{-}674$$

(a) Show that as $\rho \downarrow 0$ some of the closed-loop poles go to infinity while the others stay finite. Show that those poles that remain finite approach the left-half plane zeroes of

$$\det \left[B^T(-sI - A^T)^{-1} R_1 (sI - A)^{-1} B \right]. \qquad 3\text{-}675$$

(b) Prove that at least k closed-loop poles approach infinity, where k is the dimension of the input u. *Hint:* Let $|s| \to \infty$ to determine the maximum number of zeroes of **3-675**. Compare the proof of Theorem 1.19 (Section 1.5.3).

(c) Prove that as $\rho \to \infty$ the closed-loop poles approach the numbers $\hat{\pi}_i$, $i = 1, 2, \cdots, n$, which are the characteristic values of the matrix A mirrored into the left-half complex plane.

3.15*. *Estimation of the radius of the faraway closed-loop poles from the Bode plot* (Leake, 1965; Schultz and Melsa, 1967, Section 8.4)

Consider the problem of minimizing

$$\int_{t_0}^{\infty} [\zeta^2(t) + \rho \mu^2(t)]\, dt \qquad \text{3-676}$$

for the single-input single-output system

$$\begin{aligned} \dot{x}(t) &= Ax(t) + b\mu(t), \\ \zeta(t) &= dx(t). \end{aligned} \qquad \text{3-677}$$

Suppose that a Bode plot is available of the open-loop frequency response function $H(j\omega) = d(j\omega I - A)^{-1}b$. Show that for small ρ the radius of the faraway poles of the steady-state optimal closed-loop system can be estimated: as the frequency ω_c for which $|H(j\omega_c)| = \sqrt{\rho}$.

4 OPTIMAL LINEAR RECONSTRUCTION OF THE STATE

4.1 INTRODUCTION

All the versions of the regulator and tracking problems solved in Chapter 3 have the following basic assumption in common: *the complete state vector can be measured accurately.* This assumption is often unrealistic. The most frequent situation is that for a given system

$$\dot{x}(t) = A(t)x(t) + B(t)u(t), \qquad x(t_0) = x_0, \qquad \textbf{4-1}$$

only certain linear combinations of the state, denoted by y, can be measured:

$$y(t) = C(t)x(t). \qquad \textbf{4-2}$$

The quantity y, which is assumed to be an l-dimensional vector, with l usually less than the dimension n of the state x, will be referred to as the *observed variable*.

The purpose of this chapter is to present methods of reconstructing the state vector, or finding approximations to the state vector, from the observed variable. In particular, we wish to find a functional F,

$$x'(t) = F[y(\tau), t_0 \leq \tau \leq t], \qquad t_0 \leq t, \qquad \textbf{4-3}$$

such that $x'(t) \simeq x(t)$, where $x'(t)$ represents the *reconstructed* state. Here t_0 is the initial time of the observations. Note that $F[y(\tau), t_0 \leq \tau \leq t]$, the reconstructed $x(t)$, is a function of the *past* observations $y(\tau)$, $t_0 \leq \tau \leq t$, and does not depend upon future observations, $y(\tau)$, $\tau \geq t$. Once the state vector has been reconstructed, we shall be able to use the control laws of Chapter 3, which assume knowledge of the complete state vector, by replacing the *actual* state with the *reconstructed* state.

In Section 4.2 we introduce the *observer*, which is a dynamic system whose output approaches, as time increases, the state that must be reconstructed. Although this approach does not explicitly take into account the difficulties that arise because of the presence of noise, it seeks methods of reconstructing the state that implicitly involve a certain degree of filtering of the noise.

In Section 4.3 we introduce all the stochastic phenomena associated with the problem explicitly and quantitatively and find the *optimal observer*, also referred to as the *Kalman–Bucy filter*. The derivation of the optimal observer is based upon the fact that the optimal observer problem is "dual" to the optimal regulator problem presented in Chapter 3.

Finally, in Section 4.4 the steady-state and asymptotic properties of the Kalman–Bucy filter are studied. These results are easily obtained from optimal regulator theory using the duality of the optimal regulator and observer problems.

4.2 OBSERVERS

4.2.1 Full-Order Observers

In order to reconstruct the state x of the system **4-1** from the observed variable y as given by **4-2**, we propose a linear differential system the output of which is to be an approximation to the state x in a suitable sense. It will be investigated what structure this system should have and how it should behave. We first introduce the following terminology (Luenberger, 1966).

Definition 4.1. *The system*
$$\dot{q}(t) = F(t)q(t) + G(t)y(t) + H(t)u(t),$$
$$z(t) = K(t)q(t) + L(t)y(t) + M(t)u(t),$$
4-4

*is an **observer** for the system*
$$\dot{x}(t) = A(t)x(t) + B(t)u(t),$$
$$y(t) = C(t)x(t),$$
4-5

if for every initial state $x(t_0)$ of the system **4-5** *there exists an initial state q_0 for the system* **4-4** *such that*
$$q(t_0) = q_0 \qquad \textbf{4-6}$$
implies
$$z(t) = x(t), \qquad t \geq t_0, \qquad \textbf{4-7}$$
for all $u(t), t \geq t_0$.

We note that the observer **4-4** has the system input u and the system observed variable y as inputs, and as output the variable z. We are mainly interested in observers of a special type where the state $q(t)$ of the observer itself is to be an approximation to the system state $x(t)$:

Definition 4.2. *The n-dimensional system*
$$\dot{\hat{x}}(t) = F(t)\hat{x}(t) + G(t)y(t) + H(t)u(t) \qquad \textbf{4-8}$$

is a **full-order observer** for the n-dimensional system

$$\dot{x}(t) = A(t)x(t) + B(t)u(t), \quad \textbf{4-9a}$$

$$y(t) = C(t)x(t), \quad \textbf{4-9b}$$

if

$$\hat{x}(t_0) = x(t_0) \quad \textbf{4-10}$$

implies

$$\hat{x}(t) = x(t), \quad t \geq t_0, \quad \textbf{4-11}$$

for all $u(t)$, $t \geq t_0$.

The observer **4-8** is called a full-order observer since its state $\hat{x}$ has the same dimension as the state x of the system **4-9**. In Section 4.2.3 we consider observers of the type **4-4** whose dimension is less than that of the state x. Such observers will be called *reduced-order observers*.

We now investigate what conditions the matrices F, G, and H must satisfy so that **4-8** qualifies as an observer. We first state the result.

Theorem 4.1. *The system **4-8** is an observer for the system **4-9** if, and only if,*

$$F(t) = A(t) - K(t)C(t),$$
$$G(t) = K(t), \quad \textbf{4-12}$$
$$H(t) = B(t),$$

where $K(t)$ is an arbitrary time-varying matrix. As a result, full-order observers have the following structure:

$$\dot{\hat{x}}(t) = A(t)\hat{x}(t) + B(t)u(t) + K(t)[y(t) - C(t)\hat{x}(t)]. \quad \textbf{4-13}$$

This theorem can be proved as follows. By subtracting **4-8** from **4-9a** and using **4-9b**, the following differential equation for $x(t) - \hat{x}(t)$ is obtained:

$$\dot{x}(t) - \dot{\hat{x}}(t) = [A(t) - G(t)C(t)]x(t) - F(t)\hat{x}(t) + [B(t) - H(t)]u(t). \quad \textbf{4-14}$$

This immediately shows that $x(t) = \hat{x}(t)$ for $t \geq t_0$, for all $u(t)$, $t \geq t_0$, implies **4-12**. Conversely, if **4-12** is satisfied, it follows that

$$\dot{x}(t) - \dot{\hat{x}}(t) = [A(t) - K(t)C(t)][x(t) - \hat{x}(t)], \quad \textbf{4-15}$$

which shows that if $x(t_0) = \hat{x}(t_0)$ then $x(t) = \hat{x}(t)$ for all $t \geq t_0$, for all $u(t)$, $t \geq t_0$. This concludes the proof of the theorem.

The structure **4-13** follows by substituting **4-12** into **4-8**. Therefore, a full-order observer (see Fig. 4.1) consists simply of a model of the system with as extra driving variable a term that is proportional to the difference

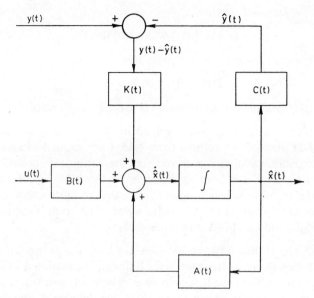

Fig. 4.1. Block diagram of a full-order observer.

$y(t) - \hat{y}(t)$, where
$$\hat{y}(t) = C(t)\hat{x}(t) \qquad \text{4-16}$$
is the observed variable as reconstructed by the observer. We call the matrix $K(t)$ the *gain matrix* of the observer. Up to this point the choice of $K(t)$ for $t \geq t_0$ is still arbitrary.

From **4-13** we see that the observer can also be represented as
$$\dot{\hat{x}}(t) = [A(t) - K(t)C(t)]\hat{x}(t) + B(t)u(t) + K(t)y(t). \qquad \text{4-17}$$
This shows that the *stability* of the observer is determined by the behavior of $A(t) - K(t)C(t)$. Of course stability of the observer is a desirable property in itself, but the following result shows that stability of the observer has further implications.

Theorem 4.2. *Consider the observer*
$$\dot{\hat{x}}(t) = A(t)\hat{x}(t) + B(t)u(t) + K(t)[y(t) - C(t)\hat{x}(t)] \qquad \text{4-18}$$
for the system
$$\begin{aligned} \dot{x}(t) &= A(t)x(t) + B(t)u(t), \\ y(t) &= C(t)x(t). \end{aligned} \qquad \text{4-19}$$
*Then the **reconstruction error***
$$e(t) = x(t) - \hat{x}(t) \qquad \text{4-20}$$

satisfies the differential equation

$$\dot{e}(t) = [A(t) - K(t)C(t)]e(t). \qquad \textbf{4-21}$$

The reconstruction error has the property that

$$e(t) \to 0 \quad \text{as } t \to \infty, \qquad \textbf{4-22}$$

for all $e(t_0)$, if, and only if, the observer is asymptotically stable.

That the reconstruction error, as defined by **4-20**, satisfies the differential equation **4-21** immediately follows from **4-15**. Comparing **4-21** and **4-17**, we see that the stability of the observer and the asymptotic behavior of the reconstruction error are both determined by the behavior of the matrix $A(t) - K(t)C(t)$. This clearly shows that the reconstruction error $e(t)$ approaches zero, irrespective of its initial value, if and only if the observer is asymptotically stable. This is a very desirable result.

Observer design thus revolves about determining the gain matrix $K(t)$ for $t \geq t_0$ such that the reconstruction error differential equation **4-21** is asymptotically stable. In the time-invariant case, where all matrices occurring in the problem formulation are constant, including the gain K, the stability of the observer follows from the locations of the characteristic values of the matrix $A - KC$. We refer to the characteristic values of $A - KC$ as the *observer poles*. In the next section we prove that, under a mildly restrictive condition (complete reconstructibility of the system), all observer poles can be arbitrarily located in the complex plane by choosing K suitably (within the restriction that complex poles occur in complex conjugate pairs).

At this point we can only offer some intuitive guidelines for a choice of K to obtain satisfactory performance of the observer. To obtain fast convergence of the reconstruction error to zero, K should be chosen so that the observer poles are quite deep in the left-half complex plane. This, however, generally must be achieved by making the gain matrix K large, which in turn makes the observer very sensitive to any observation noise that may be present, added to the observed variable $y(t)$. A compromise must be found. Section 4.3 is devoted to the problem of finding an *optimal* compromise, taking into account all the statistical aspects of the problem.

Example 4.1. *Positioning system*

In Example 2.4 (Section 2.3), we considered a positioning system described by the state differential equation

$$\dot{x}(t) = \begin{pmatrix} 0 & 1 \\ 0 & -\alpha \end{pmatrix} x(t) + \begin{pmatrix} 0 \\ \kappa \end{pmatrix} \mu(t). \qquad \textbf{4-23}$$

Here $x(t) = \text{col}\,[\xi_1(t), \xi_2(t)]$, where $\xi_1(t)$ denotes the angular displacement and $\xi_2(t)$ the angular velocity. Let us assume that the observed variable $\eta(t)$ is the angular displacement, that is,

$$\eta(t) = (1, \ 0)x(t).$$

A time-invariant observer for this system is given by

$$\dot{\hat{x}}(t) = \begin{pmatrix} 0 & 1 \\ 0 & -\alpha \end{pmatrix} \hat{x}(t) + \begin{pmatrix} 0 \\ \kappa \end{pmatrix} \mu(t) + \begin{pmatrix} k_1 \\ k_2 \end{pmatrix} [\eta(t) - (1, \ 0)\hat{x}(t)], \quad \text{4-24}$$

where the constant gains k_1 and k_2 are to be selected. The characteristic polynomial of the observer is given by

$$\det \left[\begin{pmatrix} s & 0 \\ 0 & s \end{pmatrix} - \begin{pmatrix} 0 & 1 \\ 0 & -\alpha \end{pmatrix} + \begin{pmatrix} k_1 \\ k_2 \end{pmatrix} (1, \ 0) \right] = \det \left[\begin{pmatrix} s + k_1 & -1 \\ k_2 & s + \alpha \end{pmatrix} \right]$$

$$= s^2 + (\alpha + k_1)s + k_2. \quad \text{4-25}$$

With the numerical values of Example 2.4, the characteristic values of the system **4-23** are located at 0 and $-\alpha = -4.6 \text{ s}^{-1}$. In order to make the observer fast as compared to the system itself, let us select the gains k_1 and k_2 such that the observer poles are located at $-50 \pm j50 \text{ s}^{-1}$. This yields for the gains:

$$k_1 = 95.40 \text{ s}^{-1}, \qquad k_2 = 4561 \text{ s}^{-2}. \quad \text{4-26}$$

In Fig. 4.2 we compare the output of the observer to the actual response of the system. The initial conditions of the positioning system are

$$\xi_1(0) = 0.1 \text{ rad}, \qquad \xi_2(0) = 0.5 \text{ rad/s}, \quad \text{4-27}$$

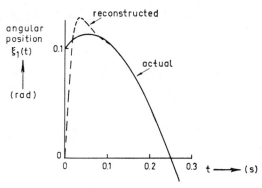

Fig. 4.2. Actual response of a positioning system and the response as reconstructed by a full-order observer.

while the input voltage is given by

$$\mu(t) = -10 \text{ V}, \quad t \geq 0. \qquad \textbf{4-28}$$

The observer has zero initial conditions. Figure 4.2 clearly shows the excellent convergence of the reconstructed angular position to its actual behavior.

4.2.2* Conditions for Pole Assignment and Stabilization of Observers

In this section we state necessary and sufficient conditions for pole assignment and stabilization of time-invariant full-order observers. We first have the following result, which is dual to Theorem 3.1 (Section 3.2.2).

Theorem 4.3. *Consider the time-invariant full-order observer*

$$\dot{\hat{x}}(t) = A\hat{x}(t) + K[y(t) - C\hat{x}(t)] + Bu(t) \qquad \textbf{4-29}$$

for the time-invariant system

$$\begin{aligned} \dot{x}(t) &= Ax(t) + Bu(t), \\ y(t) &= Cx(t). \end{aligned} \qquad \textbf{4-30}$$

Then the observer poles, that is, the characteristic values of $A - KC$, *can be arbitrarily located in the complex plane* (*within the restriction that complex characteristic values occur in complex conjugate pairs*), *by choosing the constant matrix K suitably, if and only if the system* **4-30** *is completely reconstructible.*

To prove this theorem we note that

$$\det[\lambda I - (A - KC)] = \det[\lambda I - (A^T - C^T K^T)], \qquad \textbf{4-31}$$

so that the characteristic values of $A - KC$ are identical to those of $A^T - C^T K^T$. However, by Theorem 3.1 the characteristic values of $A^T - C^T K^T$ can be arbitrarily located by choosing K appropriately if and only if the pair $\{A^T, C^T\}$ is completely controllable. From Theorem 1.41 (Section 1.8), we know that $\{A^T, C^T\}$ is completely controllable if and only if $\{A, C\}$ is completely reconstructible. This completes the proof.

If $\{A, C\}$ is not completely reconstructible, the following theorem, which is dual to Theorem 3.2 (Section 3.2.2) gives conditions for the stability of the observer.

Theorem 4.4. *Consider the time-invariant observer*

$$\dot{\hat{x}}(t) = A\hat{x}(t) + K[y(t) - C\hat{x}(t)] + Bu(t) \qquad \textbf{4-32}$$

for the time-invariant system

$$\begin{aligned} \dot{x}(t) &= Ax(t) + Bu(t), \\ y(t) &= Cx(t). \end{aligned} \qquad \textbf{4-33}$$

Then a matrix K can be found such that the observer is asymptotically stable if and only if the system **4-33** is detectable.

Detectability was defined in Section 1.7.4. The proof of this theorem follows by duality from Theorem 3.2.

4.2.3* Reduced-Order Observers

In this section we show that it is possible to find observers of dimension *less* than the dimension of the system to be observed. Such observers are called *reduced-order observers*. For simplicity we discuss only the time-invariant case. Let the system to be observed be described by

$$\dot{x}(t) = Ax(t) + Bu(t),$$
$$y(t) = Cx(t), \qquad \text{4-34}$$

where the dimension of the state $x(t)$ is n and the dimension of the observed variable $y(t)$ is given by l. Since the observation equation $y(t) = Cx(t)$ provides us with l linear equations in the unknown state $x(t)$, it is necessary to reconstruct only $n - l$ linear combinations of the components of the state. This approach was first considered by Luenberger (1964, 1966). We follow the derivation of Cumming (1969).

Assuming that C has full rank, we introduce an $(n - l)$-dimensional vector $p(t)$,

$$p(t) = C'x(t), \qquad \text{4-35}$$

such that

$$\begin{pmatrix} C \\ C' \end{pmatrix} \qquad \text{4-36}$$

is nonsingular. By the relations

$$y(t) = Cx(t),$$
$$p(t) = C'x(t), \qquad \text{4-37}$$

it follows that

$$x(t) = \begin{pmatrix} C \\ C' \end{pmatrix}^{-1} \begin{pmatrix} y(t) \\ p(t) \end{pmatrix}. \qquad \text{4-38}$$

It is convenient to write

$$\begin{pmatrix} C \\ C' \end{pmatrix}^{-1} = (L_1, L_2), \qquad \text{4-39}$$

so that

$$x(t) = L_1 y(t) + L_2 p(t). \qquad \text{4-40}$$

Thus if we reconstruct $p(t)$ and denote the reconstructed value by $\hat{p}(t)$, we

can write the reconstructed state as
$$\hat{x}(t) = L_1 y(t) + L_2 \hat{p}(t). \qquad \text{4-41}$$

An observer for $p(t)$ can be found by noting that $p(t)$ obeys the following differential equation
$$\dot{p}(t) = C'Ax(t) + C'Bu(t), \qquad \text{4-42}$$
or
$$\dot{p}(t) = C'AL_2 p(t) + C'AL_1 y(t) + C'Bu(t). \qquad \text{4-43}$$

Note that in this differential equation $y(t)$ serves as a forcing variable. If we now try to determine an observer for p by replacing p with $\hat{p}$ in **4-43** and adding a term of the form $K(t)[y(t) - C\hat{x}(t)]$, where K is a gain matrix, this is unsuccessful since from **4-41** we have $y - C\hat{x} = y - CL_1 y - CL_2 \hat{p} = y - y = 0$; apparently, y does not carry any information about p. New information must be laid bare by differentiating $y(t)$:
$$\begin{aligned}\dot{y}(t) &= CAx(t) + CBu(t) \\ &= CAL_2 p(t) + CAL_1 y(t) + CBu(t).\end{aligned} \qquad \text{4-44}$$

Equations **4-43** and **4-44** suggest the observer
$$\dot{\hat{p}}(t) = C'AL_2 \hat{p}(t) + C'AL_1 y(t) + C'Bu(t) \\ + K[\dot{y}(t) - CAL_1 y(t) - CBu(t) - CAL_2 \hat{p}(t)]. \qquad \text{4-45}$$

We leave it as an exercise to show that, if the pair $\{A, C\}$ is completely reconstructible, also the pair $\{C'AL_2, CAL_2\}$ is completely reconstructible, so that by a suitable choice of K all the poles of **4-45** can be placed at arbitrary positions (Wonham, 1970a).

In the realization of the observer, there is no need to take the derivative of $y(t)$. To show this, define
$$q(t) = \hat{p}(t) - Ky(t). \qquad \text{4-46}$$
It is easily seen that
$$\begin{aligned}\dot{q}(t) = {}& [C'AL_2 - KCAL_2]q(t) \\ & + [C'AL_2 K + C'AL_1 - KCAL_1 - KCAL_2 K]y(t) \\ & + [C'B - KCB]u(t).\end{aligned} \qquad \text{4-47}$$

This equation does not contain $\dot{y}(t)$. The reconstructed state follows from
$$\hat{x}(t) = L_2 q(t) + (L_1 + L_2 K)y(t). \qquad \text{4-48}$$

Together, **4-47** and **4-48** constitute an observer of the form **4-4**.

Since the reduced-order observer has a direct link from the observed variable $y(t)$ to the reconstructed state $\hat{x}(t)$, the estimate $\hat{x}(t)$ will be more sensitive to measurement errors in $y(t)$ than the estimate generated by a full-order observer. The question of the effects of measurement errors and system disturbances upon the observer is discussed in Section 4.3.

Example 4.2 *Positioning system*

In this example we derive a one-dimensional observer for the positioning system we considered in Example 4.1. For this system the observed variable is given by

$$\eta(t) = (1, \ 0)x(t). \qquad 4\text{-}49$$

Understandably, we choose the variable $p(t)$, which now is a scalar, as

$$p(t) = (0, \ 1)x(t), \qquad 4\text{-}50$$

so that $p(t)$ is precisely the angular velocity. It is immediately seen that $p(t)$ satisfies the differential equation

$$\dot{p}(t) = -\alpha p(t) + \kappa \mu(t). \qquad 4\text{-}51$$

Our observation equation we obtain by differentiation of $\eta(t)$:

$$\dot{\eta}(t) = \dot{\xi}_1(t) = \xi_2(t) = p(t). \qquad 4\text{-}52$$

An observer for $p(t)$ is therefore given by

$$\dot{\hat{p}}(t) = -\alpha \hat{p}(t) + \kappa \mu(t) + \lambda[\dot{\eta}(t) - \hat{p}(t)], \qquad 4\text{-}53$$

where the scalar observer gain λ is to be selected. The characteristic value of the observer is $-(\alpha + \lambda)$. To make the present design comparable to the full-order observer of Example 4.1, we choose the observer pole at the same distance from the origin as the pair of poles of Example 4.1. Thus we let $\alpha + \lambda = 50\sqrt{2} = 70.71 \text{ s}^{-1}$. With $\alpha = 4.6 \text{ s}^{-1}$ this yields for the gain

$$\lambda = 66.11 \text{ s}^{-1}. \qquad 4\text{-}54$$

The reconstructed state of the original system is given by

$$\hat{x}(t) = \begin{pmatrix} \eta(t) \\ \hat{p}(t) \end{pmatrix}, \quad t \geq 0. \qquad 4\text{-}55$$

To obtain a reduced-order observer without derivatives, we set

$$q(t) = \hat{p}(t) - \lambda \eta(t). \qquad 4\text{-}56$$

By using **4-53** it follows that $q(t)$ satisfies the differential equation

$$\dot{q}(t) = -(\alpha + \lambda)q(t) + \kappa \mu(t) - (\alpha + \lambda)\lambda \eta(t). \qquad 4\text{-}57$$

In terms of $q(t)$ the reconstructed state of the original system is given by

$$\hat{x}(t) = \begin{pmatrix} \eta(t) \\ q(t) + \lambda \eta(t) \end{pmatrix}. \qquad 4\text{-}58$$

338 Optimal Reconstruction of the State

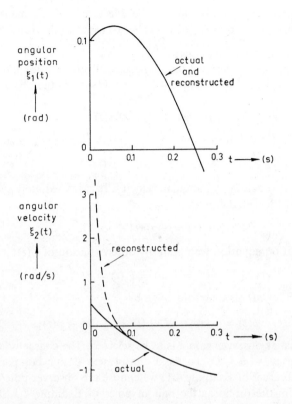

Fig. 4.3. Actual response of a positioning system and the response as reconstructed by a reduced-order observer.

In Fig. 4.3 we compare the output of the reduced-order observer described by **4-57** and **4-58** to the actual behavior of the system. The initial conditions of the system are, as in Example 4.1:

$$\xi_1(0) = 0.1 \text{ rad}, \quad \xi_2(0) = 0.5 \text{ rad/s}, \qquad \textbf{4-59}$$

while the input is given by

$$\mu(t) = -10 \text{ V}, \quad t \geq 0. \qquad \textbf{4-60}$$

The observer initial condition is

$$q(0) = 0 \text{ rad/s}. \qquad \textbf{4-61}$$

Figure 4.3 shows that the angular position is of course faithfully reproduced and that the estimated angular velocity quickly converges to the correct value, although the initial estimate is not very good.

4.3 THE OPTIMAL OBSERVER

4.3.1 A Stochastic Approach to the Observer Problem

In Section 4.2 we introduced observers. It has been seen, however, that in the selection of an observer for a given system a certain arbitrariness remains in the choice of the gain matrix K. In this section we present methods of finding the *optimal* gain matrix. To this end we must make specific assumptions concerning the disturbances and observation errors that occur in the system that is to be observed. We shall then be able to define the sense in which the observer is to be optimal.

It is assumed that the actual system equations are

$$\dot{x}(t) = A(t)x(t) + B(t)u(t) + w_1(t), \qquad \text{4-62a}$$

$$y(t) = C(t)x(t) + w_2(t). \qquad \text{4-62b}$$

Here $w_1(t)$ is termed the *state excitation noise*, while $w_2(t)$ is the *observation* or *measurement noise*. It is assumed that the joint process col $[w_1(t), w_2(t)]$ can be described as white noise with intensity

$$V(t) = \begin{pmatrix} V_1(t) & V_{12}(t) \\ V_{12}^T(t) & V_2(t) \end{pmatrix}, \qquad \text{4-63}$$

that is,

$$E\left\{ \begin{pmatrix} w_1(t_1) \\ w_2(t_1) \end{pmatrix} [w_1^T(t_2), w_2^T(t_2)] \right\} = V(t_1)\,\delta(t_1 - t_2). \qquad \text{4-64}$$

If $V_{12}(t) = 0$, the state excitation noise and the observation noise are *uncorrelated*. Later (in Section 4.3.5) we consider the possibility that $w_1(t)$ and $w_2(t)$ can not be represented as white noise processes. A case of special interest occurs when

$$V_2(t) > 0, \qquad t \geq t_0. \qquad \text{4-65}$$

This assumption means in essence that all components of the observed variable are corrupted by white noise and that it is impossible to extract from $y(t)$ information that does not contain white noise. If this condition is satisfied, we call the problem of reconstructing the state of the system **4-62** *nonsingular*.

Finally, we denote

$$E\{x(t_0)\} = \bar{x}_0, \qquad E\{[x(t_0) - \bar{x}_0][x(t_0) - \bar{x}_0]^T\} = Q_0. \qquad \text{4-66}$$

Suppose now that a full-order observer of the form

$$\dot{\hat{x}}(t) = A(t)\hat{x}(t) + B(t)u(t) + K(t)[y(t) - C(t)\hat{x}(t)] \qquad \text{4-67}$$

is connected to the system **4-62**. Then the *reconstruction error* is given by

$$e(t) = x(t) - \hat{x}(t). \quad \textbf{4-68}$$

The *mean square reconstruction error*

$$E\{e^T(t)W(t)e(t)\}, \quad \textbf{4-69}$$

with $W(t)$ a given positive-definite symmetric weighting matrix, is a measure of how well the observer reconstructs the state of the system at time t. The mean square reconstruction error is determined by the choice of $\hat{x}(t_0)$ and of $K(\tau)$, $t_0 \leq \tau \leq t$. The problem of how to choose these quantities optimally is termed the *optimal observer problem*.

Definition 4.3. *Consider the system*

$$\begin{aligned} \dot{x}(t) &= A(t)x(t) + B(t)u(t) + w_1(t), \\ y(t) &= C(t)x(t) + w_2(t), \end{aligned} \quad t \geq t_0. \quad \textbf{4-70}$$

Here col $[w_1(t), w_2(t)]$ *is a white noise process with intensity*

$$\begin{pmatrix} V_1(t) & V_{12}(t) \\ V_{12}^T(t) & V_2(t) \end{pmatrix}, \quad t \geq t_0. \quad \textbf{4-71}$$

Furthermore, the initial state $x(t_0)$ *is uncorrelated with* w_1 *and* w_2,

$$E\{x(t_0)\} = \bar{x}_0, \quad E\{[x(t_0) - \bar{x}_0][x(t_0) - \bar{x}_0]^T\} = Q_0, \quad \textbf{4-72}$$

and $u(t)$, $t \geq t_0$, *is a given input to the system. Consider the observer*

$$\dot{\hat{x}}(t) = A(t)\hat{x}(t) + (Bt)u(t) + K(t)[y(t) - C(t)\hat{x}(t)]. \quad \textbf{4-73}$$

Then the problem of finding the matrix function $K(\tau)$, $t_0 \leq \tau \leq t$, *and the initial condition* $\hat{x}(t_0)$, *so as to minimize*

$$E\{e^T(t)W(t)e(t)\}, \quad \textbf{4-74}$$

where

$$e(t) = x(t) - \hat{x}(t), \quad \textbf{4-75}$$

and where $W(t)$ *is a positive-definite symmetric weighting matrix, is termed the* **optimal observer problem.** *If*

$$V_2(t) > 0, \quad t \geq t_0, \quad \textbf{4-76}$$

the optimal observer problem is called **nonsingular.**

In Section 4.3.2 we study the nonsingular optimal observer problem where the state excitation noise and the observation noise are assumed moreover to be uncorrelated. In Section 4.3.3 we relax the condition of uncorrelatedness, while in Section 4.3.4 the singular problem is considered.

4.3.2 The Nonsingular Optimal Observer Problem with Uncorrelated State Excitation and Observation Noises

In this section we consider the nonsingular optimal observer problem where it is assumed that the state excitation noise and the observation noise are uncorrelated. This very important problem was first solved by Kalman and Bucy (Kalman and Bucy, 1961), and its solution has had a tremendous impact on optimal filtering theory. A historical account of the derivation of the so-called *Kalman–Bucy filter* is given by Sorenson (1970).

Somewhat surprisingly the derivation of the optimal observer can be based on Lemma 3.1 (Section 3.3.3). Before proceeding to this derivation, however, we introduce the following lemma, which shows how time can be reversed in any differential equation.

Lemma 4.1. *Consider the differential equations*

$$\frac{dx(t)}{dt} = f[t, x(t)], \qquad t \geq t_0, \qquad \text{4-77}$$

$$x(t_0) = x_0,$$

and

$$-\frac{dy(t)}{dt} = f[t^* - t, y(t)], \qquad t \leq t_1, \qquad \text{4-78}$$

$$y(t_1) = y_1,$$

where $t_0 < t_1$, and

$$t^* = t_0 + t_1. \qquad \text{4-79}$$

Then if

$$x_0 = y_1, \qquad \text{4-80}$$

the solutions of **4-77** *and* **4-78** *are related as follows:*

$$\begin{aligned} x(t) &= y(t^* - t), & t \geq t_0, \\ y(t) &= x(t^* - t), & t \leq t_1. \end{aligned} \qquad \text{4-81}$$

This lemma is easily proved by a change in variable from t to $t^* - t$.

We now proceed with our derivation of the optimal observer. Subtracting **4-67** from **4-62a** and using **4-62b**, we obtain the following differential equation for the reconstruction error $e(t) = x(t) - \hat{x}(t)$:

$$\dot{e}(t) = [A(t) - K(t)C(t)]e(t) + (I, -K(t))\begin{pmatrix} w_1(t) \\ w_2(t) \end{pmatrix}, \qquad \text{4-82}$$

$$e(t_0) = e_0,$$

where

$$e_0 = x(t_0) - \hat{x}(t_0), \qquad \text{4-83}$$

and where, as yet, $K(t)$, $t \geq t_0$, is an arbitrary matrix function. Let us denote by $\tilde{Q}(t)$ the variance matrix of $e(t)$, and by $\bar{e}(t)$ the mean of $e(t)$:

$$E\{e(t)\} = \bar{e}(t),$$

$$E\{[e(t) - \bar{e}(t)][e(t) - \bar{e}(t)]^T\} = \tilde{Q}(t). \qquad \text{4-84}$$

Then we write

$$E\{e(t)e^T(t)\} = \bar{e}(t)\bar{e}^T(t) + \tilde{Q}(t). \qquad \text{4-85}$$

With this, using **1-469**, the mean square reconstruction error can be expressed as

$$E\{e^T(t)W(t)e(t)\} = \bar{e}^T(t)W(t)\bar{e}(t) + \text{tr}\,[\tilde{Q}(t)W(t)]. \qquad \text{4-86}$$

The first term of this expression is obviously minimal when $\bar{e}(t) = 0$. This can be achieved by letting $\bar{e}(t_0) = 0$, since by Theorem 1.52 (Section 1.11.2) $\bar{e}(t)$ obeys the homogeneous differential equation

$$\dot{\bar{e}}(t) = [A(t) - K(t)C(t)]\bar{e}(t), \qquad t \geq t_0. \qquad \text{4-87}$$

We can make $\bar{e}(t_0) = 0$ by choosing the initial condition of the observer as

$$\hat{x}(t_0) = \bar{x}_0. \qquad \text{4-88}$$

Since the second term of **4-86** does not depend upon $\bar{e}(t)$, it can be minimized independently. From Theorem 1.52 (Section 1.11.2), we obtain the following differential equation for $\tilde{Q}(t)$:

$$\dot{\tilde{Q}}(t) = [A(t) - K(t)C(t)]\tilde{Q}(t) + \tilde{Q}(t)[A(t) - K(t)C(t)]^T$$
$$+ V_1(t) + K(t)V_2(t)K^T(t). \qquad \text{4-89}$$

The corresponding initial condition is

$$\tilde{Q}(t_0) = Q_0. \qquad \text{4-90}$$

Let us now introduce a differential equation in a matrix function $\tilde{P}(t)$, which is derived from **4-89** by reversing time (Lemma 4.1):

$$-\dot{\tilde{P}}(t) = [A^T(t^* - t) - C^T(t^* - t)K^T(t^* - t)]^T\tilde{P}(t)$$
$$+ \tilde{P}(t)[A^T(t^* - t) - C^T(t^* - t)K^T(t^* - t)]$$
$$+ V_1(t^* - t) + K(t^* - t)V_2(t^* - t)K^T(t^* - t), \qquad t \leq t_1. \qquad \text{4-91}$$

Here

$$t^* = t_0 + t_1, \qquad \text{4-92}$$

with $t_1 > t_0$. We associate with **4-91** the terminal condition

$$\tilde{P}(t_1) = Q_0. \qquad \text{4-93}$$

It immediately follows from Lemma 4.1 that

$$\tilde{Q}(t) = \tilde{P}(t^* - t), \qquad t \leq t_1. \qquad \text{4-94}$$

4.3 The Optimal Observer

Let us now apply Lemma 3.1 (Section 3.3.3) to **4-91**. This lemma shows that the matrix $\tilde{P}(t)$ is minimized if $K(t^* - \tau)$, $t \leq \tau \leq t_1$, is chosen as $K^0(t^* - \tau)$, $t \leq \tau \leq t_1$, where

$$K^0(t^* - \tau) = V_2^{-1}(t^* - \tau)C(t^* - \tau)P(\tau). \qquad \textbf{4-95}$$

In this expression $P(t)$ is the solution of **4-91** with K replaced by K^0, that is,

$$-\dot{P}(t) = V_1(t^* - t) - P(t)C^T(t^* - t)V_2^{-1}(t^* - t)C(t^* - t)P(t)$$
$$+ P(t)A^T(t^* - t) + A(t^* - t)P(t), \qquad t \leq t_1, \qquad \textbf{4-96}$$

with the terminal condition

$$P(t_1) = Q_0. \qquad \textbf{4-97}$$

The minimal value of $\tilde{P}(t)$ is $P(t)$, where the minimization is in the sense that

$$P(t) \leq \tilde{P}(t), \qquad t \leq t_1. \qquad \textbf{4-98}$$

By reversing time back again in **4-96**, we see that the variance matrix $\tilde{Q}(t)$ of $e(t)$ is minimized in the sense that

$$\tilde{Q}(t) \geq Q(t), \qquad t \geq t_0, \qquad \textbf{4-99}$$

by choosing $K(\tau) = K^0(\tau)$, $t_0 \leq \tau \leq t$, where

$$K^0(\tau) = Q(\tau)C^T(\tau)V_2^{-1}(\tau), \qquad \tau \geq t_0, \qquad \textbf{4-100}$$

and where the matrix $Q(t)$ satisfies the matrix Riccati equation

$$\dot{Q}(t) = V_1(t) - Q(t)C^T(t)V_2^{-1}(t)C(t)Q(t) + Q(t)A^T(t) + A(t)Q(t),$$
$$t \geq t_0, \qquad \textbf{4-101}$$

with the initial condition

$$Q(t_0) = Q_0. \qquad \textbf{4-102}$$

Since **4-99** implies that

$$\text{tr}\,[Q(t)W(t)] \leq \text{tr}\,[\tilde{Q}(t)W(t)] \qquad \textbf{4-103}$$

for any positive-definite symmetric matrix $W(t)$, we conclude that the gain matrix **4-100** optimizes the observer. We moreover see from **4-86** that for the optimal observer the mean square reconstruction error is given by

$$E\{e^T(t)W(t)e(t)\} = \text{tr}\,[Q(t)W(t)], \qquad \textbf{4-104}$$

while the variance matrix of $e(t)$ is $Q(t)$.

We finally remark that the result we have obtained is independent of the particular time t at which we have chosen to minimize the mean square reconstruction error. Thus if the gain is determined according to **4-100**, the mean square reconstruction error is simultaneously minimized *for all $t \geq t_0$*.

Our findings can be summarized as follows.

Theorem 4.5. *Consider the optimal observer problem of Definition* **4.3**. *Suppose that the problem is nonsingular and that the state excitation and observation noise are uncorrelated. Then the solution of the optimal observer problem is obtained by choosing for the gain matrix*

$$K^0(t) = Q(t)C^T(t)V_2^{-1}(t), \qquad t \geq t_0, \qquad \text{4-105}$$

where $Q(t)$ is the solution of the matrix Riccati equation

$$\dot{Q}(t) = A(t)Q(t) + Q(t)A^T(t) + V_1(t) - Q(t)C^T(t)V_2^{-1}(t)C(t)Q(t),$$
$$t \geq t_0, \quad \text{4-106}$$

with the initial condition

$$Q(t_0) = Q_0. \qquad \text{4-107}$$

The initial condition of the observer should be chosen as

$$\hat{x}(t_0) = \bar{x}_0. \qquad \text{4-108}$$

If **4-105** *and* **4-108** *are satisfied,*

$$E\{[x(t) - \hat{x}(t)]^T W(t)[x(t) - \hat{x}(t)]\} \qquad \text{4-109}$$

is minimized for all $t \geq t_0$. The variance matrix of the reconstruction error is given by

$$E\{[x(t) - \hat{x}(t)][x(t) - \hat{x}(t)]^T\} = Q(t), \qquad \text{4-110}$$

while the mean square reconstruction error is

$$E\{[x(t) - \hat{x}(t)]^T W(t)[x(t) - \hat{x}(t)]\} = \text{tr}\,[Q(t)W(t)]. \qquad \text{4-111}$$

It is noted that the solution of the optimal observer problem is, surprisingly, independent of the weighting matrix $W(t)$.

The optimal observer of Theorem 4.5 is known as the *Kalman–Bucy filter*. In this section we have derived this filter by first assuming that it has the form of an observer. In the original derivation of Kalman and Bucy (1961), however, it is proved that this filter is the *minimum mean square linear estimator*, that is, we cannot find another linear functional of the observations $y(\tau)$ and the input $u(\tau)$, $t_0 \leq \tau \leq t$, that produces an estimate of the state $x(t)$ with a smaller mean square reconstruction error. It can also be proved (see, e.g., Jazwinski, 1970) that if the initial state $x(t_0)$ is Gaussian, and the state excitation noise w_1 and the observation noise w_2 are Gaussian white noise processes, the Kalman–Bucy filter produces an estimate $\hat{x}(t)$ of $x(t)$ that has minimal mean square reconstruction error among *all* estimates that can be obtained by processing the data $y(\tau)$ and $u(\tau)$, $t_0 \leq \tau \leq t$.

The close relationship between the optimal *regulator* problem and the optimal *observer* problem is evident from the fact that the matrix Riccati equation for the observer variance matrix is just the time-reversed Riccati equation that holds for the regulator problem. In later sections we make further use of this relationship, which will be referred to as the *duality property*, in deriving facts about observers from facts about regulators.

The gain matrix $K^0(t)$ can be obtained by solving the matrix Riccati equation **4-106** in real time and using **4-105**. Alternatively, $K^0(t)$ can be computed in advance, stored, and played back during the state reconstruction process. It is noted that in contrast to the optimal regulator described in Chapter 3 the optimal observer can easily be implemented in real time, since **4-106** is a differential equation with given *initial* conditions, whereas the optimal regulator requires solution of a Riccati equation with given *terminal* conditions that must be solved backward in time.

In Theorem 3.3 (Section 3.3.2), we saw that the regulator Riccati equation can be obtained by solving a set of $2n \times 2n$ differential equations (where n is the dimension of the state). The same can be done with the observer Riccati equation, as is outlined in Problem 4.3.

We now briefly discuss the steady-state properties of the optimal observer. What we state here is proved in Section 4.4.3. It can be shown that under mildly restrictive conditions the solution $Q(t)$ of the observer Riccati equation **4-106** converges to a *steady-state solution* $\bar{Q}(t)$ which is independent of Q_0 as the initial time t_0 approaches $-\infty$. In the time-invariant case, where all the matrices occurring in Definition 4.3 are constant, the steady-state solution $\bar{Q}$ is, in addition, a constant matrix and is, in general, the unique non-negative-definite solution of the *algebraic observer Riccati equation*

$$0 = AQ + QA^T + V_1 - QC^TV_2^{-1}CQ. \qquad \textbf{4-112}$$

This equation is obtained from **4-106** by setting the time derivative equal to zero.

Corresponding to the steady-state solution $\bar{Q}$ of the observer Riccati equation, we obtain the *steady-state optimal observer gain matrix*

$$\bar{K}(t) = \bar{Q}(t)C^T(t)V_2^{-1}(t). \qquad \textbf{4-113}$$

It is proved in Section 4.4.3, again under mildly restrictive conditions, that the observer with $\bar{K}$ as gain matrix is, in general, asymptotically stable. We refer to this observer as the *steady-state optimal observer*. Since in the time-invariant case the steady-state observer is also time-invariant, it is very attractive to use the steady-state optimal observer since it is much easier to implement. In the time-invariant case, the steady-state optimal observer is optimal in the sense that

$$\lim_{t_0 \to -\infty} E\{e^T(t)We(t)\} = \lim_{t \to \infty} E\{e^T(t)We(t)\} \qquad \textbf{4-114}$$

is minimal with respect to all other time-invariant observers.

346   Optimal Reconstruction of the State

We conclude this section with the following discussion which is restricted to the time-invariant case. The optimal observer provides a compromise between the *speed of state reconstruction* and the *immunity to observation noise*. The balance between these two properties is determined by the magnitudes of the white noise intensities V_1 and V_2. This balance can be varied by keeping V_1 constant and setting

$$V_2 = \rho M, \qquad \text{4-115}$$

where M is a constant positive-definite symmetric matrix and ρ is a positive scalar that is varied. It is intuitively clear that decreasing ρ improves the speed of state reconstruction, since less attention can be paid to filtering the observation noise. This increase in reconstruction speed is accompanied by a shift of the observer poles further into the left-half complex plane. In cases where one is not sure of the exact values of V_1 or V_2, a good design procedure may be to assume that V_2 has the form **4-115** and vary ρ until a satisfactory observer is obtained. The limiting properties of the optimal observer as $\rho \downarrow 0$ or $\rho \to \infty$ are reviewed in Section 4.4.4.

Example 4.3. *The estimation of a "constant"*

In many practical situations variables are encountered that stay constant over relatively long periods of time and only occasionally change value. One possible approach to model such a constant is to represent it as the state of an undisturbed integrator with a stochastic initial condition. Thus let $\xi(t)$ represent the constant. Then we suppose that

$$\begin{aligned}\dot{\xi}(t) &= 0, \\ \xi(0) &= \xi_0,\end{aligned} \qquad \text{4-116}$$

where ξ_0 is a scalar stochastic variable with mean $\bar{\xi}_0$ and variance Q_0. We assume that we measure this constant with observation noise $v_2(t)$, that is, we observe

$$\eta(t) = \xi(t) + v_2(t), \qquad \text{4-117}$$

where $v_2(t)$ is assumed to be white noise with constant scalar intensity V_2.

The optimal observer for $\xi(t)$ is given by

$$\begin{aligned}\dot{\hat{\xi}}(t) &= k(t)[\eta(t) - \hat{\xi}(t)] \\ \hat{\xi}(0) &= \bar{\xi}_0,\end{aligned} \qquad \text{4-118}$$

where the scalar gain $k(t)$ is, from **4-105**, given by

$$k(t) = \frac{Q(t)}{V_2}. \qquad \text{4-119}$$

4.3 The Optimal Observer

The error variance $Q(t)$ is the solution of the Riccati equation

$$\dot{Q}(t) = -\frac{Q^2(t)}{V_2}, \qquad Q(0) = Q_0. \qquad \text{4-120}$$

Equation **4-120** can be solved explicitly:

$$Q(t) = \frac{Q_0 V_2}{V_2 + Q_0 t}, \qquad t \geq 0, \qquad \text{4-121}$$

so that

$$k(t) = \frac{Q_0}{V_2 + Q_0 t}, \qquad t \geq 0. \qquad \text{4-122}$$

We note that as $t \to \infty$ the error variance $Q(t)$ approaches zero, which means that eventually a completely accurate estimate of $\xi(t)$ becomes available. As a result, also $k(t) \to 0$, signifying that there is no point in processing any new data.

This observer is not satisfactory when the constant occasionally changes value, or in reality varies slowly. In such a case we can model the constant as the output of an integrator driven by white noise. The justification for modeling the process in this way is that integrated white noise has a very large low-frequency content. Thus we write

$$\begin{aligned}\dot{\xi}(t) &= v_1(t), \\ \eta(t) &= \xi(t) + v_2(t),\end{aligned} \qquad \text{4-123}$$

where v_1 is white noise with constant intensity V_1 and v_2 is white noise as before, independent of v_1. The steady-state optimal observer is now easily found to be given by

$$\dot{\hat{\xi}}(t) = \bar{k}[\eta(t) - \hat{\xi}(t)], \qquad \text{4-124}$$

where

$$\bar{k} = \sqrt{V_1/V_2}. \qquad \text{4-125}$$

In transfer function form we have

$$\hat{\mathbf{X}}(s) = \frac{\sqrt{V_1/V_2}}{s + \sqrt{V_1/V_2}} \mathbf{Y}(s), \qquad \text{4-126}$$

where $\hat{\mathbf{X}}(s)$ and $\mathbf{Y}(s)$ are the Laplace transforms of $\hat{\xi}(t)$ and $\eta(t)$, respectively. As can be seen, the observer is a first-order filter with unity gain at zero frequency and break frequency $\sqrt{V_1/V_2}$.

Example 4.4. *Positioning system*

In Example 2.4 (Section 2.3), we considered a positioning system which is described by the state differential equation

$$\dot{x}(t) = \begin{pmatrix} 0 & 1 \\ 0 & -\alpha \end{pmatrix} x(t) + \begin{pmatrix} 0 \\ \kappa \end{pmatrix} \mu(t). \qquad \text{4-127}$$

348 Optimal Reconstruction of the State

Here $x(t) = \text{col}\,[\xi_1(t), \xi_2(t)]$, where $\xi_1(t)$ denotes the angular displacement $\theta(t)$ and $\xi_2(t)$ the angular velocity $\dot\theta(t)$. Let us now assume, as in Example 2.4, that a disturbing torque $\tau_d(t)$ acts upon the shaft of the motor. Accordingly, the state differential equation must be modified as follows:

$$\dot x(t) = \begin{pmatrix} 0 & 1 \\ 0 & -\alpha \end{pmatrix} x(t) + \begin{pmatrix} 0 \\ \kappa \end{pmatrix} \mu(t) + \begin{pmatrix} 0 \\ \gamma \end{pmatrix} \tau_d(t), \qquad 4\text{-}128$$

where $1/\gamma$ is the rotational moment of inertia of all the rotating parts. If the fluctuations of the disturbing torque are fast as compared to the motion of the system itself, the assumption might be justified that $\tau_d(t)$ is white noise. Let us therefore suppose that $\tau_d(t)$ is white noise, with constant, scalar intensity V_d. Let us furthermore assume that the observed variable is given by

$$\eta(t) = (1, 0)x(t) + v_m(t), \qquad 4\text{-}129$$

where $v_m(t)$ is white noise with constant, scalar intensity V_m.

We compute the steady-state optimal observer for this system. The variance Riccati equation takes the form

$$\dot Q(t) = \begin{pmatrix} 0 & 1 \\ 0 & -\alpha \end{pmatrix} Q(t) + Q(t) \begin{pmatrix} 0 & 0 \\ 1 & -\alpha \end{pmatrix}$$
$$+ \begin{pmatrix} 0 & 0 \\ 0 & \gamma^2 V_d \end{pmatrix} - Q(t) \begin{pmatrix} 1 \\ 0 \end{pmatrix} \frac{1}{V_m} (1, 0) Q(t). \qquad 4\text{-}130$$

In terms of the entries $q_{ij}(t)$, $i,j = 1, 2$, of $Q(t)$, we obtain the following set of differential equations (using the fact that $q_{12}(t) = q_{21}(t)$):

$$\dot q_{11}(t) = 2 q_{12}(t) - \frac{1}{V_m} q_{11}^2(t),$$

$$\dot q_{12}(t) = q_{22}(t) - \alpha q_{12}(t) - \frac{1}{V_m} q_{11}(t) q_{12}(t), \qquad 4\text{-}131$$

$$\dot q_{22}(t) = -2\alpha q_{22}(t) + \gamma^2 V_d - \frac{1}{V_m} q_{12}^2(t).$$

It can be found that the steady-state solution of the equations as $t \to \infty$ is given by

$$\bar Q = V_m \begin{pmatrix} -\alpha + \sqrt{\alpha^2 + 2\beta} & \alpha^2 + \beta - \alpha\sqrt{\alpha^2 + 2\beta} \\ \alpha^2 + \beta - \alpha\sqrt{\alpha^2 + 2\beta} & -\alpha^3 - 2\alpha\beta + (\alpha^2 + \beta)\sqrt{\alpha^2 + 2\beta} \end{pmatrix}, \qquad 4\text{-}132$$

where

$$\beta = \gamma\sqrt{V_d/V_m}. \qquad 4\text{-}133$$

4.3 The Optimal Observer

It follows that the steady-state optimal gain matrix is given by

$$\bar{K} = \begin{pmatrix} -\alpha + \sqrt{\alpha^2 + 2\beta} \\ \alpha^2 + \beta - \alpha\sqrt{\alpha^2 + 2\beta} \end{pmatrix}. \qquad \text{4-134}$$

The characteristic polynomial of the matrix $A - \bar{K}C$ can be found to be

$$\det(sI - A + \bar{K}C) = s^2 + s\sqrt{\alpha^2 + 2\beta} + \beta, \qquad \text{4-135}$$

from which it can be derived that the poles of the steady-state optimal observer are

$$\tfrac{1}{2}(-\sqrt{\alpha^2 + 2\beta} \pm \sqrt{\alpha^2 - 2\beta}). \qquad \text{4-136}$$

Let us adopt the following numerical values:

$$\begin{aligned} \kappa &= 0.787 \text{ rad}/(\text{V s}^2), \\ \alpha &= 4.6 \text{ s}^{-1}, \\ \gamma &= 0.1 \text{ kg}^{-1} \text{ m}^{-2}, \qquad \text{4-137} \\ V_d &= 10 \text{ N}^2 \text{ m}^2 \text{ s}, \\ V_m &= 10^{-7} \text{ rad}^2 \text{ s}. \end{aligned}$$

It is supposed that the value of V_d is derived from the knowledge that the disturbing torque has an rms value of $\sqrt{1000} \simeq 31.6$ N m and that its power spectral density is constant from about -50 to 50 Hz and zero outside this frequency band. Similarly, we assume that the observation noise, which has an rms value of 0.01 rad, has a flat power spectral density function from about -500 to 500 Hz and is zero outside this frequency range. We carry out the calculations as if the noises were white with intensities as indicated in **4-137** and then see if this assumption is justified.

With the numerical values as given, the steady-state gain matrix is found to be

$$\bar{K} = \begin{pmatrix} 40.36 \\ 814.3 \end{pmatrix}. \qquad \text{4-138}$$

The observer poles are $-22.48 \pm j22.24$. These pole locations apparently provide an optimal compromise between the speed of convergence of the reconstruction error and the immunity against observation noise.

The break frequency of the optimal observer can be determined from the pole locations. The observer characteristic polynomial is

$$s^2 + s\sqrt{\alpha^2 + 2\beta} + \beta \simeq s^2 + 45s + 1000, \qquad \text{4-139}$$

350 Optimal Reconstruction of the State

which represents a second-order system with undamped natural frequency $\omega_0 = 31.6$ rad/s $\simeq 5$ Hz and relative damping of about 0.71. The undamped natural frequency is also the break frequency of the observer. Since this frequency is quite small as compared to the observation noise bandwidth of about 500 Hz and the disturbance bandwidth of about 50 Hz, we conjecture that it is safe to approximate both processes as white noise. We must compare both the disturbance bandwidth and the observation noise bandwidth to the *observer* bandwidth, since as can be seen from the error differential equation **4-82** both processes directly influence the behavior of the reconstruction error. In Example 4.5, at the end of Section 4.3.5, we compute the optimal filter without approximating the observation noise as white noise and see whether or not this approximation is justified.

The steady-state variance matrix of the reconstruction error is given by

$$\bar{Q} = \begin{pmatrix} 0.000004036 & 0.00008143 \\ 0.00008143 & 0.003661 \end{pmatrix}. \qquad \textbf{4-140}$$

By taking the square roots of the diagonal elements, it follows that the rms reconstruction error of the position is about 0.002 rad, while that of the angular velocity is about 0.06 rad/s.

We conclude this example with a discussion of the optimal observer that has been found. First, we note that the filter is completely determined by the ratio V_d/V_m, which can be seen as a sort of "signal-to-noise" ratio. The expression **4-136** shows that as this ratio increases, which means that β increases, the observer poles move further and further away. As a result, the observer becomes faster, but also more sensitive to observation noise. For $\beta = \infty$ we obtain a differentiating filter, which can be seen as follows. In transfer matrix form the observer can be represented as

$$\hat{\mathbf{X}}(s) = (sI - A + \bar{K}C)^{-1}[\bar{K}\mathbf{Y}(s) + B\mathbf{U}(s)]$$

$$= \frac{1}{s^2 + s\sqrt{\alpha^2 + 2\beta} + \beta}$$

$$\cdot \left\{ \begin{pmatrix} s(-\alpha + \sqrt{\alpha^2 + 2\beta}) + \beta \\ s(\alpha^2 + \beta - \alpha\sqrt{\alpha^2 + 2\beta}) \end{pmatrix} \mathbf{Y}(s) + \begin{pmatrix} \dfrac{\kappa}{s - \alpha + \sqrt{\alpha^2 + 2\beta}} \end{pmatrix} \mathbf{U}(s) \right\}.$$

$$\textbf{4-141}$$

Here $\hat{\mathbf{X}}(s)$, $\mathbf{Y}(s)$, and $\mathbf{U}(s)$ are the Laplace transforms of $\hat{x}(t)$, $y(t)$, and $u(t)$, respectively. As the observation noise becomes smaller and smaller, that is, $\beta \to \infty$, **4-141** converges to

$$\hat{\mathbf{X}}(s) = \begin{pmatrix} 1 \\ s \end{pmatrix} \mathbf{Y}(s). \qquad \textbf{4-142}$$

4.3 The Optimal Observer

This means that the observed variable is taken as the reconstructed angular position and that the observed variable is differentiated to obtain the reconstructed angular velocity.

4.3.3* The Nonsingular Optimal Observer Problem with Correlated State Excitation and Observation Noises

In this section the results of the preceding section are extended to the case where the state excitation noise and the measurement noise are correlated, that is, $V_{12}(t) \neq 0$, $t \geq t_0$. To determine the optimal observer, we proceed in a fashion similar to the correlated case. Again, let $\tilde{Q}(t)$ denote the variance matrix of the reconstruction error when the observer is implemented with an arbitrary gain matrix $K(t)$, $t \geq t_0$. Using Theorem 1.52 (Section 1.11.2), we obtain the following differential equation for $\tilde{Q}(t)$, which is an extended version of **4-89**:

$$\dot{\tilde{Q}}(t) = [A(t) - K(t)C(t)]\tilde{Q}(t) + \tilde{Q}(t)[A(t) - K(t)C(t)]^T$$
$$+ V_1(t) - V_{12}(t)K^T(t) - K(t)V_{12}^T(t) + K(t)V_2(t)K^T(t), \quad t \geq t_0,$$
4-143

with the initial condition

$$\tilde{Q}(t_0) = Q_0.$$
4-144

To convert the problem of finding the optimal gain matrix to a familiar problem, we reverse time in this differential equation. It then turns out that the present problem is dual to the "extended regulator problem" discussed in Problem 3.7 in which the integral criterion contains a cross-term in the state x and the input u. By using the results of Problem 3.7, it can easily be shown that the solution of the present problem is as follows (see, e.g., Wonham, 1963).

Theorem 4.6. *Consider the optimal observer problem of Definition 4.3 (Section 4.3.1). Suppose that the problem is nonsingular, that is, $V_2(t) > 0$, $t \geq t_0$. Then the solution of the optimal observer problem is achieved by choosing the gain matrix $K(t)$ of the observer* **4-73** *as*

$$K^0(t) = [Q(t)C^T(t) + V_{12}(t)]V_2^{-1}(t), \quad t \geq t_0, \qquad \textbf{4-145}$$

where $Q(t)$ is the solution of the matrix Riccati equation

$$\dot{Q}(t) = [A(t) - V_{12}(t)V_2^{-1}(t)C(t)]Q(t)$$
$$+ Q(t)[A(t) - V_{12}(t)V_2^{-1}(t)C(t)]^T$$
$$- Q(t)C^T(t)V_2^{-1}(t)C(t)Q(t)$$
$$+ V_1(t) - V_{12}(t)V_2^{-1}(t)V_{12}^T(t), \quad t \geq t_0, \qquad \textbf{4-146}$$

with the initial condition

$$Q(t_0) = Q_0. \qquad 4\text{-}147$$

The initial condition of the observer is

$$\hat{x}(t_0) = \bar{x}_0. \qquad 4\text{-}148$$

For the choices **4-145** and **4-148**, the mean square reconstruction error

$$E\{[x(t) - \hat{x}(t)]^T W(t)[x(t) - \hat{x}(t)]\} \qquad 4\text{-}149$$

is minimized for all $t \geq t_0$. The variance matrix of the reconstruction error is given by

$$E\{[x(t) - \hat{x}(t)][x(t) - \hat{x}(t)]^T\} = Q(t), \qquad 4\text{-}150$$

hence

$$E\{[x(t) - \hat{x}(t)]^T W(t)[x(t) - \hat{x}(t)]\} = \text{tr}\,[W(t)Q(t)], \quad t \geq t_0. \qquad 4\text{-}151$$

4.3.4* The Time-Invariant Singular Optimal Observer Problem

This section is devoted to the derivation of the optimal observer for the singular case, namely, the case where the matrix $V_2(t)$ is not positive-definite. To avoid the difficulties that occur when $V_2(t)$ is positive-definite during certain periods and singular during other periods, we restrict the derivation of this section to the time-invariant case, where all the matrices occurring in Definition 4.3 (Section 4.3.1) are constant. Singular observation problems arise when some of the components of the observed variable are free of observation noise, and also when the observation noise is not a white noise process, as we see in the following section. The present derivation roughly follows that of Bryson and Johansen (1965).

First, we note that when V_2 is singular the derivation of Section 4.3.2 breaks down; upon investigation it turns out that an infinite gain matrix would be required for a full-order observer as proposed. As a result, the problem formulation of Definition 4.3 is inadequate for the singular case. What we do in this section is to reduce the singular problem to a nonsingular problem (of lower dimension) and then apply the results of Sections 4.3.2 or 4.3.3.

Since V_2 is singular, we can always introduce another white noise process $w_2'(t)$, with *non*singular intensity V_2', such that

$$w_2(t) = H w_2'(t), \qquad 4\text{-}152$$

with $\dim(w_2') < \dim(w_2)$, and where H has full rank. This means that the observed variable is given by

$$y(t) = Cx(t) + H w_2'(t). \qquad 4\text{-}153$$

4.3 The Optimal Observer

With this assumption the intensity of $w_2(t)$ is given by

$$V_2 = HV'_2H^T. \qquad \text{4-154}$$

Since V_2 is singular, it is possible to decompose the observed variable into two parts: a part that is "completely noisy," and a part that is noise-free. We shall see how this decomposition is performed.

Since dim $(w'_2) <$ dim (w_2), it is always possible to find an $l \times l$ nonsingular matrix T (l is the dimension of the observed variable y) partitioned as

$$T = \begin{pmatrix} T_1 \\ T_2 \end{pmatrix}, \qquad \text{4-155}$$

such that

$$\begin{pmatrix} T_1 \\ T_2 \end{pmatrix} H = \begin{pmatrix} H_1 \\ 0 \end{pmatrix}. \qquad \text{4-156}$$

Here H_1 is square and nonsingular, and the partitioning of T has been chosen corresponding to that in the right hand side of **4-156**. Multiplying the output equation

$$y(t) = Cx(t) + Hw'_2(t) \qquad \text{4-157}$$

by T we obtain

$$y_1(t) = C_1 x(t) + H_1 w'_2(t), \qquad \text{4-158a}$$

$$y_2(t) = C_2 x(t), \qquad \text{4-158b}$$

where

$$\begin{pmatrix} y_1(t) \\ y_2(t) \end{pmatrix} = \begin{pmatrix} T_1 \\ T_2 \end{pmatrix} y(t), \quad \begin{pmatrix} C_1 \\ C_2 \end{pmatrix} = \begin{pmatrix} T_1 \\ T_2 \end{pmatrix} C. \qquad \text{4-159}$$

We see that **4-158** represents the decomposition of the observed variable $y(t)$ into a "completely noisy" part $y_1(t)$ (since $H_1 V'_2 H_1^T$ is nonsingular), and a noise-free part $y_2(t)$.

We now suppose that C_2 has full rank. If this is not the case, we can redefine $y_2(t)$ by eliminating all components that are linear combinations of other components, so that the redefined C_2 has full rank. We denote the dimension of $y_2(t)$ by k.

Equation **4-158b** will be used in two ways. First, we conclude that since $y_2(t)$ provides us with k linear equations for $x(t)$ we must reconstruct only $n - k$ (n is the dimension of x) additional linear combinations of $x(t)$. Second, since $y_2(t)$ does not contain white noise it can be differentiated in order to extract more data. Let us thus define, as we did in Section 4.2.3, an $(n - k)$-dimensional vector variable

$$p(t) = C'_2 x(t), \qquad \text{4-160}$$

354 Optimal Reconstruction of the State

where C_2' is so chosen that the $n \times n$ matrix

$$\begin{pmatrix} C_2 \\ C_2' \end{pmatrix} \qquad \text{4-161}$$

is nonsingular. From $y_2(t)$ and $p(t)$ we can reconstruct $x(t)$ exactly by the relations

$$\begin{aligned} y_2(t) &= C_2 x(t), \\ p(t) &= C_2' x(t), \end{aligned} \qquad \text{4-162}$$

or

$$x(t) = \begin{pmatrix} C_2 \\ C_2' \end{pmatrix}^{-1} \begin{pmatrix} y_2(t) \\ p(t) \end{pmatrix}. \qquad \text{4-163}$$

It is convenient to introduce the notation

$$\begin{pmatrix} C_2 \\ C_2' \end{pmatrix}^{-1} = (L_1, L_2), \qquad \text{4-164}$$

so that

$$x(t) = L_1 y_2(t) + L_2 p(t). \qquad \text{4-165}$$

Our next step is to construct an observer for $p(t)$. The reconstructed $p(t)$ will be denoted by $\hat{p}(t)$. It follows from **4-165** that $\hat{x}(t)$, the reconstructed state, is given by

$$\hat{x}(t) = L_1 y_2(t) + L_2 \hat{p}(t). \qquad \text{4-166}$$

The state differential equation for $p(t)$ is obtained by differentiation of **4-160**. It follows with **4-165**

$$\begin{aligned} \dot{p}(t) &= C_2' \dot{x}(t) = C_2' A x(t) + C_2' B u(t) + C_2' w_1(t) \\ &= C_2' A [L_1 y_2(t) + L_2 p(t)] + C_2' B u(t) + C_2' w_1(t), \end{aligned} \qquad \text{4-167}$$

or

$$\dot{p}(t) = A' p(t) + B' u(t) + B'' y_2(t) + C_2' w_1(t), \qquad \text{4-168}$$

where

$$A' = C_2' A L_2, \qquad B' = C_2' B, \qquad B'' = C_2' A L_1. \qquad \text{4-169}$$

Note that both $u(t)$ and $y_2(t)$ are forcing variables for this equation. The observations that are available are $y_1(t)$, as well as $\dot{y}_2(t)$, for which we find

$$\begin{aligned} \dot{y}_2(t) &= C_2 \dot{x}(t) = C_2 A x(t) + C_2 B u(t) + C_2 w_1(t) \\ &= C_2 A [L_1 y_2(t) + L_2 p(t)] + C_2 B u(t) + C_2 w_1(t). \end{aligned} \qquad \text{4-170}$$

For $y_1(t)$ we write

$$\begin{aligned} y_1(t) &= C_1 x(t) + H_1 w_2'(t) \\ &= C_1 [L_1 y_2(t) + L_2 p(t)] + H_1 w_2'(t). \end{aligned} \qquad \text{4-171}$$

4.3 The Optimal Observer

Combining $y_1(t)$ and $\dot{y}_2(t)$ we write for the observed variable of the system 4-168

$$y'(t) = \begin{pmatrix} y_1(t) \\ \dot{y}_2(t) \end{pmatrix} = C'p(t) + D'u(t) + D''y_2(t) + H'\begin{pmatrix} w_1(t) \\ w_2'(t) \end{pmatrix}, \quad \textbf{4-172}$$

where

$$C' = \begin{pmatrix} C_1L_2 \\ C_2AL_2 \end{pmatrix}, \quad D' = \begin{pmatrix} 0 \\ C_2B \end{pmatrix}, \quad D'' = \begin{pmatrix} C_1L_1 \\ C_2AL_1 \end{pmatrix}, \quad H' = \begin{pmatrix} 0 & H_1 \\ C_2 & 0 \end{pmatrix}.$$
4-173

Note that in the state differential equation **4-168** and in the output equation **4-172** we treat both $u(t)$ and $y_2(t)$ as *given* data. To make the problem formulation complete, we must compute the *a priori* statistical data of the auxiliary variable $p(t_0)$:

$$\hat{p}(t_0) = E\{C_2'x(t_0) \mid y_2(t_0)\} \quad \textbf{4-174}$$

and

$$Q(t_0) = E\{[p(t_0) - \hat{p}(t_0)][p(t_0) - \hat{p}(t_0)]^T \mid y_2(t_0)\}. \quad \textbf{4-175}$$

It is outlined in Problem 4.4 how these quantities can be found.

The observation problem that we now have obtained, and which is defined by **4-168**, **4-172**, **4-174**, and **4-175**, is an observation problem with correlated state excitation and observation noises. It is either singular or nonsingular. If it is nonsingular it can be solved according to Section 4.3.3, and once $\hat{p}(t)$ is available we can use **4-166** for the reconstruction of the state. If the observation problem is still singular, we repeat the entire procedure by choosing a new transformation matrix T for **4-172** and continuing as outlined. This process terminates in one of two fashions:

(a) A nonsingular observation problem is obtained.

(b) Since the dimension of the quantity to be estimated is reduced at each step, eventually a stage can be reached where the matrix C_2 in **4-162** is square and nonsingular. This means that we can solve for $x(t)$ directly and no dynamic observer is required.

We conclude this section by pointing out that if **4-168** and **4-172** define a nonsingular observer problem, in the actual realization of the optimal observer it is not necessary to take the derivative of $y_2(t)$, since later this derivative is integrated by the observer. To show this consider the following observer for $p(t)$:

$$\dot{\hat{p}}(t) = A'\hat{p}(t) + B'u(t) + B''y_2(t)$$
$$+ K(t)[y'(t) - D'u(t) - D''y_2(t) - C'\hat{p}(t)]. \quad \textbf{4-176}$$

Partitioning
$$K(t) = [K_1(t), K_2(t)], \qquad \textbf{4-177}$$
it follows for **4-176**:
$$\hat{p}(t) = [A' - K(t)C']\hat{p}(t) + B'u(t) + B''y_2(t)$$
$$+ K_1(t)y_1(t) + K_2(t)\dot{y}_2(t) - K(t)[D'u(t) + D''y_2(t)]. \qquad \textbf{4-178}$$
Now by defining
$$q(t) = \hat{p}(t) - K_2(t)y_2(t), \qquad \textbf{4-179}$$
a state differential equation for $q(t)$ can be obtained with $y_1(t)$, $y_2(t)$, and $u(t)$, but not $\dot{y}_2(t)$, as inputs. Thus, by using **4-179**, $\hat{p}(t)$ can be found without using $\dot{y}_2(t)$.

4.3.5 The Colored Noise Observation Problem

This section is devoted to the case where the state excitation noise $w_1(t)$ and the observation noise $w_2(t)$ cannot be represented as white noise processes. In this case we assume that these processes can be modeled as follows:

$$w_1(t) = C_1(t)x'(t) + w_1'(t),$$
$$w_2(t) = C_2(t)x'(t) + w_2'(t), \qquad \textbf{4-180}$$

with
$$\dot{x}'(t) = A'(t)x'(t) + w_3(t). \qquad \textbf{4-181}$$

Here $w_1'(t)$, $w_2'(t)$, and $w_3(t)$ are white noise processes that in general need not be uncorrelated. Combining **4-180** and **4-181** with the state differential and output equations
$$\dot{x}(t) = A(t)x(t) + B(t)u(t) + w_1(t),$$
$$y(t) = C(t)x(t) + w_2(t), \qquad \textbf{4-182}$$
we obtain the augmented state differential and output equations
$$\begin{pmatrix} \dot{x}(t) \\ \dot{x}'(t) \end{pmatrix} = \begin{pmatrix} A(t) & C_1(t) \\ 0 & A'(t) \end{pmatrix} \begin{pmatrix} x(t) \\ x'(t) \end{pmatrix} + \begin{pmatrix} B(t) \\ 0 \end{pmatrix} u(t) + \begin{pmatrix} w_1'(t) \\ w_3(t) \end{pmatrix},$$
$$y(t) = [C(t), C_2(t)] \begin{pmatrix} x(t) \\ x'(t) \end{pmatrix} + w_2'(t). \qquad \textbf{4-183}$$

To complete the problem formulation the mean and variance matrix of the initial augmented state col $[x(t), x'(t)]$ must be given. In many cases the white noise $w_2'(t)$ is absent, which makes the observation problem singular. If the

4.3 The Optimal Observer

problem is time-invariant, the techniques of Section 4.3.4 can then be applied. This approach is essentially that of Bryson and Johansen (1965).

We illustrate this section by means of an example.

Example 4.5. *Positioning system with colored observation noise*

In Example 4.4 we considered the positioning system with state differential equation

$$\dot{x}(t) = \begin{pmatrix} 0 & 1 \\ 0 & -\alpha \end{pmatrix} x(t) + \begin{pmatrix} 0 \\ \kappa \end{pmatrix} \mu(t) + \begin{pmatrix} 0 \\ \gamma \end{pmatrix} \tau_d(t) \qquad \text{4-184}$$

and the output equation

$$\eta(t) = (1, 0)x(t) + \nu_m(t). \qquad \text{4-185}$$

The measurement noise $\nu_m(t)$ was approximated as white noise with intensity V_m. Let us now suppose that a better approximation is to model $\nu_m(t)$ as exponentially correlated noise (see Example 1.30, Section 1.10.2) with power spectral density function

$$\Sigma_m(\omega) = \frac{2\sigma^2 \theta}{1 + \omega^2 \theta^2}. \qquad \text{4-186}$$

This means that we can write (Example 1.36, Section 1.11.4)

$$\nu_m(t) = \xi_3(t), \qquad \text{4-187}$$

where

$$\dot{\xi}_3(t) = -\frac{1}{\theta} \xi_3(t) + \omega(t). \qquad \text{4-188}$$

Here $\omega(t)$ is white noise with scalar intensity $2\sigma^2/\theta$. In Example 4.4 we assumed that $\tau_d(t)$ is also white noise with intensity V_d. In order not to complicate the problem too much, we stay with this hypothesis. The augmented problem is now represented by the state differential and output equations:

$$\begin{pmatrix} \dot{\xi}_1(t) \\ \dot{\xi}_2(t) \\ \dot{\xi}_3(t) \end{pmatrix} = \begin{pmatrix} 0 & 1 & 0 \\ 0 & -\alpha & 0 \\ 0 & 0 & -\frac{1}{\theta} \end{pmatrix} \begin{pmatrix} \xi_1(t) \\ \xi_2(t) \\ \xi_3(t) \end{pmatrix} + \begin{pmatrix} 0 \\ \kappa \\ 0 \end{pmatrix} \mu(t) + \begin{pmatrix} 0 \\ \gamma \tau_d(t) \\ \omega(t) \end{pmatrix},$$

$$\text{4-189}$$

$$y(t) = (1, 0, 1) \begin{pmatrix} \xi_1(t) \\ \xi_2(t) \\ \xi_3(t) \end{pmatrix},$$

where col $[\xi_1(t), \xi_2(t)] = x(t)$. This is obviously a singular observation problem, because the observation noise is absent. Following the argument of Section 4.3.4, we note that the output equation is already in the form **4-158**, where C_1 and H_1 are zero matrices. It is natural to choose

$$p(t) = x(t), \qquad \textbf{4-190}$$

so that

$$C_2' = \begin{pmatrix} 1 & 0 & 0 \\ 0 & 1 & 0 \end{pmatrix}. \qquad \textbf{4-191}$$

Writing

$$p(t) = \text{col}\,[\pi_1(t), \pi_2(t)], \qquad \textbf{4-192}$$

it follows by matrix inversion from

$$\begin{pmatrix} \eta(t) \\ \pi_1(t) \\ \pi_2(t) \end{pmatrix} = \begin{pmatrix} 1 & 0 & 1 \\ 1 & 0 & 0 \\ 0 & 1 & 0 \end{pmatrix} \begin{pmatrix} \xi_1(t) \\ \xi_2(t) \\ \xi_3(t) \end{pmatrix} \qquad \textbf{4-193}$$

that

$$\begin{pmatrix} \xi_1(t) \\ \xi_2(t) \\ \xi_3(t) \end{pmatrix} = \begin{pmatrix} 0 & 1 & 0 \\ 0 & 0 & 1 \\ 1 & -1 & 0 \end{pmatrix} \begin{pmatrix} \eta(t) \\ \pi_1(t) \\ \pi_2(t) \end{pmatrix}. \qquad \textbf{4-194}$$

Since $p(t) = x(t)$, it immediately follows that $p(t)$ satisfies the state differential equation

$$\dot{p}(t) = \begin{pmatrix} 0 & 1 \\ 0 & -\alpha \end{pmatrix} p(t) + \begin{pmatrix} 0 \\ \kappa \end{pmatrix} \mu(t) + \begin{pmatrix} 0 \\ \gamma \end{pmatrix} \tau_d(t). \qquad \textbf{4-195}$$

To obtain the output equation, we differentiate $\eta(t)$:

$$\dot{\eta}(t) = (1, 0)\dot{x}(t) + \dot{\xi}_3(t). \qquad \textbf{4-196}$$

Using **4-184**, **4-188**, and **4-194**, it follows that we can write

$$\dot{\eta}(t) = \left(\frac{1}{\theta}, 1\right) p(t) - \frac{1}{\theta}\eta(t) + \omega(t). \qquad \textbf{4-197}$$

Together, **4-195** and **4-197** constitute an observation problem for $p(t)$ that is nonsingular and where the state excitation and observation noises happen to be uncorrelated. The optimal observer is of the form

$$\dot{\hat{p}}(t) = \begin{pmatrix} 0 & 1 \\ 0 & -\alpha \end{pmatrix} \hat{p}(t) + \begin{pmatrix} 0 \\ \kappa \end{pmatrix} \mu(t) + K^0(t)\left[\dot{\eta}(t) + \frac{1}{\theta}\eta(t) - \left(\frac{1}{\theta}, 1\right)\hat{p}(t)\right], \qquad \textbf{4-198}$$

where the optimal gain matrix $K^0(t)$ can be computed from the appropriate Riccati equation. From **4-194** we see that the optimal estimates $\hat{x}(t)$ of the state of the plant and $\hat{\xi}_3(t)$ of the observation noise are given by

$$\hat{x}(t) = \hat{p}(t),$$
$$\hat{\xi}_3(t) = \eta(t) + (-1, 0)\hat{p}(t).$$
4-199

Let us assume the following numerical values:

$$\kappa = 0.787 \text{ rad}/(\text{Vs}^2),$$
$$\alpha = 4.6 \text{ s}^{-1},$$
$$\gamma = 0.1 \text{ kg}^{-1}\text{m}^{-2},$$
$$V_d = 10 \text{ N}^2\text{m}^2\text{s},$$
$$\theta = 5 \times 10^{-4} \text{ s},$$
$$\sigma = 0.01 \text{ rad}.$$
4-200

The numerical values for σ and θ imply that the observation noise has an rms value of 0.01 rad and a break frequency of $1/\theta = 2000$ rad/s $\simeq 320$ Hz. With these values we find for the steady-state optimal gain matrix in **4-198**

$$\bar{K}^0 = \begin{pmatrix} 0.01998 \\ 0.4031 \end{pmatrix}.$$
4-201

The variance matrix of the reconstruction error is

$$\begin{pmatrix} 0.000003955 & 0.00007981 \\ 0.00007981 & 0.003628 \end{pmatrix}.$$
4-202

Insertion of $\bar{K}^0$ for $K^0(t)$ into **4-198** immediately gives us the optimal steady-state observer for $x(t)$. An implementation that does not require differentiation of $\eta(t)$ can easily be found.

The problem just solved differs from that of Example 4.4 by the assumption that ν_m is colored noise and not white noise. The present problem reduces to that of Example 4.4 if we approximate ν_m by white noise with an intensity V_m which equals the power spectral density of the colored noise for low frequencies, that is, we set

$$V_m = 2\sigma^2\theta.$$
4-203

The numerical values in the present example and in Example 4.4 have been chosen consistently. We are now in a position to answer a question raised in Example 4.4: Are we justified in considering ν_m white noise because it has a large bandwidth, and in computing the optimal observer accordingly? In

360 Optimal Reconstruction of the State

order to deal with this question, let us compute the reconstruction error variance matrix for the present problem by using the observer found in Example 4.4. In Example 4.4 the reconstruction error obeys the differential equation

$$\dot{e}(t) = \begin{pmatrix} -k_1 & 1 \\ -k_2 & -\alpha \end{pmatrix} e(t) + \begin{pmatrix} 0 \\ \gamma \end{pmatrix} \tau_d(t) - \begin{pmatrix} k_1 \\ k_2 \end{pmatrix} \nu_m(t), \qquad \textbf{4-204}$$

where we have set $\bar{K} = \text{col}\,(k_1, k_2)$. With the aid of **4-187** and **4-188**, we obtain the augmented differential equation

$$\begin{pmatrix} \dot{\varepsilon}_1(t) \\ \dot{\varepsilon}_2(t) \\ \dot{\xi}_3(t) \end{pmatrix} = \begin{pmatrix} -k_1 & 1 & -k_1 \\ -k_2 & -\alpha & -k_2 \\ 0 & 0 & -\dfrac{1}{\theta} \end{pmatrix} \begin{pmatrix} \varepsilon_1(t) \\ \varepsilon_2(t) \\ \xi_3(t) \end{pmatrix} + \begin{pmatrix} 0 \\ \gamma \tau_d(t) \\ \omega(t) \end{pmatrix}, \qquad \textbf{4-205}$$

where $e(t) = \text{col}\,[\varepsilon_1(t), \varepsilon_2(t)]$. It follows from Theorem 1.52 (Section 1.11.2) that the variance matrix $Q(t)$ of $\text{col}\,[\varepsilon_1(t), \varepsilon_2(t), \xi_3(t)]$ satisfies the matrix differential equation

$$\dot{Q}(t) = \begin{pmatrix} -k_1 & 1 & -k_1 \\ -k_2 & -\alpha & -k_2 \\ 0 & 0 & -\dfrac{1}{\theta} \end{pmatrix} Q(t)$$

$$+ Q(t) \begin{pmatrix} -k_1 & -k_2 & 0 \\ 1 & -\alpha & 0 \\ -k_1 & -k_2 & -\dfrac{1}{\theta} \end{pmatrix} + \begin{pmatrix} 0 & 0 & 0 \\ 0 & \gamma^2 V_d & 0 \\ 0 & 0 & \dfrac{2\sigma^2}{\theta} \end{pmatrix}. \qquad \textbf{4-206}$$

Numerical solution with the numerical values **4-200** and **4-138** yields for the steady-state variance matrix of the reconstruction error $e(t)$

$$\begin{pmatrix} 0.000003995 & 0.00008062 \\ 0.00008062 & 0.003645 \end{pmatrix}. \qquad \textbf{4-207}$$

Comparison with **4-202** shows that the rms reconstruction errors that result from the white noise approximation of Example 4.4 are only very slightly greater than for the more accurate approach of the present example. This confirms the conjecture of Example 4.4 where we argued that for the optimal observer the observation noise $\nu_m(t)$ to a good approximation is white noise, so that a more refined filter designed on the assumption that $\nu_m(t)$ is actually exponentially correlated noise gives very little improvement.

4.3 The Optimal Observer

4.3.6* Innovations

Consider the optimal observer problem of Definition 4.3 and its solution as given in Sections 4.3.2, 4.3.3, and 4.3.4. In this section we discuss an interesting property of the process

$$y(t) - C(t)\hat{x}(t), \qquad t \geq t_0, \qquad \text{4-208}$$

where $\hat{x}(t)$ is the optimal reconstruction of the state at time t based upon data up to time t. In fact, we prove that this process, **4-208**, is *white noise* with intensity $V_2(t)$, which is precisely the intensity of the observation noise $w_2(t)$. This process is called the *innovation process* (Kailath, 1968), a term that can be traced back to Wiener. The quantity $y(t) - C(t)\hat{x}(t)$ can be thought of as carrying the new information contained in $y(t)$, since $y(t) - C(t)\hat{x}(t)$ is the extra driving variable that together with the model of the system constitutes the optimal observer. The innovations concept is useful in understanding the separation theorem of linear stochastic optimal control theroy (see Chapter 5). It also has applications in state reconstruction problems outside the scope of this book, in particular the so-called optimal smoothing problem (Kailath, 1968).

We limit ourselves to the situation where the state excitation noise w_1 and the observation noise w_2 are uncorrelated and have intensities $V_1(t)$ and $V_2(t)$, respectively, where $V_2(t) > 0$, $t \geq t_0$. In order to prove that $y(t) - C(t)\hat{x}(t)$ is a white noise process with intensity $V_2(t)$, we compute the covariance matrix of its integral and show that this covariance matrix is identical to the covariance matrix of the integral of a white noise process with intensity $V_2(t)$.

Let us denote by $s(t)$ the integral of $y(t) - C(t)\hat{x}(t)$, so that

$$\dot{s}(t) = y(t) - C(t)\hat{x}(t),$$
$$s(t_0) = 0. \qquad \text{4-209}$$

Furthermore,

$$e(t) = x(t) - \hat{x}(t) \qquad \text{4-210}$$

is the reconstruction error. Referring back to Section 4.3.2, we obtain from **4-209** and **4-82** the following joint state differential equation for $s(t)$ and $e(t)$:

$$\begin{pmatrix} \dot{s}(t) \\ \dot{e}(t) \end{pmatrix} = \begin{pmatrix} 0 & C(t) \\ 0 & A(t) - K^0(t)C(t) \end{pmatrix} \begin{pmatrix} s(t) \\ e(t) \end{pmatrix} + \begin{pmatrix} 0 & I \\ I & -K^0(t) \end{pmatrix} \begin{pmatrix} w_1(t) \\ w_2(t) \end{pmatrix}, \qquad \text{4-211}$$

where $K^0(t)$ is the gain of the optimal observer. Using Theorem 1.52 (Section 1.11.2), we obtain the following matrix differential equation for the variance matrix $\tilde{Q}(t)$ of col $[s(t), e(t)]$:

$$\dot{\tilde{Q}}(t) = \begin{pmatrix} 0 & C(t) \\ 0 & A(t) - K^0(t)C(t) \end{pmatrix} \tilde{Q}(t) + \tilde{Q}(t) \begin{pmatrix} 0 & 0 \\ C^T(t) & A^T(t) - C^T(t)K^{0T}(t) \end{pmatrix}$$

$$+ \begin{pmatrix} 0 & I \\ I & -K^0(t) \end{pmatrix} \begin{pmatrix} V_1(t) & 0 \\ 0 & V_2(t) \end{pmatrix} \begin{pmatrix} 0 & I \\ I & -K^{0T}(t) \end{pmatrix}, \quad \textbf{4-212}$$

with the initial condition

$$\tilde{Q}(t_0) = \begin{pmatrix} 0 & 0 \\ 0 & Q_0 \end{pmatrix}, \quad \textbf{4-213}$$

where Q_0 is the variance matrix of $x(t_0)$. Let us partition $\tilde{Q}(t)$ as follows:

$$\tilde{Q}(t) = \begin{pmatrix} Q_{11}(t) & Q_{12}(t) \\ Q_{12}^T(t) & Q_{22}(t) \end{pmatrix}. \quad \textbf{4-214}$$

Then we can rewrite the matrix differential equation **4-212** in the form

$$\dot{Q}_{11}(t) = C(t)Q_{12}^T(t) + Q_{12}(t)C^T(t) + V_2(t), \quad Q_{11}(t_0) = 0, \quad \textbf{4-215}$$

$$\dot{Q}_{12}(t) = C(t)Q_{22}(t) + Q_{12}(t)[A(t) - K^0(t)C(t)]^T - V_2(t)K^{0T}(t),$$

$$Q_{12}(t_0) = 0, \quad \textbf{4-216}$$

$$\dot{Q}_{22}(t) = [A(t) - K^0(t)C(t)]Q_{22}(t) + Q_{22}(t)[A(t) - K^0(t)C(t)]^T$$

$$+ V_1(t) + K^0(t)V_2(t)K^{0T}(t), \quad Q_{22}(t_0) = Q_0. \quad \textbf{4-217}$$

As can be seen from **4-217**, and as could also have been seen beforehand, $Q_{22}(t) = Q(t)$, where $Q(t)$ is the variance matrix of the reconstruction error. It follows with **4-105** that in **4-216** we have

$$C(t)Q_{22}(t) - V_2(t)K^{0T}(t) = 0, \quad \textbf{4-218}$$

so that **4-216** reduces to

$$\dot{Q}_{12}(t) = Q_{12}(t)[A(t) - K^0(t)C(t)]^T, \quad Q_{12}(t_0) = 0, \quad \textbf{4-219}$$

which has the solution

$$Q_{12}(t) = 0, \quad t \geq t_0. \quad \textbf{4-220}$$

Consequently, **4-215** reduces to

$$\dot{Q}_{11}(t) = V_2(t), \quad Q_{11}(t_0) = 0, \quad \textbf{4-221}$$

so that

$$Q_{11}(t) = \int_{t_0}^{t} V_2(\tau) \, d\tau. \quad \textbf{4-222}$$

4.3 The Optimal Observer

By invoking Theorem 1.52 once again, the covariance matrix of col $[s(t), e(t)]$ can be written as

$$\tilde{R}(t_1, t_2) = \begin{cases} \tilde{Q}(t_1)\tilde{\Psi}^T(t_2, t_1) & \text{for } t_2 \geq t_1, \\ \tilde{\Psi}(t_1, t_2)\tilde{Q}(t_2) & \text{for } t_1 \geq t_2, \end{cases} \quad \text{4-223}$$

where $\tilde{\Psi}(t_1, t_0)$ is the transition matrix of the system

$$\begin{pmatrix} \dot{s}(t) \\ \dot{e}(t) \end{pmatrix} = \begin{pmatrix} 0 & C(t) \\ 0 & A(t) - K^0(t)C(t) \end{pmatrix} \begin{pmatrix} s(t) \\ e(t) \end{pmatrix}. \quad \text{4-224}$$

It is easily found that this transition matrix is given by

$$\tilde{\Psi}(t_1, t_0) = \begin{pmatrix} I & \int_{t_0}^{t_1} C(t)\Psi(t, t_0)\, dt \\ 0 & \Psi(t_1, t_0) \end{pmatrix}. \quad \text{4-225}$$

where $\Psi(t_1, t_0)$ is the transition matrix of the system

$$\dot{e}(t) = [A(t) - K^0(t)C(t)]e(t). \quad \text{4-226}$$

The covariance matrix of $s(t)$ is the (1, 1)-block of $\tilde{R}(t_1, t_2)$, which can be found to be given by

$$R_s(t_1, t_2) = \int_{t_0}^{\min(t_1, t_2)} V_2(t)\, dt. \quad \text{4-227}$$

This is the covariance matrix of a process with uncorrelated increments (see Example 1.29, Section 1.10.1). Since the process $y(t) - C(t)\hat{x}(t)$ is the derivative of the process $s(t)$, it is white noise with intensity $V_2(t)$ (see Example 1.33, Section 1.11.1).

We summarize as follows.

Theorem 4.7. *Consider the solution of the nonsingular optimal observer problem with uncorrelated state excitation noise and observation noise as given in Theorem 4.5. Then the innovation process*

$$y(t) - C(t)\hat{x}(t), \quad t \geq t_0, \quad \text{4-228}$$

is a white noise process with intensity $V_2(t)$.

It can be proved that this theorem is also true for the singular optimal observer problem with correlated state excitation and observation noises.

4.4* THE DUALITY OF THE OPTIMAL OBSERVER AND THE OPTIMAL REGULATOR; STEADY-STATE PROPERTIES OF THE OPTIMAL OBSERVER

4.4.1* Introduction

In this section we study the steady-state and stability properties of the optimal observer. All of these results are based upon the properties of the optimal regulator obtained in Chapter 3. These results are derived through the *duality* of the optimal regulator and the optimal observer problem (Kalman and Bucy, 1961). Section 4.4.2 is devoted to setting forth this duality, while in Section 4.4.3 the steady-state properties of the optimal observer are discussed. Finally, in Section 4.4.4 we study the asymptotic behavior of the steady-state time-invariant optimal observer as the intensity of the observation noise goes to zero.

4.4.2* The Duality of the Optimal Regulator and the Optimal Observer Problem

The main result of this section is summarized in the following theorem.

Theorem 4.8. *Consider the optimal regulator problem (ORP) of Definition 3.2 (Section 3.3.1) and the nonsingular optimal observer problem (OOP) with uncorrelated state excitation and observation noises of Definition 4.3 (Section 4.3.1). In the observer problem let the matrix $V_1(t)$ be given by*

$$V_1(t) = G(t)V_3(t)G^T(t), \quad t \geq t_0, \qquad \text{4-229}$$

where

$$V_3(t) > 0, \quad t \geq t_0. \qquad \text{4-230}$$

Let the various matrices occurring in the definitions of the ORP and the OOP be related as follows:

$$\begin{aligned}
&A(t) \text{ of the ORP equals } A^T(t^* - t) \text{ of the OOP,} \\
&B(t) \text{ of the ORP equals } C^T(t^* - t) \text{ of the OOP,} \\
&D(t) \text{ of the ORP equals } G^T(t^* - t) \text{ of the OOP,} \\
&R_3(t) \text{ of the ORP equals } V_3(t^* - t) \text{ of the OOP,} \\
&R_2(t) \text{ of the ORP equals } V_2(t^* - t) \text{ of the OOP,} \\
&P_1 \quad \text{of the ORP equals } Q_0 \qquad \text{of the OOP,}
\end{aligned} \qquad \text{4-231}$$

all for $t \leq t_1$. Here

$$t^* = t_0 + t_1. \qquad \text{4-232}$$

Under these conditions the solutions of the optimal regulator problem (Theorem

4.4 Duality and Steady-State Properties

3.4, Section 3.3.3) and the nonsingular optimal observer problem with uncorrelated state excitation and observation noises (Theorem 4.5, Section 4.3.2) are related as follows:

(a) $P(t)$ of the ORP equals $Q(t^* - t)$ of the OOP for $t \leq t_1$;
(b) $F^0(t)$ of the ORP equals $K^{0T}(t^* - t)$ of the OOP for $t \leq t_1$;
(c) The closed-loop regulator of the ORP:

$$\dot{x}(t) = [A(t) - B(t)F^0(t)]x(t), \qquad \textbf{4-233}$$

and the unforced reconstruction error equation of the OOP:

$$\dot{e}(t) = [A(t) - K^0(t)C(t)]e(t), \qquad \textbf{4-234}$$

are dual with respect to t^* in the sense of Definition 1.23 (Section 1.8).

The proof of this theorem easily follows by comparing the regulator Riccati equation **3-130** and the observer Riccati equation **4-106**, and using time reversal (Lemma 4.1, Section 4.3.2).

In Section 4.4.3 we use the duality of the optimal regulator and the optimal observer problem to obtain the steady-state properties of the optimal observer from those of the optimal regulator. Moreover, this duality enables us to use computer programs designed for optimal regulator problems for optimal observer problems, and vice versa, by making the substitutions **4-231**.

4.4.3* Steady-State Properties of the Optimal Observer

Theorem 4.8 enables us to transfer from the regulator to the observer problem the steady-state properties (Theorem 3.5, Section 3.4.2), the steady-state stability properties (Theorem 3.6, Section 3.4.2), and various results for the time-invariant case (Theorems 3.7, Section 3.4.3, and 3.8, Section 3.4.4).

In this section we state some of the more important steady-state and stability properties. Theorem 3.5, concerning the steady-state behavior of the Riccati equation, can be rephrased as follows (Kalman and Bucy, 1961).

Theorem 4.9. *Consider the matrix Riccati equation*

$$\dot{Q}(t) = A(t)Q(t) + Q(t)A^T(t) + G(t)V_3(t)G^T(t)$$
$$- Q(t)C^T(t)V_2^{-1}(t)C(t)Q(t). \qquad \textbf{4-235}$$

Suppose that $A(t)$ is continuous and bounded, that $C(t)$, $G(t)$, $V_3(t)$, and $V_2(t)$ are piecewise continuous and bounded, and furthermore that

$$V_3(t) \geq \alpha I, \qquad V_2(t) \geq \beta I \qquad \text{for all } t, \qquad \textbf{4-236}$$

where α and β are positive constants.

(i) Then if the system

$$\dot{x}(t) = A(t)x(t) + G(t)w_3(t),$$
$$y(t) = C(t)x(t),\qquad\qquad \textbf{4-237}$$

is either
  (a) *completely reconstructible*, or
  (b) *exponentially stable*,
the solution $Q(t)$ of the Riccati equation **4-235** with the initial condition $Q(t_0) = 0$ converges to a nonnegative-definite matrix $\bar{Q}(t)$ as $t_0 \to -\infty$. $\bar{Q}(t)$ is a solution of the Riccati equation **4-235**.

(ii) Moreover, if the system **4-237** is either
  (c) *both uniformly completely reconstructible and uniformly completely controllable*, or
  (d) *exponentially stable*,
the solution $Q(t)$ of the Riccati equation **4-235** with the initial condition $Q(t_0) = Q_0$ converges to $\bar{Q}(t)$ as $t_0 \to -\infty$ for any $Q_0 \geq 0$.

The proof of this theorem immediately follows by applying the duality relations of Theorem 4.8 to Theorem 3.5, and recalling that if a system is completely reconstructible its dual is completely controllable (Theorem 1.41, Section 1.8), and that if a system is exponentially stable its dual is also exponentially stable (Theorem 1.42, Section 1.8).

We now state the dual of Theorem 3.6 (Section 3.4.2):

Theorem 4.10. *Consider the nonsingular optimal observer problem with uncorrelated state excitation and observation noises and let*

$$V_1(t) = G(t)V_3(t)G^T(t), \qquad \text{for all } t, \qquad \textbf{4-238}$$

where $V_3(t) > 0$, for all t. Suppose that the continuity, boundedness, and positive-definiteness conditions of Theorem 4.9 concerning A, C, G, V_3, and V_2 are satisfied. Then if the system **4-237** *is either*
  (a) *uniformly completely reconstructible and uniformly completely controllable*, *or*
  (b) *exponentially stable*,
the following facts hold.
(i) *The* **steady-state optimal observer**

$$\dot{\hat{x}}(t) = A(t)\hat{x}(t) + \bar{K}(t)[y(t) - C(t)\hat{x}(t)], \qquad \textbf{4-239}$$

where

$$\bar{K}(t) = \bar{Q}(t)C^T(t)V_2^{-1}(t), \qquad \textbf{4-240}$$

is exponentially stable. Here $\bar{Q}(t)$ is as defined in Theorem **4.9**.

4.4 Duality and Steady-State Properties

(ii) *The steady-state optimal observer gain $\bar{K}(t)$ minimizes*

$$\lim_{t_0 \to -\infty} E\{e^T(t)W(t)e(t)\} \quad \text{4-241}$$

for every $Q_0 \geq 0$. The minimal value of **4-241**, *which is achieved by the steady-state optimal observer, is given by*

$$\operatorname{tr}[\bar{Q}(t)W(t)]. \quad \text{4-242}$$

We also state the counterpart of Theorem 3.7 (Section 3.4.3), which is concerned with time-invariant systems.

Theorem 4.11. *Consider the time-invariant nonsingular optimal observer problem of Definition 4.3 with uncorrelated state excitation and observation noises for the system*

$$\begin{aligned} \dot{x}(t) &= Ax(t) + Gw_3(t), \\ y(t) &= Cx(t) + w_2(t). \end{aligned} \quad \text{4-243}$$

Here w_3 is white noise with intensity V_3, and w_2 has intensity V_2. It is assumed that $V_3 > 0$, $V_2 > 0$, and $Q_0 \geq 0$. The associated Riccati equation is given by

$$\dot{Q}(t) = AQ(t) + Q(t)A^T + GV_3G^T - Q(t)C^TV_2^{-1}CQ(t), \quad \text{4-244}$$

with the initial condition

$$Q(t_0) = Q_0. \quad \text{4-245}$$

(a) *Assume that $Q_0 = 0$. Then as $t_0 \to -\infty$ the solution of the Riccati equation approaches a constant steady-state value $\bar{Q}$ if and only if the system* **4-243** *possesses no poles that are at the same time unstable, unreconstructible, and controllable.*
(b) *If the system* **4-243** *is both detectable and stabilizable, the solution of the Riccati equation approaches the value $\bar{Q}$ as $t_0 \to -\infty$ for every $Q_0 \geq 0$.*
(c) *If $\bar{Q}$ exists, it is a nonnegative-definite symmetric solution of the algebraic Riccati equation*

$$0 = AQ + QA^T + GV_3G^T - QC^TV_2^{-1}CQ. \quad \text{4-246}$$

If the system **4-243** *is detectable and stabilizable, $\bar{Q}$ is the unique nonnegative-definite solution of the algebraic Riccati equation.*
(d) *If $\bar{Q}$ exists, it is positive-definite if and only if the system is completely controllable.*
(e) *If $\bar{Q}$ exists, the steady-state optimal observer*

$$\dot{\hat{x}}(t) = A\hat{x}(t) + \bar{K}[y(t) - C\hat{x}(t)], \quad \text{4-247}$$

where

$$\bar{K} = \bar{Q}C^TV_2^{-1}, \quad \text{4-248}$$

is asymptotically stable if and only if the system is detectable and stabilizable.

(f) *If the system is detectable and stabilizable, the steady-state optimal observer* **4-247** *minimizes*

$$\lim_{t_0 \to -\infty} E\{e^T(t)We(t)\} \qquad \text{4-249}$$

for all $Q_0 \geq 0$. *For the steady-state optimal observer,* **4-249** *is given by*

$$\text{tr } [\bar{Q}W]. \qquad \text{4-250}$$

We note that the conditions (b) and (c) are sufficient but not necessary.

4.4.4* Asymptotic Properties of Time-Invariant Steady-State Optimal Observers

In this section we consider the properties of the steady-state optimal filter for the time-invariant case, when the intensity of the observation noise approaches zero. This section is quite short since we are able to obtain our results immediately by "dualizing" the results of Section 3.8.

We first consider the case in which both the state excitation noise $w_3(t)$ (see **4-237**) and the observed variable are scalar. From Theorem 3.11 (Section 3.8.1), the following result is obtained almost immediately.

Theorem 4.12. *Consider the n-dimensional time-invariant system*

$$\begin{aligned}\dot{x}(t) &= Ax(t) + Bu(t) + g\omega_3(t), \\ \eta(t) &= cx(t) + \omega_2(t),\end{aligned} \qquad \text{4-251}$$

where ω_3 is scalar white noise with constant intensity V_3, ω_2 scalar white noise uncorrelated with ω_3 with positive constant intensity V_2, g a column vector, and c a row vector. Suppose that $\{A, g\}$ is stabilizable and $\{A, c\}$ detectable. Let $H(s)$ be the scalar transfer function

$$H(s) = c(sI - A)^{-1}g = \frac{\psi(s)}{\phi(s)} = \frac{\alpha \prod_{i=1}^{p}(s - \nu_i)}{\prod_{i=1}^{n}(s - \pi_i)}, \qquad \text{4-252}$$

where $\phi(s)$ is the characteristic polynomial of the system, and π_i, $i = 1, 2, \cdots, n$, its characteristic values. Then the characteristic values of the steady-state optimal observer are the left-half plane zeroes of the polynomial

$$(-1)^n \phi(s)\phi(-s)\left[1 + \frac{V_3}{V_2}H(-s)H(s)\right]. \qquad \text{4-253}$$

As a result, the following statements hold.

4.4 Duality and Steady-State Properties

(a) As $V_2/V_3 \to 0$, p of the n steady-state optimal observer poles approach the numbers $\hat{\nu}_i$, $i = 1, 2, \cdots, p$, where

$$\hat{\nu}_i = \begin{cases} \nu_i & \text{if Re } (\nu_i) \leq 0, \\ -\nu_i & \text{if Re } (\nu_i) > 0. \end{cases} \qquad 4\text{-}254$$

(b) As $V_2/V_3 \to 0$, the remaining $n - p$ observer poles asymptotically approach straight lines which intersect in the origin and make angles with the negative real axis of

$$\pm l \frac{\pi}{n-p}, \quad l = 0, 1, \cdots, \frac{n-p-1}{2}, \quad n - p \text{ odd},$$

$$\pm \frac{(l + \tfrac{1}{2})\pi}{n-p}, \quad l = 0, 1, \cdots, \frac{n-p}{2} - 1, \quad n - p \text{ even}. \qquad 4\text{-}255$$

These faraway observer poles asymptotically are at a distance

$$\omega_0 = \left(\alpha^2 \frac{V_3}{V_2}\right)^{1/[2(n-p)]} \qquad 4\text{-}256$$

from the origin.

(c) As $V_2/V_3 \to \infty$, the n observer poles approach the numbers $\hat{\pi}_i$, $i = 1, 2, \cdots, n$, where

$$\hat{\pi}_i = \begin{cases} \pi_i & \text{if Re } (\pi_i) \leq 0, \\ -\pi_i & \text{if Re } (\pi_i) > 0. \end{cases} \qquad 4\text{-}257$$

It follows from (b) that the faraway poles approach a Butterworth configuration.

For the general case we have the following results, which follow from Theorem 3.12 (Section 3.8.1).

Theorem 4.13. *Consider the n-dimensional time-invariant system*

$$\begin{aligned} \dot{x}(t) &= Ax(t) + Bu(t) + Gw_3(t), \\ y(t) &= Cx(t) + w_2(t), \end{aligned} \qquad 4\text{-}258$$

where w_3 is white noise with constant intensity V_3 and w_2 is white noise uncorrelated with w_3 with constant intensity $V_2 > 0$. Suppose that $\{A, G\}$ is stabilizable and $\{A, C\}$ detectable. Then the poles of the steady-state optimal observer are the left-half plane zeroes of the polynomial

$$(-1)^n \phi(s)\phi(-s) \det [I + V_2^{-1} H(s) V_3 H^T(-s)], \qquad 4\text{-}259$$

where $H(s)$ is the transfer matrix

$$H(s) = C(sI - A)^{-1}G, \qquad 4\text{-}260$$

and $\phi(s)$ is the characteristic polynomial of the system **4-258**. Suppose that $\dim(w_3) = \dim(y) = k$, so that $H(s)$ is a $k \times k$ transfer matrix. Let

$$\det[H(s)] = \frac{\psi(s)}{\phi(s)} = \frac{\alpha \prod_{i=1}^{p}(s - \nu_i)}{\prod_{i=1}^{n}(s - \pi_i)}, \qquad \text{4-261}$$

and assume that $\alpha \neq 0$. Also, suppose that

$$V_2 = \rho N, \qquad \text{4-262}$$

with $N > 0$ and ρ a positive scalar.

(a) Then as $\rho \downarrow 0$, p of the optimal observer poles approach the numbers $\hat{\nu}_i$, $i = 1, 2, \cdots, p$, where

$$\hat{\nu}_i = \begin{cases} \nu_i & \text{if } \operatorname{Re}(\nu_i) \leq 0, \\ -\nu_i & \text{if } \operatorname{Re}(\nu_i) > 0. \end{cases} \qquad \text{4-263}$$

The remaining observer poles go to infinity and group into several Butterworth configurations of different orders and different radii. A rough estimate of the distance of the faraway poles to the origin is

$$\left(\alpha^2 \frac{\det(V_3)}{\rho^k \det(N)} \right)^{1/[2(n-p)]}. \qquad \text{4-264}$$

(b) As $\rho \to \infty$, the n optimal observer poles approach the numbers $\hat{\pi}_i$, $i = 1, 2, \cdots, n$, where

$$\hat{\pi}_i = \begin{cases} \pi_i & \text{if } \operatorname{Re}(\pi_i) \leq 0, \\ -\pi_i & \text{if } \operatorname{Re}(\pi_i) > 0. \end{cases} \qquad \text{4-265}$$

Some information concerning the behavior of the observer poles when $\dim(w_3) \neq \dim(y)$ follows by dualizing the results of Problem 3.14.

We finally transcribe Theorem 3.14 (Section 3.8.3) as follows.

Theorem 4.14. *Consider the time-invariant system*

$$\begin{aligned} \dot{x}(t) &= Ax(t) + Gw_3(t), \\ y(t) &= Cx(t) + w_2(t), \end{aligned} \qquad \text{4-266}$$

where G and C have full rank, w_3 is white noise with constant intensity V_3 and w_2 is white noise uncorrelated with w_3 with constant nonsingular intensity $V_2 = \rho N$, $\rho > 0$, $N > 0$. Suppose that $\{A, G\}$ is stabilizable and $\{A, C\}$ detectable and let $\bar{Q}$ be the steady-state solution of the variance Riccati equation **4-244** *associated with the optimal observer problem. Then the following facts hold.*

(a) *The limit*
$$\lim_{\rho \downarrow 0} \bar{Q} = Q_1 \qquad \text{4-267}$$
exists.

(b) *Let $e_s(t)$ denote the contribution of the state excitation noise to the reconstruction error $e(t) = x(t) - \hat{x}(t)$, and $e_o(t)$ the contribution of the observation noise to $e(t)$. Then for the steady-state optimal observer the following limits hold:*

$$\lim_{\rho \downarrow 0} E\{e^T(t)We(t)\} = \text{tr}\,(Q_1 W),$$

$$\lim_{\rho \downarrow 0} E\{e_s^T(t)We_s(t)\} = \text{tr}\,(Q_1 W), \qquad \text{4-268}$$

$$\lim_{\rho \downarrow 0} E\{e_o^T(t)We_o(t)\} = 0.$$

(c) *If* dim (w_3) > dim (y), *then* $Q_1 \neq 0$.
(d) *If* dim (w_3) = dim (y), *and the numerator polynomial $\psi(s)$ of the square transfer matrix*
$$C(sI - A)^{-1}G \qquad \text{4-269}$$
is nonzero, then $Q_1 = 0$ if and only if $\psi(s)$ has zeroes with nonpositive real parts only.
(e) *If* dim (w_3) < dim (y), *then a sufficient condition for Q_1 to be the zero matrix is that there exists a rectangular matrix M such that the numerator polynomial of the square transfer matrix $MC(sI - A)^{-1}G$ is nonzero and has zeroes with nonpositive real parts only.*

This theorem shows that if no observation noise is present, completely accurate reconstruction of the state of the system is possible only if the number of components of the observed variable is at least as great as the number of components of the state excitation noise $w_1(t)$. Even if this condition is satisfied, completely faultless reconstruction is possible only if the transfer matrix from the system noise w_3 to the observed variable y possesses no right-half plane zeroes.

The following question now comes to mind. For very small values of the observation noise intensity V_2, the optimal observer has some of its poles very far away, but some other poles may remain in the neighborhood of the origin. These nearby poles cause the reconstruction error to recover relatively slowly from certain initial values. Nevertheless, Theorem 4.14 states that the reconstruction error variance matrix can be quite small. This seems to be a contradiction. The answer to this question must be that the structure of the system to be observed is so exploited that the reconstruction error cannot be driven into the subspace from which it can recover only slowly.

372   Optimal Reconstruction of the State

We conclude this section by remarking that Q_1, the limiting variance matrix for $\rho \downarrow 0$, can be computed by solving the singular optimal observer problem that results from setting $w_2(t) \equiv 0$. As it turns out, occasionally the reduced-order observation problem thus obtained involves a nondetectable system, which causes the appropriate algebraic Riccati equation to possess more than one nonnegative-definite solution. In such a case one of course has to select that solution that makes the reduced-order observer stable (asymptotically or in the sense of Lyapunov), since the full-order observer that approaches the reduced-order observer as $V_2 \to 0$ is always asymptotically stable.

The problem that is dual to computing Q_1, that is, the problem of computing

$$P_0 = \lim_{R_2 \to 0} \bar{P} \qquad 4\text{-}270$$

for the optimal deterministic regulator problem (Section 3.8.3), can be solved by formulating the dual observer problem and attacking the resulting singular optimal observer problem as outlined above. Butman (1968) gives a direct approach to the "control-free costs" linear regulator problem.

Example 4.6. *Positioning system*

In Example 4.4 (Section 4.3.2), we found that for the positioning system under consideration the steady-state solution of the error variance matrix is given by

$$\bar{Q} = V_m \begin{pmatrix} -\alpha + \sqrt{\alpha^2 + 2\beta} & \alpha^2 + \beta - \alpha\sqrt{\alpha^2 + 2\beta} \\ \alpha^2 + \beta - \alpha\sqrt{\alpha^2 + 2\beta} & -\alpha^3 - 2\alpha\beta + (\alpha^2 + \beta)\sqrt{\alpha^2 + 2\beta} \end{pmatrix}, \qquad 4\text{-}271$$

where

$$\beta = \gamma\sqrt{V_d/V_m}. \qquad 4\text{-}272$$

As $V_m \downarrow 0$, the variance matrix behaves as

$$\bar{Q} \simeq \begin{pmatrix} 2^{1/2}\gamma^{1/2}V_d^{1/4}V_m^{3/4} & \gamma V_d^{1/2}V_m^{1/2} \\ \gamma V_d^{1/2}V_m^{1/2} & 2^{1/2}\gamma^{3/2}V_d^{3/4}V_m^{1/4} \end{pmatrix}. \qquad 4\text{-}273$$

Obviously, $\bar{Q}$ approaches the zero matrix as $V_m \downarrow 0$. In Example 4.4 we found that the optimal observer poles are

$$\tfrac{1}{2}(-\sqrt{\alpha^2 + 2\beta} \pm \sqrt{\alpha^2 - 2\beta}). \qquad 4\text{-}274$$

Asymptotically, these poles behave as

$$\tfrac{1}{2}\sqrt{2}\left(\frac{V_d}{V_m}\right)^{1/4}\gamma^{1/2}(-1 \pm j), \qquad 4\text{-}275$$

which represents a second-order Butterworth configuration. All these facts accord with what we might suppose, since the system transfer function is given by

$$H(s) = c(sI - A)^{-1}g = \frac{1}{s(s + \alpha)}, \qquad \text{4-276}$$

which possesses no zeroes. As we have seen in Example 4.4, for $V_m \downarrow 0$ the optimal filter approaches the differentiating reduced-order filter

$$\hat{\xi}_1(t) = \eta(t),$$
$$\hat{\xi}_2(t) = \dot{\eta}(t). \qquad \text{4-277}$$

If no observation noise is present, this differentiating filter reconstructs the state completely accurately, no matter how large the state excitation noise.

4.5 CONCLUSIONS

In this chapter we have solved the problem of reconstructing the state of a linear differential system from incomplete and inaccurate measurements. Several versions of this problem have been discussed. The steady-state and asymptotic properties of optimal observers have been reviewed. It has been seen that some of the results of this chapter are reminiscent of those obtained in Chapter 3, and in fact we have derived several of the properties of optimal observers from the corresponding properties of optimal regulators as obtained in Chapter 3.

With the results of this chapter, we are in a position to extend the results of Chapter 3 where we considered linear state feedback control systems. We can now remove the usually unacceptable assumption that all the components of the state can always be accurately measured. This is done in Chapter 5, where we show how *output feedback control systems* can be designed by connecting the state feedback laws of Chapter 3 to the observers of the present chapter.

4.6 PROBLEMS

4.1. *An observer for the inverted pendulum positioning system*
Consider the inverted pendulum positioning system described in Example

1.1 (Section 1.2.3). The state differential equation of this system is given by

$$\dot{x}(t) = \begin{pmatrix} 0 & 1 & 0 & 0 \\ 0 & -\dfrac{F}{M} & 0 & 0 \\ 0 & 0 & 0 & 1 \\ -\dfrac{g}{L'} & 0 & \dfrac{g}{L'} & 0 \end{pmatrix} x(t) + \begin{pmatrix} 0 \\ \dfrac{1}{M} \\ 0 \\ 0 \end{pmatrix} \mu(t). \qquad 4\text{-}278$$

Suppose we choose as the observed variable the angle $\phi(t)$ that the pendulum makes with the vertical, that is, we let

$$\eta_1(t) = \left(-\dfrac{1}{L'}, 0, \dfrac{1}{L'}, 0\right) x(t). \qquad 4\text{-}279$$

Consider the problem of finding a time-invariant observer for this system.

(a) Show that it is impossible to find an asymptotically stable observer. Explain this physically.

(b) Show that if in addition to the angle $\phi(t)$ the displacement $s(t)$ of the carriage is also measured, that is, we add a component

$$\eta_2(t) = (1, 0, 0, 0) x(t) \qquad 4\text{-}280$$

to the observed variable, an asymptotically stable time-invariant observer can be found.

4.2. *Reconstruction of the angular velocity*

Consider the angular velocity control system of Example 3.3 (Section 3.3.1), which is described by the state differential equation

$$\dot{\xi}(t) = -\alpha \xi(t) + \kappa \mu(t), \qquad 4\text{-}281$$

where $\xi(t)$ is the angular velocity and $\mu(t)$ the driving voltage. Suppose that the system is disturbed by a stochastically varying torque operating on the shaft, so that we write

$$\dot{\xi}(t) = -\alpha \xi(t) + \kappa \mu(t) + \omega_1(t), \qquad 4\text{-}282$$

where $\omega_1(t)$ is exponentially correlated noise with rms value σ_1 and time constant θ_1. The observed variable is given by

$$\eta(t) = \xi(t) + \omega_2(t), \qquad 4\text{-}283$$

where ω_2 is exponentially correlated noise with rms value σ_2 and time constant θ_2. The processes ω_1 and ω_2 are uncorrelated.

The following numerical values are assumed:

$$\alpha = 0.5 \text{ s}^{-1},$$
$$\kappa = 150 \text{ rad}/(\text{V s}^2),$$
$$\sigma_1 = 54.78 \text{ rad}/\text{s}^2,$$
$$\theta_1 = 0.1 \text{ s},$$
$$\sigma_2 = 5 \text{ rad}/\text{s},$$
$$\theta_2 = 0.01 \text{ s}.$$

4-284

(a) Since the state excitation noise and the observation noise have quite large bandwidths as compared to the system bandwidth, we first attempt to find an optimal observer for the angular velocity by approximating both the state excitation noise and the observation noise as white noise processes, with intensities equal to the power spectral densities of ω_1 and ω_2 at zero frequency. Compute the steady-state optimal observer that results from this approach.

(b) To verify whether or not it is justified to represent ω_1 and ω_2 as white noise processes, model ω_1 and ω_2 as exponentially correlated noise processes, and find the augmented state differential equation that describes the angular velocity control system. Using the observer differential equation obtained under (a), obtain a three-dimensional augmented state differential equation for the reconstruction error $\varepsilon(t) = \xi(t) - \hat{\xi}(t)$ and the state variables of the processes ω_1 and ω_2. Next compute the steady-state variance of the reconstruction error and compare this number to the value that has been predicted under (a). Comment on the difference and the reason that it exists.

(c) Attempt to reach a better agreement between the predicted and the actual results by reformulating the observation problem as follows. The state excitation noise is modeled as exponentially correlated noise, but the approximation of the observation noise by white noise is maintained, since the observation noise bandwidth is very large. Compute the steady-state optimal observer for this situation and compare its predicted steady-state mean square reconstruction error with the actual value (taking into account that the observation noise is exponentially correlated noise). Comment on the results.

(d)* Determine the completely accurate solution of the optimal observer problem by modeling the observation noise as exponentially correlated noise also. Compare the performance of the resulting steady-state optimal observer to that of the observer obtained under (c) and comment.

4.3. *Solution of the observer Riccati equation.*
Consider the matrix Riccati equation

$$\dot{Q}(t) = A(t)Q(t) + Q(t)A^T(t) + V_1(t) - Q(t)C^T(t)V_2^{-1}(t)C(t)Q(t) \quad \text{4-285}$$

with the initial condition

$$Q(t_0) = Q_0. \quad \text{4-286}$$

Define $\Psi(t, t_0)$ as the $(2n \times 2n)$-dimensional [$Q(t)$ is $n \times n$] solution of

$$\frac{d}{dt}\Psi(t, t_0) = \begin{pmatrix} -A^T(t) & C^T(t)V_2^{-1}(t)C(t) \\ V_1(t) & A(t) \end{pmatrix}\Psi(t, t_0),$$ **4-287**

$\Psi(t_0, t_0) = I$.

Partition $\Psi(t, t_0)$ corresponding to the partitioning occurring in **4-287** as follows.

$$\Psi(t, t_0) = \begin{pmatrix} \Psi_{11}(t, t_0) & \Psi_{12}(t, t_0) \\ \Psi_{21}(t, t_0) & \Psi_{22}(t, t_0) \end{pmatrix}.$$ **4-288**

Show that the solution of the Riccati equation can be written as

$$Q(t) = [\Psi_{21}(t, t_0) + \Psi_{22}(t, t_0)Q_0][\Psi_{11}(t, t_0) + \Psi_{12}(t, t_0)Q_0]^{-1}.$$ **4-289**

4.4.* *Determination of a priori data for the singular optimal observer*

When computing an optimal observer for the singular observation problem as described in Section 4.3.4, we must determine the a priori data

$$\hat{p}(t_0) = E\{C_2'x(t_0) \mid y_2(t_0)\}$$ **4-290**

and

$$Q(t_0) = E\{[p(t_0) - \hat{p}(t_0)][p(t_0) - \hat{p}(t_0)]^T \mid y_2(t_0)\},$$ **4-291**

where

$$y_2(t_0) = C_2x(t_0).$$ **4-292**

We assume that

$$E\{x(t_0)\} = \bar{x}_0$$ **4-293**

and

$$E\{[x(t_0) - \bar{x}_0][x(t_0) - \bar{x}_0]^T\} = Q_0$$ **4-294**

are given. Prove that if $x(t_0)$ is Gaussian then

$$E\{x(t_0) \mid y_2(t_0)\} = \hat{x}(t_0) = \bar{x}_0 + Q_0C_2^T(C_2Q_0C_2^T)^{-1}[y_2(t_0) - C_2\bar{x}_0] \quad \textbf{4-295}$$

and

$$E\{[x(t_0) - \hat{x}(t_0)][x(t_0) - \hat{x}(t_0)]^T \mid y_2(t_0)\} = Q_0 - Q_0C_2^T(C_2Q_0C_2^T)^{-1}C_2Q_0.$$ **4-296**

Determine from these results expressions for **4-290** and **4-291**. *Hint:* Use the vector formula for a multidimensional Gaussian density function (compare **1-434**) and the expression for the inverse of a partitioned matrix as given by Noble (1969, Exercise 1.59, p. 25).

5 OPTIMAL LINEAR OUTPUT FEEDBACK CONTROL SYSTEMS

5.1 INTRODUCTION

In Chapter 3 we considered the control of linear systems described by a state differential equation of the form

$$\dot{x}(t) = A(t)x(t) + B(t)u(t). \qquad 5\text{-}1$$

An essential part of the theory of Chapter 3 is that it is assumed that the complete state vector $x(t)$ is available for measurement and feedback.

In this chapter we relax this assumption and study the much more realistic case where there is an observed variable of the form

$$y(t) = C(t)x(t), \qquad 5\text{-}2$$

which is available for measurement and feedback. Control systems where the observed variable y serves as input to the controller, and not the state x, will be called *output feedback control systems*.

In view of the results of Chapter 4, it is not surprising that the optimal output feedback controller turns out to be a combination of an observer, through which the state of the system is reconstructed, and a control law which is an instantaneous, linear function of the reconstructed state. This control law is the same control law that would have been obtained if the state had been directly available for observation.

In Section 5.2 we consider a deterministic approach to the output feedback problem and we obtain regulators through a combination of asymptotically stable observers and linear, stabilizing control laws. In Section 5.3 a stochastic approach is taken, and optimal linear feedback regulators are derived as interconnections of optimal observers and optimal linear state feedback laws. In Section 5.4 tracking problems are studied. In Section 5.5 we consider regulators and tracking systems with nonzero set points and constant disturbances. Section 5.6 concerns the sensitivity of linear optimal feedback systems to disturbances and system variations, while the chapter concludes with Section 5.7, dealing with reduced-order feedback controllers.

5.2 THE REGULATION OF LINEAR SYSTEMS WITH INCOMPLETE MEASUREMENTS

5.2.1 The Structure of Output Feedback Control Systems

In this section we take a deterministic approach to the problem of regulating a linear system with incomplete measurements. Consider the system described by the state differential equation

$$\dot{x}(t) = A(t)x(t) + B(t)u(t), \qquad \text{5-3}$$

while the observed variable is given by

$$y(t) = C(t)x(t). \qquad \text{5-4}$$

In Chapter 3 we considered control laws of the form

$$u(t) = -F(t)x(t), \qquad \text{5-5}$$

where it was assumed that the whole state $x(t)$ can be accurately measured. If the state is not directly available for measurement, a natural approach is first to construct an observer of the form

$$\dot{\hat{x}}(t) = A(t)\hat{x}(t) + B(t)u(t) + K(t)[y(t) - C(t)\hat{x}(t)], \qquad \text{5-6}$$

and then interconnect the control law with the *reconstructed* state $\hat{x}(t)$:

$$u(t) = -F(t)\hat{x}(t), \qquad \text{5-7}$$

where $F(t)$ is the same as in **5-5**. Figure 5.1 depicts the interconnection of the plant, the observer, and the control law. By substitution of the control law **5-7** into the observer equation **5-6**, the controller equations take the form

$$\begin{aligned}\dot{\hat{x}}(t) &= [A(t) - B(t)F(t) - K(t)C(t)]\hat{x}(t) + K(t)y(t),\\ u(t) &= -F(t)\hat{x}(t).\end{aligned} \qquad \text{5-8}$$

This leads to the simplified structure of Fig. 5.2.

The closed-loop system that results from interconnecting the plant with the controller is a linear system of dimension $2n$ (where n is the dimension of the state x), which can be described as

$$\begin{pmatrix}\dot{x}(t)\\ \dot{\hat{x}}(t)\end{pmatrix} = \begin{pmatrix} A(t) & -B(t)F(t) \\ K(t)C(t) & A(t) - K(t)C(t) - B(t)F(t) \end{pmatrix}\begin{pmatrix}x(t)\\ \hat{x}(t)\end{pmatrix}. \qquad \text{5-9}$$

We now analyze the stability properties of the closed-loop system. To this end we consider the state $x(t)$ and the reconstruction error

$$e(t) = x(t) - \hat{x}(t). \qquad \text{5-10}$$

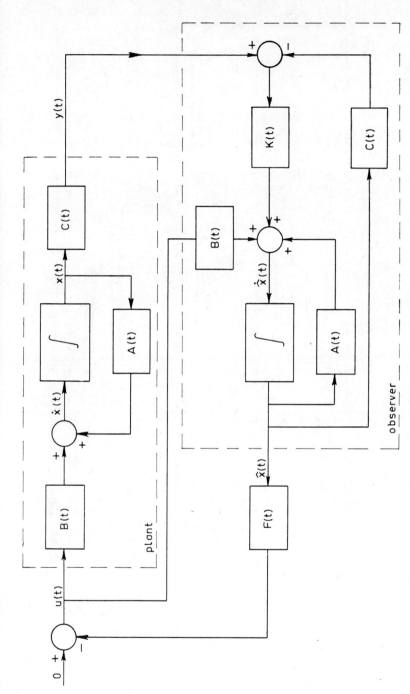

Fig. 5.1. The structure of an output feedback control system.

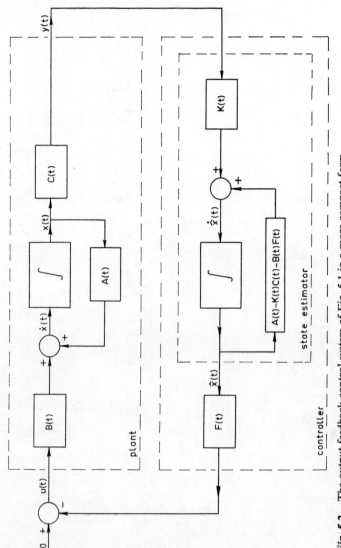

Fig. 5.2. The output feedback control system of Fig. 5.1 in a more compact form.

5.2 Regulation with Incomplete Measurements

By subtracting **5-3** and **5-6**, it easily follows with the use of **5-4** that $e(t)$ satisfies

$$\dot{e}(t) = [A(t) - K(t)C(t)]e(t). \qquad \textbf{5-11}$$

Substitution of $\hat{x}(t) = x(t) - e(t)$ into **5-3** and **5-7** yields

$$\dot{x}(t) = [A(t) - B(t)F(t)]x(t) + B(t)F(t)e(t). \qquad \textbf{5-12}$$

When considering **5-11**, it is seen that $e(t)$ converges to zero, independent of the initial state, if a gain matrix $K(t)$ can be found that makes **5-11** asymptotically stable. However, finding a gain matrix $K(t)$ that makes **5-11** stable is equivalent to determining $K(t)$ such that the observer is asymptotically stable. As we know from Chapter 4, such a gain often can be found.

Next we consider **5-12**. If $B(t)$ and $F(t)$ are bounded and $e(t) \to 0$ as $t \to \infty$, $x(t)$ will always converge to zero if the system

$$\dot{x}(t) = [A(t) - B(t)F(t)]x(t) \qquad \textbf{5-13}$$

is asymptotically stable. From Chapter 3 we know that often $F(t)$ can be determined so that **5-13** is asymptotically stable. Thus we have seen that it is usually possible to find gain matrices $F(t)$ and $K(t)$ such that Eqs. **5-11** and **5-12** constitute an asymptotically stable system. Since the system **5-9** is obtained from the system described by **5-11** and **5-12** by a nonsingular linear transformation, it follows that it is usually possible to find gain matrices $F(t)$ and $K(t)$ such that the closed-loop control systems **5-9** is stable. In the following subsection the precise conditions under which this can be done are stated.

Finally, we remark the following. Combining **5-11** and **5-12** we obtain

$$\begin{pmatrix} \dot{x}(t) \\ \dot{e}(t) \end{pmatrix} = \begin{pmatrix} A(t) - B(t)F(t) & B(t)F(t) \\ 0 & A(t) - K(t)C(t) \end{pmatrix} \begin{pmatrix} x(t) \\ e(t) \end{pmatrix}. \qquad \textbf{5-14}$$

Let us consider the time-invariant case, where all the matrices occurring in **5-14** are constant. Then the characteristic values of the system **5-14**, which are also the characteristic values of the system **5-9**, are the zeroes of

$$\det \begin{pmatrix} sI - A + BF & -BF \\ 0 & sI - A + KC \end{pmatrix}$$
$$= \det (sI - A + BF) \det (sI - A + KC). \qquad \textbf{5-15}$$

The reason that the systems **5-9** and **5-14** have the same characteristic values is that their respective state vectors are related by a nonsingular linear transformation (see Problem 1.3). Consequently, the set of closed-loop characteristic values comprises the characteristic values of $A - BF$ (the regulator poles) and the characteristic values of $A - KC$ (the *observer poles*):

Theorem 5.1. *Consider the interconnection of the time-invariant system*

$$\dot{x}(t) = Ax(t) + Bu(t),$$
$$y(t) = Cx(t), \qquad \text{5-16}$$

the time-invariant observer

$$\dot{\hat{x}}(t) = A\hat{x}(t) + Bu(t) + K[y(t) - C\hat{x}(t)], \qquad \text{5-17}$$

and the time-invariant control law

$$u(t) = -F\hat{x}(t). \qquad \text{5-18}$$

Then the characteristic values of the interconnected system consist of the regulator poles (the characteristic values of $A - BF$) together with the observer poles (the characteristic values of $A - KC$).

These results show that we can consider the problem of determining an asymptotically stable observer and an asymptotically stable state feedback control law separately, since their interconnection results in an asymptotically stable control system.

Apart from stability considerations, are we otherwise justified in separately designing the observer and the control law? In Section 5.3 we formulate a stochastic optimal regulation problem. The solution of this stochastic version of the problem leads to an affirmative answer to the question just posed.

In this section we have considered full-order observers only. It can be shown that reduced-order observers interconnected with state feedback laws also lead to closed-loop poles that consist of the observer poles together with the controller poles.

Example 5.1. *Position control system.*

Consider the positioning system described by the state differential equation (see Example 2.1, Section 2.2.2, and Example 2.4, Section 2.3)

$$\dot{x}(t) = \begin{pmatrix} 0 & 1 \\ 0 & -\alpha \end{pmatrix} x(t) + \begin{pmatrix} 0 \\ \kappa \end{pmatrix} \mu(t), \qquad \text{5-19}$$

with

$$\kappa = 0.787 \text{ rad}/(\text{V s}^2),$$
$$\alpha = 4.6 \text{ s}^{-1}. \qquad \text{5-20}$$

The control law

$$\mu(t) = -(f_1, f_2)\hat{x}(t) \qquad \text{5-21}$$

produces the regulator characteristic polynomial
$$\det(sI - A + BF) = s^2 + (\alpha + \kappa f_2)s + \kappa f_1. \qquad 5\text{-}22$$
By choosing
$$f_1 = 254.1 \text{ V/rad},$$
$$f_2 = 19.57 \text{ V s/rad}, \qquad 5\text{-}23$$
the regulator poles are placed at $-10 \pm j10 \text{ s}^{-1}$. Let us consider the observer
$$\dot{\hat{x}}(t) = \begin{pmatrix} 0 & 1 \\ 0 & -\alpha \end{pmatrix} x(t) + \begin{pmatrix} 0 \\ \kappa \end{pmatrix} \mu(t) + \begin{pmatrix} k_1 \\ k_2 \end{pmatrix} [\eta(t) - (1, \ 0)\hat{x}(t)], \qquad 5\text{-}24$$
where it is assumed that
$$\eta(t) = (1, \ 0)x(t) \qquad 5\text{-}25$$
is the observed variable. The observer characteristic polynomial is
$$\det(sI - A + KC) = s^2 + (\alpha + k_1)s + \alpha k_1 + k_2. \qquad 5\text{-}26$$
To make the observer fast as compared to the regulator, we place the observer poles at $-50 \pm j50 \text{ s}^{-1}$. This yields for the gains:
$$k_1 = 95.40 \text{ s}^{-1},$$
$$k_2 = 4561 \text{ s}^{-2}. \qquad 5\text{-}27$$

In Fig. 5.3 we sketch the response of the output feedback system to the initial state $x(0) = \text{col}(0.1, 0)$, $\hat{x}(0) = 0$. For comparison we give in Fig. 5.4 the response of the corresponding state feedback system, where the control law **5-21** is directly connected to the state. We note that in the system with an observer, the observer very quickly catches up with the actual behavior of the state. Because of the slight time lag introduced by the observer, however, a greater input is required and the response is somewhat different from that of the system without an observer.

Example 5.2. *The pendulum positioning system.*

In this example we discuss the pendulum positioning system of Example 1.1 (Section 1.2.3). The state differential equation of this system is given by
$$\dot{x}(t) = \begin{pmatrix} 0 & 1 & 0 & 0 \\ 0 & -\dfrac{F}{M} & 0 & 0 \\ 0 & 0 & 0 & 1 \\ -\dfrac{g}{L'} & 0 & \dfrac{g}{L'} & 0 \end{pmatrix} x(t) + \begin{pmatrix} 0 \\ \dfrac{1}{M} \\ 0 \\ 0 \end{pmatrix} \mu(t). \qquad 5\text{-}28$$

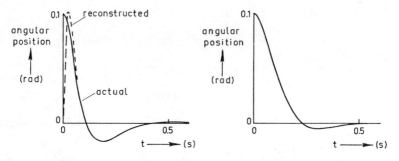

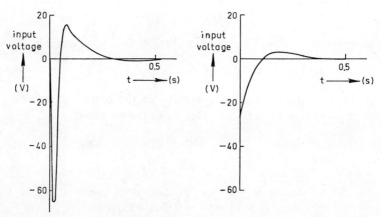

Fig. 5.3. The response and the input of the position control system with observer for $x(0) = \text{col}(0.1, 0)$; $\hat{x}(0) = \text{col}(0, 0)$.

Fig. 5.4. The response and the input of the position control system with state feedback (without observer) for $x(0) = \text{col}(0.1, 0)$.

The components of the state are

$$\begin{aligned}\xi_1(t) &= s(t), \\ \xi_2(t) &= \dot{s}(t), \\ \xi_3(t) &= s(t) + L'\phi(t), \\ \xi_4(t) &= \dot{s}(t) + L'\dot{\phi}(t).\end{aligned} \qquad 5\text{-}29$$

Here $s(t)$ is the displacement of the carriage and $\phi(t)$ the angle the pendulum makes with the vertical. We assume that both these quantities can be measured. This yields for the observed variable

$$y(t) = \begin{pmatrix}\eta_1(t) \\ \eta_2(t)\end{pmatrix} = \begin{pmatrix}1 & 0 & 0 & 0 \\ -\dfrac{1}{L'} & 0 & \dfrac{1}{L'} & 0\end{pmatrix} x(t). \qquad 5\text{-}30$$

5.2 Regulation with Incomplete Measurements

The main function of the control system is to stabilize the system. We therefore choose as the controlled variable the position of the pendulum

$$\zeta(t) = \xi_3(t) = s(t) + L'\phi(t). \qquad 5\text{-}31$$

We first select the regulator poles by solving the regulator problem with the criterion

$$\int_{t_0}^{\infty} [\zeta^2(t) + \rho\mu^2(t)] \, dt. \qquad 5\text{-}32$$

To determine an appropriate value of ρ, we select it such that the estimated radius ω_0 of the faraway poles such as given in Theorem 3.11 (Section 3.8.1) is 10 s^{-1}. This yields a settling time of roughly $10/\omega_0 = 1$ s. It follows from the numerical values of Example 1.1 that the oscillation period of the pendulum is $2\pi\sqrt{L'/g} \simeq 1.84$ s, so that we have chosen the settling time somewhat less than the oscillation period.

To compute ρ from ω_0, we must know the transfer function $H(s)$ of the system from the input force μ to the controlled variable ζ. This transfer function is given by

$$H(s) = \frac{-\dfrac{g}{L'M}}{s\left(s + \dfrac{F}{M}\right)\left(s^2 - \dfrac{g}{L'}\right)}. \qquad 5\text{-}33$$

It follows with **3-486** that

$$\omega_0 = \left[\frac{(g/L'M)^2}{\rho}\right]^{1/8}. \qquad 5\text{-}34$$

With the numerical values of Example 1.1, it can be found that we must choose

$$\rho = 10^{-6} \text{ m}^2/\text{N}^2 \qquad 5\text{-}35$$

to make ω_0 approximately 10 s^{-1}. It can be computed that the resulting steady-state gain matrix is given by

$$\bar{F} = (389.0, \quad 26.91, \quad -1389, \quad -282.4), \qquad 5\text{-}36$$

while the closed-loop poles are $-9.870 \pm j3.861$ and $-4.085 \pm j9.329$ s^{-1}. Figure 5.5 gives the response of the state feedback control system to the initial state $s(0) = 0$, $\dot{s}(0) = 0$, $\phi(0) = 0.1$ rad ($\simeq 6°$), $\dot{\phi}(0) = 0$. It is seen that the input force assumes values up to about 100 N, the carriage displacement undergoes an excursion of about 0.3 m, and the maximal pendulum displacement is about 0.08 m.

Assuming that this performance is acceptable, we now proceed to determine an observer for the system. Since we have two observed variables, there is considerable freedom in choosing the observer gain matrix in order to

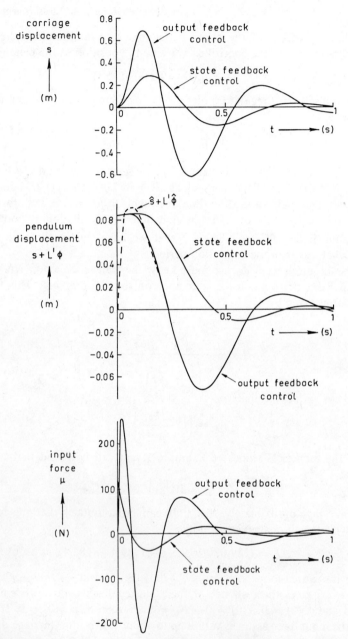

Fig. 5.5. Responses of the state feedback and output feedback pendulum-balancing systems to the initial state $x(0) = \text{col}(0, 0, 0.0842, 0)$ [the observer initial state is $\hat{x}(0) = 0$].

5.2 Regulation with Incomplete Measurements

attain a given set of observer poles. To simplify the problem we impose the restriction that the first component of the observed variable (the displacement) is used only to reconstruct the state of the carriage (i.e., ξ_1 and ξ_2), and the second component of the observed variable is used only to reconstruct the motion of the pendulum (i.e., ξ_3 and ξ_4). Thus we assume the following structure of the observer:

$$\dot{\hat{x}}(t) = \begin{pmatrix} 0 & 1 & 0 & 0 \\ 0 & -\dfrac{F}{M} & 0 & 0 \\ 0 & 0 & 0 & 1 \\ -\dfrac{g}{L'} & 0 & \dfrac{g}{L'} & 0 \end{pmatrix} \hat{x}(t) + \begin{pmatrix} 0 \\ \dfrac{1}{M} \\ 0 \\ 0 \end{pmatrix} \mu(t)$$

$$+ \begin{pmatrix} k_1 & 0 \\ k_2 & 0 \\ 0 & k_3 \\ 0 & k_4 \end{pmatrix} \left[y(t) - \begin{pmatrix} 1 & 0 & 0 & 0 \\ -\dfrac{1}{L'} & 0 & \dfrac{1}{L'} & 0 \end{pmatrix} \hat{x}(t) \right]. \qquad \text{5-37}$$

Here the gains k_1, k_2, k_3, and k_4 are to be determined. It is easily found that with the structure of **5-37** the observer characteristic polynomial is given by

$$\left[s^2 + s\left(k_1 + \dfrac{F}{M}\right) + \dfrac{F}{M} k_1 + k_2 \right]\left[s^2 + s\dfrac{k_3}{L'} + \dfrac{k_4 - g}{L'} \right]. \qquad \text{5-38}$$

It is clearly seen that one pair of poles governs the speed of reconstruction of the motion of the carriage, and the other that of the pendulum. We now choose the gains k_1 to k_4 such that both pairs of poles are somewhat further away from the origin than the regulator poles obtained above. There is no point in choosing the observer poles very far away, since the resulting high observer gains will give difficulties in the implementation without improving the control system response very much. We thus select both pairs of observer poles as

$$21.2(-1 \pm j)\ \text{s}^{-1}.$$

The distance of these poles to the origin is $30\ \text{s}^{-1}$. It can be found with the numerical values of Example 1.1 that to achieve these observer poles the gains must be chosen as

$$k_1 = 41.4, \qquad k_3 = 35.6,$$
$$k_2 = 859, \qquad k_4 = 767. \qquad \text{5-39}$$

Figure 5.5 also gives the response of the interconnection of the resulting observer with the control law and the pendulum positioning system to the same initial conditions as before, with $\hat{x}(0) = 0$. The estimate $\hat{s}(t)$ of the carriage displacement is not shown in the figure since it coincides with the actual carriage displacement right from the beginning owing to the special set of initial conditions. It is seen that the estimate $\hat{s} + L'\hat{\phi}$ of the pendulum displacement $s + L'\phi$ very quickly catches up with the correct value. Nevertheless, because of the slight time lag in the reconstruction process, the motion of the output feedback pendulum balancing system is more violent than in the state feedback case. From a practical point of view, this control system is probably not acceptable because the motion is too violent and the system moves well out of the range where the linearization is valid; very likely the pendulum will topple. A solution can be sought in decreasing ρ so as to damp the motion of the system. An alternative solution is to make the observer faster, but this may cause difficulties with noise in the system.

5.2.2* Conditions for Pole Assignment and Stabilization of Output Feedback Control Systems

In this section we state the precise conditions on the system described by 5-3 and 5-4 such that there exist an observer 5-6 and a control law 5-7 that make the closed-loop control system 5-9 asymptotically stable (G. W. Johnson, 1969; Potter and VanderVelde, 1969):

Theorem 5.2. *Consider the interconnection of the system*

$$\dot{x}(t) = A(t)x(t) + B(t)u(t),$$
$$y(t) = C(t)x(t),$$

5-40

the observer

$$\dot{\hat{x}}(t) = A(t)\hat{x}(t) + B(t)u(t) + K(t)[y(t) - C(t)\hat{x}(t)],$$

5-41

and the control law

$$u(t) = -F(t)\hat{x}(t).$$

5-42

Then sufficient conditions for the existence of gain matrices $K(t)$ and $F(t)$, $t \geq t_0$, such that the interconnected system is exponentially stable, are that the system **5-40** *be uniformly completely controllable and uniformly completely reconstructible or that it be exponentially stable. In the time-invariant situation (i.e., all matrices occurring in* **5-40**, **5-41**, *and* **5-42** *are constant), necessary and sufficient conditions for the existence of stabilizing gain matrices K and F are that the system* **5-40** *be both stabilizable and detectable. In the time-invariant case, necessary and sufficient conditions for arbitrary assignment of both the regulator and the observer poles (within the restriction that complex* *poles occur in complex conjugate pairs) are that the system be completely controllable and completely reconstructible.*

The proof of this theorem is based upon Theorems 3.1 (Section 3.2.2), 3.2 (Section 3.2.2), 3.6 (Section 3.4.2), 4.3 (Section 4.2.2), 4.4 (Section 4.2.2), and 4.10 (Section 4.4.3).

5.3 OPTIMAL LINEAR REGULATORS WITH INCOMPLETE AND NOISY MEASUREMENTS

5.3.1 Problem Formulation and Solution

In this section we formulate the optimal linear regulator problem when the observations of the system are *incomplete* and *inaccurate*, that is, the complete state vector cannot be measured, and the measurements that are available are noisy. In addition, we assume that the system is subject to stochastically varying disturbances. The precise formulation of this problem is as follows.

Definition 5.1. *Consider the system*

$$\dot{x}(t) = A(t)x(t) + B(t)u(t) + w_1(t), \qquad t \geq t_0,$$
$$x(t_0) = x_0,$$
5-43

where x_0 is a stochastic vector with mean $\bar{x}_0$ and variance matrix Q_0. The observed variable is given by

$$y(t) = C(t)x(t) + w_2(t), \qquad t \geq t_0.$$
5-44

The joint stochastic process col (w_1, w_2) *is a white noise process with intensity*

$$\begin{pmatrix} V_1(t) & V_{12}(t) \\ V_{12}^T(t) & V_2(t) \end{pmatrix}, \qquad t \geq t_0.$$
5-45

The controlled variable can be expressed as

$$z(t) = D(t)x(t), \qquad t \geq t_0.$$
5-46

Then the **stochastic linear optimal output feedback regulator problem** *is the problem of finding the functional*

$$u(t) = f[y(\tau), t_0 \leq \tau \leq t], \qquad t_0 \leq t \leq t_1,$$
5-47

such that the criterion

$$\sigma = E\left\{ \int_{t_0}^{t_1} [z^T(t)R_3(t)z(t) + u^T(t)R_2(t)u(t)] \, dt + x^T(t_1)P_1 x(t_1) \right\}$$
5-48

is minimized. Here $R_3(t)$, $R_2(t)$, and P_1 are symmetric weighting matrices such that $R_2(t) > 0$, $R_3(t) > 0$, $t_0 \leq t \leq t_1$, and $P_1 \geq 0$.

The solution of this problem is, as expected, the combination of the solutions of the stochastic optimal regulator problem of Chapter 3 (Theorem 3.9, Section 3.6.3) and the optimal reconstruction problem of Chapter 4. This rather deep result is known as the *separation principle* and is stated in the following theorem.

Theorem 5.3. *The optimal* **linear** *solution of the stochastic linear optimal output feedback regulator problem is the same as the solution of the corresponding stochastic optimal state feedback regulator problem (Theorem 3.9, Section 3.6.3) except that in the control law the state $x(t)$ is replaced with its minimum mean square linear estimator $\hat{x}(t)$, that is, the input is chosen as*

$$u(t) = -F^0(t)\hat{x}(t), \qquad 5\text{-}49$$

where $F^0(t)$ is the gain matrix given by **3-344** *and $\hat{x}(t)$ is the output of the optimal observer derived in Sections 4.3.2, 4.3.3, and 4.3.4 for the nonsingular uncorrelated, nonsingular correlated, and the singular cases, respectively.*

An outline of the proof of this theorem for the nonsingular uncorrelated case is given in Section 5.3.3. We remark that the solution as indicated is the best *linear* solution. It can be proved (Wonham, 1968b, 1970b; Fleming, 1969; Kushner, 1967, 1971) that, if the processes w_1 and w_2 are Gaussian white noise processes and the initial state x_0 is Gaussian, the optimal linear solution is *the* optimal solution (without qualification).

Restricting ourselves to the case where the problem of estimating the state is nonsingular and the state excitation and observation noises are uncorrelated, we now write out in detail the solution to the stochastic linear output feedback regulator problem. For the input we have

$$u(t) = -F^0(t)\hat{x}(t), \qquad 5\text{-}50$$

with

$$F^0(t) = R_2^{-1}(t)B^T(t)P(t). \qquad 5\text{-}51$$

Here $P(t)$ is the solution of the Riccati equation

$$-\dot{P}(t) = D^T(t)R_3(t)D(t) - P(t)B(t)R_2^{-1}(t)B^T(t)P(t) \\ + A^T(t)P(t) + P(t)A(t), \qquad 5\text{-}52$$

$P(t_1) = P_1$.

The estimate $\hat{x}(t)$ is obtained as the solution of

$$\dot{\hat{x}}(t) = A(t)\hat{x}(t) + B(t)u(t) + K^0(t)[y(t) - C(t)\hat{x}(t)],$$

$$\hat{x}(t_0) = \bar{x}_0, \qquad 5\text{-}53$$

where
$$K^0(t) = Q(t)C^T(t)V_2^{-1}(t). \quad \text{5-54}$$
The variance matrix $Q(t)$ is the solution of the Riccati equation
$$\dot{Q}(t) = V_1(t) - Q(t)C^T(t)V_2^{-1}(t)C(t)Q(t) + A(t)Q(t) + Q(t)A^T(t),$$
$$Q(t_0) = Q_0. \quad \text{5-55}$$
Figure 5.6 gives a block diagram of this stochastic optimal output feedback control system.

5.3.2 Evaluation of the Performance of Optimal Output Feedback Regulators

We proceed by analyzing the performance of optimal output feedback control systems, still limiting ourselves to the nonsingular case with uncorrelated state excitation and observation noises. The interconnection of the system **5-43**, the optimal observer **5-53**, and the control law **5-50** forms a system of dimension $2n$, where n is the dimension of the state x. Let us define, as before, the reconstruction error
$$e(t) = x(t) - \hat{x}(t). \quad \text{5-56}$$
It is easily obtained from Eqs. **5-43**, **5-53**, and **5-50** that the augmented vector col $[e(t), \hat{x}(t)]$ satisfies the differential equation
$$\begin{pmatrix} \dot{e}(t) \\ \dot{\hat{x}}(t) \end{pmatrix} = \begin{pmatrix} A(t) - K^0(t)C(t) & 0 \\ K^0(t)C(t) & A(t) - B(t)F^0(t) \end{pmatrix} \begin{pmatrix} e(t) \\ \hat{x}(t) \end{pmatrix}$$
$$+ \begin{pmatrix} I & -K^0(t) \\ 0 & K^0(t) \end{pmatrix} \begin{pmatrix} w_1(t) \\ w_2(t) \end{pmatrix}, \quad \text{5-57}$$
with the initial condition
$$\begin{pmatrix} e(t_0) \\ \hat{x}(t_0) \end{pmatrix} = \begin{pmatrix} x(t_0) - \bar{x}_0 \\ \bar{x}_0 \end{pmatrix}. \quad \text{5-58}$$
The reason that we consider col $(e, \hat{x})$ is that the variance matrix of this augmented vector is relatively easily found, as we shall see. All mean square quantities of interest can then be obtained from this variance matrix. Let us denote the variance matrix of col $[e(t), \hat{x}(t)]$ as
$$E\left\{ \begin{pmatrix} e(t) - E\{e(t)\} \\ \hat{x}(t) - E\{\hat{x}(t)\} \end{pmatrix} ([e(t) - E\{e(t)\}]^T, [\hat{x}(t) - E\{\hat{x}(t)\}]^T) \right\}$$
$$= \begin{pmatrix} Q_{11}(t) & Q_{12}(t) \\ Q_{12}^T(t) & Q_{22}(t) \end{pmatrix}. \quad \text{5-59}$$

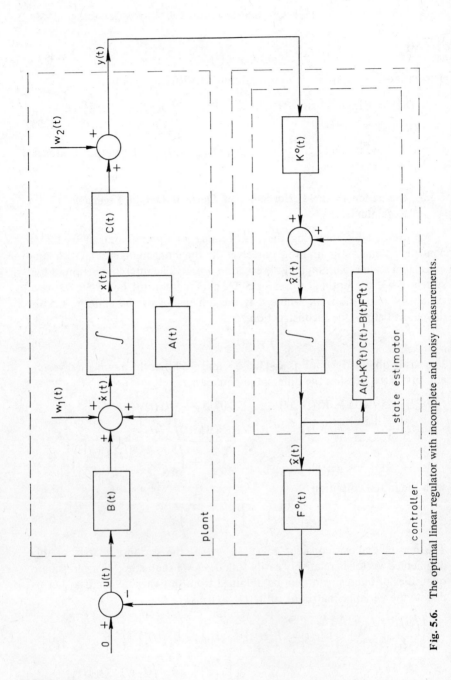

Fig. 5.6. The optimal linear regulator with incomplete and noisy measurements.

5.3 Regulators with Incomplete and Noisy Measurements

The differential equations for the matrices Q_{11}, Q_{12}, and Q_{22} can be obtained by application of Theorem 1.52 (Section 1.11.2). It easily follows that these matrices satisfy the equations:

$$\dot{Q}_{11}(t) = [A(t) - K^0(t)C(t)]Q_{11}(t) + Q_{11}(t)[A(t) - K^0(t)C(t)]^T$$
$$+ V_1(t) + K^0(t)V_2(t)K^{0T}(t),$$

$$\dot{Q}_{12}(t) = Q_{11}(t)C^T(t)K^{0T}(t) + Q_{12}(t)[A(t) - B(t)F^0(t)]^T$$
$$+ [A(t) - K^0(t)C(t)]Q_{12}(t) - K^0(t)V_2(t)K^{0T}(t), \quad \textbf{5-60}$$

$$\dot{Q}_{22}(t) = Q_{12}^T(t)C^T(t)K^{0T}(t) + Q_{22}(t)[A(t) - B(t)F^0(t)]^T + K^0(t)C(t)Q_{12}(t)$$
$$+ [A(t) - B(t)F^0(t)]Q_{22}(t) + K^0(t)V_2(t)K^{0T}(t),$$

with the initial conditions

$$Q_{11}(t_0) = Q_0, \quad Q_{12}(t_0) = 0, \quad Q_{22}(t_0) = 0. \quad \textbf{5-61}$$

When considering these equations, we immediately note that of course

$$Q_{11}(t) = Q(t), \quad t \geq t_0. \quad \textbf{5-62}$$

As a result, in the differential equation for $Q_{12}(t)$ the terms $Q_{11}(t)C^T(t)K^{0T}(t)$ and $-K^0(t)V_2(t)K^{0T}(t)$ cancel because $K^0(t) = Q(t)C^T(t)V_2^{-1}(t)$. What is left of the equation for $Q_{12}(t)$ is a homogeneous differential equation in $Q_{12}(t)$ with the initial condition $Q_{12}(t_0) = 0$, which of course has the solution

$$Q_{12}(t) = 0, \quad t \geq t_0. \quad \textbf{5-63}$$

Apparently, $e(t)$ and $\hat{x}(t)$ are uncorrelated stochastic processes. This is why we have chosen to work with the joint process col $(e, \hat{x})$. Note that $e(t)$ and $\hat{x}(t)$ are uncorrelated no matter how the input to the plant is chosen. The reason for this is that the behavior of the reconstruction error e is independent of that of the input u, and the contribution of the input $u(\tau)$, $t_0 \leq \tau \leq t$, to the reconstructed state $\hat{x}(t)$ is a known quantity which is subtracted to compute the covariance of $e(t)$ and $\hat{x}(t)$. We use this fact in the proof of the separation principle in Section 5.3.3.

The differential equation for $Q_{22}(t)$ now simplifies to

$$\dot{Q}_{22}(t) = [A(t) - B(t)F^0(t)]Q_{22}(t) + Q_{22}(t)[A(t) - B(t)F^0(t)]^T$$
$$+ K^0(t)V_2(t)K^{0T}(t), \quad \textbf{5-64}$$

with the initial condition

$$Q_{22}(t_0) = 0. \quad \textbf{5-65}$$

Once we have computed $Q_{22}(t)$, the variance matrix of the joint process col $(e, \hat{x})$ is known, and all mean square quantities or integrated mean square quantities of interest can be obtained, since

$$x(t) = e(t) + \hat{x}(t). \quad \textbf{5-66}$$

Thus we can compute the mean square regulation error as

$$\begin{aligned}E\{z^T(t)W_e(t)z(t)\} &= E\{x^T(t)D^T(t)W_e(t)D(t)x(t)\\&= \operatorname{tr}\,[D^T(t)W_e(t)D(t)E\{x(t)x^T(t)\}] \quad \text{5-67}\\&= \operatorname{tr}\,\{D^T(t)W_e(t)D(t)[\bar{x}(t)\bar{x}^T(t) + Q_{11}(t) + Q_{22}(t)]\},\end{aligned}$$

where $W_e(t)$ is the weighting matrix and $\bar{x}(t)$ is the mean of $x(t)$. Similarly, we can compute the mean square input as

$$\begin{aligned}E\{u^T(t)W_u(t)u(t)\} &= E\{\hat{x}^T(t)F^{0T}(t)W_u(t)F^0(t)\hat{x}(t)\}\\&= \operatorname{tr}\,[F^{0T}(t)W_u(t)F^0(t)E\{\hat{x}(t)\hat{x}^T(t)\}]\\&= \operatorname{tr}\,\{F^{0T}(t)W_u(t)F^0(t)[\bar{x}(t)\bar{x}^T(t) + Q_{22}(t)]\}, \quad \text{5-68}\end{aligned}$$

where $W_u(t)$ is the weighting matrix of the mean square input.

It follows that in order to compute the optimal regulator gain matrix $F^0(t)$, the optimal filter gain matrix $K^0(t)$, the mean square regulation error, and the mean square input one must solve *three* $n \times n$ matrix differential equations: the Riccati equation **5-52** to obtain $P(t)$ and from this $F^0(t)$, the Riccati equation **5-55** to determine $Q(t)$ and from this $K^0(t)$, and finally the linear matrix differential equation **5-64** to obtain the variance matrix $Q_{22}(t)$ of $\hat{x}(t)$. In the next theorem, however, we state that if the mean square regulation error and the mean square input are not required separately, but only the value of the criterion σ as given by **5-48** is required, then merely the basic Riccati equations for $P(t)$ and $Q(t)$ need be solved.

Theorem 5.4. *Consider the stochastic regulator problem of Definition 5.1. Suppose that*

$$V_2(t) > 0, \qquad V_{12}(t) = 0 \quad \text{for all } t. \qquad \text{5-69}$$

Then the following facts hold:

(a) *All mean square quantities of interest can be obtained from the variance matrix* diag $[Q(t), Q_{22}(t)]$ *of* col $[e(t), \hat{x}(t)]$, *where* $e(t) = x(t) - \hat{x}(t)$, $Q(t)$ *is the variance matrix of* $e(t)$, *and* $Q_{22}(t)$ *can be obtained as the solution of the matrix differential equation*

$$\begin{aligned}\dot{Q}_{22}(t) &= [A(t) - B(t)F^0(t)]Q_{22}(t) + Q_{22}(t)[A(t) - B(t)F^0(t)]^T\\&\quad + K^0(t)V_2(t)K^{0T}(t), \qquad t \geq t_0, \quad \text{5-70}\end{aligned}$$

$$Q_{22}(t_0) = 0.$$

(b) *The minimal value of the criterion* **5-48** *can be expressed in the following two alternative forms*

$$\sigma^0 = \bar{x}_0^T P(t_0)\bar{x}_0 + \operatorname{tr}\left\{\int_{t_0}^{t_1}[P(t)K^0(t)V_2(t)K^{0T}(t) + Q(t)R_1(t)]\,dt + P_1 Q(t_1)\right\}$$

$$\text{5-71}$$

5.3 Regulators with Incomplete and Noisy Measurements

and

$$\sigma^0 = \bar{x}_0^T P(t_0)\bar{x}_0 + \operatorname{tr}\left\{P(t_0)Q_0 + \int_{t_0}^{t_1}[P(t)V_1(t) + Q(t)F^{0T}(t)R_2(t)F^0(t)]\,dt\right\}.$$

5-72

Here we have abbreviated

$$R_1(t) = D^T(t)R_3(t)D(t),$$

5-73

and $P(t)$ and $Q(t)$ are the solutions of the Riccati equations **5-52** and **5-55**, respectively.

(c) Furthermore, if the optimal observer and regulator Riccati equations have the steady-state solutions $\bar{Q}(t)$ and $\bar{P}(t)$ as $t_0 \to -\infty$ and $t_1 \to \infty$, respectively, then the time-averaged criterion

$$\bar{\sigma} = \lim_{\substack{t_0 \to -\infty \\ t_1 \to \infty}} \frac{1}{t_1 - t_0} E\left\{\int_{t_0}^{t_1}[z^T(t)R_3(t)z(t) + u^T(t)R_2(t)u(t)]\,dt\right\},$$

5-74

if it exists, can be expressed in the alternative forms

$$\bar{\sigma} = \lim_{\substack{t_0 \to -\infty \\ t_1 \to \infty}} \frac{1}{t_1 - t_0} \operatorname{tr}\left\{\int_{t_0}^{t_1}[\bar{P}(t)\bar{K}(t)V_2(t)\bar{K}^T(t) + \bar{Q}(t)R_1(t)]\,dt\right\}$$

5-75

and

$$\bar{\sigma} = \lim_{\substack{t_0 \to -\infty \\ t \to \infty}} \frac{1}{t_1 - t_0} \operatorname{tr}\left\{\int_{t_0}^{t_1}[\bar{P}(t)V_1(t) + \bar{Q}(t)\bar{F}^T(t)R_2(t)\bar{F}(t)]\,dt\right\}.$$

5-76

Here $\bar{K}(t)$ and $\bar{F}(t)$ are the gains corresponding to the steady-state solutions $\bar{Q}(t)$ and $\bar{P}(t)$, respectively.

(d) Finally, in the time-invariant case, where $\bar{Q}(t)$ and $\bar{P}(t)$ and thus also $\bar{F}(t)$ and $\bar{K}(t)$ are constant matrices, the following expressions hold:

$$\bar{\sigma} = E\{z^T(t)R_3z(t) + u^T(t)R_2u(t)\}$$
$$= \operatorname{tr}[\bar{P}\bar{K}V_2\bar{K}^T + \bar{Q}R_1]$$ 5-77a
$$= \operatorname{tr}[\bar{P}V_1 + \bar{Q}\bar{F}^T R_2 \bar{F}].$$ 5-77b

This theorem can be proved as follows. Setting $W_e(t) = R_3(t)$ and $W_u(t) = R_2(t)$ in **5-67** and **5-68**, we write for the criterion

$$E\left\{\int_{t_0}^{t_1}[z^T(t)R_3(t)z(t) + u^T(t)R_2(t)u(t)]\,dt + x^T(t_1)P_1x(t_1)\right\}$$

$$= \int_{t_0}^{t_1}[\bar{z}^T(t)R_3(t)\bar{z}(t) + \bar{u}^T(t)R_2(t)\bar{u}(t)]\,dt + \bar{x}^T(t_1)P_1\bar{x}(t_1)$$

$$+ \operatorname{tr}\left\{\int_{t_0}^{t_1}[R_1(t)[Q(t) + Q_{22}(t)] + F^{0T}(t)R_2(t)F^0(t)Q_{22}(t)]\,dt\right.$$

$$\left. + P_1[Q(t_1) + Q_{22}(t_1)]\right\}.$$ 5-78

Let us separately consider the expression

$$\mathrm{tr}\left\{\int_{t_0}^{t_1}[R_1(t)+F^{0T}(t)R_2(t)F^0(t)]Q_{22}(t)\,dt+P_1Q_{22}(t_1)\right\}, \qquad \text{5-79}$$

where, as we know, $Q_{22}(t)$ is the solution of the matrix differential equation

$$\dot{Q}_{22}(t)=[A(t)-B(t)F^0(t)]Q_{22}(t)+Q_{22}(t)[A(t)-B(t)F^0(t)]^T$$
$$+K^0(t)V_2(t)K^{0T}(t), \qquad \text{5-80}$$
$$Q_{22}(t_0)=0.$$

It is not difficult to show (Problem 5.5) that **5-79** can be written in the form

$$\mathrm{tr}\left\{\int_{t_0}^{t_1}S(t)K^0(t)V_2(t)K^{0T}(t)\,dt\right\}, \qquad \text{5-81}$$

where $S(t)$ is the solution of the matrix differential equation

$$-\dot{S}(t)=[A(t)-B(t)F^0(t)]^T S(t)+S(t)[A(t)-B(t)F^0(t)]+R_1(t)$$
$$+F^{0T}(t)R_2(t)F^0(t), \qquad \text{5-82}$$
$$S(t_1)=P_1.$$

Obviously, the solution of this differential equation is

$$S(t)=P(t), \qquad t\le t_1. \qquad \text{5-83}$$

Combining these results, and using the fact that the first two terms of the right-hand side of **5-78** can be replaced with $\bar{x}_0^T P(t_0)\bar{x}_0$, we obtain the desired expression **5-71** from **5-78**.

The alternative expression **5-72** for the criterion can be obtained by substituting

$$R_1(t)=P(t)B(t)R_2^{-1}(t)B^T(t)P(t)-A^T(t)P(t)-P(t)A(t)-\dot{P}(t) \qquad \text{5-84}$$

into **5-71** and integrating by parts. The proofs of parts (c) and (d) of Theorem 5.4 follow from **5-71** and **5-72** by letting $t_0\to-\infty$ and $t_1\to\infty$.

Of course in any practical situation in which t_1-t_0 is large, we use the steady-state gain matrices $\bar{K}(t)$ and $\bar{F}(t)$ even when t_1-t_0 is not infinite. Particularly, we do so in the time-invariant case, where $\bar{K}$ and $\bar{F}$ are constant. From optimal regulator and observer theory and in view of Section 5.2, we know that the resulting *steady-state output feedback control system* is asymptotically stable whenever the corresponding state feedback regulator and observer are asymptotically stable.

Before concluding this section with an example, two remarks are made. First, we note that in the time-invariant steady-state case the following lower

5.3 Regulators with Incomplete and Noisy Measurements

bounds follow from **5-77a** and **5-77b**:

$$\lim_{R_2 \to 0} \bar{\sigma} \geq \text{tr}(\bar{Q}R_1), \qquad \text{5-85a}$$

$$\lim_{V_2 \to 0} \bar{\sigma} \geq \text{tr}(\bar{P}V_1). \qquad \text{5-85b}$$

These inequalities can be interpreted as follows. Even if we do not at all weight the input u, and thus do not constrain the input amplitude, the criterion $\bar{\sigma}$ still cannot be less than $\text{tr}(\bar{Q}R_1)$ according to **5-85a**. This minimum contribution to the criterion is caused by the unavoidable inaccuracy in reconstructing the state. Similarly, even when no measurement noise is present, that is, V_2 approaches zero, the criterion $\bar{\sigma}$ cannot be less then $\text{tr}(\bar{P}V_1)$. This value is not surprising since it is exactly the value of the criterion for the state feedback stochastic regulator (see Theorem 3.9, Section 3.6.3).

The second remark concerns the locations of the control system poles in the time-invariant steady-state case. In Section 5.2 we saw that the control system poles consist of the regulator poles and the observer poles. It seems a good rule of thumb that the weighting matrices R_2 and V_2 be chosen so that the regulator poles and the observer poles have distances to the origin of roughly the same order of magnitude. It seems to be wasteful to have very fast regulation when the reconstruction process is slow, and vice versa. In particular, when there is a great deal of observation noise as compared to the state excitation noise, the observer poles are relatively close to the origin and the reconstruction process is slow. When we now make the regulator just a little faster than the observer, it is to be expected that the regulator can keep up with the observer. A further increase in the speed of the regulator will merely increase the mean square input without decreasing the mean square regulation error appreciably. On the other hand, when there is very little observation noise, the limiting factor in the design will be the permissible mean square input. This will constrain the speed of the regulator, and there will be very little point in choosing an observer that is very much faster, even though the noise conditions would permit it.

Example 5.3. *Position control system.*

Let us consider the position control system discussed in many previous examples. Its state differential equation is

$$\dot{x}(t) = \begin{pmatrix} 0 & 1 \\ 0 & -\alpha \end{pmatrix} x(t) + \begin{pmatrix} 0 \\ \kappa \end{pmatrix} \mu(t). \qquad \text{5-86}$$

Here $x(t) = \text{col}[\xi_1(t), \xi_2(t)]$, with $\xi_1(t)$ the angular position and $\xi_2(t)$ the angular velocity of the system. The input variable $\mu(t)$ is the input voltage.

The controlled variable is the position, hence is given by

$$\zeta(t) = (1, \ 0)x(t). \tag{5-87}$$

In Example 3.8 (Section 3.4.1), we solved the deterministic regulator problem with the criterion

$$\int_{t_0}^{t_1} [\zeta^2(t) + \rho\mu^2(t)] \, dt. \tag{5-88}$$

With the numerical values

$$\kappa = 0.787 \text{ rad}/(\text{V s}^2),$$

$$\alpha = 4.6 \text{ s}^{-1}, \tag{5-89}$$

$$\rho = 0.00002 \text{ rad}^2/\text{V}^2,$$

we found the steady-state feedback gain matrix

$$\bar{F} = (223.6, \quad 18.69). \tag{5-90}$$

The steady-state solution of the regulator Riccati equation is given by

$$\bar{P} = \begin{pmatrix} 0.1098 & 0.005682 \\ 0.005682 & 0.0004753 \end{pmatrix}. \tag{5-91}$$

The closed-loop regulator poles are $-9.66 \pm j9.09 \text{ s}^{-1}$. From Fig. 3.9 (Section 3.4.1), we know that the settling time of the system is of the order of 0.3 s, while an initial deviation in the position of 0.1 rad causes an input voltage with an initial peak value of 25 V.

In Example 4.4 (Section 4.3.2), we assumed that the system is disturbed by an external torque on the shaft $\tau_d(t)$. This results in the following modification of the state differential equation:

$$\dot{x}(t) = \begin{pmatrix} 0 & 1 \\ 0 & -\alpha \end{pmatrix} x(t) + \begin{pmatrix} 0 \\ \kappa \end{pmatrix} \mu(t) + \begin{pmatrix} 0 \\ \gamma \end{pmatrix} \tau_d(t), \tag{5-92}$$

where $1/\gamma$ is the rotational moment of inertia of the rotating parts. It was furthermore assumed that the observed variable is given by

$$\eta(t) = (1, \ 0)x(t) + \nu_m(t), \tag{5-93}$$

where $\nu_m(t)$ represents the observation noise. This expression implies that the angular displacement is measured. Under the assumption that $\tau_d(t)$ and $\nu_m(t)$ are adequately represented as uncorrelated white noise processes with intensities

$$V_d = 10 \text{ N}^2 \text{ m}^2 \text{ s} \tag{5-94}$$

and

$$V_m = 10^{-7} \text{ rad}^2 \text{ s}, \tag{5-95}$$

5.3 Regulators with Incomplete and Noisy Measurements

respectively, we found in Example 4.4 that with $\gamma = 0.1$ kg^{-1} m^{-2} the steady-state optimal observer is given by

$$\dot{\hat{x}}(t) = \begin{pmatrix} 0 & 1 \\ 0 & -\alpha \end{pmatrix} \hat{x}(t) + \begin{pmatrix} 0 \\ \kappa \end{pmatrix} \mu(t) + \bar{K}[\eta(t) - (1, 0)\hat{x}(t)], \quad \text{5-96}$$

where the steady-state gain matrix is

$$\bar{K} = \begin{pmatrix} 40.36 \\ 814.3 \end{pmatrix}. \quad \text{5-97}$$

The observer poles are $-22.48 \pm j22.24$ s^{-1}, while the steady-state variance matrix is given by

$$\bar{Q} = \begin{pmatrix} 0.04036 \times 10^{-4} & 0.8143 \times 10^{-4} \\ 0.8143 \times 10^{-4} & 36.61 \times 10^{-4} \end{pmatrix}. \quad \text{5-98}$$

With

$$\mu(t) = -\bar{F}\hat{x}(t), \quad \text{5-99}$$

the steady-state optimal output feedback controller is described by

$$\dot{\hat{x}}(t) = \begin{pmatrix} 0 & 1 \\ 0 & -\alpha \end{pmatrix} \hat{x}(t) - \begin{pmatrix} 0 \\ \kappa \end{pmatrix} \bar{F}\hat{x}(t) + \bar{K}[\eta(t) - (1, 0)\hat{x}(t)], \quad \text{5-100}$$

$$\mu(t) = -\bar{F}\hat{x}(t).$$

It follows that

$$\lim_{t_0 \to -\infty} E\{\zeta^2(t) + \rho\mu^2(t)\} = \text{tr}(\bar{P}\bar{K}V_2\bar{K}^T + \bar{Q}R_1) = 0.00009080 \text{ rad}^2. \quad \text{5-101}$$

From this result we find the following bounds on the steady-state rms tracking error and rms input voltage:

$$\lim_{t_0 \to -\infty} \sqrt{E\{\zeta^2(t)\}} < \sqrt{0.00009080} \simeq 0.0095 \text{ rad}, \quad \text{5-102a}$$

$$\lim_{t_0 \to -\infty} E\{\rho\mu^2(t)\} < 0.00009080 \text{ rad}^2, \quad \text{5-102b}$$

so that

$$\lim_{t_0 \to -\infty} \sqrt{E\{\mu^2(t)\}} < \sqrt{\frac{0.00009080}{\rho}} \simeq 2.13 \text{ V}. \quad \text{5-103}$$

The exact values of the steady-state rms tracking error and rms input voltage must be obtained by solving for the steady-state variance matrix of the augmented state col $[x(t), \hat{x}(t)]$. As outlined in the text (Section 5.3.2), this is most efficiently done by first computing the steady-state variance matrix diag $(\bar{Q}_{11}, \bar{Q}_{22})$ of col $[e(t), \hat{x}(t)]$, which requires only the solution of an additional 2×2 linear matrix equation. It can be found that the steady-state variance matrix $\bar{\Pi}$ of col $[x(t), \hat{x}(t)]$ is given by

$$\bar{\Pi} = \begin{pmatrix} \bar{Q}_{11} + \bar{Q}_{22} & \bar{Q}_{22} \\ \bar{Q}_{22} & \bar{Q}_{22} \end{pmatrix}$$

$$= \begin{pmatrix} 0.00004562 & 0 & 0.00004158 & -0.00008145 \\ 0 & 0.006119 & -0.00008145 & 0.002458 \\ 0.00004158 & -0.00008145 & 0.00004158 & -0.00008145 \\ -0.00008145 & 0.002458 & -0.00008145 & 0.002458 \end{pmatrix}.$$

5-104

This yields for the steady-state mean square tracking error

$$\lim_{t_0 \to -\infty} E\{\zeta^2(t)\} = \operatorname{tr}\left[\bar{\Pi} \begin{pmatrix} 1 \\ 0 \\ 0 \\ 0 \end{pmatrix} (1, \ 0, \ 0, \ 0)\right] = 0.00004562 \text{ rad}^2, \qquad \text{5-105}$$

so that the rms tracking error is $\sqrt{0.00004562} \simeq 0.00674$ rad. We see that this is somewhat less than the bound **5-102**. Similarly, we obtain for the mean square input voltage

$$\lim_{t_0 \to -\infty} E\{\mu^2(t)\} = \operatorname{tr}\left[\bar{\Pi} \begin{pmatrix} 0 \\ \bar{F}^T \end{pmatrix} (0, \ \bar{F})\right] = 2.258 \text{ V}^2, \qquad \text{5-106}$$

so that the rms input voltage is about 1.5 V. It depends, of course, on the specifications of the system whether or not this performance is satisfactory.

It is noted that the regulator poles $(-9.66 \pm j9.09)$ and the observer poles $(-22.48 \pm j22.24)$ are of the same order of magnitude, which is a desirable situation. Had we found that, for example, the observer poles are very far away as compared to the regulator poles, we could have moved the observer poles closer to the origin without appreciable loss in performance.

5.3.3* Proof of the Separation Principle

In this section we prove the separation principle as stated in Theorem 5.3 for the nonsingular uncorrelated case, that is, we assume that the intensity $V_2(t)$ of the observation noise is positive-definite and that $V_{12}(t) = 0$ on $[t_0, t_1]$. It is relatively straightforward to prove that the solution as given is the best *linear* solution of the stochastic linear output feedback regulator

5.3 Regulators with Incomplete and Noisy Measurements

problem. Denoting
$$R_1(t) = D^T(t)R_3(t)D(t), \qquad \text{5-107}$$
we write

$$\begin{aligned}
E[z^T(t)&R_3(t)z(t)] \\
&= E[x^T(t)R_1(t)x(t)] \\
&= E\{[x(t) - \hat{x}(t) + \hat{x}(t)]^T R_1(t)[x(t) - \hat{x}(t) + \hat{x}(t)]\} \\
&= E\{[x(t) - \hat{x}(t)]^T R_1(t)[x(t) - \hat{x}(t)]\} \\
&\quad + 2E\{[x(t) - \hat{x}(t)]^T R_1(t)\hat{x}(t)\} + E\{\hat{x}^T(t)R_1(t)\hat{x}(t)\}. \quad \text{5-108}
\end{aligned}$$

Here $\hat{x}(t)$ is the minimum mean square linear estimator of $x(t)$ operating on $y(\tau)$ and $u(\tau)$, $t_0 \leq \tau \leq t$. From optimal observer theory we know that

$$E\{[x(t) - \hat{x}(t)]^T R_1(t)[x(t) - \hat{x}(t)]\} = \text{tr}\,[R_1(t)Q(t)], \qquad \text{5-109}$$

where $Q(t)$ is the variance matrix of the reconstruction error $x(t) - \hat{x}(t)$. Furthermore,

$$E\{[x(t) - \hat{x}(t)]^T R_1(t)\hat{x}(t)\} = \text{tr}\,[E\{[x(t) - \hat{x}(t)]\hat{x}^T(t)\}R_1(t)] = 0, \quad \text{5-110}$$

since as we have seen in Section 5.3.2 the quantities $e(t) = x(t) - \hat{x}(t)$ and $\hat{x}(t)$ are uncorrelated. Thus we find that we can write

$$\begin{aligned}
E\{x^T(t)R_1(t)x(t)\} &= \text{tr}\,[R_1(t)Q(t)] + E\{\hat{x}^T(t)R_1(t)\hat{x}(t)\}, \\
E\{x^T(t_1)P_1 x(t_1)\} &= \text{tr}\,[P_1 Q(t_1)] + E\{\hat{x}^T(t_1)P_1 \hat{x}(t_1)\}.
\end{aligned} \qquad \text{5-111}$$

Using **5-111**, we write for the criterion **5-48**:

$$E\left\{\int_{t_0}^{t_1}[\hat{x}^T(t)R_1(t)\hat{x}(t) + u^T(t)R_2(t)u(t)]\,dt + \hat{x}^T(t_1)P_1\hat{x}(t_1)\right\}$$
$$+ \text{tr}\left[\int_{t_0}^{t_1} R_1(t)Q(t)\,dt + P_1 Q(t_1)\right]. \qquad \text{5-112}$$

We observe that the last two terms in this expression are independent of the control applied to the system. Also from optimal observer theory, we know that we can write (since by assumption the reconstruction problem is nonsingular)

$$\dot{\hat{x}}(t) = A(t)\hat{x}(t) + B(t)u(t) + K^0(t)[y(t) - C(t)\hat{x}(t)], \qquad \text{5-113}$$

where $K^0(t)$ is the optimal gain matrix. However, in Section 4.3.6 we found that the innovation process $y(t) - C(t)\hat{x}(t)$ is a white noise process with intensity $V_2(t)$. Then the problem of minimizing the criterion **5-112**, with the behavior of $\hat{x}(t)$ described by **5-113**, is a stochastic linear regulator problem where the complete state can be observed, such as described in Section 3.6.1. It follows from Theorem 3.9 that the optimal *linear* solution of this state feedback stochastic regulator problem is the linear control law

$$u(t) = -F^0(t)\hat{x}(t), \qquad \text{5-114}$$

where $F^0(t)$ is given by **5-51**.

This terminates the proof of Theorem 5.3 for the case where the reconstruction problem is nonsingular and the state excitation and observation noises are uncorrelated. The proof can be extended to the singular correlated case.

5.4 LINEAR OPTIMAL TRACKING SYSTEMS WITH INCOMPLETE AND NOISY MEASUREMENTS

In Section 3.6.2 we considered tracking problems as special cases of stochastic state feedback regulator problems. Necessarily, we found control laws that require that both the state of the plant and the state of the reference variable are available. In this section we consider a similar problem, but it is assumed that only certain linear combinations of the components of the state can be measured, which moreover are contaminated with additive noise. We furthermore assume that only the reference variable itself can be measured, also contaminated with white noise.

We thus adopt the following model for the reference variable $z_r(t)$:

$$z_r(t) = D_r(t)x_r(t), \qquad \text{5-115}$$

where

$$\dot{x}_r(t) = A_r(t)x_r(t) + w_{r1}(t). \qquad \text{5-116}$$

In this expression w_{r1} is white noise with intensity $V_{r1}(t)$. It is furthermore assumed that we observe

$$y_r(t) = z_r(t) + w_{r2}(t). \qquad \text{5-117}$$

Here w_{r2} is white noise with intensity $V_{r2}(t)$.

The system to be controlled is described by the state differential equation

$$\dot{x}(t) = A(t)x(t) + B(t)u(t) + w_1(t), \qquad \text{5-118}$$

where w_1 is white noise with intensity $V_1(t)$. The system has the controlled variable

$$z(t) = D(t)x(t) \qquad \text{5-119}$$

and the observed variable

$$y(t) = C(t)x(t) + w_2(t). \qquad \text{5-120}$$

Here w_2 is white noise with intensity $V_2(t)$. We assume that $V_{r2}(t) > 0$, $V_2(t) > 0$, $t_0 \leq t \leq t_1$.

5.4 Tracking with Incomplete and Noisy Measurements

To obtain an optimization problem, we consider the criterion

$$E\left\{\int_{t_0}^{t_1}[[z(t)-z_r(t)]^T R_3(t)[z(t)-z_r(t)] + u^T(t)R_2(t)u(t)]\,dt\right\}. \quad \text{5-121}$$

Here $R_3(t) > 0$, $R_2(t) > 0$, $t_0 \leq t \leq t_1$. The first term of the integrand serves to force the controlled variable $z(t)$ to follow the reference variable $z_r(t)$, while the second term constrains the amplitudes of the input.

We now phrase the stochastic optimal tracking problem with incomplete and noisy observations as follows.

Definition 5.2. *Consider the system*

$$\dot{x}(t) = A(t)x(t) + B(t)u(t) + w_1(t), \qquad t \geq t_0, \quad \text{5-122}$$

where $x(t_0)$ is a stochastic variable with mean $\bar{x}_0$ and variance matrix Q_0, and w_1 is white noise with intensity $V_1(t)$. The controlled variable is

$$z(t) = D(t)x(t), \quad \text{5-123}$$

and the observed variable is

$$y(t) = C(t)x(t) + w_2(t), \quad \text{5-124}$$

where w_2 is white noise with intensity $V_2(t)$, with $V_2(t) > 0$, $t_0 \leq t \leq t_1$. Consider furthermore the reference variable

$$z_r(t) = D_r(t)x_r(t), \quad \text{5-125}$$

where

$$\dot{x}_r(t) = A_r(t)x_r(t) + w_{r1}(t), \qquad t \geq t_0. \quad \text{5-126}$$

Here $x_r(t_0)$ is a stochastic variable with mean $\bar{x}_{r0}$ and variance matrix Q_{r0}, and w_{r1} is white noise with intensity $V_{r1}(t)$. The observed variable for the x_r process is

$$y_r(t) = C_r(t)x_r(t) + w_{r2}(t), \quad \text{5-127}$$

*where w_{r2} is white noise with intensity $V_{r2}(t) > 0$, $t_0 \leq t \leq t_1$. Then the **optimal linear tracking problem with incomplete and noisy observations** is the problem of choosing the input to the system **5-122** as a function of $y(\tau)$ and $y_r(\tau)$, $t_0 \leq \tau \leq t$, such that the criterion*

$$E\left\{\int_{t_0}^{t_1}[[z(t)-z_r(t)]^T R_3(t)[z(t)-z_r(t)] + u^T(t)R_2(t)u(t)]\,dt\right\} \quad \text{5-128}$$

is minimized, where $R_3(t) > 0$ and $R_2(t) > 0$ for $t_0 \leq t \leq t_1$.

To solve the problem we combine the reference model and the plant in an augmented system. In terms of the augmented state $\tilde{x}(t) = \text{col}\,[x(t), x_r(t)]$, we write

$$\dot{\tilde{x}}(t) = \begin{pmatrix} A(t) & 0 \\ 0 & A_r(t) \end{pmatrix}\tilde{x}(t) + \begin{pmatrix} B(t) \\ 0 \end{pmatrix}u(t) + \begin{pmatrix} w_1(t) \\ w_{r1}(t) \end{pmatrix}. \quad \text{5-129}$$

404 Optimal Linear Output Feedback Control Systems

The observed variable for the augmented system is

$$\begin{pmatrix} y(t) \\ y_r(t) \end{pmatrix} = \begin{pmatrix} C(t) & 0 \\ 0 & C_r(t) \end{pmatrix} \tilde{x}(t) + \begin{pmatrix} w_2(t) \\ w_{r2}(t) \end{pmatrix}. \qquad 5\text{-}130$$

For the criterion we write

$$E\left\{ \int_{t_0}^{t_1} [\tilde{x}^T(t) \tilde{D}^T(t) R_3(t) \tilde{D}(t) \tilde{x}(t) + u^T(t) R_2(t) u(t)] \, dt \right\}, \qquad 5\text{-}131$$

where

$$\tilde{D}(t) = [D(t), \; -D_r(t)]. \qquad 5\text{-}132$$

The tracking problem is now in the form of a standard stochastic regulator problem and can be solved by application of Theorem 5.3. It follows that we can write

$$u(t) = -F^0(t) \begin{pmatrix} \hat{x}(t) \\ \hat{x}_r(t) \end{pmatrix}. \qquad 5\text{-}133$$

If we assume that all the white noise processes and initial values associated with the plant and the reference process are uncorrelated, two separate observers can be constructed, one for the state of the plant and one for the state of the reference process. Furthermore, we know from Section 3.6.3 that because of the special structure of the tracking problem we can write

$$F^0(t) = [F_1(t), \; -F_2(t)], \qquad 5\text{-}134$$

where the partitioning is consistent with the other partitionings, and where the feedback gain matrix $F_1(t)$ is completely independent of the properties of the reference process.

Figure 5.7 gives the block diagram of the optimal tracking system, still under the assumption that two separate observers can be used. It is seen that

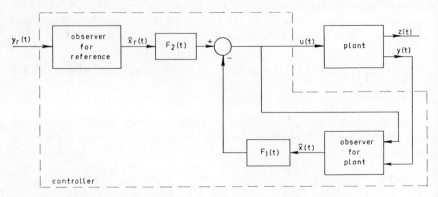

Fig. 5.7. The structure of the optimal tracking system.

5.4 Tracking with Incomplete and Noisy Measurements

the feedback link of the controller is completely independent of the properties of the reference variable.

We conclude this section with an examination of the transmission $T(s)$ of the system in the steady-state time-invariant case. A simple way to find this transfer matrix is as follows. Set $x(0) = \hat{x}(0) = 0$, and assume that the system is free of noise. It follows that $x(t) = \hat{x}(t)$ for $t \geq 0$. We can thus completely omit the plant observer in the computation of $T(s)$ and substitute $x(t)$ wherever we find $\hat{x}(t)$. We thus have the following relations:

$$\dot{x}(t) = Ax(t) + Bu(t),$$
$$z(t) = Dx(t),$$
$$u(t) = -\bar{F}_1 x(t) + \bar{F}_2 \hat{x}_r(t),$$
$$\dot{\hat{x}}(t) = A_r \hat{x}_r(t) + \bar{K}_r[y_r(t) - C_r \hat{x}_r(t)].$$
5-135

It easily follows that
$$\mathbf{Z}(s) = T(s)\mathbf{Y}_r(s),$$
5-136

where $\mathbf{Z}(s)$ and $\mathbf{Y}_r(s)$ are the Laplace transforms of $z(t)$ and $y_r(t)$, and where

$$T(s) = D(sI - A + B\bar{F}_1)^{-1} B\bar{F}_2 (sI - A_r + \bar{K}_r C_r)^{-1} \bar{K}_r.$$
5-137

In general $T(0)$ does not equal the unit matrix, so that step changes in the reference cause a steady-state error. The reason for this is that the present control system has not been designed for steps in the reference variable. If it is important that the control system have a zero steady-state error to constant references, the design method suggested in the next section should be adopted. We finally note that in the transmission only the regulator poles and the reference observer poles occur, while the plant observer poles have been canceled.

Example 5.4. *Position servo*

We return to the by now familiar positioning system. Consider the problem of designing a control system such that the angular position tracks a reference variable. For the system itself, the disturbances, and the observation noise we use the equations and numerical data of Example 5.3 (Section 5.3.2). We model the reference variable as exponentially correlated noise:

$$\zeta_r(t) = \xi_r(t),$$
5-138

with

$$\dot{\xi}_r(t) = -\frac{1}{\theta} \xi_r(t) + w_{r1}(t), \quad t \geq t_0,$$
5-139

Here w_{r1} is scalar white noise with constant intensity V_{r1}. It is assumed that the reference variable is observed with additive white noise, so that we

measure
$$\eta_r(t) = \xi_r(t) + w_{r2}(t), \quad \text{5-140}$$

where w_{r2} has constant intensity V_{r2} and is uncorrelated with w_{r1}. The steady-state optimal observer for the reference process is easily computed. It is described by

$$\dot{\hat{\xi}}_r(t) = -\frac{1}{\theta}\hat{\xi}_r(t) + \bar{K}_r[\eta_r(t) - \hat{\xi}_r(t)], \quad \text{5-141}$$

where

$$\bar{K}_r = -\frac{1}{\theta} + \sqrt{\frac{1}{\theta^2} + \frac{V_{r1}}{V_{r2}}}. \quad \text{5-142}$$

The optimization criterion is expressed as

$$E\left\{\int_{t_0}^{t_1}[[\zeta(t) - \zeta_r(t)]^2 + \rho\mu^2(t)]\,dt\right\}. \quad \text{5-143}$$

The resulting steady-state control law is given by

$$\mu(t) = -\bar{F}_1\hat{x}(t) + \bar{F}_2\hat{\xi}_r(t). \quad \text{5-144}$$

$\bar{P}_{11}$ and $\bar{F}_1$ have been computed in Example 3.8 (Section 3.4.1), in which we obtained the following results:

$$\bar{P}_{11} = \begin{pmatrix} \frac{\sqrt{\rho}}{\kappa}\sqrt{\alpha^2 + \frac{2\kappa}{\sqrt{\rho}}} & \frac{\sqrt{\rho}}{\kappa} \\ \frac{\sqrt{\rho}}{\kappa} & \frac{\rho}{\kappa^2}\left(-\alpha + \sqrt{\alpha^2 + \frac{2\kappa}{\sqrt{\rho}}}\right) \end{pmatrix}, \quad \text{5-145}$$

$$\bar{F}_1 = \left(\frac{1}{\sqrt{\rho}}, \frac{1}{\kappa}\left(-\alpha + \sqrt{\alpha^2 + \frac{2\kappa}{\sqrt{\rho}}}\right)\right).$$

Using the results of Section 3.6.3, it can be found that

$$\bar{F}_2 = \frac{\frac{\kappa}{\rho}}{\frac{\kappa}{\sqrt{\rho}} + \frac{1}{\theta^2} + \frac{1}{\theta}\left(\alpha^2 + \frac{2\kappa}{\sqrt{\rho}}\right)^{1/2}}. \quad \text{5-146}$$

Since we now have the reference observer and the regulator gains available, we can use **5-137** to calculate the transmission $T(s)$ of the closed-loop tracking

5.4 Tracking with Incomplete and Noisy Measurements

system. We obtain

$$T(s) = \frac{\kappa}{s^2 + s\left(\alpha^2 + \frac{2\kappa}{\sqrt{\rho}}\right)^{1/2} + \frac{\kappa}{\sqrt{\rho}}} \frac{\frac{\kappa}{\rho}}{\frac{\kappa}{\sqrt{\rho}} + \frac{1}{\theta^2} + \frac{1}{\theta}\left(\alpha^2 + \frac{2\kappa}{\sqrt{\rho}}\right)^{1/2}} \cdot \frac{-\frac{1}{\theta} + \left(\frac{1}{\theta^2} + \frac{V_{r1}}{V_{r2}}\right)^{1/2}}{s + \left(\frac{1}{\theta^2} + \frac{V_{r1}}{V_{r2}}\right)^{1/2}}. \quad \text{5-147}$$

We note that the break frequency of the transmission is the least of the break frequency of the closed-loop plant and the break frequency of the reference observer. The break frequency of the closed-loop plant is ω_0, where $\omega_0^2 = \kappa/\sqrt{\rho}$, while the break frequency of the reference observer is

$$\left(\frac{1}{\theta^2} + \frac{V_{r1}}{V_{r2}}\right)^{1/2}. \quad \text{5-148}$$

Which break frequency is the lowest depends upon the "signal-to-noise" ratio V_{r1}/V_{r2} of the reference variable and the value of ρ, which in turn is determined by the allowable input amplitudes to the plant. Let us first consider the effect of V_{r1}/V_{r2}. If the reference variable is accurately measured, (i.e., V_{r2} is small) the reference observer break frequency is high and the closed-loop feedback system break frequency will prevail. On the other hand, if the reference variable is inaccurately measured, the reference observer limits the total bandwidth of the system.

When we next consider the effect of the weighting factor ρ, we see that if ρ is small, that is, large input amplitudes are allowed, the closed-loop system break frequency is high and the reference observer determines the break frequency. Conversely, if ρ is large, the break frequency is limited by the closed-loop plant.

Let us assume the following numerical values for the reference process:

$$\theta = 5 \text{ s},$$
$$V_{r1} = 0.4 \text{ rad}^2/\text{s}. \quad \text{5-149}$$

This makes the reference variable break frequency 0.2 rad/s, while the reference variable rms value is 1 rad. Let us furthermore assume that the reference variable measurement noise w_{r2} is exponentially correlated noise with rms value 0.181 rad and time constant 0.025 s. This makes the break frequency of the reference variable measurement noise 40 rad/s. Since this break frequency is quite high as compared to 0.2 rad/s, we approximate the measurement noise as white noise with density

$$V_{r2} = 2(0.1)^2 0.0816 = 0.001636 \text{ rad}^2/\text{s}. \qquad \textbf{5-150}$$

With the numerical values **5-149** and **5-150**, we find for the reference observer break frequency the value

$$\left(\frac{1}{\theta^2} + \frac{V_{r1}}{V_{r2}}\right)^{1/2} \simeq 15.6 \text{ rad/s}. \qquad \textbf{5-151}$$

Since the break frequency of the reference observer is less than the break frequency of 40 rad/s of the reference measurement noise, we conclude that it is justified to approximate this measurement noise as white noise.

We finally must determine the most suitable value of the weighting factor ρ. In order to do this, we evaluate the control law for various values of ρ and compute the corresponding rms tracking errors and rms input voltages. Omitting the disturbing torque τ_d and the system measurement noise ν_m we write for the system equations

$$\begin{aligned}
\dot{x}(t) &= Ax(t) + b\mu(t), \\
\mu(t) &= -\bar{F}_1 x(t) + \bar{F}_2 \hat{\xi}_r(t), \\
\dot{\hat{\xi}}_r(t) &= -\frac{1}{\theta}\hat{\xi}_r(t) + \bar{K}_r[\eta_r(t) - \hat{\xi}_r(t)], \qquad \textbf{5-152}\\
\dot{\xi}_r(t) &= -\frac{1}{\theta}\xi_r(t) + w_{r1}(t), \\
\eta_r(t) &= \xi_r(t) + w_{r2}(t).
\end{aligned}$$

Combining all these relations we obtain the augmented differential equation

$$\begin{pmatrix} \dot{x}(t) \\ \dot{\hat{\xi}}_r(t) \\ \dot{\xi}_r(t) \end{pmatrix} = \begin{pmatrix} A - b\bar{F}_1 & b\bar{F}_2 & 0 \\ 0 & -\frac{1}{\theta} - \bar{K}_r & \bar{K}_r \\ 0 & 0 & -\frac{1}{\theta} \end{pmatrix} \begin{pmatrix} x(t) \\ \hat{\xi}_r(t) \\ \xi_r(t) \end{pmatrix} + \begin{pmatrix} 0 \\ \bar{K}_r w_{r2}(t) \\ w_{r1}(t) \end{pmatrix}. \qquad \textbf{5-153}$$

From this equation we can set up and solve the steady-state variance matrix of the augmented state col $[x(t), \hat{\xi}_r(t), \xi_r(t)]$, and from this the steady-state rms tracking error and rms input voltage can be computed. Of course we can also use the technique of Section 5.3.2. Table 5.1 lists the results for decreasing values of the weighting coefficient ρ. Note that the contributions of the reference excitation noise w_{r_1} and the reference measurement noise w_{r_2} are given separately, together with their total contribution.

If the maximally allowable input voltage is about 100 V, the weighting coefficient ρ should certainly not be chosen less than 0.00001; for this value the rms input voltage is nearly 50 V. The corresponding rms tracking error is about 0.27 rad, which is still quite a large value as compared to the rms value of the reference variable of 1 rad. If this rms value is too large, the requirements on the reference variable bandwidth must be lowered. It should be remarked, however, that the values obtained for the rms tracking error and the rms input are probably larger than the actual values encountered, since modelling stochastic processes by exponentially correlated noise usually leads to power spectral density functions that decrease much slower with increasing frequency than actual density functions.

For $\rho = 0.00001$ it can be computed from **5-152** that the zero-frequency transmission is given by $T(0) = 0.8338$. This means that the proposed control system shows a considerable steady-state error when subjected to a constant reference variable. This phenomenon occurs, first, because exponentially correlated noise has relatively much of its power at high frequencies and, second, because the term that weights the input in the optimization criterion tends to keep the input small, at the expense of the tracking accuracy. In the following section we discuss how tracking systems with a zero steady-state error can be obtained.

The rms values given in Table 5.1 do not include the contributions of the system disturbances and observation errors. Our findings in Example 5.3 suggest, however, that these contributions are negligible as compared to those of the reference variable.

5.5 REGULATORS AND TRACKING SYSTEMS WITH NONZERO SET POINTS AND CONSTANT DISTURBANCES

5.5.1 Nonzero Set Points

As we saw in Chapter 2, sometimes it is important to design tracking systems that show a zero steady-state error response to constant values of the reference variable. The design method of the preceding section can never produce such tracking systems, since the term in the optimization criterion that weights the input always forces the input to a smaller value, at the expense of a nonzero tracking error. For small weights on the input, the steady-state tracking error decreases, but it never disappears completely. In this section we approach the problem of obtaining a zero steady-state tracking error,

Table 5.1 The Effect of the Weighting Factor ρ on the Performance of the Position Servo System

ρ	Contribution of reference variable to rms tracking error (rad)	Contribution of reference measurement noise to rms tracking error (rad)	Total rms tracking error (rad)	Contribution of reference variable to rms input voltage (V)	Contribution of reference measurement noise to rms input voltage (V)	Total rms input voltage (V)
0.1	0.8720	0.0038	0.8720	1.438	0.222	1.455
0.01	0.6884	0.0125	0.6885	4.365	0.825	4.442
0.001	0.4942	0.0280	0.4950	10.32	2.69	10.67
0.0001	0.3524	0.0472	0.3556	21.84	8.15	23.31
0.00001	0.2596	0.0664	0.2680	43.03	23.08	48.82

5.5 Nonzero Set Points and Constant Disturbances

as in Section 3.7.1, from the point of view of a variable set point. Consider the system

$$\dot{x}(t) = Ax(t) + Bu(t) \qquad \text{5-154}$$

with the controlled variable

$$z(t) = Dx(t). \qquad \text{5-155}$$

In Section 3.7.1 we derived the nonzero set point optimal control law

$$u(t) = -\bar{F}x(t) + H_c^{-1}(0)z_0. \qquad \text{5-156}$$

$\bar{F}$ is the steady-state gain matrix for the criterion

$$\int_{t_0}^{\infty} [z^T(t)R_3 z(t) + u^T(t)R_2 u(t)] \, dt, \qquad \text{5-157}$$

while $H_c(s)$ is the closed-loop transfer matrix

$$H_c(s) = D(sI - A + B\bar{F})^{-1}B. \qquad \text{5-158}$$

It is assumed that the dimension of u equals that of z, and that the open-loop transfer matrix $H(s) = D(sI - A)^{-1}B$ has no zeroes at the origin. These assumptions guarantee the existence of $H_c^{-1}(0)$. Finally, z_0 is the set point for the controlled variable. The control law **5-156** causes the control system to reach the set point optimally from any initial state, and to make an optimal transition to the new set point whenever z_0 changes.

Let us now consider a stochastic version of the nonzero set point regulator problem. We assume that the plant is described by

$$\dot{x}(t) = Ax(t) + Bu(t) + w_1(t), \qquad \text{5-159}$$

where w_1 is white noise. The controlled variable again is

$$z(t) = Dx(t), \qquad \text{5-160}$$

but we introduce an observed variable

$$y(t) = Cx(t) + w_2(t), \qquad \text{5-161}$$

where w_2 is also white noise. Suppose that the set point z_0 for the controlled variable of this system is accurately known. Then the nonzero set point steady-state optimal controller for this system obviously is

$$\begin{aligned} u(t) &= -\bar{F}\hat{x}(t) + H_c^{-1}(0)z_0, \\ \dot{\hat{x}}(t) &= A\hat{x}(t) + Bu(t) + \bar{K}[y(t) - C\hat{x}(t)], \end{aligned} \qquad \text{5-162}$$

where $\bar{K}$ is the steady-state optimal observer gain and where $\bar{F}$ and $H_c(s)$ are as given before. If no state excitation noise and observation noise are present, the controlled variable will eventually approach z_0 as t increases.

The control law is optimal in the sense that the steady-state value of

$$E\{z^T(t)R_3 z(t) + u^T(t)R_2 u(t)\} \qquad 5\text{-}163$$

is minimized, where z and u are taken relative to their set points. When the set point changes, an optimal transition to the new set point is made.

The controller described by **5-162** may give quite good results when the set point z_0 is a slowly varying quantity. Unsatisfactory results may be obtained when the set point occasionally undergoes step changes. This may result in the input having too large a transient, necessitating reduction in the loop gain of the system. This in turn deteriorates the disturbance suppression properties of the system. This difficulty can be remedied by interpreting quick changes in the set point as "noise." Thus we write the control law **5-162** in the form

$$u(t) = -\bar{F}\hat{x}(t) + H_c^{-1}(0)\hat{z}_0(t), \qquad 5\text{-}164$$

where $\hat{z}_0(t)$ is the estimated set point. The observed set point, $r(t)$, is represented as

$$r(t) = z_0(t) + w_s(t), \qquad 5\text{-}165$$

where w_s is white noise and z_0 is the actual set point. In order to determine $\hat{z}_0(t)$ (compare Example 4.3, Section 4.3.2, on the estimation of a constant), we model z_0 as

$$\dot{z}_0(t) = w_0(t), \qquad 5\text{-}166$$

where w_0 is another white noise process. The steady-state optimal observer for the set point will be of the form

$$\dot{\hat{z}}_0(t) = \bar{K}_0[r(t) - \hat{z}_0(t)], \qquad 5\text{-}167$$

where $\bar{K}_0$ is the appropriate steady-state observer gain matrix.

The controller defined by **5-164** and **5-167** has the property that, if no noise is present and the observed set point $r(t)$ is constant, the controlled variable will in the steady state precisely equal $r(t)$. This follows from **5-167**, since in the steady state $\hat{z}_0(t) \equiv r(t)$ so that in **5-164** $\hat{z}_0(t)$ is replaced with $r(t)$, which in turn causes $z(t)$ to assume the value $r(t)$. It is seen that in the case where r, z_0, u, and z are scalar the *prefilter* (see Fig. 5.8) defined by **5-164** and **5-167** is nothing but a first-order filter. In the multidimensional case a generalization of this first-order filter is obtained. When the components of the uncorrelated white noise processes w_0 and w_s are assumed to be uncorrelated as well, it is easily seen that $\bar{K}_0$ is diagonal, so that the prefilter consists simply of a parallel bank of scalar first-order filters. It is suggested that the time constants of these filters be determined on the basis of the desired response to steps in the components of the reference variable and in relation to likely step sizes and permissible input amplitudes.

5.5 Nonzero Set Points and Constant Disturbances

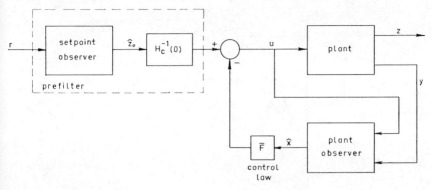

Fig. 5.8. Nonzero set point optimal controller with set point observer.

Example 5.5. *The positioning system*

In Example 5.3 (Section 5.3.2), we found a zero set point optimal controller for the positioning system. Let us determine the corresponding nonzero set point control system. We first determine the nonzero set point optimal control law. It follows from Example 3.8 (Section 3.4.1) that the closed-loop transfer function $H_c(s)$ is given by

$$H_c(s) = \frac{\kappa}{s^2 + s\left(\alpha^2 + \frac{2\kappa}{\sqrt{\rho}}\right)^{1/2} + \frac{\kappa}{\sqrt{\rho}}}. \qquad 5\text{-}168$$

Consequently, the nonzero set point control law **5-164** is

$$\mu(t) = -\bar{F}\hat{x}(t) + \frac{1}{\sqrt{\rho}}\hat{\zeta}_0(t), \qquad 5\text{-}169$$

where $\hat{\zeta}_0(t)$ is the estimated set point. Let us design for step changes in the observed set point. The observer **5-167** for the set point is of the form

$$\dot{\hat{\zeta}}_0(t) = k_0[r(t) - \hat{\zeta}_0(t)], \qquad 5\text{-}170$$

where $r(t)$ is the reference variable and k_0 a scalar gain factor. Using the numerical values of Example 5.3, we give in Fig. 5.9 the responses of the nonzero set point control system defined by **5-169** and **5-170** to a step of 1 rad in the reference variable $r(t)$ for various values of the gain k_0. Assuming that an input voltage of up to 100 V is tolerable, we see that a suitable value of k_0 is about 20 s^{-1}. The corresponding time constant of the prefilter is $1/k_0 = 0.05$ s.

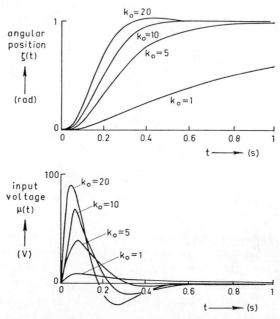

Fig. 5.9. Responses of the position control system as nonzero set point control system to a step in the set point of 1 rad for various values of the prefilter gain k_0.

5.5.2* Constant Disturbances

In the preceding section we discussed nonzero set point regulators. In the present section the question of constant disturbances is investigated, which is somewhat similar to the nonzero set point problem. The approach presented in this section is somewhat different from that in Section 3.7.2. As in Section 3.7.2, however, controllers with integrating action will be obtained.

Constant disturbances frequently occur in control problems. Often they are caused by inaccuracies in determining consistent nominal values of the input, the state, and the controlled variable. These disturbances can usually be represented through an additional constant forcing term v_0 in the state differential equation as follows:

$$\dot{x}(t) = Ax(t) + Bu(t) + v_0. \qquad \textbf{5-171}$$

As in the preceding section, we limit our discussion to the time-invariant case. For the controlled variable we write

$$z(t) = Dx(t). \qquad \textbf{5-172}$$

Let us assume, for the time being, that the complete state $x(t)$ can be

5.5 Nonzero Set Points and Constant Disturbances

observed at all times. Then we can consider the control law

$$u(t) = -\bar{F}x(t) + u_0, \qquad \text{5-173}$$

where $\bar{F}$ is a gain matrix selected according to some quadratic optimization criterion of the usual form and where the constant vector u_0 is to be chosen such that in steady-state conditions the contribution of the constant disturbance v_0 to the controlled variable z is canceled.

With the control law **5-173**, the closed-loop system equations are

$$\dot{x}(t) = (A - B\bar{F})x(t) + Bu_0 + v_0,$$
$$z(t) = Dx(t). \qquad \text{5-174}$$

Since the closed-loop system will be assumed to be asymptotically stable, the controlled variable eventually approaches a constant value, which is easily seen to be given by

$$\lim_{t \to \infty} z(t) = D(-\bar{A})^{-1}Bu_0 + D(-\bar{A})^{-1}v_0. \qquad \text{5-175}$$

Here we have abbreviated

$$\bar{A} = A - B\bar{F}. \qquad \text{5-176}$$

Does there exist a u_0 such that the steady-state value of $z(t)$ as given by **5-175** is zero? As in the nonzero set point problem, three cases must be distinguished:

(a) *The dimension of z is greater than that of u:* In this case the vector equation

$$D(-\bar{A})^{-1}Bu_0 + D(-\bar{A})^{-1}v_0 = 0 \qquad \text{5-177}$$

represents more equations than there are variables, which means that in general no solution exists. This is the case where it is attempted to control the variable $z(t)$ with an input $u(t)$ of smaller dimension and too few degrees of freedom are available.

(b) *The dimensions of u and z are the same:* In this case **5-177** can be solved for u_0 as follows:

$$u_0 = -H_c^{-1}(0)D(-\bar{A})^{-1}v_0. \qquad \text{5-178}$$

Here $H_c(s)$ is the closed-loop transfer matrix

$$H_c(s) = D(sI - A - B\bar{F})^{-1}B. \qquad \text{5-179}$$

As we know from Theorem 3.10 (Section 3.7), the inverse of $H_c(0)$ exists if the open-loop system transfer matrix $D(sI - A)^{-1}B$ has no zeroes at the origin.

(c) *The dimension of z is less than that of u:* In this case there are too many degrees of freedom and the dimension of z can be increased by adding components to the controlled variable.

In case (b), where dim (z) = dim (u), the control law

$$u(t) = -\bar{F}x(t) - H_c^{-1}(0)D(-\bar{A})^{-1}v_0 \qquad \text{5-180}$$

has the property that constant disturbances are compensated in an optimal manner. This control law, which has been given by Eklund (1969), will be referred to as the *zero-steady-state-error optimal control law*. As we have seen, it exists when dim (z) = dim (u) and the open-loop system has no zeroes at the origin.

Let us now suppose that in addition to v_0 fluctuating disturbances act upon the system as well, and that the system state can only be incompletely and inaccurately observed. We thus replace the state differential equation with

$$\dot{x}(t) = Ax(t) + Bu(t) + v_0 + w_1(t), \qquad \text{5-181}$$

where v_0 is the constant disturbance and w_1 white noise with intensity V_1. Furthermore, we assume that we have for the observed variable

$$y(t) = Cx(t) + w_2(t), \qquad \text{5-182}$$

where w_2 is white noise with intensity V_2.

In this situation the control law **5-180** must be replaced by

$$u(t) = -\bar{F}\hat{x}(t) - H_c^{-1}(0)D(-\bar{A})^{-1}\hat{v}_0, \qquad \text{5-183}$$

where $\hat{x}(t)$ and $\hat{v}_0$ are the minimum mean square estimates of $x(t)$ and v_0. An optimal observer can be obtained by modeling the constant disturbance through

$$\dot{v}_0(t) = 0. \qquad \text{5-184}$$

The resulting *steady-state* optimal observer, however, will have a zero gain matrix for updating the estimate of v_0, since according to the model **5-184** the value of v_0 never changes (compare Example 4.3, Section 4.3.2, concerning the estimation of a constant). Since in practice v_0 varies slowly, or occasionally changes value, it is better to model v_0 through

$$\dot{v}_0(t) = w_0(t), \qquad \text{5-185}$$

where the intensity V_0 of the white noise w_0 is so chosen that the increase in the fluctuations of v_0 reflects the likely variations in the slowly varying disturbance. When this model is used, the resulting steady-state optimal observer continues to track $v_0(t)$ and is of the form

$$\begin{aligned}\dot{\hat{x}}(t) &= A\hat{x}(t) + Bu(t) + \hat{v}_0(t) + \bar{K}_1[y(t) - C\hat{x}(t)], \\ \dot{\hat{v}}_0(t) &= \bar{K}_2[y(t) - C\hat{x}(t)].\end{aligned} \qquad \text{5-186}$$

5.5 Nonzero Set Points and Constant Disturbances

The control system that results from combining this observer with the control law **5-183** has the property that in the absence of other disturbances and observation noise the constant disturbance is always compensated so that a zero steady-state regulation or tracking error results (Eklund, 1969). As expected, this is achieved by "integrating action" of the controller (see Problem 2.3). The procedure of this section enables us to introduce this integrating action and at the same time improve the transient response of the control system and the suppression of fluctuating disturbances. The procedure is equally easily applied to multivariable as to single-input single-output systems.

It is not difficult to see that the procedure of this section can be combined with that of Section 5.5.1 when encountering tracking or regulating systems subject to nonzero set points as well as constant disturbances, by choosing the input as

$$u(t) = -\bar{F}\hat{x}(t) - H_c^{-1}(0)D(-\bar{A})^{-1}\hat{v}_0 + H_c^{-1}(0)\hat{z}_0. \qquad \textbf{5-187}$$

Here $\hat{z}_0$ is either the estimated set point and can be obtained as described in Section 5.5.1, or is the actual set point.

We remark that often is it is possible to trace back the constant disturbances to one or two sources. In such a case we can replace v_0 with

$$v_0 = Gv_1, \qquad \textbf{5-188}$$

where G is a given matrix and v_1 a constant disturbance of a smaller dimension than v_0. By modeling v_1 as integrated white noise, the dimension of the observer can be considerably decreased in this manner.

Example 5.6. *Integral control of the positioning system*

In this example we devise an integral control system for the positioning system. We assume that a constant disturbance can enter into the system in the form of a constant torque τ_0 on the shaft in addition to a disturbing torque τ_d which varies quickly. Thus we modify the state differential equation **5-92** of Example 5.3 (Section 5.3.2) to

$$\dot{x}(t) = \begin{pmatrix} 0 & 1 \\ 0 & -\alpha \end{pmatrix} x(t) + \begin{pmatrix} 0 \\ \kappa \end{pmatrix} \mu(t) + \begin{pmatrix} 0 \\ \gamma \end{pmatrix} \tau_d(t) + \begin{pmatrix} 0 \\ \gamma \end{pmatrix} \tau_0. \qquad \textbf{5-189}$$

As in Example 5.3, we represent the variable part of the disturbing torque as white noise with intensity V_d.

It is easily seen from **5-189** that the zero-steady-state-error optimal control law is given by

$$\mu(t) = -\bar{F}\hat{x}(t) - \frac{\gamma}{\kappa}\hat{\tau}_0, \qquad \textbf{5-190}$$

418 Optimal Linear Output Feedback Control Systems

where $\bar{F}$ is an appropriate steady-state optimal feedback gain matrix, and $\hat{\tau}_0$ is an estimate of τ_0.

To obtain an observer we model the constant part of the disturbance as

$$\dot{\tau}_0(t) = w_0(t), \qquad \text{5-191}$$

where the white noise w_0 has intensity V_0. As in Example 5.3, the observed variable is given by

$$\eta(t) = (1, 0)x(t) + v_m(t), \qquad \text{5-192}$$

where v_m is white noise with intensity V_m. The steady-state optimal observer thus has the form

$$\dot{\hat{x}}(t) = \begin{pmatrix} 0 & 1 \\ 0 & -\alpha \end{pmatrix} \hat{x}(t) + \begin{pmatrix} 0 \\ \kappa \end{pmatrix} \mu(t) + \begin{pmatrix} 0 \\ \gamma \end{pmatrix} \hat{\tau}_0(t) + \begin{pmatrix} \bar{k}_1 \\ \bar{k}_2 \end{pmatrix} [\eta(t) - (1, 0)\hat{x}(t)],$$
$$\dot{\hat{\tau}}_0(t) = \bar{k}_3[\eta(t) - (1, 0)\hat{x}(t)], \qquad \text{5-193}$$

where the scalar gains $\bar{k}_1$, $\bar{k}_2$, and $\bar{k}_3$ follow from the steady-state solution of the appropriate observer Riccati equation. With the numerical values of Example 5.3, and with the additional numerical value

$$V_0 = 60 \text{ N}^2 \text{ m}^2 \text{ s}^{-1}, \qquad \text{5-194}$$

it follows that these gains are given by

$$\bar{k}_1 = 42.74, \quad \bar{k}_2 = 913.2, \quad \bar{k}_3 = 24495. \qquad \text{5-195}$$

The assumption **5-194** implies that the rms value of the increment of τ_0 during a period of 1 s is $\sqrt{60} \simeq 7.75$ Nm. This torque is equivalent to an

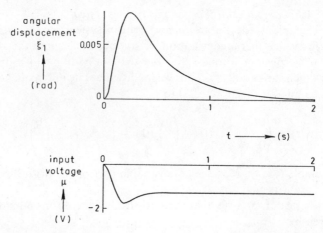

Fig. 5.10. Response of the zero steady-state error position control system to a constant torque of 10 N m on the shaft.

input voltage of nearly 1 V. The observer poles corresponding to the gains **5-195** are $-22.44 \pm j22.27$ and -2.450 s^{-1}.

By substituting the control law **5-190** into the observer equations **5-193**, it is easily found that the controller has a pole at the origin, hence exhibit integrating action, as expected. For $\bar{F}$ we choose the steady-state optimal gain matrix **5-90** derived in Example 5.3. The corresponding regulator poles are $-9.66 \pm j9.09$ s^{-1}. In Fig. 5.10 we give the response of the control system from zero initial conditions to a constant disturbance $\tau_0 = 10$ Nm. It is seen that the maximum deviation of the angular displacement caused by this constant torque is not more than about 0.008 rad.

5.6* SENSITIVITY OF TIME-INVARIANT OPTIMAL LINEAR OUTPUT FEEDBACK CONTROL SYSTEMS

In Chapter 3, Section 3.9, we saw that time-invariant linear optimal state feedback systems are insensitive to disturbances and parameter variations in the sense that the return difference matrix $J(s)$, obtained by opening the feedback loop at the state, satisfies an inequality of the form

$$J^T(-j\omega)WJ(j\omega) \geq W, \quad \text{for all real } \omega, \qquad \text{5-196}$$

where W is the weighting matrix $\bar{F}^T R_2 \bar{F}$.

In this section we see that optimal output feedback systems generally do not possess such a property, although it can be closely approximated. Consider the time-invariant system

$$\dot{x}(t) = Ax(t) + Bu(t) + w_1(t), \qquad \text{5-197}$$

where w_1 is white noise with constant intensity V_1. The observed variable is given by

$$y(t) = Cx(t) + w_2(t), \qquad \text{5-198}$$

where w_2 is white noise uncorrelated with w_1 with constant intensity V_2. The controlled variable is

$$z(t) = Dx(t), \qquad \text{5-199}$$

while the optimization criterion is specified as

$$E\left\{\int_{t_0}^{t_1} [z^T(t)R_3 z(t) + u^T(t)R_2 u(t)]\, dt\right\}, \qquad \text{5-200}$$

with R_3 and R_2 symmetric, constant, positive-definite weighting matrices.

To simplify the analysis, we assume that the controlled variable is also the observed variable (apart from the observation noise), that is, $C = D$. Then

420   Optimal Linear Output Feedback Control Systems

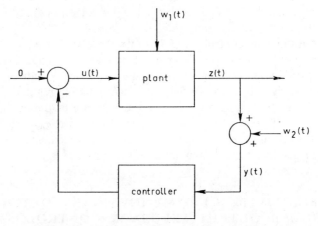

Fig. 5.11. Simplified output feedback control configuration.

we can schematically represent the control configuration as in Fig. 5.11, where observer and control law have been combined into the controller. Let us now consider the steady-state controller that results by letting $t_0 \to -\infty$ and $t_1 \to \infty$. Then the steady-state observer is described by

$$\dot{\hat{x}}(t) = A\hat{x}(t) + Bu(t) + \bar{K}[y(t) - D\hat{x}(t)], \qquad \text{5-201}$$

where $\bar{K}$ is the steady-state observer gain matrix. Laplace transformation of **5-201** and solution for the transform $\hat{\mathbf{X}}(s)$ of $\hat{x}(t)$ yields

$$\hat{\mathbf{X}}(s) = (sI - A + \bar{K}D)^{-1}[B\mathbf{U}(s) + \bar{K}\mathbf{Y}(s)], \qquad \text{5-202}$$

where $\mathbf{U}(s)$ and $\mathbf{Y}(s)$ are the Laplace transforms of $u(t)$ and $y(t)$, respectively. All initial conditions are assumed to be zero. For the input we have in terms of Laplace transforms

$$\mathbf{U}(s) = -\bar{F}\hat{\mathbf{X}}(s), \qquad \text{5-203}$$

where $\bar{F}$ is the steady-state feedback gain matrix. Substitution of **5-203** into **5-202** and solution for $\mathbf{U}(s)$ yields

$$\mathbf{U}(s) = -G(s)\mathbf{Y}(s), \qquad \text{5-204}$$

where

$$G(s) = [I + \bar{F}(sI - A + \bar{K}D)^{-1}B]^{-1}\bar{F}(sI - A + \bar{K}D)^{-1}\bar{K}. \qquad \text{5-205}$$

We now consider the return difference matrix

$$J(s) = I + H(s)G(s) \qquad \text{5-206}$$

for the control system, where

$$H(s) = D(sI - A)^{-1}B \qquad \text{5-207}$$

5.6 Sensitivity

is the plant transfer matrix. Generally, there does not exist a nonnegative-definite weighting matrix W such that an inequality of the form

$$J^T(-j\omega)WJ(j\omega) \geq W \qquad \text{5-208}$$

is satisfied for all real frequencies ω. Indeed, it can easily be proved (see Problem 5.6) that in the single-input single-output case **5-208** is *never* satisfied for all ω when $W > 0$. Of course the inequality **5-208** must hold in some useful frequency range, adapted to the frequency band of the disturbances acting upon the plant, since it follows from the optimality of the controller that the specific disturbances for which the control system has been designed are attenuated.

We now prove, however, that under certain conditions satisfaction of **5-208** for all frequencies can be obtained asymptotically. Consider the algebraic Riccati equation

$$0 = D^T R_3 D - \bar{P}BR_2^{-1}B^T\bar{P} + A^T\bar{P} + \bar{P}A, \qquad \text{5-209}$$

which must be solved to obtain the regulation gain $\bar{F} = R_2^{-1}B^T\bar{P}$. Suppose that

$$R_2 = \rho N, \qquad \text{5-210}$$

where ρ is a positive scalar and N a positive-definite matrix. Then it follows from Theorem 3.14 (Section 3.8.3) that if $\dim(z) = \dim(u)$, and the open-loop transfer matrix $H(s) = D(sI - A)^{-1}B$ has zeroes with nonpositive real parts only, as $\rho \downarrow 0$ the desired solution $\bar{P}$ of **5-209** approaches the zero matrix. This implies that

$$\lim_{\rho \downarrow 0} \bar{P}B \frac{1}{\rho} N^{-1}B^T\bar{P} = D^T R_3 D, \qquad \text{5-211}$$

or

$$\lim_{\rho \downarrow 0} \rho \bar{F}^T N \bar{F} = D^T R_3 D. \qquad \text{5-212}$$

Now the general solution of the matrix equation $X^T X = M^T M$, where X and M have equal dimensions, can be written in the form $X = UM$, where U is an arbitrary unitary matrix, that is, U satisfies $U^T U = I$. We therefore conclude from **5-212** that as $\rho \downarrow 0$ the gain matrix $\bar{F}$ asymptotically behaves as

$$\bar{F} \to \frac{1}{\sqrt{\rho}} N^{-1/2} U R_3^{1/2} D. \qquad \text{5-213}$$

As a result,

$$G(s) \to [D(sI - A + \bar{K}D)^{-1}B]^{-1}D(sI - A + \bar{K}D)^{-1}\bar{K}, \qquad \text{5-214}$$

as $\rho \downarrow 0$. It is not difficult to prove that

$$[D(sI - A + \bar{K}D)^{-1}B]^{-1}D(sI - A + \bar{K}D)^{-1}\bar{K}$$
$$= [D(sI - A)^{-1}B]^{-1}D(sI - A)^{-1}\bar{K}. \quad \text{5-215}$$

With this it follows for the return difference matrix $J(s)$ of the configuration of Fig. 5.11 that as $R_2 \to 0$

$$J(s) \to J_0(s), \quad \text{5-216}$$

where

$$J_0(s) = I + D(sI - A)^{-1}\bar{K}. \quad \text{5-217}$$

We now derive an inequality for the asymptotic return difference matrix $J_0(s)$. The steady-state variance matrix $\bar{Q}$ satisfies the algebraic Riccati equation

$$0 = V_1 - \bar{Q}D^TV_2^{-1}D\bar{Q} + A\bar{Q} + \bar{Q}A^T, \quad \text{5-218}$$

assuming that the state excitation noise and observation noise are uncorrelated, that $V_2 > 0$, and that the Riccati differential equation possesses a steady-state solution. We can now go through manipulations very similar to those in Section 3.9, where we dealt with the sensitivity of the state feedback regulator. Addition and subtraction of $s\bar{Q}$ and rearrangement yield

$$0 = V_1 - \bar{Q}D^TV_2^{-1}D\bar{Q} - (sI - A)\bar{Q} - \bar{Q}(-sI - A^T). \quad \text{5-219}$$

Premultiplication by $D(sI - A)^{-1}$ and postmultiplication by $(-sI - A^T)^{-1}D^T$ give

$$0 = D(sI - A)^{-1}(V_1 - \bar{Q}D^TV_2^{-1}D\bar{Q})(-sI - A^T)^{-1}D^T$$
$$- D\bar{Q}(-sI - A^T)^{-1}D^T - D(sI - A)^{-1}\bar{Q}D^T. \quad \text{5-220}$$

By adding and subtracting an extra term V_2, this expression can be rearranged into the form

$$[I + D(sI - A)^{-1}\bar{Q}D^TV_2^{-1}]V_2[I + V_2^{-1}D\bar{Q}(-sI - A^T)^{-1}D^T]$$
$$= V_2 + D(sI - A)^{-1}V_1(-sI - A^T)^{-1}D^T. \quad \text{5-221}$$

Since $\bar{Q}D^TV_2^{-1} = \bar{K}$ we immediately recognize that this expression implies the equality

$$J_0(s)V_2J_0^T(-s) = V_2 + D(sI - A)^{-1}V_1(-sI - A^T)^{-1}D^T. \quad \text{5-222}$$

Substituting $s = j\omega$ we see that the second term on the right-hand side is a nonnegative-definite Hermitian matrix; thus we have

$$J_0(j\omega)V_2J_0^T(-j\omega) \geq V_2 \quad \text{for all real } \omega. \quad \text{5-223}$$

It follows from Theorem 2.2 (Section 2.10) that

$$S_0^T(-j\omega)V_2^{-1}S_0(j\omega) \leq V_2^{-1} \quad \text{for a real } \omega, \quad \text{5-224}$$

where $S_0(s)$ is the asymptotic sensitivity matrix:
$$S_0(s) = J_0^{-1}(s). \qquad \text{5-225}$$
We also have
$$J_0^T(-j\omega)V_2^{-1}J_0(j\omega) \geq V_2^{-1}.$$
We thus have the following result (Kwakernaak, 1969).

Theorem 5.5. *Consider the steady-state time-invariant stochastic optimal output feedback regulator. Suppose that the observed variable is also the controlled variable, that is,*
$$\begin{aligned} y(t) &= Dx(t) + w_2(t), \\ z(t) &= Dx(t). \end{aligned} \qquad \text{5-226}$$

Also assume that the state excitation noise $w_1(t)$ and the observation noise $w_2(t)$ are uncorrelated, that the observation problem is nonsingular, that is, $V_2 > 0$, and that the steady-state output feedback regulator is asymptotically stable. Then if dim (u) = dim (z), and the open-loop transfer matrix $H(s) = D(sI - A)^{-1}B$ possesses no right-half plane zeroes, the return difference matrix of the closed-loop system asymptotically approaches $J_0(s)$ as $R_2 \to 0$, where
$$J_0(s) = I + D(sI - A)^{-1}\bar{K}. \qquad \text{5-227}$$

$\bar{K}$ is the steady-state observer gain matrix. The asymptotic return difference matrix satisfies the relation
$$J_0(s)V_2J_0^T(-s) = V_2 + D(sI - A)^{-1}V_1(-sI - A^T)^{-1}D^T. \qquad \text{5-228}$$

The asymptotic return difference matrix $J_0(s)$ and its inverse, the asymptotic sensitivity matrix $S_0(s) = J_0^{-1}(s)$, satisfy the inequalities

$$\begin{aligned} J_0(j\omega)V_2J_0^T(-j\omega) &\geq V_2 & \text{for all real } \omega, \\ S_0^T(-j\omega)V_2^{-1}S_0(j\omega) &\leq V_2^{-1} & \text{for all real } \omega, \\ J_0^T(-j\omega)V_2^{-1}J_0(j\omega) &\geq V_2^{-1} & \text{for all real } \omega. \end{aligned} \qquad \text{5-229}$$

This theorem shows that asymptotically the sensitivity matrix of the output feedback regulator system satisfies an inequality of the form **5-196**, which means that in the asymptotic control system disturbances are always reduced as compared to the open-loop steady-state equivalent control system no matter what the power spectral density matrix of the disturbances. It also means that the asymptotic control system reduces the effect of all (sufficiently small) plant variations as compared to the open-loop steady-state equivalent The following points are worth noting:

(i) The weighting matrix in the sensitivity criterion is V_2^{-1}. This is not surprising. Let us assume for simplicity that V_2 is diagonal. Then if one of the diagonal elements of V_2 is small, the corresponding component of the observed variable can be accurately measured, which means that the gain in the corresponding feedback loop can be allowed to be large. This will have a favorable effect on the suppression of disturbances and plant variations at this output, which in turn is reflected by a large weighting coefficient in the sensitivity criterion.

(ii) The theorem is not valid for systems that possess open-loop zeroes in the right-half plane.

(iii) In practical cases it is never possible to choose R_2 very small. This means that the sensitivity criterion is violated over a certain frequency range. Examples show that this is usually the case in the high-frequency region. It is to be expected that the sensitivity reduction is not spoiled too badly when R_2 is chosen so small that the faraway regulator poles are much further away from the origin than the observer poles.

(iv) The right-hand side of **5-228** can be evaluated directly without solving Riccati equations. It can be used to determine the behavior of the return difference matrix, in particular in the single-input single-output case.

(v) It can be shown (Kwakernaak, 1969), that a result similar to Theorem 5.5 holds when

$$y(t) = \begin{pmatrix} D \\ M \end{pmatrix} x(t) + w_2(t), \qquad \text{5-230}$$

that is, $y(t)$ *includes* the controlled variable $z(t)$.

Example 5.7. *Position control system*

Again we consider the positioning system described by the state differential equation

$$\dot{x}(t) = \begin{pmatrix} 0 & 1 \\ 0 & -\alpha \end{pmatrix} x(t) + \begin{pmatrix} 0 \\ \kappa \end{pmatrix} \mu(t) + \begin{pmatrix} 0 \\ \gamma \end{pmatrix} \tau_d(t). \qquad \text{5-231}$$

Here $\tau_d(t)$ is white noise with intensity V_d. The observed variable is

$$\eta(t) = (1, 0)x(t) + \nu_m(t), \qquad \text{5-232}$$

where $\nu_m(t)$ is white noise with intensity V_m. The controlled variable is

$$\zeta(t) = (1, 0)x(t). \qquad \text{5-233}$$

The system satisfies the assumptions of Theorem 5.5, since the controlled variable is the observed variable, the state excitation and observation noise are assumed to be uncorrelated, and the open-loop transfer function,

$$H(s) = \frac{\kappa}{s(s + \alpha)}, \qquad \text{5-234}$$

5.6 Sensitivity

possesses no right-half plane zeroes. To compute the asymptotic return difference $J_0(s)$, we evaluate **5-228**, which easily yields

$$J_0(s)J_0(-s) = 1 + \frac{\gamma^2 V_d/V_m}{-s^2(-s^2+\alpha^2)}$$

$$= \frac{s^4 - \alpha^2 s^2 + \gamma^2 V_d/V_m}{-s^2(-s^2+\alpha^2)}. \quad \text{5-235}$$

Substitution of $s = j\omega$ provides us with the relation

$$|J_0(j\omega)|^2 = \frac{\omega^4 + \alpha^2\omega^2 + \gamma^2 V_d/V_m}{\omega^2(\omega^2+\alpha^2)} \quad \text{5-236}$$

or

$$|S_0(j\omega)|^2 = \frac{\omega^2(\omega^2+\alpha^2)}{\omega^4 + \alpha^2\omega^2 + \gamma^2 V_d/V_m}, \quad \text{5-237}$$

which shows that $|S_0(j\omega)| < 1$ for all real ω.

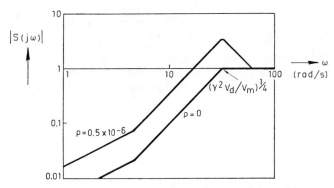

Fig. 5.12. Asymptotic Bode plots of the sensitivity function of the position control system for $\rho = 0$ and $\rho = 0.5 \times 10^{-6}$.

Figure 5.12 gives an asymptotic Bode plot of $|S(j\omega)|$ which shows that the limiting controller provides protection against all disturbances and parameter variations up to a frequency of about $(\gamma^2 V_d/V_m)^{1/4}$. With the numerical values

$$\begin{aligned} \gamma &= 0.1 \text{ kg}^{-1} \text{ m}^{-2}, \\ V_d &= 10 \text{ N}^2 \text{ m}^2 \text{ s}, \\ V_m &= 10^{-7} \text{ rad}^2 \text{ s}, \end{aligned} \quad \text{5-238}$$

this break frequency is about 31.6 rad/s.

The frequency range over which disturbance protection is obtained is reduced when the weighting factor ρ in the criterion

$$E\left\{\int_{t_0}^{t_1} [\zeta^2(t) + \rho\mu^2(t)]\, dt\right\} \qquad 5\text{-}239$$

is chosen greater than zero. It is reasonable to assume that the disturbance reduction is not affected so long as the regulator break frequency is much larger than the observer break frequency. Since the regulator break frequency (Example 5.3, Section 5.3.2) is $(\kappa/\sqrt{\rho})^{1/2}$, we conclude that with $\kappa = 0.787$ rad/(Vs^2) the value of ρ should be 0.5×10^{-6} or less (for this value of ρ the regulator break frequency is 33.4 rad/s). It can be computed, using Theorem 5.4 (Section 5.3.2), that with this value of ρ we have

$$\lim_{t\to\infty} E\{\zeta^2(t) + \rho\mu^2(t)\} = 0.00001906 \text{ rad}^2. \qquad 5\text{-}240$$

It follows that the rms input voltage is bounded by

$$\sqrt{E\{\mu^2(t)\}} \leq \sqrt{\frac{0.00001906}{\rho}} \simeq 6.17 \text{ V}, \qquad 5\text{-}241$$

which is quite an acceptable value when input amplitudes of up to 100 V are permissible. It can be calculated that the sensitivity function of the steady-state controller for this value of ρ is given by

$$S(s) = \frac{s(s + 4.6)(s^2 + 87.9s + 3859)}{(s^2 + 47.5s + 1125)(s^2 + 44.96s + 1000)}. \qquad 5\text{-}242$$

The asymptotic Bode plot of $|S(j\omega)|$ is given in Fig. 5.12 as well and is compared to the plot for $\rho = 0$. It is seen that the disturbance attenuation cutoff frequency is shifted from about 30 to about 20 rad/s, while disturbances in the frequency range near 30 rad/s are slightly amplified instead of attenuated. By making ρ smaller than 0.5×10^{-6}, the asymptotic sensitivity function can be more closely approximated.

Using the methods of Section 5.5.1, it is easy to determine the nonzero set point optimal controller for this system. Figure 5.13 gives the response of the resulting nonzero set point output feedback control system to a step of 0.1 rad in the set point of the angular position, from zero initial conditions, for the nominal parameter values, and for two sets of off-nominal values. As in Example 3.25 (Section 3.9), the off-nominal values of the plant constants α and κ are assumed to be caused by changes in the inertial load of the dc motor. It is seen that the effect of the parameter changes is moderate.

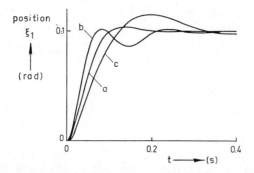

Fig. 5.13. The effect of parameter variations on the response of the output feedback position control system. (*a*) Nominal load; (*b*) inertial load $\frac{2}{3}$ of nominal; (*c*) inertial load $\frac{3}{2}$ of nominal.

5.7* LINEAR OPTIMAL OUTPUT FEEDBACK CONTROLLERS OF REDUCED DIMENSIONS

5.7.1* Introduction

In Section 5.3.1 we obtained the solution of the stochastic linear optimal output feedback regulator problem. It is immediately clear that the dimension of the controller by itself equals the dimension of the plant, since the optimal observer has the dimension of the plant. This may be a severe drawback of the design methods suggested, since in some cases a controller of much lower dimension would render quite satisfactory, although not optimal, performance. Moreover, the dimension of the mathematical model of a system is a number that very much depends on the accuracy of the model. The model may incorporate some marginal effects that drastically increase the dimension of the model without much improvement in the accuracy of the model. When this is the case, there seems to be no reason why the dimension of the controller should also be increased.

Motivated by the fact that the complexity and cost of the controller increase with its dimension, we intend to investigate in this section methods for obtaining controllers of lower dimensions than those prescribed by the methods of Section 5.3. One obvious way to approach the problem of designing controllers of low dimension is to describe the plant by a cruder mathematical model, of lower dimension. Methods are available (see e.g., Mitra, 1967; Chen and Shieh, 1968b; Davison, 1968a; Aoki, 1968; Kuppurajulu and Elangovan, 1970; Fossard, 1970; Chidambara and Schainker, 1971) for reducing the dimension of the model while retaining only the "significant modes" of the model. In this case the methods of Section 5.3 result in controllers of lower dimension. There are instances, however, in which it is not easy to achieve a reduction of the dimension of the plant. There are also situations where dimension reduction by neglecting the "parasitic" effects leads to the design of a controller that makes the actual control system unstable (Sannuti and Kokotović, 1969).

Our approach to the problem of designing low-dimensional controllers is therefore as follows. We use mathematical models for the systems which are as accurate as possible, without hesitating to include marginal effects that may or may not have significance. However we limit the dimension of the controller to some fixed number m, less than n, where n is the dimension of the plant model. In fact, we attempt to select the smallest m that still produces a satisfactory control system. We feel that this method is more dependable than that of reducing the dimension of the plant. This approach was originally suggested by Newton, Gould, and Kaiser (1957), and was further pursued by Sage and Eisenberg (1966), Sims and Melsa (1970), Johnson and Athans (1970), and others.

5.7.2* Controllers of Reduced Dimensions

Consider the system described by the equations

$$\dot{x}(t) = A(t)x(t) + B(t)u(t) + w_1(t), \qquad x(t_0) = x_0,$$
$$y(t) = C(t)x(t) + w_2(t),$$

5-243

where, as usual, $x(t)$ is an n-dimensional state vector, $u(t)$ is a k-dimensional input variable, $y(t)$ is an l-dimensional observed variable, and w_1 and w_2 are white noise processes. The joint process col (w_1, w_2) has the intensity $V(t)$. It is furthermore assumed that the initial state x_0 is a stochastic vector, uncorrelated with w_1 and w_2, with mean $\bar{x}_0$ and variance matrix Q_0.

We now consider a controller for the system given above described by

$$\dot{q}(t) = L(t)q(t) + K(t)y(t), \qquad q(t_0) = q_0,$$
$$u(t) = -F(t)q(t),$$

5-244

where q is the m-dimensional state vector of the controller. The observed variable y serves as input to the controller, and the input to the plant u is the output of the controller. It is noted that we do not allow a direct link in the controller. The reason is that a direct link causes the white observation noise w_2 to penetrate directly into the input variable u, which results in infinite input amplitudes since white noise has infinite amplitudes.

We are now in a position to formulate the linear optimal output feedback control problem for controllers of reduced dimensions (Kwakernaak and Sivan, 1971a):

5.7 Controllers of Reduced Dimensions

Definition 5.3. *Consider the system* **5-243** *with the statistical data given. Then the* **optimal output feedback control problem for a controller of reduced dimension** *is to find, for a given integer m, with $1 \leq m \leq n$, and a given final time t_1, matrix functions $L(t)$, $K(t)$, and $F(t)$, $t_0 \leq t \leq t_1$, and the probability distribution of q_0, so as to minimize σ_m, where*

$$\sigma_m = E\left\{\int_{t_0}^{t_1}[x^T(t)R_1(t)x(t) + u^T(t)R_2(t)u(t)]\,dt\right\}. \qquad 5\text{-}245$$

Here $R_1(t)$ and $R_2(t)$, $t_0 \leq t \leq t_1$, are given matrices, nonnegative-definite and positive-definite, respectively, for all t.

In the special case in which $m = n$, the solution to this problem follows from Theorem 5.3 which states that $F(t)$ and $K(t)$ in **5-244** are the optimal regulator and observer gains, respectively, and

$$L(t) = A(t) - B(t)F(t) - K(t)C(t). \qquad 5\text{-}246$$

It is easy to recognize that σ_m, $m = 1, 2, \cdots$, forms a monotonically nonincreasing sequence of numbers, that is,

$$\sigma_1 \geq \sigma_2 \geq \sigma_3 \geq \cdots, \qquad 5\text{-}247$$

since an m-dimensional controller is a special case of an $(m + 1)$-dimensional controller. Also, for $m \geq n$ the value of σ_m no longer decreases, since we know from Theorem 5.3 that the optimal controller (without restriction on its dimension) has the dimension n; thus we have

$$\sigma_1 \geq \sigma_2 \geq \sigma_3 \geq \cdots \geq \sigma_{n-1} \geq \sigma_n = \sigma_{n+1} = \sigma_{n+2} = \cdots. \qquad 5\text{-}248$$

One way to approach the problem of Definition 5.3 is to convert it to a deterministic dynamic optimization problem. This can be done as follows. Let us combine the plant equation **5-243** with the controller equation **5-244**. The control system is then described by the augmented state differential equation

$$\begin{pmatrix}\dot{x}(t)\\ \dot{q}(t)\end{pmatrix} = \begin{pmatrix}A(t) & -B(t)F(t)\\ K(t)C(t) & L(t)\end{pmatrix}\begin{pmatrix}x(t)\\ q(t)\end{pmatrix} + \begin{pmatrix}I & 0\\ 0 & K(t)\end{pmatrix}\begin{pmatrix}w_1(t)\\ w_2(t)\end{pmatrix}. \qquad 5\text{-}249$$

We now introduce the second-order joint moment matrix

$$S(t) = E\left\{\begin{pmatrix}x(t)\\ q(t)\end{pmatrix}(x^T(t), q^T(t))\right\}. \qquad 5\text{-}250$$

It follows from Theorem 1.52 (Section 1.11.2) that $S(t)$ is the solution of the matrix differential equation

$$\dot{S}(t) = M(t)S(t) + S(t)M^T(t) + N(t)V(t)N^T(t),$$
$$S(t_0) = S_0, \qquad \text{5-251}$$

where

$$M(t) = \begin{pmatrix} A(t) & -B(t)F(t) \\ K(t)C(t) & L(t) \end{pmatrix}, \quad N(t) = \begin{pmatrix} I & 0 \\ 0 & K(t) \end{pmatrix},$$
$$S_0 = E\left\{\begin{pmatrix} x_0 x_0^T & x_0 q_0^T \\ q_0 x_0^T & q_0 q_0^T \end{pmatrix}\right\}. \qquad \text{5-252}$$

Using the matrix function $S(t)$, the criterion **5-245** can be rewritten in the form

$$\sigma_m = \operatorname{tr}\left\{\int_{t_0}^{t_1} [S_{11}(t)R_1(t) + S_{22}(t)F^T(t)R_2(t)F(t)]\,dt\right\}, \qquad \text{5-253}$$

where $S_{11}(t)$ and $S_{22}(t)$ are the $n \times n$ and $m \times m$ diagonal blocks of $S(t)$, respectively.

The problem of determining the optimal behaviors of the matrix functions $L(t)$, $F(t)$, and $K(t)$ and the probability distribution of q_0 has now been reduced to the problem of choosing these matrix functions and S_0 such that σ_m as given by **5-253** is minimized, where the matrix function $S(t)$ follows from **5-251**. Application of dynamic optimization techniques to this problem (Sims and Melsa, 1970) results in a two-point boundary value problem for nonlinear matrix differential equations; this problem can be quite formidable from a computational point of view.

In order to simplify the problem, we now confine ourselves to time-invariant systems and formulate a steady-state version of the problem that is numerically more tractable and, moreover, is more easily implemented. Let us thus assume that the matrices A, B, C, V, R_1, and R_2 are constant. Furthermore, we also restrict the choice of controller to time-invariant controllers with constant matrices L, K, and F. Assuming that the interconnection of plant and controller is asymptotically stable, the limit

$$\bar{\sigma}_m = \lim_{t_0 \to -\infty} E\{x^T(t)R_1 x(t) + u^T(t)R_2 u(t)\} \qquad \text{5-254}$$

will exist. As before, the subscript m refers to the dimension of the controller. We now consider the problem of choosing the constant matrices L, K, and F (of prescribed dimensions) such that $\bar{\sigma}_m$ is minimized.

As before, we can argue that

$$\bar{\sigma}_1 \geq \bar{\sigma}_2 \geq \bar{\sigma}_3 \geq \cdots \geq \bar{\sigma}_{n-1} \geq \bar{\sigma}_n = \bar{\sigma}_{n+1} = \bar{\sigma}_{n+2} = \cdots. \qquad \text{5-255}$$

The minimal value that can ever be obtained is achieved for $m = n$, since as we know from Theorem 5.4 (Section 5.3.2) the criterion **5-254** is minimized

5.7 Controllers of Reduced Dimensions

by the interconnection of the steady-state optimal observer with the steady-state optimal control law.

The problem of minimizing 5-254 with respect to L, K, and F can be converted into a mathematical programming problem as follows. Since by assumption the closed-loop control system is asymptotically stable, that is, the constant matrix M has all its characteristic values strictly within the left-half complex plane, as $t_0 \to -\infty$ the variance matrix $S(t)$ of the augmented state approaches a constant steady-state value $\bar{S}$ that is the unique solution of the linear matrix equation

$$M\bar{S} + \bar{S}M^T + NVN^T = 0. \qquad \text{5-256}$$

Also, $\bar{\sigma}_m$ can be expressed as

$$\bar{\sigma}_m = \text{tr}\,(\bar{S}_{11}R_1 + \bar{S}_{22}F^T R_2 F), \qquad \text{5-257}$$

where $\bar{S}_{11}$ and $\bar{S}_{22}$ are the $n \times n$ and $m \times m$ diagonal blocks of $\bar{S}$, respectively.

Thus the problem of solving the steady-state version of the linear time-invariant optimal feedback control problem for controllers of reduced dimension is reduced to determining constant matrices L, K, and F of prescribed dimensions that minimize

$$\bar{\sigma}_m = \text{tr}\,(\bar{S}_{11}R_1 + \bar{S}_{22}F^T R_2 F), \qquad \text{5-258}$$

and satisfy the constraints

(i) $$M\bar{S} + \bar{S}M^T + NVN^T = 0, \qquad \text{5-259a}$$

(ii) $$\text{Re}\,[\lambda_i(M)] < 0, \quad i = 1, 2, \cdots, n + m. \qquad \text{5-259b}$$

Here the $\lambda_i(M)$, $i = 1, 2, \cdots, n + m$, denote the characteristic values of the matrix M, and Re stands for "the real part of."

It is noted that the problem of finding time-varying matrices $L(t)$, $K(t)$, and $F(t)$, $t_0 \leq t \leq t_1$, that minimize the criterion σ_m always has a solution as long as the matrix $A(t)$ is continuous, and all other matrices occurring in the problem formulation are piecewise continuous. The steady-state version of the problem, however, that is, the problem of minimizing $\bar{\sigma}_m$ with respect to the constant matrices L, K, and F, has a solution only if for the given dimension m of the controller there exist matrices L, K, and F such that the compound matrix M is asymptotically stable. For $m = n$ necessary and sufficient conditions on the matrices A, B, and C so that there exist matrices L, K, and F that render M asymptotically stable are that $\{A, B\}$ be stabilizable and $\{A, C\}$ detectable (Section 5.2.2). For $m < n$ such conditions are not known, although it is known what is the least dimension of the controller such that all closed-loop poles can be arbitrarily assigned (see, e.g., Brash and Pearson, 1970).

432 Optimal Linear Output Feedback Control Systems

In the following subsection some guidelines for the numerical determination of the matrices L, K, and F are given. We conclude this section with a note on the selection of the proper dimension of the controller. Assume that for given R_1 and R_2 the optimization problem has been solved for $m = 1, 2, \cdots, n$, and that $\bar{\sigma}_1, \bar{\sigma}_2, \cdots, \bar{\sigma}_n$ have been computed. Is it really meaningful to compare the values of $\bar{\sigma}_1, \bar{\sigma}_2, \cdots, \bar{\sigma}_n$, and thus decide upon the most desirable of m as the number that gives a sufficiently small value of $\bar{\sigma}_m$? The answer is that this is probably not meaningful since the designs all have different mean square inputs. The maximally allowable mean square input, however, is a prescribed number, which is not related to the complexity of the controller selected. Therefore, a more meaningful comparison results when for each m the weighting matrix R_2 is so adjusted that the maximally allowable mean square input is obtained. This can be achieved by letting

$$R_2 = \rho_m R_{20}, \qquad \text{5-260}$$

where ρ_m is a positive scalar and R_{20} a positive-definite weighting matrix which determines the relative importance of the components of the input. Then we rephrase our problem as follows. For given m, R_1, and R_{20}, minimize the criterion

$$\bar{\sigma}_m = \text{tr}\,(\bar{S}_{11}R_1 + \rho_m \bar{S}_{22}F^T R_{20}F), \qquad \text{5-261}$$

with respect to the constant matrices L, K, and F, subject to the constraints (i) and (ii), where ρ_m is so chosen that

$$\text{tr}\,(\bar{S}_{22}F^T R_{20}F) \qquad \text{5-262}$$

equals the given maximally allowable mean square input.

5.7.3* Numerical Determination of Optimal Controllers of Reduced Dimensions

In this section some results are given that are useful in obtaining an efficient computer program for the solution of the steady-state version of the linear time-invariant optimal output feedback control problem for a controller of reduced dimension as outlined in the preceding subsection. In particular, we describe a method for computing the gradient of the objective function (in this case $\bar{\sigma}_m$) with respect to the unknown parameters (in this case the entries of the matrices L, K, and F). This gradient can be used in any standard function minimization algorithm employing gradients, such as the conjugate gradient method or the Powell-Fletcher technique [see, e.g., Pierre (1969) or Beveridge and Schechter (1970) for extensive reviews of unconstrained optimization methods].

Gradient methods are particularly useful for solving the present function minimization problem, since the gradient can easily be computed, as we shall see. Moreover, meeting constraint (ii), which expresses that the control

5.7 Controllers of Reduced Dimensions

system be asymptotically stable, is quite simple when care is taken to choose the starting values of L, K, and F such that (ii) is satisfied, and we move with sufficiently small steps along the search directions prescribed. This is because as the boundary of the region where the control system is stable is approached, the criterion becomes infinite, and this provides a natural barrier against moving out of the stability region.

A remark on the representation of the controller is in order at this point. Clearly, the value of the criterion $\bar{\sigma}_m$ is determined only by the *external* representation of the controller, that is, its transfer matrix $F(sI - L)^{-1}K$, or, equivalently, its impulse response matrix $F \exp[L(t - \tau)]K$. It is well-known that for a given external representation many *internal* representations (in the form of a state differential equation together with an output equation) are possible. Therefore, when the optimization problem is set up starting from an internal representation of the controller, as we prefer to do, and all the entries of the matrices L, K, and F are taken as free parameters, the minimizing values of L, K, and F are not at all unique. This may give numerical difficulties. Moreover, the dimension of the function minimization problem is unnecessarily increased. These difficulties can be overcome by choosing a canonical representation of the controller equations. For example, when the controller is a single-input system, the phase canonical form of the state equations (see Section 1.9) has the minimal number of free parameters. Similarly, when the controller is a single-output system, the dual phase canonical form (see also Section 1.9) has the minimal number of free parameters. For multiinput multioutput systems related canonical forms can be used (Bucy and Ackermann, 1970). It is noted, however, that considerable reduction in the number of free parameters can often be achieved by imposing structural constraints on the controller, for example, by blocking certain feedback paths that can be expected to be of minor significance.

We discuss finally the evaluation of the gradient of $\bar{\sigma}_m$ with respect to the entries of L, K, and F. Let γ be one of the free parameters. Then introducing the matrix

$$R = \begin{pmatrix} R_1 & 0 \\ 0 & FR_2F^T \end{pmatrix}, \qquad 5\text{-}263$$

the gradient of $\bar{\sigma}_m$ with respect to γ can be written as

$$\frac{\partial \bar{\sigma}_m}{\partial \gamma} = \frac{\partial}{\partial \gamma}[\text{tr}\,(\bar{S}R)] = \text{tr}\left(\frac{\partial \bar{S}}{\partial \gamma}R + \bar{S}\frac{\partial R}{\partial \gamma}\right). \qquad 5\text{-}264$$

Furthermore, taking the partial derivative of **5-259a** with respect to the same parameter we find that

$$\frac{\partial M}{\partial \gamma}\bar{S} + M\frac{\partial \bar{S}}{\partial \gamma} + \frac{\partial \bar{S}}{\partial \gamma}M^T + \bar{S}\frac{\partial M^T}{\partial \gamma} + \frac{\partial}{\partial \gamma}(NVN^T) = 0. \qquad 5\text{-}265$$

434 Optimal Linear Output Feedback Control Systems

At this point it is convenient to introduce a linear matrix equation which is adjoint to **5-259a** and is given by

$$M^T \bar{U} + \bar{U}M + R = 0. \qquad \textbf{5-266}$$

Using the fact that for any matrices A, B, and C of compatible dimensions tr $(AB) = $ tr (BA) and tr $(C) = $ tr (C^T), we write with the aid of **5-265** and **5-266** for **5-264**

$$\frac{\partial \bar{\sigma}_m}{\partial \gamma} = \text{tr} \left[\frac{\partial \bar{S}}{\partial \gamma} (-M^T \bar{U} - \bar{U}M) + \bar{S} \frac{\partial R}{\partial \gamma} \right]$$

$$= \text{tr} \left[\bar{U} \left(-M \frac{\partial \bar{S}}{\partial \gamma} - \frac{\partial \bar{S}}{\partial \gamma} M^T \right) + \bar{S} \frac{\partial R}{\partial \gamma} \right]$$

$$= \text{tr} \left\{ \bar{U} \left[\frac{\partial M}{\partial \gamma} \bar{S} + \bar{S} \frac{\partial M^T}{\partial \gamma} + \frac{\partial}{\partial \gamma} (NVN^T) \right] + \bar{S} \frac{\partial R}{\partial \gamma} \right\}$$

$$= \text{tr} \left[2 \frac{\partial M}{\partial \gamma} \bar{S} \bar{U} + \bar{U} \frac{\partial}{\partial \gamma} (NVN^T) + \bar{S} \frac{\partial R}{\partial \gamma} \right]. \qquad \textbf{5-267}$$

Thus in order to compute the gradient of the objective function $\bar{\sigma}_m$ with respect to γ, one of the free parameters, the two linear matrix equations **5-259a** and **2-266** must be solved for $\bar{S}$ and $\bar{U}$, respectively, and the resulting values must be inserted into **5-267**. When a different parameter is considered, the bulk of the computational effort, which consists of solving the two matrix equations, need not be repeated. In Section 1.11.3 we discussed numerical methods for solving linear matrix equations of the type at hand.

Example 5.8. *Position control system*

In this example we design a position control system with a constraint on the dimension of the controller. The system to be controlled is the dc motor of Example 5.3 (Section 5.3.2), which is described by the state differential and observed variable equations

$$\dot{x}(t) = \begin{pmatrix} 0 & 1 \\ 0 & -\alpha \end{pmatrix} x(t) + \begin{pmatrix} 0 \\ \kappa \end{pmatrix} \mu(t) + \begin{pmatrix} 0 \\ \gamma \end{pmatrix} \tau_d(t), \qquad \textbf{5-268}$$

$$\eta(t) = (1, 0)x(t) + v_m(t),$$

where τ_d and v_m are described as white noise processes with intensities V_d and V_m, respectively. As in Example 5.3, we choose the criterion to be minimized as

$$\lim_{t_0 \to -\infty} E\{\zeta^2(t) + \rho \mu^2(t)\}, \qquad \textbf{5-269}$$

where $\zeta(t) = (1, 0)x(t)$ is the controlled variable. As we saw in Example 5.3, the optimal controller without limitations on its dimension is of dimension

5.7 Controllers of Reduced Dimensions

two. The only possible controller without a direct link of smaller dimension is a first-order controller, described by the scalar equations

$$\dot{q}(t) = \delta q(t) + \eta(t),$$
$$\mu(t) = -\varepsilon q(t). \quad \text{5-270}$$

Here we have taken the coefficient of $\eta(t)$ equal to 1, without loss of generality. The problem to be solved thus is: Find the constants δ and ε that minimize the criterion **5-269**.

In Example 5.3 we used the following numerical values:

$$\kappa = 0.787 \text{ rad}/(\text{V s}^2), \quad \alpha = 4.6 \text{ s}^{-1}, \quad \gamma = 0.1 \text{ kg}^{-1} \text{ m}^{-2},$$
$$V_d = 10 \text{ N}^2 \text{ m}^2 \text{ s}, \quad V_m = 10^{-7} \text{ rad}^2 \text{ s}. \quad \text{5-271}$$

For $\rho = 0.00002 \text{ rad}^2/\text{V}^2$ we found an optimal controller characterized by the data in the first column of Table 5.2.

Table 5.2 A Comparison of the Performances of the Position Control System with Controllers of Dimensions One and Two

	Second-order optimal controller with $\rho = 0.00002$	First-order optimal controller with $\rho = 0.00002$	First-order optimal controller with rms input 1.5 V
Rms input voltage (V)	1.5	1.77	1.5
Rms regulating error (rad)	0.00674	0.00947	0.0106
$E\{\zeta^2(t)\} + \rho E\{\mu^2(t)\}$ (rad^2)	$9.08 + 10^{-5}$	15.2×10^{-5}	15.8×10^{-5}
Closed-loop poles (s^{-1})	$-9.66 \pm j9.09$ $-22.48 \pm j22.24$	-400 $-2.13 \pm j11.3$	-350 $-2.15 \pm j9.92$

It is not difficult to find the parameters of the first-order controller **5-270** that minimize the criterion **5-269**. In the present case explicit expressions for the rms regulating error and input voltage can be found. Numerical or analytical evaluation of the optimal parameter values for $\rho = 0.00002$ rad^2/V^2 leads to

$$\delta = -400 \text{ s}^{-1}, \quad \varepsilon = 6.75 \times 10^4 \text{ V}/(\text{rad s}). \quad \text{5-272}$$

The performance of the resulting controller is listed in the second column of Table 5.2. It is observed that this controller results in an rms input voltage that is larger than that for the second-order optimal controller. By slightly increasing ρ a first-order controller is obtained with the same rms input voltage as the second-order controller. The third column of Table 5.2 gives the performance of this controller. It is characterized by the parameters

$$\delta = -350 \text{ s}^{-1}, \quad \varepsilon = 4.65 \times 10^4 \text{ V/(rad s)}. \qquad \text{5-273}$$

A comparison of the data of Table 5.2 shows that the first-order optimal controller has an rms regulating error that is about 1.5 times that of the second-order controller. Whether or not this is acceptable depends on the system specifications. We note that the locations of the dominating closed-loop poles at $-2.15 \pm j9.92$ of the reduced-order control system are not at all close to the locations of the dominant poles at $-9.66 \pm j9.09$ of the second-order system. Finally, we observe that the first-order controller transfer function is

$$G(s) = \frac{\varepsilon}{s - \delta} = \frac{4.65 \times 10^4}{s + 350} \text{ V/rad.} \qquad \text{5-274}$$

This controller has a very large bandwidth. Unless the bandwidth of the observation noise (which we approximated as white noise but in practice has a limited bandwidth) is larger than the bandwidth of the controller, the controller may as well be replaced with a constant gain of

$$\frac{4.65 \times 10^4}{350} \simeq 133 \text{ V/rad.} \qquad \text{5-275}$$

This suggests, however, that the optimization procedure probably should be repeated, representing the observation noise with its proper bandwidth, and searching for a zero-order controller (consisting of a constant gain).

5.8 CONCLUSIONS

In this final chapter on the design of continuous-time optimal linear feedback systems, we have seen how the results of the preceding chapters can be combined to yield optimal output feedback control systems. We have also analyzed the properties of such systems. Table 5.3 summarizes the main properties and characteristics of linear optimal output feedback control system designs of full order. Almost all of the items listed can be considered favorable features except the last two.

We first discuss the aspects of digital computation. Linear optimal control system design usually requires the use of a digital computer, but this hardly constitutes an objection because of the widespread availability of computing facilities. In fact, the need for digital computation can be converted into an

Table 5.3 Characteristics of Linear Optimal Output Feedback Control System Designs

Design characteristic	Characteristic judged favorable (+), indifferent (□), or unfavorable (−)
Stability is guaranteed	+
A good response from initial conditions and to a reference variable can be obtained	+
Information about the closed-loop poles is available	+
The input amplitude or, equivalently, the loop gain, is easily controlled	+
Good protection against disturbances can be obtained	+
Adequate protection against observation noise can be obtained	+
The control system offers protection against plant variations	+
Digital computation is usually necessary for control system design	□
The control system may turn out to be rather complex	−

advantage, since it is possible to develop computer programs that largely automate the control system design procedure and at the same time produce a great deal of detailed information about the proposed design. Table 5.4 lists several subroutines that could be contained in a computer program package for the design and analysis of time-invariant, continuous-time linear optimal control systems. Apart from the subroutines listed, such a package should contain programs for coordinating the subroutines and handling the data.

The last item in the list of Table 5.3, concerning the complexity of linear output feedback controllers, raises a substantial objection. In Section 5.7 we discussed methods for obtaining controllers of reduced complexity. At present, too little experience with such design methods is available, however, to conclude that this approach solves the complexity problem.

Altogether, the perspective that linear optimal control theory offers for the solution of real, everyday, complex linear control problems is very favorable. It truly appears that this theory is a worthy successor to traditional control theory.

438 Optimal Linear Output Feedback Control Systems

Table 5.4 Computer Subroutines for a Linear Optimal Control System Design and Analysis Package

Subroutine task	For discussion and references see
Computation of the exponential of a matrix	Section 1.3.2
Simulation of a time-invariant linear system	Section 1.3.2
Computation of the transfer matrix and characteristic values of a linear time-invariant system	Section 1.5.1
Computation of the zeroes of a square transfer matrix	Section 1.5.3
Simulation of a linear time-invariant system driven by white noise	Section 1.11.2
Solution of the linear matrix equation $M_1 X + X M_2^T = M_3$	Section 1.11.3
Solution of the algebraic Riccati equation and computation of the corresponding closed-loop regulator or observer poles	Section 3.5
Numerical determination of an optimal controller of reduced dimension	Section 5.7.3

5.9 PROBLEMS

5.1. *Angular velocity regulation system*
Consider the angular velocity system described by the state differential equation

$$\dot{\xi}(t) = -\alpha \xi(t) + \kappa \mu(t) + w_1(t). \qquad 5\text{-}276$$

Here ξ is the angular velocity, μ the driving voltage, and the disturbance w_1 is represented as white noise with intensity N. The controlled variable is the angular velocity:

$$\zeta(t) = \xi(t). \qquad 5\text{-}277$$

The observed variable is also the angular velocity:

$$\eta(t) = \xi(t) + w_2(t), \qquad 5\text{-}278$$

where w_2 is represented as white noise with intensity M. The following numerical values are assumed:

$$\alpha = 0.5 \text{ s}^{-1},$$
$$\kappa = 150 \text{ rad}/(\text{Vs}^2),$$
$$N = 600 \text{ rad}^2/\text{s}^3, \quad \quad \text{5-279}$$
$$M = 0.5 \text{ rad}^2/\text{s}.$$

Suppose that the angular velocity system is to be made into a regulator system, which keeps the angular velocity at a constant value. Determine the optimal output feedback regulator such that the rms input is 10 V. Compute the rms regulating error and compare this to the rms regulating error when no control is applied.

5.2. *Angular velocity tracking system*

Suppose that the system of Problem 5.1 is to be made into an angular velocity tracking system. For the reference variable we assume exponentially correlated noise with time constant θ and rms value σ. Furthermore, we assume that the reference variable is measured with additive white noise with intensity M_r. Compute the optimal tracking system. Assume the numerical values

$$\theta = 1 \text{ s},$$
$$\sigma = 30 \text{ rad/s}, \quad \quad \text{5-280}$$
$$M_r = 0.8 \text{ rad}^2/\text{s}^3.$$

Determine the optimal tracking system such that the total rms input is 10 V. Compute the total rms tracking error and compare this to the rms value of the reference variable.

5.3. *Nonzero set point angular velocity control system*

The tracking system of Problem 5.2 does not have the property that a constant value of the reference variable causes a zero steady-state tracking error. To obtain such a controller, design a nonzero set point controller as suggested in Section 5.5.1. For the state feedback law, choose the one obtained in Problem 5.1. Choose the prefilter such that a step of 30 rad/s in the reference variable causes a peak input voltage of 10 V or less. Compare the resulting design to that of Problem 5.2.

5.4.* *Integral control of the angular velocity regulating system*

Consider the angular velocity control system as described in Problem 5.1. Suppose that in addition to the time-varying disturbance represented by $w_1(t)$ there is also a constant disturbance $v_0(t)$ operating upon the dc motor, so that the state differential equation takes the form

$$\dot{\xi}(t) = -\alpha \xi(t) + \kappa \mu(t) + w_1(t) + v_0(t). \quad \quad \text{5-281}$$

The observed variable is given by **5-278**, while the numerical values **5-279** are assumed. The controlled variable is given by **5-277**. Design for the present situation a zero-steady-state-error controller as described in Section 5.5.2. To this end, assume that $v_0(t)$ is represented as integrated white noise and choose the intensity of this white noise as 250 rad²/s³. Compute the response of the resulting integral control system to a step of 50 rad/s² in the constant disturbance v_0 from steady-state conditions and comment on this response. What is the effect of increasing or decreasing the assumed white noise intensity of 250 rad²/s³?

5.5.* *Adjoint matrix differential equations*
Consider the matrix differential equation

$$\dot{Q}(t) = A(t)Q(t) + Q(t)A^T(t) + R(t), \qquad Q(t_0) = Q_0, \qquad \text{5-282}$$

together with the linear functional

$$\operatorname{tr}\left[\int_{t_0}^{t_1} Q(t)S(t)\, dt + Q(t_1)P_1\right]. \qquad \text{5-283}$$

Prove that **5-283** equals

$$\operatorname{tr}\left[\int_{t_0}^{t_1} P(t)R(t)\, dt + P(t_0)Q_0\right], \qquad \text{5-284}$$

where $P(t)$ is the solution of the adjoint matrix differential equation

$$-\dot{P}(t) = P(t)A(t) + A^T(t)P(t) + S(t), \qquad P(t_1) = P_1. \qquad \text{5-285}$$

5.6.* *A property of scalar sensitivity functions*
In Section 5.6 we remarked that optimal linear output feedback systems generally do not possess the property that disturbances are attenuated at all frequencies as compared to the equivalent open-loop system. For single-input single-output systems this follows from the following theorem (Bode, 1945; Westcott, 1952).

Consider a single-input single-output linear time-invariant system with transfer function $H(s)$. Let the controller transfer function (see Fig. 5.14) be given by $G(s)$ so that the control system loop gain function is

$$L(s) = H(s)G(s), \qquad \text{5-286}$$

and the sensitivity function is

$$S(s) = \frac{1}{1 + L(s)}. \qquad \text{5-287}$$

Let v denote the difference of the degree of the denominator of $L(s)$ and that of its numerator. Assume that the control system is asymptotically stable.

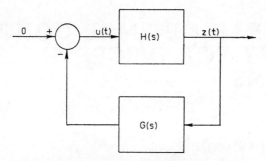

Fig. 5.14. A time-invariant linear feedback system.

Then

$$\int_{-\infty}^{\infty} \ln [|S(j\omega)|] \, d\omega = \begin{cases} \pm \infty & \text{for } \nu = 0, \\ -\gamma \dfrac{\pi}{2} & \text{for } \nu = 1, \\ 0 & \text{for } \nu \geq 2, \end{cases} \qquad 5\text{-}288$$

where

$$\gamma = \lim_{s \to \infty} sL(s). \qquad 5\text{-}289$$

Prove this result. Conclude that for plants and controllers without direct links the inequality

$$|S(j\omega)| \leq 1 \qquad 5\text{-}290$$

cannot hold for all ω. *Hint:* Integrate $\ln[S(s)]$ along a contour that consists of part of the imaginary axis closed with a semicircle in the right-half complex s-plane and let the radius of the semicircle go to infinity.

6* LINEAR OPTIMAL CONTROL THEORY FOR DISCRETE-TIME SYSTEMS

6.1 INTRODUCTION

In the first five chapters of this book, we treated in considerable detail linear control theory for continuous-time systems. In this chapter we give a condensed review of the same theory for discrete-time systems. Since the theory of linear discrete-time systems very closely parallels the theory of linear continuous-time systems, many of the results are similar. For this reason the comments in the text are brief, except in those cases where the results for discrete-time systems deviate markedly from the continuous-time situation. For the same reason many proofs are omitted.

Discrete-time systems can be classified into two types:

1. Inherently discrete-time systems, such as digital computers, digital filters, monetary systems, and inventory systems. In such systems it makes sense to consider the system at discrete instants of time only, and what happens in between is irrelevant.
2. Discrete-time systems that result from considering continuous-time systems at discrete instants of time only. This may be done for reasons of convenience (e.g., when analyzing a continuous-time system on a digital computer), or may arise naturally when the continuous-time system is interconnected with inherently discrete-time systems (such as digital controllers or digital process control computers).

Discrete-time linear optimal control theory is of great interest because of its application in computer control.

6.2 THEORY OF LINEAR DISCRETE-TIME SYSTEMS

6.2.1 Introduction

In this section the theory of linear discrete-time systems is briefly reviewed. The section is organized along the lines of Chapter 1. Many of the results stated in this section are more extensively discussed by Freeman (1965).

6.2.2 State Description of Linear Discrete-Time Systems

It sometimes happens that when dealing with a physical system it is relevant not to observe the system behavior at all instants of time t but only at a sequence of instants t_i, $i = 0, 1, 2, \cdots$. Often in such cases it is possible to characterize the system behavior by quantities defined at those instants only. For such systems the natural equivalent of the state differential equation is the *state difference equation*

$$x(i + 1) = f[x(i), u(i), i], \qquad \text{6-1}$$

where $x(i)$ is the state and $u(i)$ the input at time t_i. Similarly, we assume that the output at time t_i is given by the *output equation*

$$y(i) = g[x(i), u(i), i]. \qquad \text{6-2}$$

Linear discrete-time systems are described by state difference equations of the form

$$x(i + 1) = A(i)x(i) + B(i)u(i), \qquad \text{6-3}$$

where $A(i)$ and $B(i)$ are matrices of appropriate dimensions. The corresponding output equation is

$$y(i) = C(i)x(i) + D(i)u(i). \qquad \text{6-4}$$

If the matrices A, B, C, and D are independent of i, the system is *time-invariant*.

Example 6.1. *Savings bank account*

Let the scalar quantity $x(n)$ be the balance of a savings bank account at the beginning of the n-th month, and let α be the monthly interest rate. Also, let the scalar quantity $u(n)$ be the total of deposits and withdrawals during the n-th month. Assuming that the interest is computed monthly on the basis of the balance at the beginning of the month, the sequence $x(n), n = 0, 1, 2, \cdots$, satisfies the linear difference equation

$$x(n + 1) = (1 + \alpha)x(n) + u(n), \qquad n = 0, 1, 2, \cdots,$$
$$x(0) = x_0, \qquad \text{6-5}$$

where x_0 is the initial balance. These equations describe a linear time-invariant discrete-time system.

6.2.3 Interconnections of Discrete-Time and Continuous-Time Systems

Systems that consist of an interconnection of a discrete-time system and a continuous-time system are frequently encountered. An example of particular interest occurs when a digital computer is used to control a continuous-time plant. Whenever such interconnections exist, there must be some type of *interface system* that takes care of the communication between the discrete-time and continuous-time systems. We consider two particularly simple types

Discrete-Time Systems

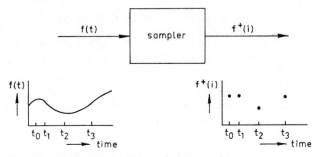

Fig. 6.1. Continuous-to-discrete-time conversion.

of interface systems, namely, *continuous-to-discrete-time* (*C-to-D*) *converters* and *discrete-to-continuous-time* (*D-to-C*) *converters*.

A C-to-D converter, also called a *sampler* (see Fig. 6.1), is a device with a continuous-time function $f(t)$, $t \geq t_0$, as input, and the sequence of real numbers $f^+(i)$, $i = 0, 1, 2, \cdots$, at times t_i, $i = 0, 1, 2, \cdots$, as output, where the following relation holds:

$$f^+(i) = f(t_i), \qquad i = 0, 1, 2, \cdots. \qquad \text{6-6}$$

The sequence of time instants t_i, $i = 0, 1, 2, \cdots$, with $t_0 < t_1 < t_2 < \cdots$, is given. In the present section we use the superscript $+$ to distinguish sequences from the corresponding continuous-time functions.

A D-to-C converter is a device that accepts a sequence of numbers $f^+(i)$, $i = 0, 1, 2, \cdots$, at given instants $t_i, i = 0, 1, 2, \cdots$, with $t_0 < t_1 < t_2 < \cdots$, and produces a continuous-time function $f(t)$, $t \geq t_0$, according to a well-defined prescription. We consider only a very simple type of D-to-C converter known as a *zero-order hold*. Other converters are described in the literature (see, e.g., Saucedo and Schiring, 1968). A zero-order hold (see Fig. 6.2) is described by the relation

$$f(t) = f^+(i), \qquad t_i \leq t < t_{i+1}, \qquad i = 0, 1, 2, \cdots. \qquad \text{6-7}$$

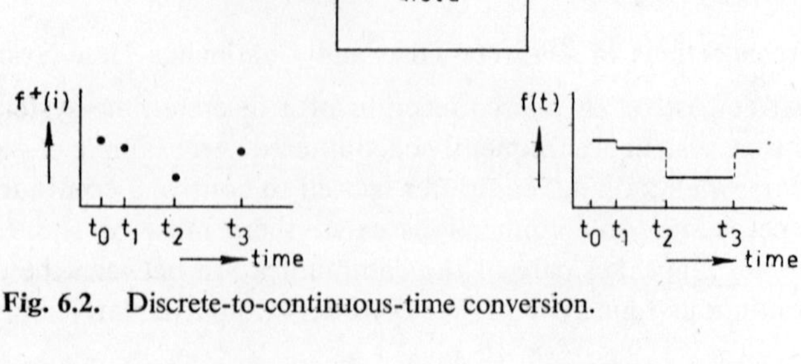

Fig. 6.2. Discrete-to-continuous-time conversion.

6.2 Linear Discrete-Time Systems

Figure 6.3 illustrates a typical example of an interconnection of discrete-time and continuous-time systems. In order to analyze such a system, it is often convenient to represent the continuous-time system together with the D-to-C converter and the C-to-D converter by an *equivalent discrete-time system*. To see how this equivalent discrete-time system can be found in a specific case, suppose that the D-to-C converter is a zero-order hold and that the C-to-D converter is a sampler. We furthermore assume that the continuous-time system of Fig. 6.3 is a linear system with state differential equation

$$\dot{x}(t) = A(t)x(t) + B(t)u(t), \qquad \text{6-8}$$

and output equation

$$y(t) = C(t)x(t) + D(t)u(t). \qquad \text{6-9}$$

Since we use a zero-order hold,

$$u(t) = u(t_i), \qquad t_i \leq t < t_{i+1}, \qquad i = 0, 1, 2, \cdots. \qquad \text{6-10}$$

Then from **1-61** we can write for the state of the system at time t_{i+1}

$$x(t_{i+1}) = \Phi(t_{i+1}, t_i)x(t_i) + \left[\int_{t_i}^{t_{i+1}} \Phi(t_{i+1}, \tau)B(\tau)\,d\tau\right]u(t_i), \qquad \text{6-11}$$

where $\Phi(t, t_0)$ is the transition matrix of the system **6-8**. This is a linear state difference equation of the type **6-3**. In deriving the corresponding output equation, we allow the possibility that the instants at which the output is sampled do not coincide with the instants at which the input is adjusted. Thus we consider the *output associated with the i-th sampling interval*, which is given by

$$y(t_i'), \qquad \text{6-12}$$

where

$$t_i \leq t_i' < t_{i+1}, \qquad \text{6-13}$$

for $i = 0, 1, 2, \cdots$. Then we write

$$y(t_i') = C(t_i')\Phi(t_i', t_i)x(t_i) + \left[C(t_i')\int_{t_i}^{t_i'} \Phi(t_i', \tau)B(\tau)\,d\tau\right]u(t_i) + D(t_i')u(t_i). \qquad \text{6-14}$$

Now replacing $x(t_i)$ by $x^+(i)$, $u(t_i)$ by $u^+(i)$, and $y(t_i')$ by $y^+(i)$, we write the system equations in the form

$$\begin{aligned} x^+(i+1) &= A_d(i)x^+(i) + B_d(i)u^+(i), \\ y^+(i) &= C_d(i)x^+(i) + D_d(i)u^+(i), \qquad i = 0, 1, 2, \cdots, \end{aligned} \qquad \text{6-15}$$

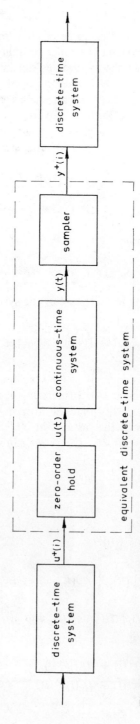

Fig. 6.3. Interconnection of discrete-time and continuous-time systems.

where

$$A_d(i) = \Phi(t_{i+1}, t_i),$$

$$B_d(i) = \int_{t_i}^{t_{i+1}} \Phi(t_{i+1}, \tau) B(\tau)\, d\tau,$$

$$C_d(i) = C(t_i')\Phi(t_i', t_i),$$

$$D_d(i) = C(t_i') \int_{t_i}^{t_i'} \Phi(t_i', \tau) B(\tau)\, d\tau + D(t_i').$$

6-16

We note that the discrete-time system defined by **6-15** has a direct link even if the continuous-time system does not have one because $D_d(i)$ can be different from zero even when $D(t_i')$ is zero. The direct link is absent, however, if $D(t) \equiv 0$ and the instants t_i' coincide with the instants t_i, that is, $t_i' = t_i$, $i = 0, 1, 2, \cdots$.

In the special case in which the sampling instants are equally spaced:

$$t_{i+1} - t_i = \Delta, \qquad \text{6-17}$$

and

$$t_i' - t_i = \Delta', \qquad \text{6-18}$$

while the system **6-8**, **6-9** is time-invariant, the discrete-time system **6-15** is also time-invariant, and

$$A_d = e^{A\Delta}, \qquad B_d = \left(\int_0^{\Delta} e^{A\tau}\, d\tau \right) B,$$

$$C_d = Ce^{A\Delta'}, \qquad D_d = C \left(\int_0^{\Delta'} e^{A\tau}\, d\tau \right) B + D.$$

6-19

We call Δ the *sampling period* and $1/\Delta$ the *sampling rate*.

Once we have obtained the discrete-time equations that represent the continuous-time system together with the converters, we are in a position to study the interconnection of the system with other discrete-time systems.

Example 6.2. *Digital positioning system*

Consider the continuous-time positioning system of Example 2.4 (Section 2.3) which is described by the state differential equation

$$\dot{x}(t) = \begin{pmatrix} 0 & 1 \\ 0 & -\alpha \end{pmatrix} x(t) + \begin{pmatrix} 0 \\ \kappa \end{pmatrix} \mu(t). \qquad \text{6-20}$$

Suppose that this system is part of a control system that is commanded by a digital computer (Fig. 6.4). The zero-order hold produces a piecewise constant input $\mu(t)$ that changes value at equidistant instants of time separated by

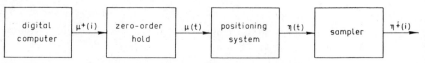

Fig. 6.4. A digital positioning system.

intervals of length Δ. The transition matrix of the system **6-20** is

$$\Phi(t, t_0) = \begin{pmatrix} 1 & \dfrac{1}{\alpha}[1 - e^{-\alpha(t-t_0)}] \\ 0 & e^{-\alpha(t-t_0)} \end{pmatrix}. \qquad \text{6-21}$$

From this it is easily found that the discrete-time description of the positioning system is given by

$$x^+(i+1) = Ax^+(i) + b\mu^+(i), \qquad \text{6-22}$$

where

$$A = \begin{pmatrix} 1 & \dfrac{1}{\alpha}(1 - e^{-\alpha\Delta}) \\ 0 & e^{-\alpha\Delta} \end{pmatrix} \qquad \text{6-23}$$

and

$$b = \begin{pmatrix} \dfrac{\kappa}{\alpha}\left(\Delta - \dfrac{1}{\alpha} + \dfrac{1}{\alpha}e^{-\alpha\Delta}\right) \\ \dfrac{\kappa}{\alpha}(1 - e^{-\alpha\Delta}) \end{pmatrix}. \qquad \text{6-24}$$

Note that we have replaced $x(t_i)$ by $x^+(i)$ and $\mu(t_i)$ by $\mu^+(i)$.

With the numerical values

$$\begin{aligned} \alpha &= 4.6 \text{ s}^{-1}, \\ \kappa &= 0.787 \text{ rad/(V s}^2), \\ \Delta &= 0.1 \text{ s}, \end{aligned} \qquad \text{6-25}$$

we obtain for the state difference equation

$$x^+(i+1) = \begin{pmatrix} 1 & 0.08015 \\ 0 & 0.6313 \end{pmatrix} x^+(i) + \begin{pmatrix} 0.003396 \\ 0.06308 \end{pmatrix} \mu^+(i). \qquad \text{6-26}$$

Let us suppose that the output variable $\eta(t)$ of the continuous-time system, where

$$\eta(t) = (1, \quad 0)x(t), \qquad \text{6-27}$$

is sampled at the instants t_i, $i = 0, 1, 2, \cdots$. Then the output equation for the discrete-time system clearly is

$$\eta^+(i) = (1,\ 0)x^+(i), \qquad \text{6-28}$$

where we have replaced $\eta(t_i)$ with $\eta^+(i)$.

Example 6.3. *Stirred tank*

Consider the stirred tank of Example 1.2 (Section 1.2.3) and suppose that it forms part of a process commanded by a process control computer. As a result, the valve settings change at discrete instants only and remain constant in between. It is assumed that these instants are separated by time intervals of constant length Δ. The continuous-time system is described by the state differential equation

$$\dot{x}(t) = \begin{pmatrix} -\dfrac{1}{2\theta} & 0 \\ 0 & -\dfrac{1}{\theta} \end{pmatrix} x(t) + \begin{pmatrix} 1 & 1 \\ \dfrac{c_1 - c_0}{V_0} & \dfrac{c_2 - c_0}{V_0} \end{pmatrix} u(t). \qquad \text{6-29}$$

It is easily found that the discrete-time description is

$$x^+(i+1) = Ax^+(i) + Bu^+(i),$$

where

$$A = \begin{pmatrix} e^{-\Delta/(2\theta)} & 0 \\ 0 & e^{-\Delta/\theta} \end{pmatrix},$$

$$B = \begin{pmatrix} 2\theta(1 - e^{-\Delta/(2\theta)}) & 2\theta(1 - e^{-\Delta/(2\theta)}) \\ \dfrac{\theta(c_1 - c_0)}{V_0}(1 - e^{-\Delta/\theta}) & \dfrac{\theta(c_2 - c_0)}{V_0}(1 - e^{-\Delta/\theta}) \end{pmatrix}. \qquad \text{6-30}$$

With the numerical data of Example 1.2, we find

$$A = \begin{pmatrix} 0.9512 & 0 \\ 0 & 0.9048 \end{pmatrix},$$

$$B = \begin{pmatrix} 4.877 & 4.877 \\ -1.1895 & 3.569 \end{pmatrix}, \qquad \text{6-31}$$

where we have chosen

$$\Delta = 5 \text{ s}. \qquad \text{6-32}$$

Example 6.4. *Stirred tank with time delay*

As an example of a system with a time delay, we again consider the stirred tank but with a slightly different arrangement, as indicated in Fig. 6.5. Here

450   Discrete-Time Systems

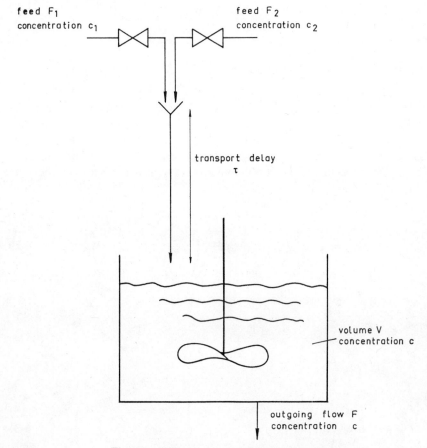

Fig. 6.5. Stirred tank with modified configuration.

the feeds are mixed *before* they flow into the tank. This would not make any difference in the dynamic behavior of the system if it were not for a transport delay τ that occurs in the common section of the pipe. Rewriting the mass balances and repeating the linearization, we find that the system equations now are

$$\dot{\xi}_1(t) = -\frac{1}{2\theta}\xi_1(t) + \mu_1(t) + \mu_2(t),$$

$$\dot{\xi}_2(t) = -\frac{1}{\theta}\xi_2(t) + \frac{c_1 - c_0}{V_0}\mu_1(t-\tau) + \frac{c_2 - c_0}{V_0}\mu_2(t-\tau),$$

6-33

where the symbols have the same meanings as in Example 1.2 (Section

1.2.3). In vector form we write

$$\dot{x}(t) = \begin{pmatrix} -\dfrac{1}{2\theta} & 0 \\ 0 & -\dfrac{1}{\theta} \end{pmatrix} x(t) + \begin{pmatrix} 1 & 1 \\ 0 & 0 \end{pmatrix} u(t) + \begin{pmatrix} 0 & 0 \\ \dfrac{c_1 - c_0}{V_0} & \dfrac{c_2 - c_0}{V_0} \end{pmatrix} u(t - \tau).$$

6-34

Note that changes in the feeds have an immediate effect on the volume but a delayed effect on the concentration.

We now suppose that the tank is part of a computer controlled process so that the valve settings change only at fixed instants separated by intervals of length Δ. For convenience we assume that the delay time τ is an exact multiple $k\Delta$ of the sampling period Δ. This means that the state difference equation of the resulting discrete-time system is of the form

$$x^+(i + 1) = Ax^+(i) + B_1 u^+(i) + B_2 u^+(i - k).$$

6-35

It can be found that with the numerical data of Example 1.2 and a sampling period

$$\Delta = 5 \text{ s},$$

6-36

A is as given by **6-31**, while

$$B_1 = \begin{pmatrix} 4.877 & 4.877 \\ 0 & 0 \end{pmatrix}, \quad B_2 = \begin{pmatrix} 0 & 0 \\ -1.1895 & 3.569 \end{pmatrix}.$$

6-37

It is not difficult to bring the difference equation **6-35** into standard state difference equation form. We illustrate this for the case $k = 1$. This means that the effect of changes in the valve settings are delayed by one sampling interval. To evaluate the effect of valve setting changes, we must therefore remember the settings of one interval ago. Thus we define an augmented state vector

$$x'(i) = \begin{pmatrix} \xi_1^+(i) \\ \xi_2^+(i) \\ \mu_1^+(i - 1) \\ \mu_2^+(i - 1) \end{pmatrix}.$$

6-38

By using this definition it is easily found that in terms of the augmented state the system is described by the state difference equation

$$x'(i + 1) = A'x'(i) + B'u^+(i),$$

6-39

where

$$A' = \begin{pmatrix} 0.9512 & 0 & 0 & 0 \\ 0 & 0.9048 & -1.1895 & 3.569 \\ 0 & 0 & 0 & 0 \\ 0 & 0 & 0 & 0 \end{pmatrix},$$

$$B' = \begin{pmatrix} 4.877 & 4.877 \\ 0 & 0 \\ 1 & 0 \\ 0 & 1 \end{pmatrix}.$$

6-40

We point out that the matrix A' has two characteristic values equal to zero. Discrete-time systems obtained by operating finite-dimensional time-invariant linear differential systems with a piecewise constant input never have zero characteristic values, since for such systems $A_d = \exp(A\Delta)$, which is always a nonsingular matrix.

6.2.4 Solution of State Difference Equations

For the solution of state difference equations, we have the following theorem, completely analogous to Theorems 1.1 and 1.3 (Section 1.3).

Theorem 6.1. *Consider the state difference equation*

$$x(i+1) = A(i)x(i) + B(i)u(i). \qquad 6\text{-}41$$

The solution of this equation can be expressed as

$$x(i) = \Phi(i, i_0)x(i_0) + \sum_{j=i_0}^{i-1} \Phi(i, j+1)B(j)u(j), \qquad i \geq i_0 + 1, \qquad 6\text{-}42$$

where $\Phi(i, i_0)$, $i \geq i_0$, *is the matrix*

$$\Phi(i, i_0) = \begin{cases} A(i-1)A(i-2)\cdots A(i_0) & \text{for } i \geq i_0 + 1, \\ I & \text{for } i = i_0. \end{cases} \qquad 6\text{-}43$$

*The **transition matrix** $\Phi(i, i_0)$ is the solution of the difference equation*

$$\Phi(i+1, i_0) = A(i)\Phi(i, i_0), \qquad i \geq i_0,$$
$$\Phi(i_0, i_0) = I. \qquad 6\text{-}44$$

If $A(i)$ does not depend upon i,

$$\Phi(i, i_0) = A^{i-i_0}. \qquad 6\text{-}45$$

6.2 Linear Discrete-Time Systems

Suppose that the system has an output

$$y(i) = C(i)x(i). \qquad \text{6-46}$$

If the initial state is zero, that is, $x(i_0) = 0$, we can write with the aid of 6-42:

$$y(i) = \sum_{j=i_0}^{i} K(i,j)u(j), \quad i \geq i_0. \qquad \text{6-47}$$

Here

$$K(i,j) = \begin{cases} C(i)\Phi(i,j+1)B(j), & j \leq i-1, \\ 0, & j = i, \end{cases} \qquad \text{6-48}$$

will be termed the *pulse response matrix* of the system. Note that for time-invariant systems K depends upon $i - j$ only. If the system has a *direct link*, that is, the output is given by

$$y(i) = C(i)x(i) + D(i)u(i), \qquad \text{6-49}$$

the output can be represented in the form

$$y(i) = \sum_{j=i_0}^{i} K(i,j)u(j), \quad i \geq i_0, \qquad \text{6-50}$$

where

$$K(i,j) = \begin{cases} C(i)\Phi(i,j+1)B(j) & \text{for } j \leq i-1, \\ D(i) & \text{for } j = i. \end{cases} \qquad \text{6-51}$$

Also in the case of time-invariant discrete-time linear systems, diagonalization of the matrix A is sometimes useful. We summarize the facts.

Theorem 6.2. *Consider the time-invariant state difference equation*

$$x(i+1) = Ax(i). \qquad \text{6-52}$$

Suppose that the matrix A has n distinct characteristic values $\lambda_1, \lambda_2, \cdots, \lambda_n$ with corresponding characteristic vectors $e_1, e_2, \cdots, e_n$. Define the $n \times n$ matrices

$$\begin{aligned} T &= (e_1, e_2, \cdots, e_n), \\ \Lambda &= \text{diag}\,(\lambda_1, \lambda_2, \cdots, \lambda_n). \end{aligned} \qquad \text{6-53}$$

Then the transition matrix of the state difference equation 6-41 can be written as

$$\Phi(i, i_0) = A^{i-i_0} = T\Lambda^{i-i_0}T^{-1}. \qquad \text{6-54}$$

Suppose that the inverse matrix T^{-1} is represented as

$$T^{-1} = \begin{pmatrix} f_1 \\ f_2 \\ \cdot \\ \cdot \\ \cdot \\ f_n \end{pmatrix}, \qquad 6\text{-}55$$

where $f_1, f_2, \cdots, f_n$ are row vectors. Then the solution of the difference equation **6-52** can be expressed as

$$x(i) = \sum_{j=1}^{n} \lambda_j^{i-i_0} e_j f_j x_0, \qquad 6\text{-}56$$

where $x_0 = x(i_0)$.

Expression **6-56** shows that the behavior of the system can be described as a composition of expanding (for $|\lambda_j| > 1$), sustained (for $|\lambda_j| = 1$), or contracting (for $|\lambda_j| < 1$) motions along the characteristic vectors $e_1, e_2, \cdots, e_n$ of the matrix A.

6.2.5 Stability

In Section 1.4 we defined the following forms of stability for continuous-time systems: stability in the sense of Lyapunov; asymptotic stability; asymptotic stability in the large; and exponential stability. All the definitions for the continuous-time case carry over to the discrete-time case if the continuous time variable t is replaced with the discrete time variable i. Time-invariant discrete-time linear systems can be tested for stability according to the following results.

Theorem 6.3. *The time-invariant linear discrete-time system*

$$x(i+1) = Ax(i) \qquad 6\text{-}57$$

is stable in the sense of Lyapunov if and only if
(a) *all the characteristic values of A have moduli not greater than 1, and*
(b) *to any characteristic value with modulus equal to 1 and multiplicity m there correspond exactly m characteristic vectors of the matrix A.*

The proof of this theorem when A has no multiple characteristic values is easily seen by inspecting **6-56**.

Theorem 6.4. *The time-invariant linear discrete-time system*

$$x(i+1) = Ax(i) \qquad 6\text{-}58$$

is asymptotically stable if and only if all of the characteristic values of A have moduli strictly less than 1.

6.2 Linear Discrete-Time Systems

Theorem 6.5. *The time-invariant linear discrete-time system*

$$x(i + 1) = Ax(i), \qquad \text{6-59}$$

is exponentially stable if and only if it is asymptotically stable.

We see that the role that the left-half complex plane plays in the analysis of continuous-time systems is taken by the inside of the unit circle for discrete-time systems. Similarly, the right-half plane is replaced with the outside of the unit circle and the imaginary axis by the unit circle itself.

Completely analogously to continuous-time systems, we define the stable subspace of a linear discrete-time system as follows.

Definition 6.1. *Consider the n-dimensional time-invariant linear discrete-time system*

$$x(i + 1) = Ax(i). \qquad \text{6-60}$$

*Suppose that A has n distinct characteristic values. Then we define the **stable subspace** of this system as the real linear subspace spanned by those characteristic vectors of A that correspond to characteristic values with moduli strictly less than 1. Similarly, the **unstable subspace** of the system is the real subspace spanned by those characteristic vectors of A that correspond to characteristic values with moduli equal to or greater than 1.*

For systems where the characteristic values of A are not all distinct, we have:

Definition 6.2. *Consider the n-dimensional time-invariant linear discrete-time system*

$$x(i + 1) = Ax(i). \qquad \text{6-61}$$

Let $\mathcal{N}_j$ be the null space of $(A - \lambda_j I)^{m_j}$, where λ_j is a characteristic value of A and m_j the multiplicity of this characteristic value in the characteristic polynomial of A. Then we define the stable subspace of the system as the real subspace of the direct sum of those null spaces $\mathcal{N}_j$ that correspond to characteristic values of A with moduli strictly less than 1. Similarly, the unstable subspace is the real subspace of the direct sum of those null spaces $\mathcal{N}_j$ that correspond to characteristic values of A with moduli greater than or equal to 1.

Example 6.5. *Digital positioning system*

It is easily found that the characteristic values of the digital positioning system of Example 6.2 (Section 6.2.3) are 1 and $\exp(-\alpha\Delta)$. As a result, the system is stable in the sense of Lyapunov but not asymptotically stable.

6.2.6 Transform Analysis of Linear Discrete-Time Systems

The natural equivalent of the Laplace transform for continuous-time variables is the z-transform for discrete-time sequences. We define the *z-transform*

$\mathbf{V}(z)$ of a sequence of vectors $v(i)$, $i = 0, 1, 2, \cdots$, as follows

$$\mathbf{V}(z) = \sum_{i=0}^{\infty} z^{-i} v(i), \qquad \text{6-62}$$

where z is a complex variable. This transform is defined for those values of z for which the sum converges.

To understand the application of the z-transform to the analysis of linear time-invariant discrete-time systems, consider the state difference equation

$$x(i + 1) = Ax(i) + Bu(i). \qquad \text{6-63}$$

Multiplication of both sides of **6-63** by z^{-i} and summation over $i = 0, 1, 2, \cdots$ yields

$$z\mathbf{X}(z) - zx(0) = A\mathbf{X}(z) + B\mathbf{U}(z), \qquad \text{6-64}$$

where $\mathbf{X}(z)$ is the z-transform of $x(i)$, $i = 0, 1, 2, \cdots$, and $\mathbf{U}(z)$ that of $u(i)$, $i = 0, 1, 2, \cdots$. Solution for $\mathbf{X}(z)$ gives

$$\mathbf{X}(z) = (zI - A)^{-1} B\mathbf{U}(z) + (zI - A)^{-1} zx(0). \qquad \text{6-65}$$

In the evaluation of $(zI - A)^{-1}$, Leverrier's algorithm (Theorem 1.18, Section 1.5.1) may be useful. Suppose that an output $y(i)$ is given by

$$y(i) = Cx(i) + Du(i). \qquad \text{6-66}$$

Transformation of this expression and substitution of **6-65** yields for $x(0) = 0$

$$\mathbf{Y}(z) = H(z)\mathbf{U}(z), \qquad \text{6-67}$$

where $\mathbf{Y}(z)$ is the z-transform of $y(i)$, $i = 0, 1, 2, \cdots$, and

$$H(z) = C(zI - A)^{-1} B + D \qquad \text{6-68}$$

is the *z-transfer matrix* of the system.

For the *inverse transformation* of z-transforms, there exist several methods for which we refer the reader to the literature (see, e.g., Saucedo and Schiring, 1968).

It is easily proved that the z-transform transfer matrix $H(z)$ is the z-transform of the pulse response matrix of the system. More precisely, let the pulse transfer matrix of time-invariant system be given by $K(i - j)$ (with a slight inconsistency in the notation). Then

$$H(z) = \sum_{i=0}^{\infty} z^{-i} K(i). \qquad \text{6-69}$$

We note that $H(z)$ is generally of the form

$$H(z) = \frac{P(z)}{\det(zI - A)}, \qquad \text{6-70}$$

where $P(z)$ is a polynomial matrix in z. The poles of the transfer matrix $H(z)$ are clearly the characteristic values of the matrix A, unless a factor of the form $z - \lambda_j$ cancels in all entries of $H(z)$, where λ_j is a characteristic value of A.

Just as in Section 1.5.3, if $H(z)$ is a square matrix, we have

$$\det [H(z)] = \frac{\psi(z)}{\phi(z)}, \qquad \text{6-71}$$

where $\phi(z)$ is the characteristic polynomial $\phi(z) = \det(zI - A)$ and $\psi(z)$ is a polynomial in z. We call the roots of $\psi(z)$ the *zeroes* of the system.

The *frequency response* of discrete-time systems can conveniently be investigated with the aid of the z-transfer matrix. Suppose that we have a complex-valued input of the form

$$u(i) = u_m e^{j\theta i}, \qquad i = 0, 1, 2, \cdots, \qquad \text{6-72}$$

where $j = \sqrt{-1}$. We refer to the quantity θ as the *normalized angular frequency*. Let us first attempt to find a particular solution to the state difference equation **6-63** of the form

$$x_p(i) = x_m e^{j\theta i}, \qquad i = 0, 1, 2, \cdots. \qquad \text{6-73}$$

It is easily found that this particular solution is given by

$$x_p(i) = (e^{j\theta}I - A)^{-1} B u_m e^{j\theta i}, \qquad i = 0, 1, 2, \cdots. \qquad \text{6-74}$$

The general solution of the *homogeneous* difference equation is

$$x_h(i) = A^i a, \qquad \text{6-75}$$

where a is an arbitrary constant vector. The general solution of the inhomogeneous state difference equation is therefore

$$x(i) = A^i a + (e^{j\theta}I - A)^{-1} B u_m e^{j\theta i}, \qquad i = 0, 1, 2, \cdots. \qquad \text{6-76}$$

If the system is asymptotically stable, the first term vanishes as $i \to \infty$; then the second term corresponds to the *steady-state response* of the state to the input **6-72**. The corresponding steady-state response of the output **6-66** is given by

$$y(i) = C(e^{j\theta}I - A)^{-1} B u_m e^{j\theta i} + D u_m e^{j\theta i}$$
$$= H(e^{j\theta}) u_m e^{j\theta i}, \qquad \text{6-77}$$

where $H(z)$ is the transfer matrix of the system.

We see that the response of the system to inputs of the type **6-72** is determined by the behavior of the z-transfer matrix for values of z on the unit circle. The steady-state response to real "sinusoidal inputs," that is, inputs of the

form
$$u(i) = \alpha \cos(i\theta) + \beta \sin(i\theta), \quad i = 0, 1, 2, \cdots, \qquad 6\text{-}78$$

can be ascertained from the moduli and arguments of the entries of $H(e^{j\theta})$. The steady-state response of an asymptotically stable discrete-time system with z-transfer matrix $H(z)$ to a constant input
$$u(i) = u_m, \quad i = 0, 1, 2, \cdots, \qquad 6\text{-}79$$
is given by
$$\lim_{i \to \infty} y(i) = H(1)u_m. \qquad 6\text{-}80$$

In the special case in which the discrete-time system is actually an equivalent description of a continuous-time system with zero-order hold and sampler, we let
$$\theta = \omega\Delta, \qquad 6\text{-}81$$
where Δ is the sampling period. The harmonic input
$$u(i) = e^{j\theta i}u_m = e^{j\omega\Delta i}u_m, \quad i = 0, 1, 2, \cdots, \qquad 6\text{-}82$$
is now the discrete-time version of the continuous-time harmonic function
$$e^{j\omega t}u_m, \quad t \geq 0, \qquad 6\text{-}83$$
from which **6-82** is obtained by sampling at equidistant instants with sampling rate $1/\Delta$.

For sufficiently small values of the angular frequency ω, the frequency response $H(e^{j\omega\Delta})$ of the discrete-time version of the system approximates the frequency response matrix of the continuous-time system. It is noted that $H(e^{j\omega\Delta})$ is periodic in ω with period $2\pi/\Delta$. This is caused by the phenomenon of *aliasing*; because of the sampling procedure, high-frequency signals are indistinguishable from low-frequency signals.

Example 6.6. *Digital positioning system*

Consider the digital positioning system of Example 6.2 (Section 6.2.3) and suppose that the position is chosen as the output:
$$y(i) = (1, \ 0)x(i). \qquad 6\text{-}84$$
It is easily found that the z-transfer function is given by
$$H(z) = \frac{0.003396z + 0.002912}{(z-1)(z-0.6313)}. \qquad 6\text{-}85$$

Figure 6.6 shows a plot of the modulus and the argument of $H(e^{j\omega\Delta})$, where $\Delta = 0.1$ s. In the same figure the corresponding plots are given of the frequency response function of the original continuous-time system, which

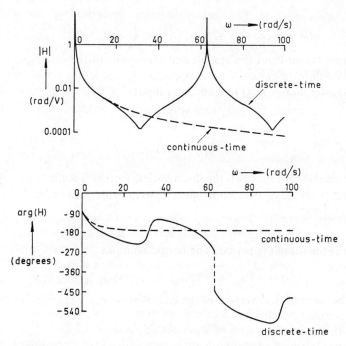

Fig. 6.6. The frequency response functions of the continuous-time and the discrete-time positioning systems.

is given by

$$\frac{0.787}{j\omega(j\omega + 4.6)}. \qquad \text{6-86}$$

We observe that for low frequencies (up to about 15 rad/s) the continuous-time and the discrete-time frequency response function have about the same modulus but that the discrete-time version has a larger phase shift. The plot also illustrates the aliasing phenomenon.

6.2.7 Controllability

In Section 1.6 we defined controllability for continuous-time systems. This definition carries over to the discrete-time case if the discrete-time variable i is substituted for the continuous-time variable t. For the controllability of time-invariant linear discrete-time systems, we have the following result which is surprisingly similar to the continuous-time equivalent.

Theorem 6.6. *The n-dimensional linear time-invariant discrete-time system with state difference equation*

$$x(i + 1) = Ax(i) + Bu(i) \qquad \text{6-87}$$

is completely controllable if and only if the column vectors of the **controllability matrix**

$$P = (B, AB, A^2B, \cdots, A^{n-1}B) \qquad 6\text{-}88$$

span the n-dimensional space.

For a proof we refer the reader to, for example, Kalman, Falb, and Arbib (1969). At this point, the following comment is in order. Frequently, complete controllability is defined as the property that any initial state can be reduced to the zero state in a finite number of steps (or in a finite length of time in the continuous-time case). According to this definition, the system with the state difference equation

$$x(i+1) = 0 \qquad 6\text{-}89$$

is completely controllable, although obviously it is not controllable in any intuitive sense. This is why we have chosen to define controllability by the requirement that the system can be brought from the zero state to any nonzero state in a finite time. In the continuous-time case it makes little difference which definition is used, but in the discrete-time it does. The reason is that in the latter case the transition matrix $\Phi(i, i_0)$, as given by **6-43**, can be singular, caused by the fact that one or more of the matrices $A(j)$ can be singular (see, e.g., the system of Example 6.4, Section 6.2.3).

The complete controllability of time-varying linear discrete-time systems can be tested as follows.

Theorem 6.7. *The linear discrete-time system*

$$x(i+1) = A(i)x(i) + B(i)u(i) \qquad 6\text{-}90$$

is completely controllable if and only if for every i_0 there exists an $i_1 \geq i_0 + 1$ such that the symmetric nonnegative-definite matrix

$$W(i_0, i_1) = \sum_{i=i_0}^{i_1-1} \Phi(i_1, i+1)B(i)B^T(i)\Phi^T(i_1, i+1) \qquad 6\text{-}91$$

is nonsingular. Here $\Phi(i, i_0)$ is the transition matrix of the system.

Uniform controllability is defined as follows.

Definition 6.3. *The time-varying system* **6-90** *is* **uniformly completely controllable** *if there exist an integer $k \geq 1$ and positive constants $\alpha_0, \alpha_1, \beta_0$, and β_1 such that*

(a) $W(i_0, i_0 + k) > 0 \quad$ *for all i_0;* $\qquad$ 6-92
(b) $\alpha_0 I \leq W^{-1}(i_0, i_0 + k) \leq \alpha_1 I \quad$ *for all i_0;* $\qquad$ 6-93
(c) $\beta_0 I \leq \Phi^T(i_0 + k, i_0)W^{-1}(i_0, i_0 + k)\Phi(i_0 + k, i_0) \leq \beta_1 I$
$\hfill$ *for all i_0.* $\quad$ 6-94

Here $W(i_0, i_1)$ is the matrix **6-91**, and $\Phi(i, i_0)$ is the transition matrix of the system.

It is noted that this definition is slightly different from the corresponding continuous-time definition. This is caused by the fact that in the discrete-time case we have avoided defining the transition matrix $\Phi(i, i_0)$ for $i < i_0$. This would involve the inverses of the matrices $A(j)$, which do not necessarily exist.

For time-invariant systems we have:

Theorem 6.8. *The time-invariant linear discrete-time system*

$$x(i+1) = Ax(i) + Bu(i) \qquad 6\text{-}95$$

is uniformly completely controllable if and only if it is completely controllable.

For time-invariant systems it is useful to define the concept of controllable subspace.

Definition 6.4. *The **controllable subspace** of the linear time-invariant discrete-time system*

$$x(i+1) = Ax(i) + Bu(i) \qquad 6\text{-}96$$

is the linear subspace consisting of the states that can be reached from the zero state within a finite number of steps.

The following characterization of the controllable subspace is quite convenient.

Theorem 6.9. *The controllable subspace of the n-dimensional time-invariant linear discrete-time system*

$$x(i+1) = Ax(i) + Bu(i) \qquad 6\text{-}97$$

is the linear subspace spanned by the column vectors of the controllability matrix P.

Discrete-time systems, too, can be decomposed into a controllable and an uncontrollable part.

Theorem 6.10. *Consider the n-dimensional linear time-invariant discrete-time system*

$$x(i+1) = Ax(i) + Bu(i). \qquad 6\text{-}98$$

Form a nonsingular transformation matrix $T = (T_1, T_2)$, where the columns of T_1 form a basis for the controllable subspace of the system, and the column vectors of T_2 together with those of T_1 span the whole n-dimensional space. Define the transformed state variable

$$x'(i) = T^{-1}x(i). \qquad 6\text{-}99$$

Then the transformed state variable satisfies the state difference equation

$$x'(i + 1) = \begin{pmatrix} A'_{11} & A'_{12} \\ 0 & A'_{22} \end{pmatrix} x'(i) + \begin{pmatrix} B'_1 \\ 0 \end{pmatrix} u(i), \qquad 6\text{-}100$$

where the pair $\{A'_{11}, B'_1\}$ is completely controllable.

Here the terminology "the pair $\{A, B\}$ is completely controllable" is shorthand for "the system $x(i + 1) = Ax(i) + Bu(i)$ is completely controllable."

Also stabilizability can be defined for discrete-time systems.

Definition 6.5. *The linear time-invariant discrete-time system*

$$x(i + 1) = Ax(i) + Bu(i) \qquad 6\text{-}101$$

*is **stabilizable** if its unstable subspace is contained in its controllable subspace.*

Stabilizability may be tested as follows.

Theorem 6.11. *Suppose that the linear time-invariant discrete-time system*

$$x(i + 1) = Ax(i) + Bu(i) \qquad 6\text{-}102$$

is transformed according to Theorem 6.10 into the form **6-100**. *Then the system is stabilizable if and only if all the characteristic values of the matrix A'_{22} have moduli strictly less than 1.*

Analogously to the continuous-time case, we define the characteristic values of the matrix A'_{11} as the *controllable poles* of the sytem, and the remaining poles as the *uncontrollable poles*. Thus a system is stabilizable if and only if all its uncontrollable poles are stable (where a *stable pole* is defined as a characteristic value of the system with modulus strictly less than 1).

6.2.8 Reconstructibility

The definition of reconstructibility given in Section 1.7 can be applied to discrete-time systems if the continuous time variable t is replaced by the discrete variable i. The reconstructibility of a time-invariant linear discrete-time system can be tested as follows.

Theorem 6.12. *The n-dimensional time-invariant linear discrete-time system*

$$\begin{aligned} x(i + 1) &= Ax(i) + Bu(i), \\ y(i) &= Cx(i), \end{aligned} \qquad 6\text{-}103$$

is completely reconstructible if and only if the row vectors of the **reconstructibility matrix**

$$Q = \begin{pmatrix} C \\ CA \\ CA^2 \\ \cdot \\ \cdot \\ \cdot \\ CA^{n-1} \end{pmatrix} \qquad 6\text{-}104$$

span the whole n-dimensional space.

A proof of this theorem can be found in Meditch (1969). For general, time-varying systems the following test applies.

Theorem 6.13. *The linear discrete-time system*

$$\begin{aligned} x(i+1) &= A(i)x(i) + B(i)u(i), \\ y(i) &= C(i)x(i) \end{aligned} \qquad 6\text{-}105$$

is completely reconstructible if and only if for every i_1 there exists an $i_0 \leq i_1 - 1$ such that the symmetric nonnegative-definite matrix

$$M(i_0, i_1) = \sum_{i=i_0+1}^{i_1} \Phi^T(i, i_0 + 1) C^T(i) C(i) \Phi(i, i_0 + 1) \qquad 6\text{-}106$$

is nonsingular. Here $\Phi(i, i_0)$ is the transition matrix of the system.

A proof of this theorem is given by Meditch (1969).

Uniform complete reconstructibility can be defined as follows.

Definition 6.6. *The time-varying system* **6-105** *is* **uniformly completely reconstructible** *if there exist an integer $k \geq 1$ and positive constants α_0, α_1, β_0, and β_1 such that*

(a) $M(i_1 - k, i_1) > 0 \quad$ for all i_1;   6-107
(b) $\alpha_0 I \leq M^{-1}(i_1 - k, i_1) \leq \alpha_1 I \quad$ for all i_1;   6-108
(c) $\beta_0 I \leq \Phi(i_1, i_1 - k) M^{-1}(i_1 - k, i_1) \Phi^T(i_1, i_1 - k) \leq \beta_1 I$
                for all i_1.   6-109

Here $M(i_0, i_1)$ is the matrix **6-106** and $\Phi(i, i_0)$ is the transition matrix of the system.

We are forced to introduce the inverse of $M(i_0, i_1)$ in order to avoid defining $\Phi(i, i_0)$ for i less than i_0.

For time-invariant systems we have:

Theorem 6.14. *The time-invariant linear discrete-time system*

$$x(i+1) = Ax(i), \qquad y(i) = Cx(i) \qquad \text{6-110}$$

is uniformly completely reconstructible if and only if it is completely reconstructible.

For time-invariant systems we introduce the concept of unreconstructible subspace.

Definition 6.7. *The **unreconstructible subspace** of the n-dimensional linear time-invariant discrete-time system*

$$\begin{aligned} x(i+1) &= Ax(i) + Bu(i), \\ y(i) &= Cx(i) \end{aligned} \qquad \text{6-111}$$

is the linear subspace consisting of the states x_0 for which

$$y(i; x_0, i_0, 0) = 0, \qquad i \geq i_0. \qquad \text{6-112}$$

Here **6-112** denotes the response of the output variable y of the system to the initial state $x(i_0) = x_0$, with $u(i) = 0$, $i \geq i_0$. The following theorem gives more information about the unreconstructible subspace.

Theorem 6.15. *The unreconstructible subspace of the linear time-invariant discrete-time system*

$$\begin{aligned} x(i+1) &= Ax(i) + Bu(i), \\ y(i) &= Cx(i) \end{aligned} \qquad \text{6-113}$$

is the null space of the reconstructibility matrix Q.

Using the concept of an unreconstructible subspace, discrete-time linear systems can also be decomposed into a reconstructible and an unreconstructible part.

Theorem 6.16. *Consider the linear time-invariant discrete-time system*

$$\begin{aligned} x(i+1) &= Ax(i) + Bu(i), \\ y(i) &= Cx(i). \end{aligned} \qquad \text{6-114}$$

Form the nonsingular transformation matrix

$$U = \begin{pmatrix} U_1 \\ U_2 \end{pmatrix}, \qquad \text{6-115}$$

where the rows of U_1 form a basis for the subspace which is spanned by the rows of the reconstructibility matrix Q of the system. U_2 is so chosen that its rows together with those of U_1 span the whole n-dimensional space. Define the transformed state variable

$$x'(t) = Ux(t). \qquad \text{6-116}$$

Then in terms of the transformed state variable the system can be represented by the state difference equation

$$x'(i+1) = \begin{pmatrix} A'_{11} & 0 \\ A'_{21} & A'_{22} \end{pmatrix} x'(i) + \begin{pmatrix} B'_1 \\ B'_2 \end{pmatrix} u(i), \qquad \textbf{6-117}$$

$$y(i) = (C'_1, \ 0) x'(i),$$

where the pair $\{A'_{11}, C'_1\}$ is completely reconstructible.

Here the terminology "the pair $\{A, C\}$ is completely reconstructible" means that the system $x(i+1) = Ax(i)$, $y(i) = Cx(i)$ is completely reconstructible. A *detectable* discrete-time system is defined as follows.

Definition 6.8. *The linear time-invariant discrete-time system*

$$\begin{aligned} x(i+1) &= Ax(i) + Bu(i), \\ y(i) &= Cx(i), \end{aligned} \qquad \textbf{6-118}$$

is **detectable** *if its unreconstructible subspace is contained within its stable subspace.*

One way of testing for detectability is through the following result.

Theorem 6.17. *Consider the linear time-invariant discrete-time system*

$$\begin{aligned} x(i+1) &= Ax(i) + Bu(i), \\ y(i) &= Cx(i). \end{aligned} \qquad \textbf{6-119}$$

Suppose that it is transformed according to Theorem 6.16 into the form **6-117**. *Then the system is detectable if and only if all the characteristic values of the matrix A'_{22} have moduli strictly less than one.*

Analogously to the continuous-time case, we define the characteristic values of the matrix A'_{11} as the *reconstructible poles*, and the characteristic values of A'_{22} as the *unreconstructible poles* of the system. Then a system is detectable if and only if all its unreconstructible poles are stable.

6.2.9 Duality

As in the continuous-time case, discrete-time regulator and filtering theory turn out to be related through duality. It is convenient to introduce the following definition.

Definition 6.9. *Consider the linear discrete-time system*

$$\begin{aligned} x(i+1) &= A(i)x(i) + B(i)u(i), \\ y(i) &= C(i)x(i). \end{aligned} \qquad \textbf{6-120}$$

In addition, consider the system

$$x^*(i+1) = A^T(i^* - i)x^*(i) + C^T(i^* - i)u^*(i),$$
$$y^*(i) = B^T(i^* - i)x^*(i), \qquad \textbf{6-121}$$

where i^* is an arbitrary fixed integer. Then the system **6-121** is termed the **dual** of the system **6-120** with respect to i^*.

Obviously, we have the following.

Theorem 6.18. *The dual of the system* **6-121** *with respect to* i^* *is the original system* **6-120**.

Controllability and reconstructibility of systems and their duals are related as follows.

Theorem 6.19. *Consider the system* **6-120** *and its dual* **6-121**:

(a) *The system* **6-120** *is completely controllable if and only if its dual is completely reconstructible.*
(b) *The system* **6-120** *is completely reconstructible if and only if its dual is completely controllable.*
(c) *Assume that* **6-120** *is time-invariant. Then* **6-120** *is stabilizable if and only if* **6-121** *is detectable.*
(d) *Assume that* **6-120** *is time-invariant. Then* **6-120** *is detectable if and only if* **6-121** *is stabilizable.*

The proof of this theorem is analogous to that of Theorem 1.41 (Section 1.8).

6.2.10 Phase-Variable Canonical Forms

Just as for continuous-time systems, phase-variable canonical forms can be defined for discrete-time systems. For single-input systems we have the following definition.

Definition 6.10. *A single-input time-invariant linear discrete-time system is in* **phase-variable canonical form** *if it is represented in the form*

$$x(i+1) = \begin{pmatrix} 0 & 1 & 0 & \cdots & 0 \\ 0 & 0 & 1 & 0 & \cdots & 0 \\ \cdots & & & & & \cdots \\ 0 & \cdots & & & 0 & 1 \\ -\alpha_0 & -\alpha_1 & \cdots & & & -\alpha_{n-1} \end{pmatrix} x(i) + \begin{pmatrix} 0 \\ 0 \\ \cdots \\ 0 \\ 1 \end{pmatrix} \mu(i),$$

$$y(i) = Cx(i). \qquad \textbf{6-122}$$

Here the α_i, $i = 0, 1, \cdots, n - 1$ are the coefficients of the characteristic polynomial

$$\sum_{i=0}^{n} \alpha_i z^i \qquad \text{6-123}$$

of the system, where $\alpha_n = 1$. Any completely controllable time-invariant linear discrete-time system can be transformed into this form by the prescription of Theorem 1.43 (Section 1.9).

Similarly we introduce for single-output systems the following definition.

Definition 6.11. *A single-output time-invariant linear discrete-time system is in **dual phase-variable canonical form** if it is represented as follows*

$$x(i+1) = \begin{pmatrix} 0 & 0 & 0 & \cdots & 0 & -\alpha_0 \\ 1 & 0 & 0 & \cdots & 0 & -\alpha_1 \\ 0 & 1 & 0 & \cdots & 0 & -\alpha_2 \\ \cdots & \cdots & \cdots & \cdots & \cdots & \cdots \\ 0 & \cdots & \cdots & 0 & 1 & -\alpha_{n-1} \end{pmatrix} x(i) + Bu(i),$$

$$\eta(i) = (0, \ 0, \cdots, 0, \ 1) x(i). \qquad \text{6-124}$$

6.2.11 Discrete-Time Vector Stochastic Processes

In this section we give a very brief discussion of discrete-time vector stochastic processes, which is a different name for infinite sequences of stochastic vector variables of the form $v(i)$, $i = \cdots, -1, 0, 1, 2, \cdots$. Discrete-time vector stochastic processes can be characterized by specifying all joint probability distributions

$$P\{v(i_1) \leq v_1, v(i_2) \leq v_2, \cdots, v(i_m) \leq v_m\} \qquad \text{6-125}$$

for all real $v_1, v_2, \cdots, v_m$, for all integers $i_1, i_2, \cdots, i_m$, and all integers m. If

$$P\{v(i_1) \leq v_1, v(i_2) \leq v_2, \cdots, v(i_m) \leq v_m\}$$
$$= P\{v(i_1 + k) \leq v_1, v(i_2 + k) \leq v_2, \cdots, v(i_m + k) \leq v_m\} \qquad \text{6-126}$$

for all real $v_1, v_2, \cdots, v_m$, for all integers $i_1, i_2, \cdots, i_m$, and for any integers m and k the process is called *stationary*. If the joint distributions **6-126** are all multidimensional Gaussian distributions, the process is termed *Gaussian*. We furthermore define:

Definition 6.12. *Consider the discrete-time vector stochastic process $v(i)$. Then we call*

$$m(i) = E\{v(i)\} \qquad \text{6-127}$$

the **mean** of the process,

$$C_v(i,j) = E\{v(i)v^T(j)\} \qquad \text{6-128}$$

the **second-order joint moment matrix**, and

$$R_v(i,j) = E\{[v(i) - m(i)][v(j) - m(j)]^T\} \qquad \text{6-129}$$

the **covariance matrix** of the process. Finally,

$$Q(i) = E\{[v(i) - m(i)][v(i) - m(i)]^T\} = R_v(i,i) \qquad \text{6-130}$$

is the **variance matrix** and $C_v(i,i)$ the **second-order moment matrix** of the process.

If the process v is stationary, its mean and variance matrix are independent of i, and its joint moment matrix $C_v(i,j)$ and its covariance matrix $R_v(i,j)$ depend upon $i - j$ only. A process that is not stationary, but that has the property that its mean is constant, its second-order moment matrix is finite for all i and its second-order joint moment matrix and covariance matrix depend on $i - j$ only, is called *wide-sense stationary*.

For wide-sense stationary discrete-time processes, we define the following.

Definition 6.13. *The* **power spectral density matrix** $\Sigma_v(\theta)$, $-\pi \leq \theta < \pi$, *of a wide-sense stationary discrete-time process v is defined as*

$$\Sigma_v(\theta) = \sum_{i=-\infty}^{\infty} z^{-i} R_v(i), \quad z = e^{j\theta}, \quad -\pi \leq \theta < \pi, \qquad \text{6-131}$$

if it exists, where $R_v(i - k)$ is the covariance matrix of the process and where $j = \sqrt{-1}$.

The name power spectral density matrix stems from its close connection with the identically named quantity for continuous-time stochastic processes. The following fact sheds some light on this.

Theorem 6.20. *Let v be a wide-sense stationary zero mean discrete-time stochastic process with power spectral density matrix $\Sigma_v(\theta)$. Then*

$$E\{v(i)v^T(i)\} = R_v(0) = \frac{1}{2\pi} \int_{-\pi}^{\pi} \Sigma_v(\theta)\, d\theta. \qquad \text{6-132}$$

A nonrigorous proof is as follows. We write

$$\begin{aligned}
\frac{1}{2\pi} \int_{-\pi}^{\pi} \Sigma_v(\theta)\, d\theta &= \frac{1}{2\pi} \int_{-\pi}^{\pi} \left(\sum_{i=-\infty}^{\infty} R_v(i) e^{-j\theta i} \right) d\theta \\
&= \sum_{i=-\infty}^{\infty} R_v(i) \left(\frac{1}{2\pi} \int_{-\pi}^{\pi} e^{-j\theta i}\, d\theta \right) \\
&= R_v(0), \qquad \text{6-133}
\end{aligned}$$

since

$$\int_{-\pi}^{\pi} e^{-j\theta i}\, d\theta = \begin{cases} 2\pi & \text{for } i = 0, \\ 0 & \text{otherwise.} \end{cases} \qquad 6\text{-}134$$

Power spectral density matrices are especially useful when analyzing the response of time-invariant linear discrete-time systems when a realization of a discrete-time stochastic process serves as the input. We have the following result.

Theorem 6.21. *Consider an asymptotically stable time-invariant linear discrete-time system with z-transfer matrix $H(z)$. Let the input to the system be a realization of a wide-sense stationary discrete-time stochastic process u with power spectral density matrix $\Sigma_u(\theta)$, which is applied from time $-\infty$ on. Then the output y is a realization of a wide-sense stationary discrete-time stochastic process with power spectral density matrix*

$$\Sigma_y(\theta) = H(e^{j\theta})\Sigma_u(\theta)H^T(e^{-j\theta}), \qquad -\pi \leq \theta < \pi. \qquad 6\text{-}135$$

Example 6.7. *Sequence of mutually uncorrelated variables*

Suppose that the stochastic process $v(i)$, $i = \cdots, -1, 0, 1, 2, \cdots$, consists of a sequence of mutually uncorrelated, zero-mean, vector-valued stochastic variables with constant variance matices Q. Then the covariance matrix of the process is given by

$$R_v(i - j) = \begin{cases} Q & \text{for } i = j, \\ 0 & \text{for } i \neq j. \end{cases} \qquad 6\text{-}136$$

This is a wide-sense stationary process. Its power spectral density matrix is

$$\Sigma_v(\theta) = Q. \qquad 6\text{-}137$$

This process is the discrete-time equivalent of white noise.

Example 6.8. *Exponentially correlated noise*

Consider the scalar wide-sense stationary, zero-mean discrete-time stochastic process v with covariance function

$$R_v(i - k) = \sigma^2 \exp\left(-\left|\frac{(i - k)\Delta}{T}\right|\right). \qquad 6\text{-}138$$

We refer to Δ as the sampling period and to T as the time constant of the process. The power spectral density function of the process is easily found to be

$$\Sigma_v(\theta) = \frac{\sigma^2(1 - e^{-2\Delta/T})}{(e^{j\theta} - e^{-\Delta/T})(e^{-j\theta} - e^{-\Delta/T})}, \qquad -\pi \leq \theta < \pi. \qquad 6\text{-}139$$

6.2.12 Linear Discrete-Time Systems Driven by White Noise

In the context of linear discrete-time systems, we often describe disturbances and other stochastically varying phenomena as the outputs of linear discrete-time systems of the form

$$x(i + 1) = A(i)x(i) + B(i)w(i),$$
$$y(i) = C(i)x(i). \qquad \text{6-140}$$

Here $x(i)$ is the state variable, $y(i)$ the output variable, and $w(i)$, $i = \cdots, -1, 0, 1, 2, \cdots$, a sequence of mutually uncorrelated, zero-mean, vector-valued stochastic vectors with variance matrix

$$E\{w(i)w^T(i)\} = V(i). \qquad \text{6-141}$$

As we saw in Example 6.7, the process w shows resemblance to the white noise process we considered in the continuous-time case, and we therefore refer to the process w as *discrete-time white noise*. We call $V(i)$ the variance matrix of the process. When $V(i)$ does not depend upon i, the discrete-time white noise process is wide-sense stationary. When $w(i)$ has a Gaussian probability distribution for each i, we refer to w as a *Gaussian discrete-time white noise process*.

Processes described by **6-140** may arise when continuous-time processes described as the outputs of continuous-time systems driven by white noise are sampled. Let the continuous-time variable $x(t)$ be described by

$$\dot{x}(t) = A(t)x(t) + B(t)w(t), \qquad \text{6-142}$$

where w is white noise with intensity $V(t)$. Then if t_i, $i = 0, 1, 2, \cdots$, is a sequence of sampling instants, we can write from **1-61**:

$$x(t_{i+1}) = \Phi(t_{i+1}, t_i)x(t_i) + \int_{t_i}^{t_{i+1}} \Phi(t_{i+1}, \tau)B(\tau)w(\tau)\, d\tau, \qquad \text{6-143}$$

where $\Phi(t, t_0)$ is the transition matrix of the differential system **6-142**. Now using the integration rules of Theorem 1.51 (Section 1.11.1) it can be seen that the quantities

$$\int_{t_i}^{t_{i+1}} \Phi(t_{i+1}, \tau)B(\tau)w(\tau)\, d\tau, \qquad \text{6-144}$$

$i = 0, 1, 2, \cdots$, form a sequence of zero mean, mutually uncorrelated stochastic variables with variance matrices

$$\int_{t_i}^{t_{i+1}} \Phi(t_{i+1}, \tau)B(\tau)V(\tau)B^T(\tau)\Phi^T(t_{i+1}, \tau)\, d\tau. \qquad \text{6-145}$$

It is observed that **6-143** is in the form **6-140**.

It is sometimes of interest to compute the variance matrix of the stochastic process x described by **6-140**. The following result is easily verified.

Theorem 6.22. *Let the stochastic discrete-time process x be the solution of the linear stochastic difference equation*

$$x(i+1) = A(i)x(i) + B(i)w(i), \qquad \text{6-146}$$

where $w(i)$, $i = -1, 0, 1, 2, \cdots$, is a sequence of mutually uncorrelated zero-mean, vector-valued stochastic variables with variance matrices $V(i)$. Suppose that $x(i_0) = x_0$ has mean m_0 and variance matrix Q_0. Then the mean of $x(i)$

$$m(i) = E\{x(i)\}, \qquad \text{6-147}$$

and the variance matrix of $x(i)$,

$$Q(i) = E\{[x(i) - m(i)][x(i) - m(i)]^T\}, \qquad \text{6-148}$$

can be given as follows. The mean is

$$m(i) = \Phi(i, i_0)m_0, \qquad i \geq i_0, \qquad \text{6-149}$$

where $\Phi(i, i_0)$ is the transition matrix of the difference equation **6-146**, *while $Q(i)$ is the solution of the matrix difference equation*

$$Q(i+1) = A(i)Q(i)A^T(i) + B(i)V(i)B^T(i), \qquad i = i_0, i_0+1, \cdots,$$
$$Q(i_0) = Q_0. \qquad \text{6-150}$$

When the matrices A, B, and V are constant, the following can be stated about the steady-state behavior of the stochastic process x.

Theorem 6.23. *Let the discrete-time stochastic process x be the solution of the stochastic difference equation*

$$x(i+1) = Ax(i) + Bw(i),$$
$$x(i_0) = x_0. \qquad \text{6-151}$$

where A and B are constant and where the uncorrelated sequence of zero-mean stochastic variables w has a constant variance matrix V. Then if all the characteristic values of A have moduli strictly less than 1, and $i_0 \to -\infty$, the covariance matrix of the process tends to an asymptotic value $\bar{R}_x(i,j)$ which depends on $i - j$ only. The corresponding asymptotic variance matrix $\bar{Q}$ is the unique solution of the matrix equation

$$\bar{Q} = A\bar{Q}A^T + BVB^T. \qquad \text{6-152}$$

In later sections we will be interested in quadratic expressions. The following results are useful.

Theorem 6.24. *Let the process x be the solution of*

$$x(i+1) = A(i)x(i) + B(i)w(i),$$
$$x(i_0) = x_0, \qquad \text{6-153}$$

where the $w(i)$ are a sequence of mutually uncorrelated zero mean stochastic variables with variance matrices $V(i)$. Let $R(i)$ be a given sequence of nonnegative-definite symmetric matrices. Then

$$E\left\{\sum_{i=i_0}^{i_1} x^T(i)R(i)x(i)\right\} = \operatorname{tr}\left[E\{x_0 x_0^T\}P(i_0) + \sum_{i=i_0}^{i_1-1} B(i)V(i)B^T(i)P(i+1)\right], \quad \text{6-154}$$

where the nonnegative-definite symmetric matrices $P(i)$ are the solution of the matrix difference equation

$$P(i) = A^T(i)P(i+1)A(i) + R(i), \qquad i = i_1 - 1, i_1 - 2, \cdots, i_0,$$
$$P(i_1) = R(i_1). \qquad \text{6-155}$$

If A and R are constant, and all the characteristic values of A have moduli strictly less than 1, $P(i)$ approaches a constant value $\bar{P}$ as $i_1 \to \infty$, where $\bar{P}$ is the unique solution of the matrix equation

$$\bar{P} = A^T \bar{P} A + R. \qquad \text{6-156}$$

One method for obtaining the solutions to the linear matrix equations **6-152** and **6-156** is repeated application of **6-150** or **6-155**. Berger (1971) gives another method. Power (1969) gives a transformation that brings equations of the type **6-152** or **6-156** into the form

$$M_1 X + X M_2^T = N_3, \qquad \text{6-157}$$

or vice versa, so that methods of solution available for one of these equations can also be used for the other (see Section 1.11.3 for equations of the type **6-157**).

A special case occurs when all stochastic variables involved are Gaussian.

Theorem 6.25. *Consider the stochastic discrete-time process x described by*

$$x(i+1) = A(i)x(i) + B(i)w(i),$$
$$x(i_0) = x_0. \qquad \text{6-158}$$

Then if the mutually uncorrelated stochastic variables $w(i)$ are Gaussian and the initial state x_0 is Gaussian, x is a Gaussian process.

Example 6.9. *Exponentially correlated noise*
Consider the stochastic process described by the scalar difference equation

$$\xi(i+1) = \alpha \xi(i) + \omega(i), \qquad \xi(i_0) = \xi_0, \qquad i_0 \to -\infty, \qquad \text{6-159}$$

where the $\omega(i)$ form a sequence of scalar uncorrelated stochastic variables with variance σ_ω^2 and where $|\alpha| < 1$. We consider ξ the output of a time-invariant discrete-time system with z-transfer function

$$\frac{1}{z - \alpha} \qquad \text{6-160}$$

6.2 Linear Discrete-Time Systems

and with the sequence ω as input. Since the power spectral density function of ω is

$$\Sigma_\omega(\theta) = \sigma_\omega^2, \qquad \text{6-161}$$

we find for the spectral density matrix of ξ, according to **6-135**,

$$\Sigma_\xi(\theta) = \frac{\sigma_\omega^2}{(e^{j\theta} - \alpha)(e^{-j\theta} - \alpha)}. \qquad \text{6-162}$$

We observe that **6-162** and **6-139** have identical appearances; therefore, **6-159** generates exponentially correlated noise. The steady-state variance σ_ξ^2 of the process ξ follows from **6-152**; in this case we have

$$\sigma_\xi^2 = \alpha^2 \sigma_\xi^2 + \sigma_\omega^2 \qquad \text{6-163}$$

or

$$\sigma_\xi^2 = \frac{\sigma_\omega^2}{1 - \alpha^2}. \qquad \text{6-164}$$

Example 6.10. *Stirred tank with disturbances*

In Example 1.37 (Section 1.11.4), we considered a continuous-time model of the stirred tank with disturbances included. The stochastic state differential equation is given by

$$\dot{x}(t) = \begin{pmatrix} -\dfrac{1}{2\theta} & 0 & 0 & 0 \\ 0 & -\dfrac{1}{\theta} & \dfrac{F_{10}}{V_0} & \dfrac{F_{20}}{V_0} \\ 0 & 0 & -\dfrac{1}{\theta_1} & 0 \\ 0 & 0 & 0 & -\dfrac{1}{\theta_2} \end{pmatrix} x(t) + \begin{pmatrix} 1 & 1 \\ \dfrac{c_{10} - c_0}{V_0} & \dfrac{c_{20} - c_0}{V_0} \\ 0 & 0 \\ 0 & 0 \end{pmatrix} u(t)$$

$$+ \begin{pmatrix} 0 & 0 \\ 0 & 0 \\ 1 & 0 \\ 0 & 1 \end{pmatrix} w(t), \qquad \text{6-165}$$

where w is white noise with intensity

$$V = \begin{pmatrix} \dfrac{2\sigma_1^2}{\theta_1} & 0 \\ 0 & \dfrac{2\sigma_2^2}{\theta_2} \end{pmatrix}. \qquad \text{6-166}$$

Here the components of the state are, respectively, the incremental volume of fluid, the incremental concentration in the tank, the incremental concentration of the feed F_1, and the incremental concentration of the feed F_2. The variations in the concentrations of the feeds are represented as exponentially correlated noise processes with rms values σ_1 and σ_2 and time constants θ_1 and θ_2, respectively.

When we assume that the system is controlled by a process computer so that the valve settings change at instants separated by intervals Δ, the discrete-time version of the system description can be found according to the method described in the beginning of this section. Since this leads to somewhat involved expressions, we give only the outcome for the numerical values of Example 1.37 supplemented with the following values:

$$\sigma_1 = 0.1 \text{ kmol/m}^3,$$
$$\sigma_2 = 0.2 \text{ kmol/m}^3,$$
$$\theta_1 = 40 \text{ s}, \qquad \text{6-167}$$
$$\theta_2 = 50 \text{ s},$$
$$\Delta = 5 \text{ s}.$$

With this the stochastic state difference equation is

$$x(i+1) = \begin{pmatrix} 0.9512 & 0 & 0 & 0 \\ 0 & 0.9048 & 0.0669 & 0.02262 \\ 0 & 0 & 0.8825 & 0 \\ 0 & 0 & 0 & 0.9048 \end{pmatrix} x(i)$$

$$+ \begin{pmatrix} 4.877 & 4.877 \\ -1.1895 & 3.569 \\ 0 & 0 \\ 0 & 0 \end{pmatrix} u(i) + w(i), \quad \text{6-168}$$

where $w(i)$, $i \geq i_0$, is a sequence of uncorrelated zero-mean stochastic vectors with variance matrix

$$\begin{pmatrix} 0 & 0 & 0 & 0 \\ 0 & 0.00004886 & 0.00009375 & 0.0001 \\ 0 & 0.00009375 & 0.002212 & 0 \\ 0 & 0.0001 & 0 & 0.007252 \end{pmatrix}. \quad \textbf{6-169}$$

By repeated application of **6-150**, it is possible to find the steady-state value $\bar{Q}$ of the variance matrix of the state. Numerically, we obtain

$$\bar{Q} = \begin{pmatrix} 0 & 0 & 0 & 0 \\ 0 & 0.00390 & 0.00339 & 0.00504 \\ 0 & 0.00339 & 0.0100 & 0 \\ 0 & 0.00504 & 0 & 0.0400 \end{pmatrix}. \quad \textbf{6-170}$$

This means that the rms value of the variations in the tank volume is zero (this is obvious, since the concentration variations do not affect the flows), the rms value of the concentration in the tank is $\sqrt{0.00390} \simeq 0.0625$ kmol/m³, and the rms values of the concentrations of the incoming feeds are 0.1 kmol/m³ and 0.2 kmol/m³, respectively. The latter two values are of course precisely σ_1 and σ_2.

6.3 ANALYSIS OF LINEAR DISCRETE-TIME CONTROL SYSTEMS

6.3.1 Introduction

In this section a brief review is given of the analysis of linear discrete-time control systems. The section closely parallels Chapter 2.

6.3.2 Discrete-Time Linear Control Systems

In this section we briefly describe discrete-time control problems, introduce the equations that will be used to characterize plant and controller, define the notions of the mean square tracking error and mean square input, and state the basic design objective. First, we introduce the *plant*, which is the system to be controlled and which is represented as a linear discrete-time system characterized by the equations

$$x(i + 1) = A(i)x(i) + B(i)u(i) + v_p(i),$$
$$x(i_0) = x_0,$$
$$y(i) = C(i)x(i) + E_1(i)u(i) + v_m(i),$$
$$z(i) = D(i)x(i) + E_2(i)u(i),$$
$$\text{for } i = i_0, i_0 + 1, \cdots.$$

6-171

Here x is the *state* of the plant, x_0 the *initial state*, u the *input variable*, y the *observed variable*, and z the *controlled variable*. Furthermore v_p represents the *disturbance variable* and v_m the *observation noise*. Finally, we associate with the plant a *reference variable* $r(i)$, $i = i_0, i_0 + 1, \cdots$. It is noted that in contrast to the continuous-time case we allow both the observed variable and the controlled variable to have a direct link from the plant input. The reason is that direct links easily arise in discrete-time systems obtained by sampling continuous-time systems where the sampling instants of the output variables do not coincide with the instants at which the input variable changes value (see Section 6.2.3). As in the continuous-time case, we consider separately *tracking problems*, where the controlled variable $z(i)$ is to follow a time-varying reference variable $r(i)$, and *regulator problems*, where the reference variable is constant or slowly varying.

Analogously to the continuous-time case, we consider *closed-loop* and *open-loop controllers*. The general closed-loop controller is taken as a linear discrete-time system described by the state difference equation and the output equation

$$q(i + 1) = L(i)q(i) + K_r(i)r(i) - K_f(i)y(i),$$
$$u(i) = F(i)q(i) + H_r(i)r(i) - H_f(i)y(i).$$

6-172

We note that these equations imply that the controller is able to process the input data $r(i)$ and $y(i)$ instantaneously while generating the plant input $u(i)$. If there actually are appreciable processing delays, such as may be the case in computer control when high sampling rates are used, we assume that these delays have been accounted for when setting up the plant equations (see Section 6.2.3).

The general open-loop controller follows from **6-172** with K_f and H_f identical to zero.

Closely following the continuous-time theory, we judge the performance of a control system, open- or closed-loop, in terms of its *mean square tracking error* and its *mean square input*. The mean square tracking error is defined as

$$C_e(i) = E\{e^T(i)W_e(i)e(i)\},$$

6-173

6.3 Linear Discrete-Time Control Systems

where

$$e(i) = z(i) - r(i). \qquad \text{6-174}$$

$W_e(i)$ is a nonnegative-definite symmetric weighting matrix. Similarly, the mean square input is defined as

$$C_u(i) = E\{u^T(i)W_u(i)u(i)\}, \qquad \text{6-175}$$

where $W_u(i)$ is another nonnegative-definite weighting matrix. Our *basic objective* in designing a control system is to *reduce the mean square tracking error as much as possible, while at the same time keeping the mean square input down to a reasonable value.*

As in the continuous-time case, a requirement of primary importance is contained in the following design rule.

Design Objective 6.1. *A control system should be asymptotically stable.*

Discrete-time control systems, just as continuous-time control systems, have the property that an unstable plant can be stabilized by closed-loop control but never by open-loop control.

Example 6.11. *Digital position control system with proportional feedback*

As an example, we consider the digital positioning system of Example 6.2 (Section 6.2.3). This system is described by the state difference equation

$$x(i+1) = \begin{pmatrix} 1 & 0.08015 \\ 0 & 0.6313 \end{pmatrix} x(i) + \begin{pmatrix} 0.003396 \\ 0.06308 \end{pmatrix} \mu(i). \qquad \text{6-176}$$

Here the first component $\xi_1(i)$ of $x(i)$ is the angular position, and the second component $\xi_2(i)$ the angular velocity. Furthermore, $\mu(i)$ is the input voltage. Suppose that this system is made into a position servo by using proportional feedback as indicated in Fig. 6.7. Here the controlled variable $\zeta(i)$ is the position, and the input voltage is determined by the relation

$$\mu(i) = \lambda[r(i) - \zeta(i)]. \qquad \text{6-177}$$

In this expression $r(i)$ is the reference variable and λ a gain constant. We assume that there are no processing delays, so that the sampling instant of the

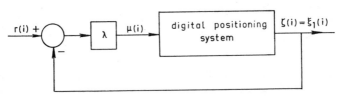

Fig. 6.7. A digital positioning system with proportional feedback.

output variable coincides with the instant at which a new control interval is initiated. Thus we have

$$\zeta(i) = (1, \ 0)x(i). \qquad 6\text{-}178$$

In Example 6.6 (Section 6.2.6), it was found that the open-loop z-transfer function of the plant is given by

$$H(z) = \frac{0.003396(z + 0.8575)}{(z - 1)(z - 0.6313)}. \qquad 6\text{-}179$$

By using this it is easily found that the characteristic polynomial of the closed-loop system is given by

$$(z - 1)(z - 0.6313) + 0.003396\lambda(z + 0.8575). \qquad 6\text{-}180$$

In Fig. 6.8 the loci of the closed-loop roots are sketched. It is seen that when λ changes from 100 to 150 V/rad the closed-loop poles leave the unit circle,

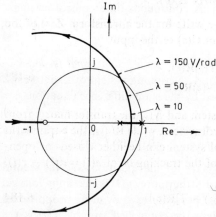

Fig. 6.8. The root loci of the digital position control system. ×, Open-loop poles; ○, open-loop zero.

hence the closed-loop system becomes unstable. Furthermore, it is to be expected that, in the stable region, as λ increases the system becomes more and more oscillatory since the closed-loop poles approach the unit circle more and more closely. To avoid resonance effects, while maximizing λ, the value of λ should be chosen somewhere between 10 and 50 V/rad.

6.3.3 The Steady-State and the Transient Analysis of the Tracking Properties

In this section the response of a linear discrete-time control system to the reference variable is studied. Both the steady-state response and the transient response are considered. The following assumptions are made.

1. *Design Objective 6.1 is satisfied, that is, the control system is asymptotically stable.*
2. *The control system is time-invariant and the weighting matrices W_e and W_u are constant.*
3. *The disturbance variable v_p and the observation noise v_m are identical to zero.*
4. *The reference variable can be represented as*

$$r(i) = r_0 + r_v(i), \qquad i = i_0, i_0 + 1, \cdots, \qquad \text{6-181}$$

where the **constant part** r_0 is a stochastic vector with second-order moment matrix

$$E\{r_0 r_0^T\} = R_0, \qquad \text{6-182}$$

and the **variable part** r_v is a wide-sense stationary zero-mean vector stochastic process with power spectral density matrix $\Sigma_r(\theta)$.

Assuming zero initial conditions, we write for the z-transform $\mathbf{Z}(z)$ of the controlled variable and the z-transform $\mathbf{U}(z)$ of the input

$$\begin{aligned} \mathbf{Z}(z) &= T(z)\mathbf{R}(z), \\ \mathbf{U}(z) &= N(z)\mathbf{R}(z). \end{aligned} \qquad \text{6-183}$$

Here $T(z)$ is the *transmission* of the system and $N(z)$ the transfer matrix from reference variable to input of the control system, while $\mathbf{R}(z)$ is the z-transform of the reference variable. The control system can be either closed- or open-loop. Thus if $\mathbf{E}(z)$ is the z-transform of the tracking error $e(i) = z(i) - r(i)$, we have

$$\mathbf{E}(z) = [T(z) - I]\mathbf{R}(z). \qquad \text{6-184}$$

To derive expressions for the steady-state mean square tracking error and input, we study the contributions of the constant part and the variable part of the reference variable separately. The constant part of the reference variable yields a steady-state response of the tracking error and the input as follows:

$$\begin{aligned} \lim_{i \to \infty} e(i) &= [T(1) - I]r_0, \\ \lim_{i \to \infty} u(i) &= N(1)r_0. \end{aligned} \qquad \text{6-185}$$

From Section 6.2.11 it follows that in steady-state conditions the response of the tracking error to the variable part of the reference variable has the power spectral density matrix

$$[T(e^{j\theta}) - I]\Sigma_r(\theta)[T(e^{-j\theta}) - I]^T. \qquad \text{6-186}$$

Consequently, the *steady-state mean square tracking error* can be expressed as

$$C_{e\infty} = \lim_{i \to \infty} C_e(i)$$

$$= E\{r_0^T[T(1) - I]^T W_e[T(1) - I]r_0\}$$

$$+ \text{tr}\left\{\frac{1}{2\pi}\int_{-\pi}^{\pi}[T(e^{j\theta}) - I]\Sigma_r(\theta)[T(e^{-j\theta}) - I]^T W_e \, d\theta\right\}. \quad \textbf{6-187}$$

This expression can be rewritten as

$$C_{e\infty} = \text{tr}\left\{[T(1) - I]^T W_e[T(1) - I]R_0\right.$$

$$\left.+ \frac{1}{2\pi}\int_{-\pi}^{\pi}[T(e^{-j\theta}) - I]^T W_e[T(e^{j\theta}) - I]\Sigma_r(\theta) \, d\theta\right\}. \quad \textbf{6-188}$$

Similarly, the *steady-state mean square input* can be expressed in the form

$$C_{u\infty} = \lim_{i \to \infty} C_u(i)$$

$$= \text{tr}\left\{N^T(1)W_u N(1)R_0 + \frac{1}{2\pi}\int_{-\pi}^{\pi} N^T(e^{-j\theta})W_u N(e^{j\theta})\Sigma_r(\theta) \, d\theta\right\}. \quad \textbf{6-189}$$

Before further analyzing these expressions, we introduce the following additional assumption.

5. *The constant part and the variable part of the reference variable have uncorrelated components, that is, both R_0 and $\Sigma_r(\theta)$ are diagonal and can be written in the form*

$$R_0 = \text{diag}(R_{0,1}, R_{0,2}, \cdots, R_{0,p}),$$
$$\Sigma_r(\theta) = \text{diag}[\Sigma_{r,1}(\theta), \Sigma_{r,2}(\theta), \cdots, \Sigma_{r,p}(\theta)], \quad \textbf{6-190}$$

where p is the dimension of the reference variable and the controlled variable.

With this assumption we write for **6-188**:

$$C_{e\infty} = \sum_{i=1}^{p} R_{0,i}\{[T(1) - I]^T W_e[T(1) - I]\}_{ii}$$

$$+ \frac{1}{2\pi}\sum_{i=1}^{p}\int_{-\pi}^{\pi}\Sigma_{r,i}(\theta)\{[T(e^{-j\theta}) - I]^T W_e[T(e^{j\theta}) - I]\}_{ii} \, d\theta, \quad \textbf{6-191}$$

where $\{M\}_{ii}$ denotes the i-th diagonal entry of the matrix M. Following Chapter 2, we now introduce the following notions.

Definition 6.14. *Let $\rho(i)$, $i = \cdots, -1, 0, 1, 2, \cdots$, be a scalar wide-sense stationary discrete-time stochastic process with power spectral density function $\Sigma_\rho(\theta)$. Then the **normalized frequency band** Θ of this process is defined as the* set of normalized frequencies θ, $0 \leq \theta \leq \pi$, for which

$$\Sigma_\rho(e^{j\theta}) \geq \alpha. \qquad 6\text{-}192$$

Here α is so chosen that the frequency band contains a given fraction $1 - \varepsilon$, where ε is small with respect to 1, of half the power of the process, that is,

$$\int_{\theta \in \Theta} \Sigma_\rho(\theta)\, d\theta = (1 - \varepsilon) \int_0^\pi \Sigma_\rho(\theta)\, d\theta. \qquad 6\text{-}193$$

As in Chapter 2, when the frequency band is an interval $[\theta_1, \theta_2]$, we define $\theta_2 - \theta_1$ as the *normalized bandwidth* of the process. When the frequency band is an interval $[0, \theta_c]$, we define θ_c as the *normalized cutoff frequency* of the process.

In the special case where the discrete-time process is derived from a continuous-time process by sampling, the (not normalized) bandwidth and cutoff frequency follow from the corresponding normalized quantities by the relation

$$\omega = \theta/\Delta, \qquad 6\text{-}194$$

where Δ is the sampling period and ω the (not normalized) angular frequency.

Before returning to our discussion of the steady-state mean square tracking error we introduce another concept.

Definition 6.15. *Let $T(z)$ be the transmission of an asymptotically stable time-invariant linear discrete-time control system. Then we define the **normalized frequency band of the i-th link** of the control system as the set of normalized frequencies θ, $0 \leq \theta \leq \pi$, for which*

$$\{[T(e^{-j\theta}) - I]^T W_e [T(e^{j\theta}) - I]\}_{ii} \leq \varepsilon^2 W_{e,ii}. \qquad 6\text{-}195$$

Here ε is a given number which is small with respect to 1, W_e is the weighting matrix for the mean square tracking error, and $W_{e,ii}$ the i-th diagonal entry of W_e.

Here as well we speak of the *bandwidth* and the *cutoff frequency* of the i-th link, if they exist. If the discrete-time system is derived from a continuous-time system by sampling, the (not normalized) bandwidth and cutoff frequency can be obtained by the relation **6-194**.

We can now phrase the following advice, which follows from a consideration of **6-191**.

Design Objective 6.2. *Let $T(z)$ be the $p \times p$ transmission of an asymptotically stable time-invariant linear discrete-time control system, for which both the constant and the variable part of the reference variable have uncorrelated components. Then in order to obtain a small steady-state mean square tracking error, the frequency band of each of the p links should contain the frequency* band of the corresponding component of the reference variable. If the i-th component of the reference variable, $i = 1, 2, \cdots, p$, is likely to have a nonzero constant part, $\{[T(1) - I]^T W_e [T(1) - I]\}_{ii}$ should be small, preferably zero.

Let us now consider the steady-state mean square input as given by **6-189**. Under assumption 5 this expression can be rewritten as

$$C_{u\infty} = \sum_{i=1}^{p} R_{0,i} \{N^T(1) W_u N(1)\}_{ii}$$

$$+ \frac{1}{2\pi} \sum_{i=1}^{p} \int_{\theta=-\pi}^{\pi} \Sigma_{r,i}(\theta) \{N^T(e^{-j\theta}) W_u N(e^{j\theta})\}_{ii} \, d\theta. \qquad \textbf{6-196}$$

Since $C_{u\infty}$ should not be made too large, we extract the following advice.

Design Objective 6.3. *In order to obtain a small steady-state mean square input in an asymptotically stable time-invariant linear discrete-time control system with a p-dimensional reference variable with uncorrelated components,*

$$\{N^T(e^{-j\theta}) W_u N(e^{j\theta})\}_{ii} \qquad \textbf{6-197}$$

should be made small over the normalized frequency band of the i-th component of the reference variable, for $i = 1, 2, \cdots, p$.

As in Chapter 2, we do not impose restrictions on the first term of **6-196** because only the fluctuations of the input variable about its set point need be considered.

We conclude this section with a discussion of the *transient* behavior of the response of the control system to the reference variable. As in the continuous-time case, we define the *settling time* of the mean square tracking error, the mean square input, or any other quantity, as *the time it takes this quantity to reach its steady-state value within a specified accuracy*. This settling time can be expressed as a number of intervals, or in seconds when the sampling interval is known. Obviously, it is desirable that the mean square tracking error of a control system settle down to its steady-state value as soon as possible after start-up or after upsets. We thus have the following design rule.

Design Objective 6.4. *The settling time of the mean square tracking error of a discrete-time control system should be as short as possible.*

The transient behavior of the mean square tracking error, the mean square input, and other quantities of interest can be computed in a manner similar to the continuous-time approach. For the various stochastic processes that influence the evolution of the control system, mathematical models are assumed in the form of discrete-time systems driven by discrete-time white noise. The variance matrix of the state of the system that results by augmenting the control system difference equation with these models can be computed according to Theorem 6.22 (Section 6.2.12). This variance matrix yields all the data required. The example at the end of this section illustrates the procedure. Often, however, a satisfactory estimate of the settling time of a given quantity can be obtained by evaluating the transient behavior of the response of the control system to the constant part of the reference variable alone; this then becomes a simple matter of computing step responses.

For time-invariant control systems, information about the settling time can often be derived from the location of the closed-loop characteristic values of the system. From Section 6.2.4 we know that all responses are linear combinations of functions of the form λ^i, $i = i_0, i_0 + 1, \cdots$, where λ is a characteristic value. Since the time it takes $|\lambda|^i$ to reach 1 % of its initial value of 1 is (assuming that $|\lambda| < 1$)

$$\frac{2}{\log_{10}\left(\frac{1}{|\lambda|}\right)} \qquad \text{6-198}$$

time intervals, an estimate of the 1 % settling time of an asymptotically stable linear time-invariant discrete-time control system is

$$\max_i \left\{ \frac{2}{\log_{10}\left(\frac{1}{|\lambda_i|}\right)} \right\} \qquad \text{6-199}$$

time intervals, where λ_l, $l = 1, 2, \cdots, n$, are the characteristic values of the control system. As with continuous-time systems, this formula may give misleading results inasmuch as some of the characteristic values may not appear in the response of certain variables.

We conclude this section by pointing out that when a discrete-time control system is used to describe a sampled continuous-time system the settling time as obtained from the discrete-time description may give a completely erroneous impression of the settling time for the continuous-time system. This is because it occasionally happens that a sampled system exhibits quite satisfactory behavior at the sampling instants, while *between* the sampling instants large overshoots appear that do not settle down for a long time. We shall meet examples of such situations in later sections.

Example 6.12. *Digital position control system with proportional feedback*

We illustrate the results of this section for a single-input single-output system only, for which we take the digital position control system of Example 6.11. Here the steady-state tracking properties can be analyzed by considering the scalar transmission $T(z)$, which is easily computed and turns out to be given by

$$T(z) = \frac{0.003396\lambda(z + 0.8575)}{(z - 1)(z - 0.6313) + 0.003396\lambda(z + 0.8575)}.\qquad \text{6-200}$$

In Fig. 6.9 plots are given of $|T(e^{j\omega\Delta})|$ for $\Delta = 0.1$ s, and for values of λ between 5 and 100 V/rad. It is seen from these plots that the most favorable value of λ is about 15 V/rad; for this value the system bandwidth is maximal without the occurrence of undesirable resonance effects.

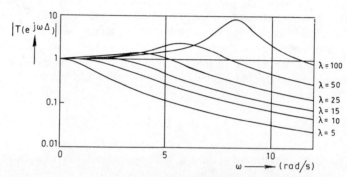

Fig. 6.9. The transmissions of the digital position control system for various values of the gain factor λ.

To compute the mean square tracking error and the mean square input voltage, we assume that the reference variable can be described by the model

$$r(i + 1) = 0.9802 r(i) + w(i). \qquad \text{6-201}$$

Here w forms a sequence of scalar uncorrelated stochastic variables with variance 0.0392 rad². With a sampling interval of 0.1 s, this represents a sampled exponentially correlated noise process with a time constant of 5 s. The steady-state rms value of r can be found to be 1 rad (see Example 6.9).

With the simple feedback scheme of Example 6.11, the input to the plant is given by

$$\mu(i) = \lambda r(i) - \lambda \xi_1(i), \qquad \text{6-202}$$

which results in the closed-loop difference equation

$$x(i + 1) = \begin{pmatrix} 0.94906 & 0.08015 \\ -0.9462 & 0.6313 \end{pmatrix} x(i) + \begin{pmatrix} 0.05094 \\ 0.9462 \end{pmatrix} r(i). \qquad \text{6-203}$$

Here the value $\lambda = 15$ V/rad has been substituted. Augmenting this equation

6.3 Linear Discrete-Time Control Systems

with **6-201**, we obtain

$$\begin{pmatrix} \xi_1(i+1) \\ \xi_2(i+1) \\ r(i+1) \end{pmatrix} = \begin{pmatrix} 0.94906 & 0.08015 & 0.05094 \\ -0.9462 & 0.6313 & 0.9462 \\ 0 & 0 & 0.9802 \end{pmatrix} \begin{pmatrix} \xi_1(i) \\ \xi_2(i) \\ r(i) \end{pmatrix} + \begin{pmatrix} 0 \\ 0 \\ 1 \end{pmatrix} w(i). \qquad \text{6-204}$$

We now define the variance matrix

$$Q(i) = E\left\{ \begin{pmatrix} \xi_1(i) \\ \xi_2(i) \\ r(i) \end{pmatrix} (\xi_1(i), \xi_2(i), r(i)) \right\}. \qquad \text{6-205}$$

Here it is assumed that $E\{x(i_0)\} = 0$ and $E\{r(i_0)\} = 0$, so that $x(i)$ and $r(i)$ have zero means for all i. Denoting the entries of $Q(i)$ as $Q_{jk}(i)$, $j, k = 1, 2, 3$, the mean square tracking error can be expressed as

$$\begin{aligned} C_e(i) &= E\{[\xi_1(i) - r(i)]^2\} \\ &= E\{\xi_1^2(i)\} - 2E\{\xi_1(i)r(i)\} + E\{r^2(i)\} \\ &= Q_{11}(i) - 2Q_{13}(i) + Q_{33}(i) \\ &= \operatorname{tr}\left\{ Q(i) \begin{pmatrix} 1 & 0 & -1 \\ 0 & 0 & 0 \\ -1 & 0 & 1 \end{pmatrix} \right\}. \end{aligned} \qquad \text{6-206}$$

For the mean square input, we have

$$C_u(i) = E\{\mu^2(i)\} = E\{\lambda^2[r(i) - \xi_1(i)]^2\} = \lambda^2 C_e(i). \qquad \text{6-207}$$

For the variance matrix $Q(i)$, we obtain from Theorem 6.22 the matrix difference equation

$$Q(i+1) = MQ(i)M^T + NVN^T, \qquad \text{6-208}$$

where M is the 3×3 matrix and N the 3×1 matrix in **6-204**. V is the variance of $w(i)$. For the initial condition of this matrix difference equation, we choose

$$Q(0) = \begin{pmatrix} 0 & 0 & 0 \\ 0 & 0 & 0 \\ 0 & 0 & 1 \end{pmatrix}. \qquad \text{6-209}$$

This choice of $Q(0)$ implies that at $i = 0$ the plant is at rest, while the initial variance of the reference variable equals the steady-state variance 1 rad². Figure 6.10 pictures the evolution of the rms tracking error and the rms input voltage. It is seen that the settling time is somewhere between 10 and 20 sampling intervals.

It is also seen that the steady-state rms tracking error is nearly 0.4 rad, which is quite a large value. This means that the reference variable is not very well tracked. To explain this we note that continuous-time exponentially correlated noise with a time constant of 5 s (from which the reference variable is

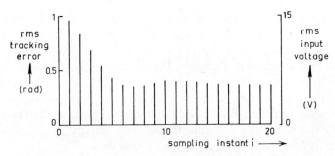

Fig. 6.10. Rms tracking error and rms input voltage for the digital position control system.

derived) has a 1 % cutoff frequency of $63.66/5 = 12.7$ rad/s (see Section 2.5.2). The digital position servo is too slow to track this reference variable properly since its 1 % cutoff frequency is perhaps 1 rad/s. We also see, however, that the steady-state rms input voltage is about 4 V. By assuming that the maximally allowable rms input voltage is 25 V, it is clear that there is considerable room for improvement.

Finally, in Fig. 6.11 we show the response of the position digital system to a step of 1 rad in the reference variable. This plot confirms that the settling time of the tracking error is somewhere between 10 and 20 time intervals,

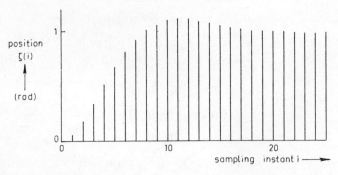

Fig. 6.11. The response of the digital position control system to a step in the reference variable of 1 rad.

depending upon the accuracy required. From the root locus of Fig. 6.8, we see that the distance of the closed-loop poles from the origin is about 0.8. The corresponding estimated 1% settling time according to **6-199** is 20.6 time intervals.

6.3.4 Further Aspects of Linear Discrete-Time Control System Performance

In this section we briefly discuss other aspects of the performance of linear discrete-time control systems. They are: *the effect of disturbances; the effect of observation noise; and the effect of plant parameter uncertainty*. We can carry out an analysis very similar to that for the continuous-time case. We very briefly summarize the results of this analysis. To describe the effect of the disturbances on the mean square tracking error in the single-input single-output case, it turns out to be useful to introduce the *sensitivity function*

$$S(z) = \frac{1}{1 + H(z)G(z)}, \qquad \text{6-210}$$

where

$$H(z) = D(zI - A)^{-1}B + E \qquad \text{6-211}$$

is the open-loop transfer function of the plant, and

$$G(z) = F(zI - L)^{-1}K_f + H_f \qquad \text{6-212}$$

is the transfer function of the feedback link of the controller. Here it is assumed that the controlled variable of the plant is also the observed variable, that is, in **6-171** $C = D$ and $E_1 = E_2 = E$. To reduce the effect of the disturbances, it turns out that $|S(e^{j\theta})|$ must be made small over the frequency band of the equivalent disturbance at the controlled variable. If

$$|S(e^{j\theta})| \leq 1 \qquad \text{for all } 0 \leq \theta < \pi, \qquad \text{6-213}$$

the closed-loop system always reduces the effect of disturbances, no matter what their statistical properties are. If constant disturbances are to be suppressed, $S(1)$ should be made small (this statement is not true without qualification if the matrix A has a characteristic value at 1). In the case of a multiinput multioutput system, the sensitivity function **6-210** is replaced with the *sensitivity matrix*

$$S(z) = [I + H(z)G(z)]^{-1}, \qquad \text{6-214}$$

and the condition **6-213** is replaced with the condition

$$S^T(e^{-j\theta})W_e S(e^{j\theta}) \leq W_e \qquad \text{for all } 0 \leq \theta < \pi, \qquad \text{6-215}$$

where W_e is the weighting matrix of the mean square tracking error.

In the scalar case, making $S(e^{j\theta})$ small over a prescribed frequency band can be achieved by making the controller transfer function $G(e^{j\theta})$ large over that frequency band. This conflicts, however, with the requirement that the mean square input be restricted, that the effect of the observation noise be restrained, and, possibly, with the requirement of stability. A compromise must be found.

The condition that $S(e^{j\theta})$ be small over as large a frequency band as possible also ensures that the closed-loop system receives protection against parameter variations. Here the condition **6-213**, or **6-215** in the multivariable case, guarantees that the effect of small parameter variations in the closed-loop system is always less than in an equivalent open-loop system.

6.4 OPTIMAL LINEAR DISCRETE-TIME STATE FEEDBACK CONTROL SYSTEMS

6.4.1 Introduction

In this section a review is given of linear optimal control theory for discrete-time systems, where it is assumed that the state of the system can be completely and accurately observed at all times. As in the continuous-time case, much of the attention is focused upon the regulator problem, although the tracking problem is discussed as well. The section is organized along the lines of Chapter 3.

6.4.2 Stability Improvement by State Feedback

In Section 3.2 we proved that a continuous-time linear system can be stabilized by an appropriate feedback law if the system is completely controllable or stabilizable. The same is true for discrete-time systems.

Theorem 6.26. *Let*

$$x(i+1) = Ax(i) + Bu(i) \qquad \textbf{6-216}$$

represent a time-invariant linear discrete-time system. Consider the time-invariant control law

$$u(i) = -Fx(i). \qquad \textbf{6-217}$$

Then the closed-loop characteristic values, that is, the characteristic values of $A - BF$, can be arbitrarily located in the complex plane (within the restriction that complex characteristic values occur in complex conjugate pairs) by choosing F suitably if and only if **6-216** *is completely controllable. It is possible to choose F such that the closed-loop system is stable if and only if* **6-216** *is stabilizable.*

Since the proof of the theorem depends entirely on the properties of the matrix $A - BF$, it is essentially identical to that for continuous-time systems. Moreover, the computational methods of assigning closed-loop poles are the same as those for continuous-time systems.

A case of special interest occurs when all closed-loop characteristic values are assigned to the origin. The characteristic polynomial of $A - BF$ then is of the form

$$\det(\lambda I - A + BF) = \lambda^n, \qquad \textbf{6-218}$$

where n is the dimension of the system. Since according to the Cayley-Hamilton theorem every matrix satisfies its own characteristic equation, we must have

$$(A - BF)^n = 0. \qquad \textbf{6-219}$$

In matrix theory it is said that this matrix is *nilpotent* with index n. Let us consider what implications this has. The state at the instant i can be expressed as

$$x(i) = (A - BF)^i x(0). \qquad \textbf{6-220}$$

This shows that, if **6-219** is satisfied, any initial state $x(0)$ is reduced to the zero state at or before the instant n, that is, in n steps or less (Cadzow, 1968; Farison and Fu, 1970). We say that a system with this property exhibits a *state deadbeat response*. In Section 6.4.7 we encounter systems with *output deadbeat responses*.

The preceding shows that the state of any completely controllable time-invariant discrete-time system can be forced to the zero state in at most n steps, where n is the dimension of the system. It may very well be, however, that the control law that assigns all closed-loop poles to the origin leads to excessively large input amplitudes or to an undesirable transient behavior.

We summarize the present results as follows.

Theorem 6.27. *Let the state difference equation*

$$x(i+1) = Ax(i) + Bu(i) \qquad \textbf{6-221}$$

represent a completely controllable, time-invariant, n-dimensional, linear discrete-time system. Then any initial state can be reduced to the zero state in at most n steps, that is, for every $x(0)$ there exists an input that makes $x(n) = 0$. This can be achieved through the time-invariant feedback law

$$u(i) = -Fx(i), \qquad \textbf{6-222}$$

where F is so chosen that the matrix $A - BF$ has all its characteristic values at the origin.

Example 6.13. *Digital position control system*

The digital positioning system of Example 6.2 (Section 6.2.3) is described by the state difference equation

$$x(i+1) = \begin{pmatrix} 1 & 0.08015 \\ 0 & 0.6313 \end{pmatrix} x(i) + \begin{pmatrix} 0.003396 \\ 0.06308 \end{pmatrix} \mu(i). \qquad \text{6-223}$$

The system has the characteristic polynomial

$$(z-1)(z-0.6313) = z^2 - 1.6313z + 0.6313. \qquad \text{6-224}$$

In phase-variable canonical form the system can therefore be represented as

$$x'(i+1) = \begin{pmatrix} 0 & 1 \\ -0.6313 & 1.6313 \end{pmatrix} x'(i) + \begin{pmatrix} 0 \\ 1 \end{pmatrix} \mu(i). \qquad \text{6-225}$$

The transformed state $x'(i)$ is related to the original state $x(i)$ by $x(i) = Tx'(i)$, where by Theorem 1.43 (Section 1.9) the matrix T can be found to be

$$T = \begin{pmatrix} 0.002912 & 0.003396 \\ -0.06308 & 0.06308 \end{pmatrix}. \qquad \text{6-226}$$

It is immediately seen that in terms of the transformed state the state deadbeat control law is given by

$$\mu(i) = -(-0.6313, 1.6313)x'(i). \qquad \text{6-227}$$

In terms of the original state, we have

$$\mu(i) = -(-0.6313, 1.6313)T^{-1}x(i), \qquad \text{6-228}$$

or

$$\mu(i) = -(158.5, 17.33)x(i). \qquad \text{6-229}$$

In Fig. 6.12 the complete response of the deadbeat digital position control system to an initial condition $x(0) = \text{col}(0.1, 0)$ is sketched, not only at the sampling instants, but also at the intermediate times. This response has been obtained by simulating the continuous-time positioning system while it is controlled with piecewise constant inputs obtained from the discrete-time control law **6-229**. It is seen that the system is completely at rest after two sampling periods.

6.4.3 The Linear Discrete-Time Optimal Regulator Problem

Analogously to the continuous-time problem, we define the discrete-time regulator problem as follows.

Definition 6.16. *Consider the discrete-time linear system*

$$x(i+1) = A(i)x(i) + B(i)u(i), \qquad \text{6-230}$$

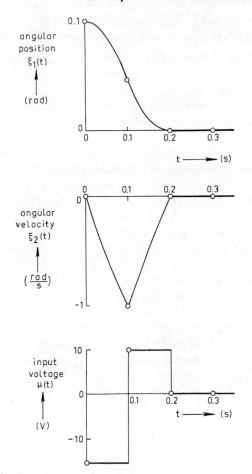

Fig. 6.12. State deadbeat response of the digital position control system.

where
$$x(i_0) = x_0, \qquad \text{6-231}$$
with the controlled variable
$$z(i) = D(i)x(i). \qquad \text{6-232}$$
Consider as well the criterion
$$\sum_{i=i_0}^{i_1-1}[z^T(i+1)R_3(i+1)z(i+1) + u^T(i)R_2(i)u(i)] + x^T(i_1)P_1x(i_1), \quad \text{6-233}$$
where $R_3(i+1) > 0$ and $R_2(i) > 0$ for $i = i_0, i_0 + 1, \cdots, i_1 - 1$, and $P_1 \geq 0$. Then the problem of determining the input $u(i)$ for $i = i_0, i_0 + 1, \cdots, i_1 - 1$, is called the **discrete-time deterministic linear optimal regulator problem**.

*If all matrices occurring in the problem formulation are constant, we refer to it as the **time-invariant discrete-time linear optimal regulator problem**.*

It is noted that the two terms following the summation sign in the criterion do not have the same index. This is motivated as follows. The initial value of the controlled variable $z(i_0)$ depends entirely upon the initial state $x(i_0)$ and cannot be changed. Therefore there is no point in including a term with $z(i_0)$ in the criterion. Similarly, the final value of the input $u(i_1)$ affects only the system behavior beyond the terminal instant i_1; therefore the term involving $u(i_1)$ can be excluded as well. For an extended criterion, where the criterion contains a cross-term, see Problem 6.1.

It is also noted that the controlled variable does not contain a direct link in the problem formulation of Definition 6.16, although as we saw in Section 6.2.3 such a direct link easily arises when a continuous-time system is discretized. The omission of a direct link can be motivated by the fact that usually some freedom exists in selecting the controlled variable, so that often it is justifiable to make the instants at which the controlled variable is to be controlled coincide with the sampling instants. In this case no direct link enters into the controlled variable (see Section 6.2.3). Regulator problems where the controlled variable does have a direct link, however, are easily converted to the formulation of Problem 6.1.

In deriving the optimal control law, our approach is different from the continuous-time case where we used elementary calculus of variations; here we invoke dynamic programming (Bellman, 1957; Kalman and Koepcke, 1958). Let us define the scalar function $\sigma[x(i), i]$ as follows:

$$\sigma[x(i), i] = \begin{cases} \min_{u(i), \cdots, u(i_1-1)} \Big\{ \sum_{j=i}^{i_1-1} [z^T(j+1)R_3(j+1)z(j+1) \\ \qquad + u^T(j)R_2(j)u(j)] + x^T(i_1)P_1 x(i_1) \Big\} \\ \qquad \text{for } i = i_0, i_0+1, \cdots, i_1-1, \\ x^T(i_1)P_1 x(i_1) \qquad \text{for } i = i_1. \end{cases} \qquad \text{6-234}$$

We see that $\sigma[x(i), i]$ represents the minimal value of the criterion, computed over the period $i, i+1, \cdots, i_1$, when at the instant i the system is in the state $x(i)$. We derive an iterative equation for this function. Consider the instant $i-1$. Then if the input $u(i-1)$ is arbitrarily selected, but $u(i), u(i+1), \cdots, u(i_1-1)$ are chosen optimally with respect to the state at time i, we can write for the criterion over the period $i-1, i, \cdots, i_1$:

$$\sum_{j=i-1}^{i_1-1} [z^T(j+1)R_3(j+1)z(j+1) + u^T(j)R_2(j)u(j)] + x^T(i_1)P_1 x(i_1)$$
$$= [z^T(i)R_3(i)z(i) + u^T(i-1)R_2(i-1)u(i-1)] + \sigma[x(i), i]. \qquad \text{6-235}$$

6.4 Optimal Discrete-Time State Feedback

Obviously, to determine $u^0(i-1)$, the optimal input at time $i-1$, we must choose $u(i-1)$ so that the expression

$$z^T(i)R_3z(i) + u^T(i-1)R_2u(i-1) + \sigma[x(i), i] \qquad \text{6-236}$$

is minimized. The minimal value of **6-236** must of course be the minimal value of the criterion evaluated over the control periods $i-1, i, \cdots, i_1-1$. Consequently, we have the equality

$$\sigma[x(i-1), i-1)] = \min_{u(i-1)} \{z^T(i)R_3(i)z(i)$$
$$+ u^T(i-1)R_2(i-1)u(i-1) + \sigma[x(i), i]\}. \qquad \text{6-237}$$

By using **6-230** and **6-232** and rationalizing the notation, this expression takes the form

$$\sigma(x, i-1) = \min_u \{[A(i-1)x + B(i-1)u]^T R_1(i)[A(i-1)x + B(i-1)u]$$
$$+ u^T R_2(i-1)u + \sigma([A(i-1)x + B(i-1)u], i)\}, \qquad \text{6-238}$$

where

$$R_1(i) = D^T(i)R_3(i)D(i). \qquad \text{6-239}$$

This is an iterative equation in the function $\sigma(x, i)$. It can be solved in the order $\sigma(x, i_1), \sigma(x, i_1 - 1), \sigma(x, i_1 - 2), \cdots$, since $\sigma(x, i_1)$ is given by **6-234**. Let us attempt to find a solution of the form

$$\sigma(x, i) = x^T P(i)x, \qquad \text{6-240}$$

where $P(i)$, $i = i_0, i_0 + 1, \cdots, i_1$, is a sequence of matrices to be determined. From **6-234** we immediately see that

$$P(i_1) = P_1. \qquad \text{6-241}$$

Substitution of **6-240** into **6-238** and minimization shows that the optimal input is given by

$$u(i-1) = -F(i-1)x(i-1), \qquad i = i_0 + 1, \cdots, i_1, \qquad \text{6-242}$$

where the gain matrix $F(i-1)$ follows from

$$F(i-1) = \{R_2(i-1) + B^T(i-1)[R_1(i) + P(i)]B(i-1)\}^{-1}$$
$$\cdot B^T(i-1)[R_1(i) + P(i)]A(i-1). \qquad \text{6-243}$$

The inverse matrix in this expression always exists since $R_2(i-1) > 0$ and a nonnegative-definite matrix is added. Substitution of **6-242** into **6-238** yields with **6-243** the following difference equation in $P(i)$:

$$P(i-1) = A^T(i-1)[R_1(i) + P(i)][A(i-1) - B(i-1)F(i-1)],$$
$$i = i_0 + 1, \cdots, i_1. \qquad \text{6-244}$$

It is easily verified that the right-hand side is a symmetric matrix.
We sum up these results as follows.

Theorem 6.28. *Consider the discrete-time deterministic linear optimal regulator problem. The optimal input is given by*

$$u(i) = -F(i)x(i), \quad i = i_0, i_0 + 1, \cdots, i_1 - 1, \qquad \text{6-245}$$

where

$$F(i) = \{R_2(i) + B^T(i)[R_1(i+1) + P(i+1)]B(i)\}^{-1}$$
$$\cdot B^T(i)[R_1(i+1) + P(i+1)]A(i). \qquad \text{6-246}$$

Here the inverse always exists and

$$R_1(i) = D^T(i)R_3(i)D(i), \quad i = i_0 + 1, i_0 + 2, \cdots, i_1. \qquad \text{6-247}$$

The sequence of matrices $P(i)$, $i = i_0, i_0 + 1, \cdots, i_1 - 1$, satisfies the matrix difference equation

$$P(i) = A^T(i)[R_1(i+1) + P(i+1)][A(i) - B(i)F(i)],$$
$$i = i_0, i_0 + 1, \cdots, i_1 - 1, \qquad \text{6-248}$$

with the terminal condition

$$P(i_1) = P_1. \qquad \text{6-249}$$

The value of the criterion **6-233** *achieved with this control law is given by*

$$x^T(i_0)P(i_0)x(i_0). \qquad \text{6-250}$$

We note that the difference equation **6-248** is conveniently solved backward, where first $F(i)$ is computed from $P(i+1)$ through **6-246**, and then $P(i)$ from $P(i+1)$ and $F(i)$ through **6-248**. This presents no difficulties when the aid of a digital computer is invoked. Equation **6-248** is the equivalent of the continuous-time Riccati equation.

It is not difficult to show that under the conditions of Definition 6.16 the solution of the discrete-time deterministic linear optimal regulator problem as given in Theorem 6.28 always exists and is unique.

Example 6.14. *Digital position control system*
Let us consider the digital positioning system of Example 6.2 (Section 6.2.3). We take as the controlled variable the position, that is, we let

$$\zeta(i) = (1, 0)x(i). \qquad \text{6-251}$$

6.4 Optimal Discrete-Time State Feedback

The following criterion is selected. Minimize

$$\sum_{i=0}^{i_1-1} [\zeta^2(i+1) + \rho\mu^2(i)]. \qquad 6\text{-}252$$

Table 6.1 shows the behavior of the gain vector $F(i)$ for $i_1 = 10$ and $\rho = 0.00002$. We see that as i decreases, $F(i)$ approaches a steady-state value

$$\bar{F} = (110.4, 12.66). \qquad 6\text{-}253$$

The response of the corresponding steady-state closed-loop system to the initial state $x(0) = \text{col}(0.1, 0)$ is given in Fig. 6.13.

Table 6.1 Behavior of the Feedback Gain Vector $F(i)$ for the Digital Position Control System

i	$F(i)$
9	(107.7, 8.63)
8	(114.0, 12.66)
7	(109.4, 12.58)
6	(110.3, 12.64)
5	(110.4, 12.66)
4	(110.4, 12.66)
3	(110.4, 12.66)
2	(110.4, 12.66)
1	(110.4, 12.66)
0	(110.4, 12.66)

6.4.4 Steady-State Solution of the Discrete-Time Regulator Problem

In this section we study the case where the control period extends from i_0 to infinity. The following results are in essence identical to those for the continuous-time case.

Theorem 6.29. *Consider the discrete-time deterministic linear optimal regulator problem and its solution as given in Theorem 6.28. Assume that $A(i)$, $B(i)$, $R_1(i + 1)$, and $R_2(i)$ are bounded for $i \geq i_0$, and suppose that*

$$R_3(i+1) \geq \alpha I, \qquad R_2(i) \geq \beta I, \qquad i \geq i_0, \qquad 6\text{-}254$$

where α and β are positive constants.
(i) *Then if the system 6-230 is either*
  (a) *completely controllable, or*
  (b) *exponentially stable,*

496 Discrete-Time Systems

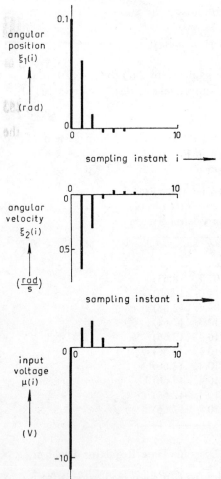

Fig. 6.13. Response of the optimal digital position control system to the initial condition $x(0) = \text{col}(0.1, 0)$.

the solution $P(i)$ of the difference equations **6-246** and **6-248** with the terminal condition $P(i_1) = 0$ converges to a nonnegative-definite sequence of matrices $\bar{P}(i)$ as $i_1 \to \infty$, which is a solution of the difference equations **6-246** and **6-248**.
(ii) *Moreover, if the system* **6-230**, **6-232** *is either*
   (c) *both uniformly completely controllable and uniformly completely reconstructible, or*
   (d) *exponentially stable,*
the solution $P(i)$ of the difference equations **6-246** and **6-248** with the terminal condition $P(i_1) = P_1$ converges to $\bar{P}(i)$ as $i_1 \to \infty$ for any $P_1 \geq 0$.

6.4 Optimal Discrete-Time State Feedback

The stability of the steady-state control law that corresponds to the steady-state solution $\bar{P}$ is ascertained from the following result.

Theorem 6.30. *Consider the discrete-time deterministic linear optimal regulator problem and suppose that the assumptions of Theorem 6.29 concerning A, B, R_1, R_3, and R_2 are satisfied. Then if the system* **6-230**, **6-232** *is either*

(a) *uniformly completely controllable and uniformly completely reconstructible, or*

(b) *exponentially stable,*

the following facts hold.

(i) *The steady-state optimal control law*

$$u(i) = -\bar{F}(i)x(i), \qquad \text{6-255}$$

where $\bar{F}(i)$ is obtained by substituting $\bar{P}(i)$ for $P(i)$ in **6-246**, *is exponentially stable.*

(ii) *The steady-state optimal control law* **6-255** *minimizes*

$$\lim_{i_1 \to \infty} \left\{ \sum_{i=i_0}^{i_1} [z^T(i+1)R_3(i+1)z(i+1) + u^T(i)R_2(i)u(i)] + x^T(i_1)P_1 x(i_1) \right\} \qquad \text{6-256}$$

for all $P_1 \geq 0$. The minimal value of **6-256**, *which is achieved by the steady-state optimal control law, is given by*

$$x^T(i_0)\bar{P}(i_0)x(i_0). \qquad \text{6-257}$$

The proofs of these theorems can be given along the lines of Kalman's proofs (Kalman, 1960) for continuous-time systems. The duals of these theorems (for reconstruction) are considered by Deyst and Price (1968). In the time-invariant case, the following facts hold (Caines and Mayne, 1970, 1971).

Theorem 6.31. *Consider the time-invariant discrete-time linear optimal regulator problem. Then if the system is both stabilizable and detectable the following facts hold.*

(i) *The solution $P(i)$ of the difference equations* **6-246** *and* **6-248** *with the terminal condition $P(i_1) = P_1$ converges to a constant steady-state solution $\bar{P}$ as $i_1 \to \infty$ for any $P_1 \geq 0$.*

(ii) *The steady-state optimal control law is time-invariant and asymptotically stable.*

(iii) *The steady-state optimal control law minimizes* **6-256** *for all $P_1 \geq 0$. The minimal value of this expression is given by*

$$x^T(i_0)\bar{P}x(i_0). \qquad \text{6-258}$$

In conclusion, we derive a result that is useful when studying the closed-loop pole locations of the steady-state time-invariant optimal regulator. Define the quantity

$$p(i) = [R_1(i+1) + P(i+1)]x(i+1),$$
$$i = i_0, i_0+1, \cdots, i_1-1, \quad \textbf{6-259}$$

where R_1 and P are as given in Theorem 6.28. We derive a difference equation for $p(i)$. From the terminal condition **6-249**, it immediately follows that

$$p(i_1-1) = [R_1(i_1) + P_1]x(i_1). \qquad \textbf{6-260}$$

Furthermore, we have with the aid of **6-248**

$$\begin{aligned}p(i-1) &= R_1(i)x(i) + P(i)x(i) \\ &= R_1(i)x(i) + A^T(i)[R_1(i+1) + P(i+1)][A(i) - B(i)F(i)]x(i) \\ &= R_1(i)x(i) + A^T(i)[R_1(i+1) + P(i+1)]x(i+1) \\ &= R_1(i)x(i) + A^T(i)p(i).\end{aligned} \qquad \textbf{6-261}$$

Finally, we express $u^0(i)$ in terms of $p(i)$. Consider the following string of equalities

$$\begin{aligned}-R_2^{-1}(i)B^T(i)p(i) &= -R_2^{-1}(i)B^T(i)[R_1(i+1) + P(i+1)]x(i+1) \\ &= -R_2^{-1}(i)B^T(i)[R_1(i+1) + P(i+1)][A(i)x(i) + B(i)u^0(i)] \\ &= -R_2^{-1}(i)B^T(i)[R_1(i+1) + P(i+1)]A(i)x(i) \\ &\quad - R_2^{-1}(i)B^T(i)[R_1(i+1) + P(i+1)]B(i)u^0(i).\end{aligned} \qquad \textbf{6-262}$$

Now from **6-246** it follows that

$$\begin{aligned}B^T(i)[R_1(i+1) &+ P(i+1)]A(i)x(i) \\ &= \{R_2(i) + B^T(i)[R_1(i+1) + P(i+1)]B(i)\}F(i)x(i) \\ &= -\{R_2(i) + B^T(i)[R_1(i+1) + P(i+1)]B(i)\}u^0(i).\end{aligned} \qquad \textbf{6-263}$$

Substitution of this into **6-262** yields

$$-R_2^{-1}(i)B^T(i)p(i) = u^0(i). \qquad \textbf{6-264}$$

Inserting $u^0(i)$ as given here into the state difference equation, we obtain the following two-point boundary-value problem

$$\begin{aligned}x(i+1) &= A(i)x(i) - B(i)R_2^{-1}(i)B^T(i)p(i), & i &= i_0, i_0+1, \cdots, i_1-1, \\ p(i-1) &= R_1(i)x(i) + A^T(i)p(i), & i &= i_0+1, i_0+2, \cdots, i_1-1 \\ x(i_0) &= x_0, \\ p(i_1-1) &= [R_1(i_1) + P_1]x(i_1). & &\textbf{6-265}\end{aligned}$$

6.4 Optimal Discrete-Time State Feedback

We could have derived these equations directly by a variational approach to the discrete-time regulator problem, analogously to the continuous-time version.

Let us now consider the time-invariant steady-state case. Then $p(i)$ is defined by

$$p(i) = (R_1 + \bar{P})x(i+1), \qquad i = i_0, i_0 + 1, \cdots. \qquad \textbf{6-266}$$

In the time-invariant case the difference equations **6-265** take the form

$$\begin{aligned} x(i+1) &= Ax(i) - BR_2^{-1}B^T p(i), & i &= i_0, i_0 + 1, \cdots, \\ p(i-1) &= R_1 x(i) + A^T p(i), & i &= i_0 + 1, i_0 + 2, \cdots. \end{aligned} \qquad \textbf{6-267}$$

Without loss of generality we take $i_0 = 0$; thus we rewrite **6-267** as

$$\begin{aligned} x(i+1) &= Ax(i) - BR_2^{-1}B^T p(i), & i &= 0, 1, 2, \cdots, \\ p(i) &= R_1 x(i+1) + A^T p(i+1), & i &= 0, 1, 2, \cdots. \end{aligned} \qquad \textbf{6-268}$$

We study these difference equations by z-transformation. Application of the z-transformation to both equations yields

$$\begin{aligned} zX(z) - zx_0 &= AX(z) - BR_2^{-1}B^T P(z), \\ P(z) &= zR_1 X(z) - zR_1 x_0 + zA^T P(z) - zA^T p_0, \end{aligned} \qquad \textbf{6-269}$$

where $x_0 = x(0)$, $p_0 = p(0)$, and $X(z)$ and $P(z)$ are the z-transforms of x and p, respectively. Solving for $X(z)$ and $P(z)$, we write

$$\begin{pmatrix} X(z) \\ P(z) \end{pmatrix} = \begin{pmatrix} zI - A & BR_2^{-1}B^T \\ -R_1 & z^{-1}I - A^T \end{pmatrix}^{-1} \begin{pmatrix} zx_0 \\ -R_1 x_0 - A^T p_0 \end{pmatrix}. \qquad \textbf{6-270}$$

When considering this expression, we note that each component of $X(z)$ and $P(z)$ is a rational function in z with singularities at those values of z where

$$\det \begin{pmatrix} zI - A & BR_2^{-1}B^T \\ -R_1 & z^{-1}I - A^T \end{pmatrix} = 0. \qquad \textbf{6-271}$$

Let z_j, $j = 1, 2, \cdots$, denote the roots of this expression, the left-hand side of which is a polynomial in z and $1/z$. If z_j is a root, $1/z_j$ also is a root. Moreover, zero can never be a root of **6-271** and there are at most $2n$ roots (n is the dimension of the state x). It follows that both $x(i)$ and $p(i)$ can be described as linear combinations of expressions of the form $z_j^i, iz_j^i, i^2 z_j^i, \cdots$, for all values of j. Terms of the form $i^k z_j^i$, $k = 0, 1, \cdots, l - 1$, occur when z_j has multiplicity l. Now we know that under suitable conditions stated in Theorem 6.31 the steady-state response of the closed-loop regulator is asymptotically stable. This means that the initial conditions of the difference equations **6-268** are such that the coefficients of the terms in $x(i)$ with powers of z_j with $|z_j| \geq 1$ are zero. Consequently, $x(i)$ is a linear combination of powers of those roots z_j for which $|z_j| < 1$. This means that these roots are characteristic values of the closed-loop regulator. Now, since **6-271** may have less than $2n$ roots, there may be less than n roots with moduli strictly less than 1 (it is seen in Section 6.4.7 that this is the case only when A has one or more characteristic values zero). This leads to the conclusion that the remaining characteristic values of the closed-loop regulator are zero, since z appears in the denominators of the expression on the right-hand side of **6-270** after inversion of the matrix.

We will need these results later (Section 6.4.7) to analyze the behavior of the closed-loop characteristic values. We summarize as follows.

Theorem 6.32. *Consider the time-invariant discrete-time deterministic linear optimal regulator problem. Suppose that the n-dimensional system*

$$x(i + 1) = Ax(i) + Bu(i),$$
$$z(i) = Dx(i), \qquad \text{6-272}$$

is stabilizable and detectable. Let $z_j, j = 1, 2, \cdots, r$, *with* $r \leq n$, *denote those roots of*

$$\det \begin{pmatrix} zI - A & BR_2^{-1}B^T \\ -D^T R_3 D & z^{-1}I - A^T \end{pmatrix} = 0 \qquad \text{6-273}$$

that have moduli strictly less than 1. Then $z_j, j = 1, 2, \cdots, r$, *constitute* r *of the characteristic values of the closed-loop steady-state optimal regulator. The remaining* $n - r$ *characteristic values are zero.*

Using an approach related to that of this section, Vaughan (1970) gives a method for finding the steady-state solution of the regulator problem by diagonalization.

Example 6.15. *Stirred tank*

Consider the problem of regulating the stirred tank of Example 6.3 (Section 6.2.3) which is described by the state difference equation

$$x(i + 1) = \begin{pmatrix} 0.9512 & 0 \\ 0 & 0.9048 \end{pmatrix} x(i) + \begin{pmatrix} 4.877 & 4.877 \\ -1.1895 & 3.569 \end{pmatrix} u(i). \qquad \text{6-274}$$

We choose as controlled variables the outgoing flow and the concentration, that is,

$$z(i) = \begin{pmatrix} 0.01 & 0 \\ 0 & 1 \end{pmatrix} x(i). \qquad \text{6-275}$$

The criterion is given by

$$\sum_{i=0}^{\infty} [z^T(i + 1)R_3 z(i + 1) + u^T(i)R_2 u(i)]. \qquad \text{6-276}$$

6.4 Optimal Discrete-Time State Feedback

Exactly as in the continuous-time case of Example 3.9 (Section 3.4.1), we choose for the weighting matrices

$$R_3 = \begin{pmatrix} 50 & 0 \\ 0 & 0.02 \end{pmatrix} \quad \text{and} \quad R_2 = \rho \begin{pmatrix} \frac{1}{3} & 0 \\ 0 & 3 \end{pmatrix}, \qquad \text{6-277}$$

where ρ is a scalar constant to be determined.

The steady-state feedback gain matrix can be found by repeated application of **6-246** and **6-248**. For $\rho = 1$ numerical computation yields

$$\bar{F} = \begin{pmatrix} 0.07125 & -0.7029 \\ 0.01357 & 0.04548 \end{pmatrix}. \qquad \text{6-278}$$

The closed-loop characteristic values are $0.5982 \pm j0.08988$. Figure 6.14 shows the response of the closed-loop system to the initial conditions $x(0) = \text{col}\,(0.1, 0)$ and $x(0) = \text{col}\,(0, 0.1)$. The response is quite similar to that of the corresponding continuous-time regulator as given in Fig. 3.11 (Section 3.4.1).

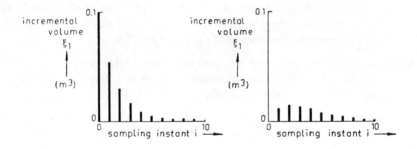

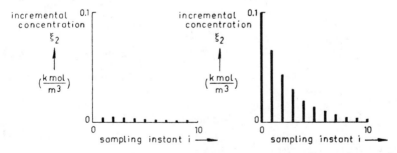

Fig. 6.14. Closed-loop responses of the regulated stirred tank, discrete-time version. Left column: Responses of volume and concentration to the initial conditions $\xi_1(0) = 0.1$ m³ and $\xi_2(0) = 0$ kmol/m³. Right column: Responses of volume and concentration to the initial conditions $\xi_1(0) = 0$ m³ and $\xi_2(0) = 0.1$ kmol/m³.

6.4.5 The Stochastic Discrete-Time Linear Optimal Regulator

The stochastic discrete-time linear optimal regulator problem is formulated as follows.

Definition 6.17. *Consider the discrete-time linear system*
$$x(i + 1) = A(i)x(i) + B(i)u(i) + w(i),$$
$$x(i_0) = x_0, \qquad 6\text{-}279$$
where $w(i)$, $i = i_0, i_0 + 1, \cdots, i_1 - 1$, *constitutes a sequence of uncorrelated, zero-mean stochastic variables with variance matrices* $V(i)$, $i = i_0, \cdots, i_1 - 1$. *Let*
$$z(i) = D(i)x(i) \qquad 6\text{-}280$$
be the controlled variable. Then the problem of minimizing the criterion
$$E\left\{\sum_{i=i_0}^{i_1-1}[z^T(i+1)R_3(i+1)z(i+1) + u^T(i)R_2(i)u(i)] + x^T(i_1)P_1 x(i_1)\right\}, \qquad 6\text{-}281$$
where $R_3(i + 1) > 0$, $R_2(i) > 0$ *for* $i = i_0, \cdots, i_1 - 1$ *and* $P_1 \geq 0$, *is termed the* **stochastic discrete-time linear optimal regulator problem.** *If all the matrices in the problem formulation are constant, we refer to it as the* **time-invariant stochastic discrete-time linear optimal regulator problem.**

As in the continuous-time case, the solution of the stochastic regulator problem is identical to that of the deterministic equivalent (Åström, Koepcke, and Tung, 1962; Tou, 1964; Kushner, 1971).

Theorem 6.33. *The criterion* 6-281 *of the stochastic discrete-time linear optimal regulator problem is minimized by choosing the input according to the control law*
$$u(i) = -F(i)x(i), \qquad i = i_0, i_0 + 1, \cdots, i_1 - 1, \qquad 6\text{-}282$$
where
$$F(i) = \{R_2(i) + B^T(i)[R_1(i + 1) + P(i + 1)]B(i)\}^{-1}$$
$$\cdot B^T(i)[R_1(i + 1) + P(i + 1)]A(i). \qquad 6\text{-}283$$
The sequence of matrices $P(i)$, $i = i_0, \cdots, i_1 - 1$, *is the solution of the matrix difference equation*
$$P(i) = A^T(i)[R_1(i + 1) + P(i + 1)][A(i) - B(i)F(i)],$$
$$i = i_0, i_0 + 1, \cdots, i_1 - 1, \qquad 6\text{-}284$$
with the terminal condition
$$P(i_1) = P_1. \qquad 6\text{-}285$$
Here
$$R_1(i) = D^T(i)R_3(i)D(i). \qquad 6\text{-}286$$

6.4 Optimal Discrete-Time State Feedback

The value of the criterion **6-281** *achieved with this control law is given by*

$$x_0^T P(i_0) x_0 + \sum_{j=i_0+1}^{i_1} \operatorname{tr}\{V(j-1)[P(j) + R_1(j)]\}. \qquad \textbf{6-287}$$

This theorem can be proved by a relatively straightforward extension of the dynamic programming argument of Section 6.4.3. We note that Theorem 6.33 gives the linear control law **6-282** as *the* optimal solution, without further qualification. This is in contrast to the continuous-time case (Theorem 3.9, Section 3.6.3), where we restricted ourself to linear control laws.

As in the continuous-time case, the stochastic regulator problem encompasses regulator problems with disturbances, tracking problems, and tracking problems with disturbances. Here as well, the structure of the solutions of each of these special versions of the problem is such that the feedback gain from the state of the plant is not affected by the properties of the disturbances of the reference variable (see Problems 6.2 and 6.3).

Here too we can investigate in what sense the steady-state control law is optimal. As in the continuous-time case, it can be surmised that, if it exists, the steady-state control law minimizes

$$\lim_{N\to\infty} \frac{1}{N} E\left\{ \sum_{i=i_0}^{i_0+N-1} [z^T(i+1) R_3(i+1) z(i+1) + u^T(i) R_2(i) u(i)] \right\} \qquad \textbf{6-288}$$

(assuming that this expression exists for the steady-state optimal control law) with respect to all linear control laws for which this expressions exists. The minimal value of **6-288** is given by

$$\lim_{N\to\infty} \frac{1}{N} \sum_{j=i_0+1}^{i_0+N} \operatorname{tr}\{[R_1(j) + \bar{P}(j)] V(j-1)\}, \qquad \textbf{6-289}$$

where $\bar{P}(j), j \geq i_0$, is the steady-state solution of **6-284**. In the time-invariant case, the steady-state control law moreover minimizes

$$\lim_{i_0\to-\infty} E\{z^T(i+1) R_3 z(i+1) + u^T(i) R_2 u(i)\} \qquad \textbf{6-290}$$

with respect to all time-invariant control laws. The minimal value of **6-290** is given by

$$\operatorname{tr}[(R_1 + \bar{P}) V]. \qquad \textbf{6-291}$$

Kushner (1971) discusses these facts.

Example 6.16. *Stirred tank with disturbances*

In Example 6.10 (Section 6.2.12), we modeled the stirred tank with disturbances in the incoming concentrations through the stochastic difference equation **6-168**. If we choose for the components of the controlled variable the outgoing flow and the concentration in the tank, we have

$$z(i) = \begin{pmatrix} 0.01 & 0 & 0 & 0 \\ 0 & 1 & 0 & 0 \end{pmatrix} x(i). \qquad \text{6-292}$$

We consider the criterion

$$E\left\{ \sum_{i=0}^{N-1} [z^T(i+1)R_3 z(i+1) + u^T(i)R_2 u(i)] \right\}, \qquad \text{6-293}$$

where the weighting matrices R_3 and R_2 are selected as in Example 6.15. For $\rho = 1$ numerical computation yields the steady-state feedback gain matrix

$$\bar{F} = \begin{pmatrix} 0.07125 & -0.07029 & -0.009772 & -0.003381 \\ 0.01357 & 0.04548 & 0.008671 & 0.003052 \end{pmatrix}. \qquad \text{6-294}$$

Comparison with the solution of Example 6.15 shows that, as in the continuous-time case, the feedback link of the control law (represented by the first two columns of $\bar{F}$) is not affected by introducing the disturbances into the model (see Problem 6.2).

The steady-state rms values of the outgoing flow, the concentration, and the incoming flows can be computed by setting up the closed-loop system state difference equation and solving for $\bar{Q}$, the steady-state variance matrix of the state of the augmented system.

6.4.6 Linear Discrete-Time Regulators with Nonzero Set Points and Constant Disturbances

In this section we study linear discrete-time regulators with nonzero set points and constant disturbances. We limit ourselves to time-invariant systems and first consider nonzero set point regulators. Suppose that the system

$$x(i+1) = Ax(i) + Bu(i),$$
$$z(i) = Dx(i), \qquad \text{6-295}$$

must be operated about the set point

$$z(i) = z_0, \qquad \text{6-296}$$

where z_0 is a given constant vector. As in the continuous-time case of Section 3.7.1, we introduce the shifted state, input, and controlled variables. Then the steady-state control law that returns the system from any initial condition to the set point optimally, in the sense that a criterion of the form

$$\sum_{i=i_0}^{\infty} [z'^T(i+1)R_3 z'(i+1) + u'^T(i)R_2 u'(i)] \qquad \text{6-297}$$

6.4 Optimal Discrete-Time State Feedback

is minimized, is of the form
$$u'(i) = -\bar{F}x'(i), \qquad \text{6-298}$$
where u', x', and z' are the shifted input, state, and controlled variables, respectively, and where $\bar{F}$ is the steady-state feedback gain matrix. In terms of the original system variables, this control law must take the form
$$u(i) = -\bar{F}x(i) + u'_0, \qquad \text{6-299}$$
where u'_0 is a constant vector. With this control law the closed-loop system is described by
$$\begin{aligned} x(i+1) &= \bar{A}x(i) + Bu'_0, \\ z(i) &= Dx(i), \end{aligned} \qquad \text{6-300}$$
where
$$\bar{A} = A - B\bar{F}. \qquad \text{6-301}$$
Assuming that the closed-loop system is asymptotically stable, the controlled variable will approach a constant steady-state value
$$\lim_{i \to \infty} z(i) = H_c(1)u'_0, \qquad \text{6-302}$$
where $H_c(z)$ is the *closed-loop transfer matrix*
$$H_c(z) = D(zI - \bar{A})^{-1}B. \qquad \text{6-303}$$

The expression **6-302** shows that a zero steady-state error is obtained when u'_0 is chosen as
$$u'_0 = H_c^{-1}(1)z_0, \qquad \text{6-304}$$
provided the inverse exists, where it is assumed that $\dim(u) = \dim(z)$. We call the control law
$$u(i) = -\bar{F}x(i) + H_c^{-1}(1)z_r(i) \qquad \text{6-305}$$
the *nonzero set point optimal control law*.

We see that the existence of this control law is determined by the existence of the inverse of $H_c(1)$. Completely analogously to the continuous-time case, it can be shown that
$$\det[H_c(z)] = \frac{\psi(z)}{\phi_c(z)}, \qquad \text{6-306}$$
where $\phi_c(z)$ is the closed-loop characteristic polynomial
$$\phi_c(z) = \det(zI - A + B\bar{F}), \qquad \text{6-307}$$
and where $\psi(z)$ is the open-loop numerator polynomial; that is, $\psi(z)$ follows from
$$\det[H(z)] = \frac{\psi(z)}{\phi(z)}. \qquad \text{6-308}$$

Here
$$H(z) = D(zI - A)^{-1}B \qquad \text{6-309}$$
is the open-loop transfer matrix and
$$\phi(z) = \det(zI - A) \qquad \text{6-310}$$
is the open-loop characteristic polynomial. The relation **6-306** shows that $H_c^{-1}(1)$ exists provided $\psi(1) \neq 0$. Since $H(e^{j\theta})$ describes the frequency response of the open-loop system, this condition is equivalent to requiring that the open-loop frequency response matrix have a numerator polynomial that does not vanish at $\theta = 0$.

We summarize as follows.

Theorem 6.34. *Consider the time-invariant discrete-time linear system*
$$x(i + 1) = Ax(i) + Bu(i),$$
$$z(i) = Dx(i), \qquad \text{6-311}$$
where $\dim(z) = \dim(u)$. *Consider any asymptotically stable time-invariant control law*
$$u(i) = -Fx(i) + u_0'. \qquad \text{6-312}$$
Let $H(z)$ be the open-loop transfer matrix
$$H(z) = D(zI - A)^{-1}B \qquad \text{6-313}$$
and $H_c(z)$ the closed-loop transfer matrix
$$H_c(z) = D(zI - A + BF)^{-1}B. \qquad \text{6-314}$$
Then $H_c(1)$ is nonsingular and the controlled variable $z(i)$ can under steady-state conditions be maintained at any constant set point z_0 by choosing
$$u_0' = H_c^{-1}(1)z_0 \qquad \text{6-315}$$
if and only if $H(z)$ has a nonzero numerator polynomial that has no zeroes at $z = 1$.

It is noted that this theorem holds not only for the optimal control law, but for any stable control law.

Next we very briefly consider regulators with constant disturbances. We suppose that the plant is described by the state difference and output equations
$$x(i + 1) = Ax(i) + Bu(i) + v_0,$$
$$z(i) = Dx(i), \qquad \text{6-316}$$
where v_0 is a constant vector. Shifting the state and input variables, we reach

6.4 Optimal Discrete-Time State Feedback

the conclusion that the control law that returns the shifted state optimally to zero must be of the form

$$u(i) = -\bar{F}x(i) + u_0', \qquad \text{6-317}$$

where u_0' is a suitable constant vector. The steady-state response of the controlled variable with this control law is given by

$$\lim_{i \to \infty} z(i) = H_c(1)u_0' + D(I - \bar{A})^{-1}v_0, \qquad \text{6-318}$$

where $H_c(z) = D(zI - A + B\bar{F})^{-1}B$. It is possible to make the steady-state response **6-318** equal to zero by choosing

$$u_0' = -H_c^{-1}(1)D(I - \bar{A})^{-1}v_0. \qquad \text{6-319}$$

provided dim (z) = dim (u) and $H_c(1)$ is nonsingular. Thus the *zero-steady-state-error optimal control law* is given by

$$u(i) = -\bar{F}x(i) - H_c^{-1}(1)D(I - \bar{A})^{-1}v_0. \qquad \text{6-320}$$

The conditions for the existence of $H_c^{-1}(1)$ are given in Theorem 6.34.

The disadvantage of the control law **6-320** is that its application requires accurate measurement of the constant disturbance v_0. This difficulty can be circumvented by appending to the system an "integral state" q (compare Section 3.7.2), defined by the difference relation

$$q(i+1) = q(i) + z(i), \qquad i \geq i_0, \qquad \text{6-321}$$

with $q(i_0)$ given. Then it can easily be seen that any asymptotically stable control law of the form

$$u(i) = -F_1 x(i) - F_2 q(i) \qquad \text{6-322}$$

suppresses the effect of constant disturbances on the controlled variable, that is, $z(i)$ assumes the value zero in steady-state conditions no matter what the value of v_0 is in **6-316**. Necessary and sufficient conditions for the existence of such an asymptotically stable control law are that the system **6-316** be stabilizable, and [assuming that dim (u) = dim (z)] that the open-loop transfer matrix possess no zeroes at the origin.

Example 6.17. *Digital position control system*

In Example 6.6 (Section 6.2.6), we saw that the digital positioning system of Example 6.2 (Section 6.2.3) has the transfer function

$$H(z) = \frac{0.003396(z + 0.8575)}{(z-1)(z-0.6313)}. \qquad \text{6-323}$$

Because the numerator polynomial of this transfer function does not have a zero at $z = 1$, a nonzero set point optimal controller can be obtained. In

Example 6.14 (Section 6.4.3), we obtained the steady-state feedback gain vector $\bar{F} = (110.4, 12.66)$. It is easily verified that the corresponding nonzero set point optimal control law is given by

$$\mu(i) = -\bar{F}x(i) + 110.4\zeta_0, \qquad \text{6-324}$$

where ζ_0 is the (scalar) set point. Figure 6.15 shows the response of the closed-loop system to a step in the set point, not only at the sampling instants but also at intermediate times, obtained by simulation of the

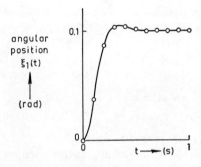

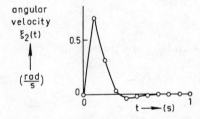

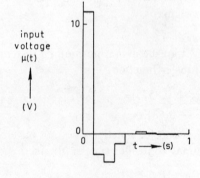

Fig. 6.15. Responses of the digital position control system to a step of 0.1 rad in the set point.

6.4 Optimal Discrete-Time State Feedback

continuous-time system. The system exhibits an excellent response, not quite as fast as the deadbeat response of Fig. 6.12, but with smaller input amplitudes.

6.4.7 Asymptotic Properties of Time-Invariant Optimal Control Laws

In this section we study the asymptotic properties of time-invariant steady-state optimal control laws when in the criterion the weighting matrix R_2 is replaced with

$$R_2 = \rho N, \qquad \text{6-325}$$

where $\rho \downarrow 0$. Let us first consider the behavior of the closed-loop poles. In Theorem 6.32 (Section 6.4.4) we saw that the nonzero closed-loop characteristic values are those roots of the equation

$$\det \begin{pmatrix} zI - A & BR_2^{-1}B^T \\ -R_1 & z^{-1}I - A^T \end{pmatrix} = 0 \qquad \text{6-326}$$

that have moduli less than 1, where $R_1 = D^T R_3 D$. Using Lemmas 1.2 (Section 1.5.4) and 1.1 (Section 1.5.3), we write

$$\det \begin{pmatrix} zI - A & BR_2^{-1}B^T \\ -R_1 & z^{-1}I - A^T \end{pmatrix}$$
$$= \det(zI - A) \det[z^{-1}I - A^T + R_1(zI - A)^{-1}BR_2^{-1}B^T]$$
$$= \det(zI - A) \det(z^{-1}I - A^T)$$
$$\quad \cdot \det[I + R_1(zI - A)^{-1}BR_2^{-1}B^T(z^{-1}I - A^T)^{-1}]$$
$$= \det(zI - A) \det(z^{-1}I - A^T)$$
$$\quad \cdot \det[I + R_2^{-1}B^T(z^{-1}I - A^T)^{-1}R_1(zI - A)^{-1}B]$$
$$= \det(zI - A) \det(z^{-1}I - A^T)$$
$$\quad \cdot \det\left[I + \frac{1}{\rho}N^{-1}B^T(z^{-1}I - A^T)^{-1}D^T R_3 D(zI - A)^{-1}B\right]$$
$$= \phi(z)\phi(z^{-1}) \det\left[I + \frac{1}{\rho}N^{-1}H^T(z^{-1})R_3 H(z)\right], \qquad \text{6-327}$$

where
$$\phi(z) = \det(zI - A) \qquad \text{6-328}$$

is the open-loop characteristic polynomial, and

$$H(z) = D(zI - A)^{-1}B \qquad \text{6-329}$$

is the open-loop transfer matrix.

To study the behavior of the closed-loop characteristic values, let us first consider the single-input single-output case. We assume that the scalar transfer function $H(z)$ can be written as

$$H(z) = \frac{\psi(z)}{\phi(z)}, \qquad \text{6-330}$$

where

$$\phi(z) = z^{n-q} \prod_{i=1}^{q} (z - \pi_i), \qquad \pi_i \neq 0, \qquad i = 1, 2, \cdots, q, \qquad \text{6-331}$$

with $q \leq n$, is the characteristic polynomial of the system, and where

$$\psi(z) = \alpha z^{s-p} \prod_{i=1}^{p} (z - \nu_i), \qquad \nu_i \neq 0, \qquad i = 1, 2, \cdots, p, \qquad \text{6-332}$$

with $p \leq s \leq n - 1$, is the numerator polynomial of the system. Then **6-327** takes the form (assuming $R_3 = 1$ and $N = 1$):

$$\prod_{i=1}^{q} (z - \pi_i)\left(\frac{1}{z} - \pi_i\right) + \frac{\alpha^2}{\rho} \prod_{i=0}^{p} (z - \nu_i)\left(\frac{1}{z} - \nu_i\right). \qquad \text{6-333}$$

To apply standard root locus techniques, we bring this expression into the form

$$\prod_{i=1}^{q} (z - \pi_i)\left(z - \frac{1}{\pi_i}\right) + \frac{\alpha^2 \prod_{i=1}^{p} (-\nu_i)}{\rho \prod_{i=1}^{q} (-\pi_i)} z^{q-p} \prod_{i=1}^{p} (z - \nu_i)\left(z - \frac{1}{\nu_i}\right). \qquad \text{6-334}$$

We conclude the following concerning the loci of the $2q$ roots of this expression, where we assume that $q \geq p$ (see Problem 6.4 for the case $q < p$).

1. The $2q$ loci originate for $\rho = \infty$ at π_i and $1/\pi_i$, $i = 1, 2, \cdots, q$.
2. As $\rho \downarrow 0$, the loci behave as follows.
   (a) p roots approach the zeroes ν_i, $i = 1, 2, \cdots, p$;
   (b) p roots approach the inverse zeroes $1/\nu_i$, $i = 1, 2, \cdots, p$;
   (c) $q - p$ roots approach 0;
   (d) the remaining $q - p$ roots approach infinity.
3. Those roots that go to infinity as $\rho \downarrow 0$ asymptotically are at a distance

$$\left| \frac{\alpha^2 \prod_{i=1}^{p} \nu_i}{\rho \prod_{i=1}^{q} \pi_i} \right|^{1/(q-p)} \qquad \text{6-335}$$

6.4 Optimal Discrete-Time State Feedback

from the origin. Consequently, those roots that go to zero are asymptotically at a distance

$$\left| \frac{\rho}{\alpha^2} \frac{\prod_{i=1}^{q} \pi_i}{\prod_{i=1}^{p} \nu_i} \right|^{1/(q-p)} \qquad \text{6-336}$$

from the origin.

Information about the optimal closed-loop poles is obtained by selecting those roots that have moduli less than 1. We conclude the following.

Theorem 6.35. *Consider the steady-state solution of the time-invariant single-input single-output discrete-time linear regulator problem. Let the open-loop transfer function be given by*

$$H(z) = \frac{\alpha z^{s-p} \prod_{i=1}^{p} (z - \nu_i)}{z^{n-q} \prod_{i=1}^{q} (z - \pi_i)}, \qquad \alpha \neq 0, \qquad \text{6-337}$$

where the $\pi_i \neq 0$, $i = 1, 2, \cdots, q$, are the nonzero open-loop characteristic values, and $\nu_i \neq 0$, $i = 1, 2, \cdots, p$, the nonzero zeroes. Suppose that $n \geq q \geq p$, $n - 1 \geq s \geq p$ and that in the criterion 6-233 we have $R_3 = 1$ and $R_2 = \rho$. Then the following holds.

(a) *Of the n closed-loop characteristic values $n - q$ are always at the origin.*

(b) *As $\rho \downarrow 0$, of the q remaining closed-loop characteristic values p approach the numbers $\hat{\nu}_i$, $i = 1, 2, \cdots, p$, where*

$$\hat{\nu}_i = \begin{cases} \nu_i & \text{if } |\nu_i| \leq 1, \\ \dfrac{1}{\nu_i} & \text{if } |\nu_i| > 1. \end{cases} \qquad \text{6-338}$$

(c) *As $\rho \downarrow 0$, the $q - p$ other closed-loop characteristic values go to zero. These closed-loop poles asymptotically are at a distance*

$$\left| \frac{\rho}{\alpha^2} \frac{\prod_{i=1}^{q} \pi_i}{\prod_{i=1}^{p} \nu_i} \right|^{1/(q-p)} \qquad \text{6-339}$$

from the origin.

(d) *As $\rho \to \infty$, the q nonzero closed-loop characteristic values approach the numbers $\hat{\pi}_i$, $i = 1, 2, \cdots, q$, where*

$$\hat{\pi}_i = \begin{cases} \pi_i & \text{if } |\pi_i| \leq 1, \\ \dfrac{1}{\pi_i} & \text{if } |\pi_i| > 1. \end{cases} \qquad \text{6-340}$$

Let us now consider the behavior of the nonzero set point optimal control law derived in Section 6.4.6. For a single-input single-output system, it is easily seen that the system transfer function from the (scalar) set point $\zeta_0(i)$ (now assumed to be variable) to the controlled variable $\zeta(i)$ is given by

$$T(z) = \frac{H_c(z)}{H_c(1)}, \qquad \text{6-341}$$

where $H_c(z)$ is the closed-loop transfer function. As in the continuous-time case (Section 3.8.2), it is easily verified that we can write

$$H_c(z) = \frac{\psi(z)}{\phi_c(z)}, \qquad \text{6-342}$$

where $\psi(z)$ is the open-loop transfer function numerator polynomial and $\phi_c(z)$ the closed-loop characteristic polynomial. For $\psi(z)$ we have

$$\psi(z) = \alpha z^{s-p} \prod_{i=1}^{p}(z - \nu_i), \qquad \text{6-343}$$

while in the limit $\rho \downarrow 0$ we write for the closed-loop characteristic polynomial

$$\phi_c(z) = z^{n-p} \prod_{i=1}^{p}(z - \hat{\nu}_i). \qquad \text{6-344}$$

Substitution into **6-342** and **6-341** shows that in the limit $\rho \downarrow 0$ the control system transfer function can be written as

$$T_0(z) = \frac{1}{z^{n-s}} \prod_{i=1}^{p}\left(\frac{z - \nu_i}{z - \hat{\nu}_i}\right) \prod_{k=1}^{p}\left(\frac{1 - \hat{\nu}_k}{1 - \nu_k}\right). \qquad \text{6-345}$$

Now if the open-loop transfer function has no zeroes outside the unit circle. the limiting control system transfer function reduces to

$$T_0(z) = \frac{1}{z^{n-s}}. \qquad \text{6-346}$$

This represents a pure delay, that is, the controlled variable and the variable set point are related as follows:

$$\zeta(i) = \zeta_0[i - (n - s)]. \qquad \text{6-347}$$

We summarize as follows.

Theorem 6.36. *Consider the nonzero set point optimal control law, as described in Section 6.4.6., for a single-input single-output system. Let $R_3 = 1$ and $R_2 = \rho$. Then as $\rho \downarrow 0$, the control system transmission (that is, the transfer function of the closed-loop system from the set point to the controlled*

6.4 Optimal Discrete-Time State Feedback

variable) approaches

$$T_0(z) = \frac{1}{z^{n-s}} \prod_{i=1}^{p} \left(\frac{z - \nu_i}{z - \hat{\nu}_i}\right) \prod_{k=1}^{p} \left(\frac{1 - \hat{\nu}_k}{1 - \nu_k}\right), \qquad 6\text{-}348$$

where the $\hat{\nu}_i, i = 1, 2, \cdots, p$ are derived from the nonzero open-loop zeroes $\nu_i, i = 1, 2, \cdots, p$, as indicated in **6-338**, and where n is the dimension of the system and s the degree of the numerator polynomial of the system. If the open-loop transfer function has no zeroes outside the unit circle, the limiting system transfer function is

$$T_0(z) = \frac{1}{z^{n-s}}, \qquad 6\text{-}349$$

which represents a pure delay.

We see that, if the open-loop system has no zeroes outside the unit circle, the limiting closed-loop system has the property that the response of the controlled variable to a step in the set point achieves a zero tracking error after $n - s$ time intervals. We refer to this as *output deadbeat response*.

We now discuss the asymptotic behavior of the closed-loop characteristic values for multiinput systems. Referring back to **6-327**, we consider the roots of

$$\phi(z)\phi(z^{-1}) \det\left[I + \frac{1}{\rho} N^{-1} H^T(z^{-1}) R_3 H(z)\right]. \qquad 6\text{-}350$$

Apparently, for $\rho = \infty$ those roots of this expression that are finite are the roots of

$$\phi(z)\phi(z^{-1}). \qquad 6\text{-}351$$

Let us write

$$\phi(z) = z^{n-q} \prod_{i=1}^{q} (z - \pi_i), \qquad 6\text{-}352$$

and assume that $\pi_i \neq 0, i = 1, 2, \cdots, q$. Then we have

$$\phi(z)\phi(z^{-1}) = \prod_{i=1}^{q} (z - \pi_i)(z^{-1} - \pi_i), \qquad 6\text{-}353$$

which shows that $2q$ root loci of **6-350** originate for $\rho = \infty$ at the nonzero characteristic values of the open-loop system and their inverses.

Let us now consider the roots of **6-350** as $\rho \downarrow 0$. Clearly, those roots that stay finite approach the zeroes of

$$\phi(z)\phi(z^{-1}) \det [H^T(z^{-1}) R_3 H(z)]. \qquad 6\text{-}354$$

Let us now assume that the input and the controlled variable have the same dimensions, so that $H(z)$ is a square transfer matrix, with

$$\det [H(z)] = \frac{\psi(z)}{\phi(z)}. \qquad \text{6-355}$$

Then the zeroes of **6-354** are the zeroes of

$$\psi(z^{-1})\psi(z). \qquad \text{6-356}$$

Let us write the numerator polynomial $\psi(z)$ in the form

$$\psi(z) = \alpha z^{s-p} \prod_{i=1}^{p} (z - \nu_i), \qquad \text{6-357}$$

where $\nu_i \neq 0$, $i = 1, 2, \cdots, p$. Then **6-356** can be written as

$$\alpha^2 \prod_{i=1}^{p} (z - \nu_i)(z^{-1} - \nu_i). \qquad \text{6-358}$$

This shows that $2p$ root loci of **6-350** terminate for $\rho = 0$ at the nonzero zeroes ν_i, $i = 1, 2, \cdots, p$, and the inverse zeroes $1/\nu_i$, $1 = 1, 2, \cdots, p$.

Let us suppose that $q \geq p$ (for the case $q < p$, see Problem 6.4). Then there are $2q$ root loci of **6-350**, which originate for $\rho = \infty$ at the nonzero open-loop poles and their inverses. As we have seen, $2p$ loci terminate for $\rho = 0$ at the nonzero open-loop zeroes and their inverses. Of the remaining $2q - 2p$ loci, $q - p$ must go to infinity as $\rho \downarrow 0$, while the other $q - p$ loci approach the origin.

The nonzero closed-loop poles are those roots of **6-350** that lie inside the unit circle. We conclude the following.

Theorem 6.37. *Consider the steady-state solution of the time-invariant regulator problem. Suppose that* $\dim (u) = \dim (z)$ *and let* $H(z)$ *be the open-loop transfer matrix*

$$H(z) = D(zI - A)^{-1}B. \qquad \text{6-359}$$

Furthermore, let

$$\det [H(z)] = \frac{\psi(z)}{\phi(z)}, \qquad \text{6-360}$$

where

$$\phi(z) = z^{n-q} \prod_{i=1}^{q} (z - \pi_i), \qquad \text{6-361}$$

with $\pi_i \neq 0$, $i = 1, 2, \cdots, q$, *is the open-loop characteristic polynomial. In addition, suppose that*

$$\psi(z) = \alpha z^{s-p} \prod_{i=1}^{p} (z - \nu_i), \qquad \text{6-362}$$

6.4 Optimal Discrete-Time State Feedback

with $p \leq q$, and where $v_i \neq 0$, $i = 1, 2, \cdots, p$. Finally, set $R_2 = \rho N$ where $N > 0$ and ρ is a positive scalar. Then we have the following.

(a) Of the n closed-loop poles, $n - q$ always are at the origin.

(b) As $\rho \downarrow 0$, of the remaining q closed-loop poles, p approach the numbers $\hat{v}_i$, $i = 1, 2, \cdots, p$, where

$$\hat{v}_i = \begin{cases} v_i & \text{if } |v_i| \leq 1, \\ \dfrac{1}{v_i} & \text{if } |v_i| > 1. \end{cases} \qquad 6\text{-}363$$

(c) As $\rho \downarrow 0$, the $q - p$ other closed-loop poles go to zero.

(d) As $\rho \to \infty$, the q nonzero closed-loop poles approach the numbers $\hat{\pi}_i$, $i = 1, 2, \cdots, q$, where

$$\hat{\pi}_i = \begin{cases} \pi_i & \text{if } |\pi_i| \leq 1, \\ \dfrac{1}{\pi_i} & \text{if } |\pi_i| > 1. \end{cases} \qquad 6\text{-}364$$

We note that contrary to the continuous-time case the closed-loop poles remain finite as the weighting matrix R_2 approaches the zero matrix. Similarly, the feedback gain matrix $\bar{F}$ also remains finite. Often, but not always, the limiting feedback gain matrix can be found by setting $R_2 = 0$ in the difference equations **6-246** and **6-248** and iterating until the steady-state value is found (see the examples, and also Pearson, 1965; Rappaport and Silverman, 1971).

For the response of the closed-loop system with this limiting feedback law, the following is to be expected. As we have seen, the limiting closed-loop system asymptotically has $n - p$ characteristic values at the origin. If the open-loop zeroes are all inside the unit circle, they cancel the corresponding limiting closed-loop poles. This means that the response is determined by the $n - p$ poles at the origin, resulting in a deadbeat response of the controlled variable after $n - p$ steps. We call this an output deadbeat response, in contrast to the state deadbeat response discussed in Section 6.4.2. If a system exhibits an output deadbeat response, the output reaches the desired value exactly after a finite number of steps, but the system as a whole may remain in motion for quite a long time, as one of the examples at the end of this section illustrates. If the open-loop system has zeroes outside the unit circle, the cancellation effect does not occur and as a result the limiting regulator does not exhibit a deadbeat response.

It is noted that these remarks are conjectures, based on analogy with the continuous-time case. A complete theory is missing as yet. The examples at the end of the section confirm the conjectures. An essential difference between the discrete-time theory and the continuous-time theory is that in the discrete-time case the steady-state solution $\bar{P}$ of the matrix equation **6-24'** generally does not approach the zero matrix as R_2 goes to zero, even if the open-loop transfer matrix possesses no zeroes outside the unit circle.

Example 6.18. *Digital position control system*

Let us consider the digital positioning system of Example 6.2 (Section 6.2.3). From Example 6.6 (Section 6.2.6), we know that the open-loop transfer function is

$$H(z) = \frac{0.003396(z + 0.8575)}{(z - 1)(z - 0.6313)}. \qquad \textbf{6-365}$$

It follows from Theorem 6.37 that the optimal closed-loop poles approach 0 and -0.8575 as $\rho \downarrow 0$. It is not difficult to find the loci of the closed-loop characteristic values. Expression **6-334** takes for this system the form

$$(z - 1)(z - 0.6313)(z - 1)(z - 1.584)$$
$$+ \frac{0.00001566}{\rho} z(z + 0.8575)(z + 1.166). \qquad \textbf{6-366}$$

The loci of the roots of this expression are sketched in Fig. 6.16. Those loci that lie inside the unit circle are the loci of the closed-loop poles. It can be

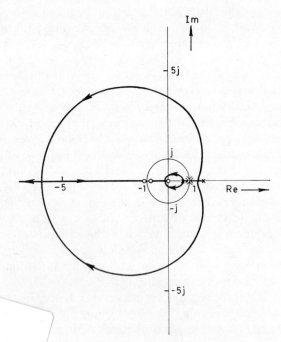

closed-loop poles and the inverse closed-loop poles for the digital

found that the limiting feedback gain matrix $\bar{F}_0$ for $\rho = 0$ is given by

$$\bar{F}_0 = (294.5, \quad 23.60). \qquad 6\text{-}367$$

Let us determine the corresponding nonzero set point optimal control law. We have for the limiting closed-loop transfer function

$$H_c(z) = \frac{\psi(z)}{\phi_c(z)} = \frac{0.003396(z + 0.8575)}{z(z + 0.8575)} = \frac{0.003396}{z}. \qquad 6\text{-}368$$

Consequently, $H_c(1) = 0.003396$ and the nonzero set point optimal control law is

$$\mu(i) = -\bar{F}_0 x(i) + 294.5\zeta_0(i). \qquad 6\text{-}369$$

Figure 6.17 gives the response of the system to a step in the set point, not only at the sampling instants but also at intermediate times. Comparing with the state deadbeat response of the same system as derived in Example 6.13, we observe the following.

(a) When considering only the response of the angular position at the sampling instants, the system shows an output deadbeat response after one sampling interval. In between the response exhibits a bad overshoot, however, and the actual settling time is in the order of 2 s, rather than 0.1 s.

(b) The input amplitude and the angular velocity assume large values.

These disadvantages are characteristic for output deadbeat control systems. Better results are achieved by not letting ρ go to zero. For $\rho = 0.00002$ the closed-loop poles are at 0.2288 ± 0.3184. The step response of the corresponding closed-loop system is given in Example 6.17 (Fig. 6.15) and is obviously much better than that of Fig. 6.17.

The disadvantages of the output deadbeat response are less pronounced when a larger sampling interval Δ is chosen. This causes the open-loop zero at -0.8575 to move closer to the origin; as a result the output deadbeat control system as a whole comes to rest much faster. For an alternative solution, which explicitly takes into account the behavior of the system between the sampling instants, see Problem 6.5.

Example 6.19. *Stirred tank with time delay*

Consider the stirred tank with time delay of Example 6.4 (Section 6.2.3). As the components of the controlled variable we choose the outgoing flow and concentration; hence

$$z(i) = \begin{pmatrix} 0.01 & 0 & 0 & 0 \\ 0 & 1 & 0 & 0 \end{pmatrix} x(i). \qquad 6\text{-}370$$

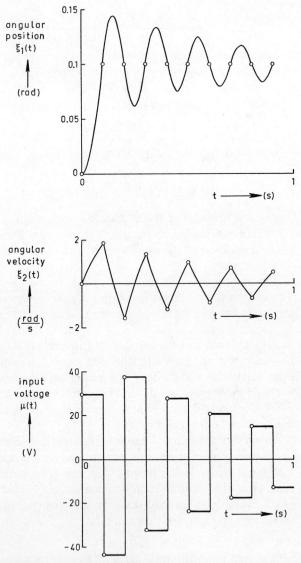

Fig. 6.17. Response of the output deadbeat digital position control system to a step in the set point of 0.1 rad.

6.4 Optimal Discrete-Time State Feedback

It can be found that the open-loop transfer matrix of the system is

$$H(z) = \begin{pmatrix} \dfrac{4.877}{z - 0.9512} & \dfrac{4.877}{z - 0.9512} \\ \dfrac{-1.1895}{z(z - 0.9048)} & \dfrac{3.569}{z(z - 0.9048)} \end{pmatrix}. \qquad \text{6-371}$$

The determinant of the transfer matrix is

$$\det[H(z)] = \frac{26.62}{z(z - 0.9512)(z - 0.9048)}. \qquad \text{6-372}$$

Because the open-loop characteristic polynomial is given by

$$\phi(z) = z^2(z - 0.9512)(z - 0.9048), \qquad \text{6-373}$$

the numerator polynomial of the transfer matrix is

$$\psi(z) = 26.62z. \qquad \text{6-374}$$

As a result, two closed-loop poles are always at the origin. The loci of the two other poles originate for $\rho = \infty$ at 0.9512 and 0.9048, respectively, and both approach the origin as $\rho \downarrow 0$. This means that in this case the output deadbeat control law is also a state deadbeat control law.

Let us consider the criterion

$$\sum_{i=0}^{\infty} [z^T(i+1)R_3 z(i+1) + u^T(i)R_2 u(i)], \qquad \text{6-375}$$

where, as in previous examples,

$$R_3 = \begin{pmatrix} 50 & 0 \\ 0 & 0.02 \end{pmatrix} \quad \text{and} \quad R_2 = \rho \begin{pmatrix} \tfrac{1}{3} & 0 \\ 0 & 3 \end{pmatrix}. \qquad \text{6-376}$$

When one attempts to compute the limiting feedback law for $\rho = 0$ by setting $R_2 = 0$ in the difference equation for $P(i)$ and $F(i)$, difficulties occur because for certain choices of P_1 the matrix

$$R_2 + B^T[R_1 + P(i+1)]B \qquad \text{6-377}$$

becomes singular at the first iteration. This can be avoided by choosing a very small value for ρ (e.g., $\rho = 10^{-6}$). By using this technique numerical computation yields the limiting feedback gain matrix

$$\bar{F}_0 = \begin{pmatrix} 0.1463 & -0.1720 & 0.2262 & -0.6786 \\ 0.04875 & 0.1720 & -0.2262 & 0.6786 \end{pmatrix}. \qquad \text{6-378}$$

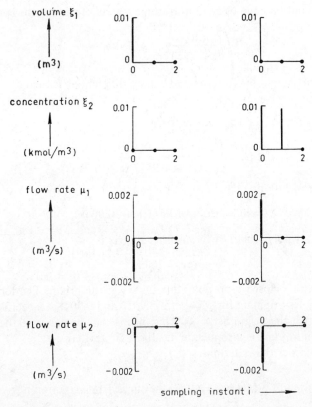

Fig. 6.18. Deadbeat response of the stirred tank with time delay. Left column: Responses of volume, concentration, feed no. 1, and feed no. 2 to the initial condition $\xi_1(0) = 0.01$ m³, while all other components of the initial state are zero. Right column: Responses of volume, concentration, feed no. 1, and feed no. 2 to the initial condition $\xi_2(0) = 0.01$ kmol/m³, while all other components of the initial state are zero.

In Fig. 6.18 the deadbeat response to two initial conditions is sketched. It is observed that initial errors in the volume ξ_1 are reduced to zero in one sampling period. For the concentration ξ_2 two sampling periods are required; this is because of the inherent delay in the system.

6.4.8 Sensitivity

In Section 3.9 we saw that the continuous-time time-invariant closed-loop regulator possesses the property that it always decreases the effect of disturbances and parameter variations as compared to the open-loop system. It is shown in this section by a counter example that this is not generally the case for discrete-time systems. The same example shows, however, that protection over a wide range of frequencies can still be obtained.

Example 6.20. *Digital angular velocity control*

Consider the angular velocity control system of Example 3.3 (Section 3.3.1), which is described by the scalar state differential equation

$$\dot{\xi}(t) = -\alpha \xi(t) + \kappa \mu(t). \qquad \text{6-379}$$

Let us assume that the input is piecewise constant over intervals of duration Δ. Then the resulting discrete-time system is described by

$$\xi(i+1) = e^{-\alpha \Delta} \xi(i) + \frac{\kappa}{\alpha}(1 - e^{-\alpha \Delta})\mu(i), \qquad \text{6-380}$$

where we have replaced $\xi(i\Delta)$ with $\xi(i)$ and $\mu(i\Delta)$ with $\mu(i)$. With the numerical values $\alpha = 0.5 \text{ s}^{-1}$, $\kappa = 150 \text{ rad}/(\text{V s}^2)$, and $\Delta = 0.1$ s, we obtain

$$\xi(i+1) = 0.9512 \xi(i) + 14.64 \mu(i). \qquad \text{6-381}$$

The controlled variable $\zeta(i)$ is the angular velocity $\xi(i)$, that is,

$$\zeta(i) = \xi(i). \qquad \text{6-382}$$

Let us consider the problem of minimizing

$$\sum_{i=0}^{\infty} [\zeta^2(i+1) + \rho \mu^2(i)]. \qquad \text{6-383}$$

It is easily found that with $\rho = 1000$ the steady-state solution is given by

$$\bar{P} = 1.456,$$
$$\bar{F} = 0.02240. \qquad \text{6-384}$$

The return difference of the closed-loop system is

$$J(z) = I + (zI - A)^{-1}B\bar{F}, \qquad \text{6-385}$$

which can be found to be

$$J(z) = \frac{z - 0.6232}{z - 0.9512}. \qquad \text{6-386}$$

To determine the behavior of $J(z)$ for z on the unit circle, set

$$z = e^{j\omega\Delta},$$

where $\Delta = 0.1$ s is the sampling interval. With this we find

$$|J(e^{j\omega\Delta})|^2 = \frac{1.388 - 1.246 \cos(\omega\Delta)}{1.905 - 1.902 \cos(\omega\Delta)}. \qquad \text{6-387}$$

522 Discrete-Time Systems

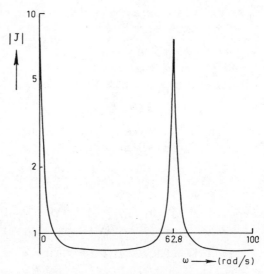

Fig. 6.19. Behavior of the return difference for a first-order discrete-time regulator.

Figure 6.19 gives a plot of the behavior of $|J(e^{j\omega\Delta})|$. We see that sensitivity reduction is achieved for low frequencies up to about 7 rad/s, but by no means for all frequencies. If the significant disturbances occur within the frequency band up to 7 rad/s, however, the sensitivity reduction may very well be adequate.

6.5 OPTIMAL LINEAR RECONSTRUCTION OF THE STATE OF LINEAR DISCRETE-TIME SYSTEMS

6.5.1 Introduction

This section is devoted to a review of the optimal reconstruction of the state of linear discrete-time systems. The section parallels Chapter 4.

6.5.2 The Formulation of Linear Discrete-Time Reconstruction Problems

In this section we discuss the formulation of linear discrete-time reconstruction problems. We pay special attention to this question since there are certain differences from the continuous-time case. As before, we take the point of view that the linear discrete-time system under consideration is obtained by operating a linear continuous-time system with a piecewise constant input, as indicated in Fig. 6.20. The instants at which the input changes value are given by $t_i, i = 0, 1, 2, \cdots$, which we call the *control*

6.5 Optimal Reconstruction of the State

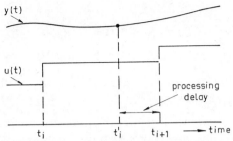

Fig. 6.20. Relationship of control actuation instant t_i and observation instant t_i'.

instants. These instants form the basic time grid. We furthermore introduce the *observation instants* t_i', $i = 0, 1, 2, \cdots$, which are the instants at which the observed variable $y(t)$ of the continuous-time system is sampled. It is assumed that the observation instant t_i' always *precedes* the control instant t_{i+1}. The difference $t_{i+1} - t_i'$ will be called the *processing delay*; in the case of a control system, it is the time that is available to process the observation $y(t_i')$ in order to determine the input $u(t_{i+1})$.

Suppose that the continuous-time system is described by

$$\dot{x}(t) = A(t)x(t) + B(t)u(t) + w_1(t), \qquad t \geq t_0, \qquad \textbf{6-388}$$

where w_1 is white noise with time-varying intensity $V_1(t)$. We furthermore assume that the observed variable is given by

$$y(t_i') = C(t_i')x(t_i') + w_2(t_i'), \qquad i = 0, 1, 2, \cdots, \qquad \textbf{6-389}$$

where the $w_2(t_i')$, $i = 0, 1, 2, \cdots$, form a sequence of uncorrelated stochastic vectors. To obtain the discrete-time description of the system, we write

$$x(t_{i+1}) = \Phi(t_{i+1}, t_i)x(t_i) + \left[\int_{t_i}^{t_{i+1}} \Phi(t_{i+1}, \tau)B(\tau)\,d\tau\right]u(t_i)$$

$$+ \int_{t_i}^{t_{i+1}} \Phi(t_{i+1}, \tau)w_1(\tau)\,d\tau, \qquad \textbf{6-390}$$

and

$$y(t_i') = C(t_i')\Phi(t_i', t_i)x(t_i) + \left[C(t_i')\int_{t_i}^{t_i'} \Phi(t_i', \tau)B(\tau)\,d\tau\right]u(t_i)$$

$$+ C(t_i')\int_{t_i}^{t_i'} \Phi(t_i', \tau)w_1(\tau)\,d\tau + w_2(t_i'), \qquad \textbf{6-391}$$

where in both cases $i = 0, 1, 2, \cdots$, and where $\Phi(t, t_0)$ is the transition matrix of the system **6-388**. We see that the two equations **6-390** and **6-391** are of

the form
$$x^+(i+1) = A_d(i)x^+(i) + B_d(i)u^+(i) + w_1^+(i),$$
$$y^+(i) = C_d(i)x^+(i) + E_d(i)u^+(i) + w_2^+(i). \qquad \text{6-392}$$

This method of setting up the discrete-time version of the problem has the following characteristics.

1. In the discrete-time version of the reconstruction problem, we assume that $y^+(i)$ is the latest observation that can be processed to obtain a reconstructed value for $x^+(i+1)$.
2. The output equation generally contains a *direct link*. As can be seen from **6-391**, the direct link is absent [i.e., $E_d(i) = 0$] when the processing delay takes up the whole interval (t_i, t_{i+1}).
3. Even if in the continuous-time problem the state excitation noise w_1 and the observation noise w_2 are uncorrelated, the state excitation noise w_1^+ and the observation noise w_2^+ of the discrete-time version of the problem will be *correlated*, because, as can be seen from **6-390**, **6-391**, and **6-392**, both $w_1^+(i)$ and $w_2^+(i)$ depend upon $w_1(t)$ for $t_i \leq t \leq t_i'$. Clearly, $w_1^+(i)$ and $w_2^+(i)$ are uncorrelated only if $t_i' = t_i$, that is, if the processing delay takes up the whole interval (t_i, t_{i+1}).

Example 6.21. *The digital positioning system*

Let us consider the digital positioning system of Example 6.2 (Section 6.2.3). It has been assumed that the sampling period is Δ. We now assume that the observed variable is the angular displacement ξ_1, so that in the continuous-time version
$$C = (1, 0). \qquad \text{6-393}$$

We moreover assume that there is a processing delay Δ_d, so that the observations are taken at an interval Δ_d before the instants at which control actuation takes place. Disregarding the noises that are possibly present, it is easily found with the use of **6-391** that the observation equation takes the form

$$\eta^+(i) = \left[1, \frac{1}{\alpha}(1 - e^{-\alpha\Delta'})\right]x^+(i) + \frac{\kappa}{\alpha}\left(\Delta' - \frac{1}{\alpha} + \frac{1}{\alpha}e^{-\alpha\Delta'}\right)\mu^+(i), \qquad \text{6-394}$$

where
$$\Delta' = \Delta - \Delta_d. \qquad \text{6-395}$$

With the numerical value
$$\Delta_d = 0.02 \text{ s}, \qquad \text{6-396}$$

we obtain for the observation equation
$$\eta^+(i) = (1, 0.06608)x^+(i) + 0.002381\mu^+(i). \qquad \text{6-397}$$

6.5.3 Discrete-Time Observers

In this section we consider dynamical systems that are able to reconstruct the state of another system that is being observed.

Definition 6.18. *The system*

$$\hat{x}(i+1) = \hat{A}(i)\hat{x}(i) + \hat{B}(i)u(i) + \hat{C}(i)y(i) \qquad \text{6-398}$$

*is a **full-order observer** for the system*

$$\begin{aligned} x(i+1) &= A(i)x(i) + B(i)u(i), \\ y(i) &= C(i)x(i) + E(i)u(i), \end{aligned} \qquad \text{6-399}$$

if

$$\hat{x}(i_0) = x(i_0) \qquad \text{6-400}$$

implies

$$\hat{x}(i) = x(i), \qquad i \geq i_0, \qquad \text{6-401}$$

for all $u(i)$, $i \geq i_0$.

It is noted that consistent with the reasoning of Section 6.5.2 the latest observation that the observer processes for obtaining $x(i+1)$ is $y(i)$. The following theorem gives more information about the structure of an observer.

Theorem 6.38. *The system **6-398** is a full order observer for the system **6-399** if and only if*

$$\begin{aligned} \hat{A}(i) &= A(i) - K(i)C(i), \\ \hat{B}(i) &= B(i) - K(i)E(i), \\ \hat{C}(i) &= K(i), \end{aligned} \qquad \text{6-402}$$

all for $i \geq i_0$, where $K(i)$ is an arbitrary time-varying matrix.

This theorem is easily proved by subtracting the state difference equations **6-399** and **6-398**. With **6-402** the observer can be represented as follows:

$$\hat{x}(i+1) = A(i)\hat{x}(i) + B(i)u(i) + K(i)[y(i) - C(i)\hat{x}(i) - E(i)u(i)]. \qquad \text{6-403}$$

The observer consists of a model of the system, with as extra driving variable an input which is proportional to the difference $y(i) - \hat{y}(i)$ of the observed variable $y(i)$ and its predicted value

$$\hat{y}(i) = C(i)\hat{x}(i) + E(i)u(i). \qquad \text{6-404}$$

We now discuss the stability of the observer and the behavior of the reconstruction error $e(i) = x(i) - \hat{x}(i)$.

Theorem 6.39. *Consider the observer* **6-398** *for the system* **6-399**. *Then the reconstruction error*

$$e(i) = x(i) - \hat{x}(i) \qquad \text{6-405}$$

satisfies the difference equation

$$e(i+1) = [A(i) - K(i)C(i)]e(i), \qquad i \geq i_0. \qquad \text{6-406}$$

The reconstruction error has the property that

$$e(i) \to 0, \quad as \quad i \to \infty, \qquad \text{6-407}$$

for all $e(i_0)$, *if and only if the observer is asymptotically stable.*

The difference equation **6-406** is easily found by subtracting the state difference equations in **6-399** and **6-398**. The behavior of $A(i) - K(i)C(i)$ determines both the stability of the observer and the behavior of the reconstruction error; hence the second part of the theorem.

As in the continuous-time case, we now consider the question: When does there exist a gain matrix K that stabilizes the observer and thus ensures that the reconstruction error will always eventually approach zero? Limiting ourselves to time-invariant systems, we have the following result.

Theorem 6.40. *Consider the time-invariant observer*

$$\hat{x}(i+1) = A\hat{x}(i) + Bu(i) + K[y(i) - C\hat{x}(i) - Eu(i)] \qquad \text{6-408}$$

for the time-invariant system

$$\begin{aligned} x(i+1) &= Ax(i) + Bu(i), \\ y(i) &= Cx(i) + Eu(i). \end{aligned} \qquad \text{6-409}$$

Then the observer poles (that is, the characteristic values of $A - KC$*) can be arbitrarily located in the complex plane (within the restriction that complex poles occur in complex conjugate pairs) by suitably choosing the gain matrix* K *if and only if the system* **6-409** *is completely reconstructible.*

The proof of this theorem immediately follows from the continuous-time equivalent (Theorem 4.3, Section 4.2.2). For systems that are only detectable, we have the following result.

Theorem 6.41. *Consider the time-invariant observer* **6-408** *for the time-invariant system* **6-409**. *Then a gain matrix* K *can be found such that the observer is asymptotically stable if and only if the system* **6-409** *is detectable.*

A case of special interest occurs when the observer poles are all located at the origin, that is, all the characteristic values of $A - KC$ are zero. Then the characteristic polynomial of $A - KC$ is given by

$$\det[\lambda I - (A - KC)] = \lambda^n, \qquad \textbf{6-410}$$

so that by the Cayley–Hamilton theorem

$$(A - KC)^n = 0. \qquad \textbf{6-411}$$

It follows by repeated application of the difference equation **6-406** for the reconstruction error that now

$$e(n) = (A - KC)^n e(0) = 0 \qquad \textbf{6-412}$$

for every $e(0)$, which means that every initial value of the reconstruction error is reduced to zero in at most n steps. In analogy with deadbeat control laws, we refer to observers with this property as *deadbeat observers*. Such observers produce a completely accurate reconstruction of the state after at most n steps.

Finally, we point out that if the system **6-409** has a scalar observed variable y, a unique solution of the gain matrix K is obtained for a given set of observer poles. In the case of multioutput systems, however, in general many different gain matrices exist that result in the same set of observer poles.

The observers considered so far in this section are systems of the same dimension as the system to be observed. Because of the output equation $y(i) = C(i)x(i) + E(i)u(i)$, we have available m equations in the unknown state $x(i)$ (assuming that y has dimension m); clearly, it must be possible to construct a reduced-order observer of dimension $n - m$ to reconstruct $x(i)$ completely. This observer can be constructed more or less analogously to the continuous-time case (Section 4.2.3).

Example 6.22. *Digital positioning system*
Consider the digital positioning system of Example 6.2 (Section 6.2.3), which is described by the state difference equation

$$x(i+1) = \begin{pmatrix} 1 & 0.08015 \\ 0 & 0.6313 \end{pmatrix} x(i) + \begin{pmatrix} 0.003396 \\ 0.06308 \end{pmatrix} \mu(i). \qquad \textbf{6-413}$$

As in Example 6.21, we assume that the observed variable is the angular position but that there is a processing delay of 0.02 s. This yields for the observed variable:

$$\eta(i) = (1, 0.06608)x(i) + 0.002381\mu(i). \qquad \textbf{6-414}$$

It is easily verified that the system is completely reconstructible so that

528 Discrete-Time Systems

Theorem 6.40 applies. Let us write $K = \text{col }(k_1, k_2)$. Then we find

$$A - KC = \begin{pmatrix} 1 - k_1 & 0.08015 - 0.06608k_1 \\ -k_2 & 0.6313 - 0.06608k_2 \end{pmatrix}. \qquad \text{6-415}$$

This matrix has the characteristic polynomial

$$z^2 + (-1.6313 + k_1 + 0.06608k_2)z + (0.6313 - 0.6313k_1 + 0.01407k_2). \qquad \text{6-416}$$

We obtain a deadbeat observer by setting

$$-1.6313 + k_1 + 0.06608k_2 = 0,$$
$$0.6313 - 0.6313k_1 + 0.01407k_2 = 0. \qquad \text{6-417}$$

This results in the gain matrix

$$K = \begin{pmatrix} 1.159 \\ 7.143 \end{pmatrix}. \qquad \text{6-418}$$

An observer with this gain reduces any initial reconstruction error to zero in at most two steps.

6.5.4 Optimal Discrete-Time Linear Observers

In this section we study discrete-time observers that are *optimal* in a well-defined sense. To this end we assume that the system under consideration is affected by disturbances and that the observations are contaminated by observation noise. We then find observers such that the reconstructed state is optimal in the sense that the mean square reconstruction error is minimized. We formulate our problem as follows.

Definition 6.19. *Consider the system*

$$x(i + 1) = A(i)x(i) + B(i)u(i) + w_1(i),$$
$$y(i) = C(i)x(i) + E(i)u(i) + w_2(i), \qquad i \geq i_0. \qquad \text{6-419}$$

Here col $[w_1(i), w_2(i)]$, $i \geq i_0$, *forms a sequence of zero-mean, uncorrelated vector stochastic variables with variance matrices*

$$\begin{pmatrix} V_1(i) & V_{12}(i) \\ V_{12}^T(i) & V_2(i) \end{pmatrix}, \qquad i \geq i_0. \qquad \text{6-420}$$

Furthermore, $x(i_0)$ *is a vector stochastic variable, uncorrelated with* w_1 *and* w_2, *with*

$$E\{x(i_0)\} = \bar{x}_0, \qquad E\{[x(i_0) - \bar{x}_0][x(i_0) - \bar{x}_0]^T\} = Q_0. \qquad \text{6-421}$$

6.5 Optimal Reconstruction of the State

Consider the observer

$$\hat{x}(i+1) = A(i)\hat{x}(i) + B(i)u(i) + K(i)[y(i) - C(i)\hat{x}(i) - E(i)u(i)] \quad \text{6-422}$$

for this system. Then the problem of finding the sequence of matrices $K^0(i_0)$, $K^0(i_0+1), \cdots, K^0(i-1)$, *and the initial condition* $\hat{x}(i_0)$, *so as to minimize*

$$E\{e^T(i)W(i)e(i)\}, \quad \text{6-423}$$

where $e(i) = x(i) - \hat{x}(i)$, *and where* $W(i)$ *is a positive-definite symmetric weighting matrix, is termed the **discrete-time optimal observer problem**. If*

$$V_2(i) > 0, \quad i \geq i_0,$$

*the optimal observer problem is called **nonsingular**.*

To solve the discrete-time optimal observer problem, we first establish the difference equation that is satisfied by the reconstruction error $e(i)$. Subtraction of the system state difference equation **6-419** and the observer equation **6-422** yields

$$e(i+1) = [A(i) - K(i)C(i)]e(i) + w_1(i) - K(i)w_2(i), \quad i \geq i_0. \quad \text{6-424}$$

Let us now denote by $\tilde{Q}(i)$ the variance matrix of $e(i)$, and by $\bar{e}(i)$ the mean of $e(i)$. Then we write

$$E\{e(i)e^T(i)\} = \tilde{Q}(i) + \bar{e}(i)\bar{e}^T(i), \quad \text{6-425}$$

so that

$$E\{e^T(i)W(i)e(i)\} = \bar{e}^T(i)W(i)\bar{e}(i) + \text{tr}\,[\tilde{Q}(i)W(i)]. \quad \text{6-426}$$

The first term of this expression is obviously minimized by making $\bar{e}(i) = 0$. This can be achieved by letting $\bar{e}(i_0) = 0$, which in turn is done by choosing

$$\hat{x}(i_0) = \bar{x}_0. \quad \text{6-427}$$

The second term in **6-425** can be minimized independently of the first term. With the aid of Theorem 6.22 (Section 6.2.12), it follows from **6-424** that $\tilde{Q}$ satisfies the recurrence relation

$$\tilde{Q}(i+1) = [A(i) - K(i)C(i)]\tilde{Q}(i)[A(i) - K(i)C(i)]^T \\ + V_1(i) - V_{12}(i)K^T(i) - K(i)V_{12}^T(i) + K(i)V_2(i)K^T(i),$$

$$i \geq i_0, \quad \text{6-428}$$

with

$$\tilde{Q}(i_0) = Q_0. \quad \text{6-429}$$

Repeated application of this recurrence relation will give us $\tilde{Q}(i+1)$ as a function of $K(i), K(i-1), \cdots, K(i_0)$. Let us now consider the problem of minimizing tr $[\tilde{Q}(i+1)W(i+1)]$ with respect to $K(i_0), K(i_0+1), \cdots, K(i)$. This is equivalent to minimizing $\tilde{Q}(i+1)$, that is, finding a sequence of matrices $K^0(i_0), K^0(i_0+1), \cdots, K^0(i)$ such that for the corresponding value $Q(i+1)$ of $\tilde{Q}(i+1)$ we have $Q(i+1) \leq \tilde{Q}(i+1)$. Now **6-428** gives us $\tilde{Q}(i+1)$ as a function of $K(i)$ and $\tilde{Q}(i)$, where $\tilde{Q}(i)$ is a function of $K(i_0), \cdots, K(i-1)$. Clearly, for given $K(i)$, $\tilde{Q}(i+1)$ is a monotone function of $\tilde{Q}(i)$, that is, if $Q(i) \leq \tilde{Q}(i)$ then $Q(i+1) \leq \tilde{Q}(i+1)$, where $Q(i+1)$ is obtained from $Q(i)$ by **6-428**. Therefore, $\tilde{Q}(i+1)$ can be minimized by first minimizing $\tilde{Q}(i)$ with respect to $K(i_0), K(i_0+1), \cdots, K(i-1)$, substituting the minimal value $Q(i)$ of $\tilde{Q}(i)$ into **6-428**, and then minimizing $\tilde{Q}(i+1)$ with respect to $K(i)$.

Let us suppose that the minimal value $Q(i)$ of $\tilde{Q}(i)$ has been found. Substituting $Q(i)$ for $\tilde{Q}(i)$ into **6-428** and completing the square, we obtain

$$\tilde{Q}(i+1) = [K - (AQC^T + V_{12})(V_2 + CQC^T)^{-1}](V_2 + CQC^T)$$
$$\cdot [K - (AQC^T + V_{12})(V_2 + CQC^T)^{-1}]^T$$
$$- (AQC^T + V_{12})(V_2 + CQC^T)^{-1}(CQA^T + V_{12}^T)$$
$$+ AQA^T + V_1, \qquad \textbf{6-430}$$

where for brevity we have omitted the arguments i on the right-hand side and where it has been assumed that

$$V_2(i) + C(i)Q(i)C^T(i) \qquad \textbf{6-431}$$

is nonsingular. This assumption is always justified in the nonsingular observer problem, where $V_2(i) > 0$. When considering **6-430**, we note that $\tilde{Q}(i+1)$ is minimized with respect to $K(i)$ if we choose $K(i)$ as $K^0(i)$, where

$$K^0(i) = [A(i)Q(i)C^T(i) + V_{12}(i)][V_2(i) + C(i)Q(i)C^T(i)]^{-1}. \qquad \textbf{6-432}$$

The corresponding value of $\tilde{Q}(i+1)$ is given by

$$Q(i+1) = [A(i) - K^0(i)C(i)]Q(i)A^T(i) + V_1(i) - K^0(i)V_{12}^T(i), \qquad \textbf{6-433}$$

with

$$Q(i_0) = Q_0. \qquad \textbf{6-434}$$

The relations **6-432** and **6-433** together with the initial condition **6-434** enable us to compute the sequence of gain matrices recurrently, starting with $K(i_0)$.

We summarize our conclusions as follows.

Theorem 6.42. *The optimal gain matrices $K^0(i)$, $i \geq i_0$, for the nonsingular optimal observer problem can be obtained from the recurrence relations*

$$K^0(i) = [A(i)Q(i)C^T(i) + V_{12}(i)][V_2(i) + C(i)Q(i)C^T(i)]^{-1},$$
$$Q(i+1) = [A(i) - K^0(i)C(i)]Q(i)A^T(i) + V_1(i) - K^0(i)V_{12}^T(i), \qquad \textbf{6-435}$$

both for $i \geq i_0$, with the initial condition

$$Q(i_0) = Q_0. \qquad \textbf{6-436}$$

The initial condition of the observer should be chosen as

$$\hat{x}(i_0) = \bar{x}_0. \qquad \textbf{6-437}$$

The matrix $Q(i)$ is the variance matrix of the reconstruction error $e(i) = x(i) - \hat{x}(i)$. For the optimal observer the mean square reconstruction error is given by

$$E\{e^T(i)W(i)e(i)\} = \operatorname{tr}[Q(i)W(i)]. \qquad \textbf{6-438}$$

Singular optimal observation problems can be handled in a manner that is more or less analogous to the continuous-time case (Brammer, 1968; Tse and Athans, 1970). Discrete-time observation problems where the state excitation noise and the observation noise are colored rather than white noise processes (Jazwinski, 1970) can be reduced to singular or nonsingular optimal observer problems.

We remark finally that in the literature a version of the discrete-time linear optimal observer problem is usually given that is different from the one considered here in that it is assumed that $y(i+1)$ rather than $y(i)$ is the latest observation available for reconstructing $x(i+1)$. In Problem 6.6 it is shown how the solution of this alternative version of the problem can be derived from the present version.

In this section we have considered optimal observers. As in the continuous-time case, it can be proved (see, e.g., Meditch, 1969) that the optimal observer is actually the *minimum mean square linear estimator* of $x(i+1)$ given the data $u(j)$ and $y(j)$, $j = i_0, i_0 + 1, \cdots, i$; that is, we cannot find any other linear operator on these data that yields an estimate with a smaller mean square reconstruction error. Moreover, if the initial state x_0 is Gaussian, and the white noise sequences w_1 and w_2 are jointly Gaussian, the optimal observer is *the* minimum mean square estimator of $x(i+1)$ given $u(j)$, $y(j)$, $j = i_0, i_0 + 1, \cdots, i$; that is, it is impossible to determine any other estimator operating on these data that has a smaller mean square reconstruction error (see, e.g., Jazwinski, 1970).

Example 6.23. *Stirred tank with disturbances*

In Example 6.10 (Section 6.2.12), we considered a discrete-time version of the stirred tank. The plant is described by the state difference equation

$$x(i+1) = \begin{pmatrix} 0.9512 & 0 & 0 & 0 \\ 0 & 0.9048 & 0.0669 & 0.02262 \\ 0 & 0 & 0.8825 & 0 \\ 0 & 0 & 0 & 0.9048 \end{pmatrix} x(i)$$

$$+ \begin{pmatrix} 4.877 & 4.877 \\ -1.1895 & 3.569 \\ 0 & 0 \\ 0 & 0 \end{pmatrix} u(i) + w_1(i), \qquad \textbf{6-439}$$

where $w_1(i)$, $i \geq i_0$, is a sequence of uncorrelated zero-mean stochastic variables with the variance matrix **6-169**. The components of the state are the incremental volume of the fluid in the tank, the incremental concentration in the tank, and the incremental concentrations of the two incoming feeds. We assume that we can observe at each instant of time i the incremental volume, as well as the incremental concentration in the tank. Both observations are contaminated with uncorrelated, zero-mean observation errors with standard deviations of 0.001 m³ and 0.001 kmol/m³, respectively. Furthermore, we assume that the whole sampling interval is used to process the data, so that the observation equation takes the form

$$y(i) = \begin{pmatrix} 1 & 0 & 0 & 0 \\ 0 & 1 & 0 & 0 \end{pmatrix} x(i) + w_2(i), \qquad \textbf{6-440}$$

where $w_2(i)$, $i \geq i_0$, have the variance matrix

$$\begin{pmatrix} 10^{-6} & 0 \\ 0 & 10^{-6} \end{pmatrix}. \qquad \textbf{6-441}$$

The processes w_1 and w_2 are uncorrelated. In Example 6.10 we found that the steady-state variance matrix of the state of the system is given by

$$\begin{pmatrix} 0 & 0 & 0 & 0 \\ 0 & 0.00369 & 0.00339 & 0.00504 \\ 0 & 0.00339 & 0.0100 & 0 \\ 0 & 0.00504 & 0 & 0.0400 \end{pmatrix}. \qquad \textbf{6-442}$$

Using this variance matrix as the initial variance matrix $Q(0) = Q_0$, the recurrence relations **6-435** can be solved. Figure 6.21 gives the evolution of the rms reconstruction errors of the last three components of the state as obtained from the evolution of $Q(i)$, $i \geq 0$. The rms reconstruction error of the first component of the state, the volume, of course remains zero all the time, since the volume does not fluctuate and thus we know its value exactly at all times.

It is seen from the plots that the concentrations of the feeds cannot be reconstructed very accurately because the rms reconstruction errors approach steady-state values that are hardly less than the rms values of the fluctuations in the concentrations of the feeds themselves. The rms reconstruction error of the concentration of the tank approaches a steady-state value of about 0.0083 kmol/m³. The reason that this error is larger than the standard deviation of 0.001 kmol/m³ of the observation error is the presence of the

6.5 Optimal Reconstruction of the State

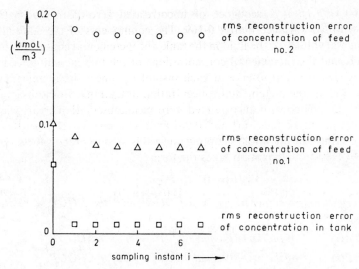

Fig. 6.21. Behavior of the rms reconstruction errors for the stirred tank with disturbances.

processing delay—the observer must predict the concentration a full sampling interval ahead.

6.5.5 Innovations

In this section we state the following fact, which is more or less analogous to the corresponding continuous-time result.

Theorem 6.43. *Consider the optimal observer of Theorem 6.42. Then the innovation process*

$$y(i) - E(i)u(i) - C(i)\hat{x}(i), \qquad i \geq i_0, \qquad \textbf{6-443}$$

is a sequence of zero-mean uncorrelated stochastic vectors with variance matrices

$$C(i)Q(i)C^T(i) + V_2(i), \qquad i \geq i_0. \qquad \textbf{6-444}$$

That the innovation sequence is discrete-time white noise can be proved analogously to the continuous-time case. That the variance matrix of **6-443** is given by **6-444** follows by inspection.

6.5.6 Duality of the Optimal Observer and Regulator Problems; Steady-State Properties of the Optimal Observer

In this subsection we expose the duality of the linear discrete-time optimal regulator and observer problems. Here the following results are available.

Theorem 6.44. *Consider the linear discrete-time optimal regulator problem (DORP) of Definition 6.16 (Section 6.4.3) and the linear discrete-time optimal observer problem (DOOP) of Definition 6.19 (Section 6.5.4). Let in the observer problem $V_1(i)$ be given by*

$$V_1(i) = G(i)V_3(i)G^T(i), \quad i \geq i_0, \qquad \text{6-445}$$

where

$$V_3(i) > 0, \quad i \geq i_0. \qquad \text{6-446}$$

Suppose also that the state excitation noise and the observation noise are uncorrelated in the DOOP, that is,

$$V_{12}(i) = 0, \quad i \geq i_0. \qquad \text{6-447}$$

Let the various matrices occurring in the DORP and the DOOP be related as follows:

$A(i)$	*of the DORP equals* $A^T(i^* - i)$ *of the DOOP,*
$B(i)$	*of the DORP equals* $C^T(i^* - i)$ *of the DOOP,*
$D(i+1)$	*of the DORP equals* $G^T(i^* - i)$ *of the DOOP,*
$R_3(i+1)$	*of the DORP equals* $V_3(i^* - i)$ *of the DOOP,*
$R_2(i)$	*of the DORP equals* $V_2(i^* - i)$ *of the DOOP,*
P_1	*of the DORP equals* Q_0 *of the DOOP,*

all for $i \leq i_1 - 1$. *Here*

$$i^* = i_0 + i_1 - 1. \qquad \text{6-448}$$

Under these conditions the solutions of the DORP (Theorem 6.28, Section 6.4.3) and the DOOP (Theorem 6.42, Section 6.5.4) are related as follows.

(a) $P(i+1)$ *of the DORP equals* $Q(i^* - 1) - V_1(i^* - i)$ *of the DOOP for* $i \leq i_1 - 1$;
(b) $F(i)$ *of the DORP equals* $K^{0T}(i^* - i)$ *of the DOOP for* $i \leq i_1 - 1$;
(c) *The closed-loop regulator of the DORP,*

$$x(i+1) = [A(i) - B(i)F(i)]x(i), \qquad \text{6-449}$$

and the unforced reconstruction error equation of the DOOP,

$$e(i+1) = [A(i) - K^0(i)C(i)]e(i), \qquad \text{6-450}$$

are dual with respect to i^* *in the sense of Definition 6.9.*

The proof of this theorem follows by a comparison of the recursive matrix equations that determine the solutions of the regulator and observer problems. Because of duality, computer programs for regulator problems can be used for observer problems, and vice versa. Moreover, by using duality it is very

6.5 Optimal Reconstruction of the State

simple to derive the following results concerning the steady-state properties of the nonsingular optimal observer with uncorrelated state excitation and observation noises from the corresponding properties of the optimal regulator.

Theorem 6.45. *Consider the nonsingular optimal observer problem with uncorrelated state excitation and observation noises of Definition 6.19 (Section 6.5.4). Assume that $A(i)$, $C(i)$, $V_1(i) = G(i)V_3(i)G^T(i)$ and $V_2(i)$ are bounded for all i, and that*

$$V_3(i) \geq \alpha I, \qquad V_2(i) \geq \beta I, \qquad \text{for all } i, \qquad \text{6-451}$$

where α and β are positive constants.

(i) *Then if the system* **6-419** *is either*
  (a) *completely reconstructible, or*
  (b) *exponentially stable,*
and the initial variance $Q_0 = 0$, the variance $Q(i)$ of the reconstruction error converges to a steady-state solution $\bar{Q}(i)$ as $i_0 \to -\infty$, which satisfies the matrix difference equations **6-435**.

(ii) *Moreover, if the system*

$$x(i+1) = A(i)x(i) + G(i)w_3(i), \qquad y(i) = C(i)x(i), \qquad \text{6-452}$$

is either
  (c) *both uniformly completely reconstructible and uniformly completely controllable (from w_3), or*
  (d) *exponentially stable,*
the variance $Q(i)$ of the reconstruction error converges to $\bar{Q}(i)$ for $i_0 \to -\infty$ for any initial variance $Q_0 \geq 0$.

(iii) *If either condition (c) or (d) holds, the steady-state optimal observer, which is obtained by using the gain matrix $\bar{K}$ corresponding to the steady-state variance $\bar{Q}$, is exponentially stable.*

(iv) *Finally, if either condition (c) or (d) holds, the steady-state observer minimizes*

$$\lim_{i_0 \to -\infty} E\{e^T(i)W(i)e(i)\} \qquad \text{6-453}$$

for every initial variance Q_0. The minimal value of **6-453**, *which is achieved by the steady-state optimal observer, is given by*

$$\text{tr } [\bar{Q}(i)W(i)]. \qquad \text{6-454}$$

Similarly, it follows by "dualizing" Theorem 6.31 (Section 6.4.4) that, in the time-invariant nonsingular optimal observer problem with uncorrelated state excitation and observation noises, the properties mentioned under (ii), (iii), and (iv) hold provided the system **6-452** is both detectable and stabilizable.

536 Discrete-Time Systems

We leave it as an exercise for the reader to state the dual of Theorem 6.37 (Section 6.4.7) concerning the asymptotic behavior of the regulator poles.

6.6 OPTIMAL LINEAR DISCRETE-TIME OUTPUT FEEDBACK SYSTEMS

6.6.1 Introduction

In this section we consider the design of optimal linear discrete-time control systems where the state of the plant cannot be completely and accurately observed, so that an observer must be connected. This section parallels Chapter 5.

6.6.2 The Regulation of Systems with Incomplete Measurements

Consider a linear discrete-time system described by the state difference equation

$$x(i + 1) = A(i)x(i) + B(i)u(i), \qquad \textbf{6-455}$$

with the controlled variable

$$z(i) = D(i)x(i). \qquad \textbf{6-456}$$

In Section 6.4 we considered controlling this system with state feedback control laws of the form

$$u(i) = -F(i)x(i). \qquad \textbf{6-457}$$

Very often it is not possible to measure the complete state accurately, however, but only an observed variable of the form

$$y(i) = C(i)x(i) + E(i)u(i) \qquad \textbf{6-458}$$

is available. Assuming, as before, that $y(i)$ is the latest observation available for reconstructing $x(i + 1)$, we can connect an observer to this system of the form

$$\hat{x}(i + 1) = A(i)\hat{x}(i) + B(i)u(i) + K(i)[y(i) - E(i)u(i) - C(i)\hat{x}(i)]. \qquad \textbf{6-459}$$

Then a most natural thing to do is to replace the state x in **6-457** with its reconstructed value $\hat{x}$:

$$u(i) = -F(i)\hat{x}(i). \qquad \textbf{6-460}$$

We first consider the stability of the interconnection of the plant given by **6-455** and **6-458**, the observer **6-459**, and the control law **6-460**. We have the following result, completely analogous to the continuous-time result of Theorem 5.2 (Section 5.2.2).

6.6 Optimal Output Feedback Systems

Theorem 6.46. *Consider the interconnection of the system described by* **6-455** *and* **6-458**, *the observer* **6-459**, *and the control law* **6-460**. *Then sufficient conditions for the existence of gain matrices F(i) and K(i), $i \geq i_0$, such that the interconnected system is exponentially stable are that the system described by* **6-455** *and* **6-458** *be uniformly completely controllable and uniformly completely reconstructible, or that it be exponentially stable. In the time-invariant case (i.e., all matrices occurring in* **6-455**, **6-458**, **6-459**, *and* **6-460** *are constant) necessary and sufficient conditions for the existence of stabilizing gain matrices K and F are that the system given by* **6-455** *and* **6-458** *be both stabilizable and detectable. Moreover, in the time-invariant case, necessary and sufficient conditions for arbitrarily assigning all the closed-loop poles in the complex plane (within the restriction that complex poles occur in complex conjugate pairs) by suitably choosing the gain matrices K and F are that the system be both completely reconstructible and completely controllable.*

The proof of this theorem follows by recognizing that the reconstruction error

$$e(i) = x(i) - \hat{x}(i) \qquad \text{6-461}$$

satisfies the difference equation

$$e(i + 1) = [A(i) - K(i)C(i)]e(i). \qquad \text{6-462}$$

Substitution of $\hat{x}(i) = x(i) + e(i)$ into **6-460** yields for **6-455**

$$x(i + 1) = [A(i) - B(i)F(i)]x(i) + B(i)F(i)e(i). \qquad \text{6-463}$$

Theorem 6.46 then follows by application of Theorem 6.29 (Section 6.4.4), Theorem 6.45 (Section 6.5.4), Theorem 6.26 (Section 6.4.2), and Theorem 6.41 (Section 6.5.3). We moreover see from **6-462** and **6-463** that in the time-invariant case the characteristic values of the interconnected system comprise the characteristic values of $A - BF$ (the *regulator poles*) and the characteristic values of $A - KC$ (the *observer poles*).

A case of special interest occurs when in the time-invariant case all the regulator poles as well as the observer poles are assigned to the origin. Then we know from Section 6.5.3 that the observer will reconstruct the state completely accurately in at most n steps (assuming that n is the dimension of the state x), and it follows from Section 6.4.2 that after this the regulator will drive the system to the zero state in at most another n steps. Thus we have obtained an output feedback control system that reduces any initial state to the origin in at most $2n$ steps. We call such systems *output feedback state deadbeat control systems*.

Example 6.24. *Digital position output feedback state deadbeat control system*
Let us consider the digital positioning system of Example 6.2 (Section 6.2.3). In Example 6.13 (Section 6.3.3) we derived the state deadbeat control

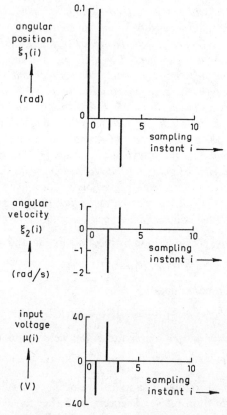

Fig. 6.22. Response of the output feedback state deadbeat position control system from the initial state col$[x(0), \hat{x}(0)]$ = col(0.1, 0, 0, 0). The responses are shown at the sampling instants only and not at intermediate times.

law for this system, while in Example 6.22 (Section 6.5.3) we found the deadbeat observer. In Fig. 6.22 we give the response of the interconnection of deadbeat control law, the deadbeat observer, and the system to the initial state

$$x(0) = \text{col }(0.1, \quad 0), \qquad \hat{x}(0) = 0. \qquad \textbf{6-464}$$

It is seen that the initial state is reduced to the zero state in four steps. Comparison with the state feedback deadbeat response of the same system, as depicted in Fig. 6.12 (Section 6.3.3), shows that the output feedback control system exhibits relatively large excursions of the state before it returns to the zero state, and requires larger input amplitudes.

6.6 Optimal Output Feedback Systems

6.6.3 Optimal Linear Discrete-Time Regulators with Incomplete and Noisy Measurements

We begin this section by defining the central problem.

Definition 6.20. *Consider the linear discrete-time system*

$$x(i+1) = A(i)x(i) + B(i)u(i) + w_1(i),$$
$$x(i_0) = x_0, \quad i \geq i_0, \quad \text{6-465}$$

where x_0 is a stochastic vector with mean $\bar{x}_0$ and variance matrix Q_0. The observed variable of the system is

$$y(i) = C(i)x(i) + E(i)u(i) + w_2(i). \quad \text{6-466}$$

The variables col $[w_1(i), w_2(i)]$ *form a sequence of uncorrelated stochastic vectors, uncorrelated with x_0, with zero means and variance matrices*

$$E\left\{\begin{pmatrix} w_1(i) \\ w_2(i) \end{pmatrix} (w_1^T(i), w_2^T(i))\right\} = \begin{pmatrix} V_1(i) & V_{12}(i) \\ V_{12}^T(i) & V_2(i) \end{pmatrix}, \quad i \geq i_0. \quad \text{6-467}$$

The controlled variable can be expressed as

$$z(i) = D(i)x(i). \quad \text{6-468}$$

Then the **stochastic linear discrete-time optimal output feedback regulator problem** *is the problem of finding the functional*

$$u(i) = f[y(i_0), y(i_0+1), \cdots, y(i-1), i], \quad i_0 \leq i \leq i_1 - 1, \quad \text{6-469}$$

such that the criterion

$$\sigma = E\left\{\sum_{i=i_0}^{i_1-1} [z^T(i+1)R_3(i+1)z(i+1) + u^T(i)R_2(i)u(i)] + x^T(i_1)P_1x(i_1)\right\} \quad \text{6-470}$$

is minimized. Here $R_3(i+1) > 0$ and $R_2(i) > 0$ for $i_0 \leq i \leq i_1 - 1$, and $P_1 \geq 0$.

As in the continuous-time case, the solution of this problem satisfies the separation principle (Gunckel and Franklin, 1963; Åström, 1970; Kushner, 1971).

Theorem 6.47. *The solution of the stochastic linear discrete-time optimal output feedback problem is as follows. The optimal input is given by*

$$u(i) = -F(i)\hat{x}(i), \quad i_0 \leq i \leq i_1 - 1, \quad \text{6-471}$$

where $F(i), i_0 \leq i \leq i_1 - 1$, is the sequence of gain matrices for the deterministic optimal regulator as given in Theorem 6.28 (Section 6.4.3). Furthermore, $\hat{x}(i)$

is the minimum mean-square linear estimator of $x(i)$ given $y(j)$, $i_0 \leq j \leq i-1$; $\hat{x}(i)$ for the nonsingular case [i.e., $V_2(i) > 0$, $i_0 \leq i \leq i_1 - 1$] can be obtained as the output of the optimal observer as described in Theorem 6.42 (Section 6.5.4).

We note that this theorem states *the* optimal solution to the stochastic linear discrete-time optimal output feedback problem and not just the optimal *linear* solution, as in the continuous-time equivalent of the present theorem (Theorem 5.3, Section 5.3.1). Theorem 6.47 can be proved analogously to the continuous-time equivalent.

We now consider the computation of the criterion **6-470**, where we restrict ourselves to the nonsingular case. The closed-loop control system is described by the relations

$$x(i+1) = A(i)x(i) + B(i)u(i) + w_1(i),$$
$$y(i) = C(i)x(i) + E(i)u(i) + w_2(i), \qquad \textbf{6-472}$$
$$u(i) = -F(i)\hat{x}(i),$$
$$\hat{x}(i+1) = A(i)\hat{x}(i) + B(i)u(i) + K(i)[y(i) - E(i)u(i) - C(i)\hat{x}(i)].$$

In terms of the reconstruction error,

$$e(i) = x(i) - \hat{x}(i), \qquad \textbf{6-473}$$

and the observer state $\hat{x}(i)$, **6-472** can be rewritten in the form

$$\begin{pmatrix} e(i+1) \\ \hat{x}(i+1) \end{pmatrix} = \begin{pmatrix} A(i) - K(i)C(i) & 0 \\ K(i)C(i) & A(i) - B(i)F(i) \end{pmatrix} \begin{pmatrix} e(i) \\ \hat{x}(i) \end{pmatrix}$$
$$+ \begin{pmatrix} I & -K(i) \\ 0 & K(i) \end{pmatrix} \begin{pmatrix} w_1(i) \\ w_2(i) \end{pmatrix}, \qquad \textbf{6-474}$$

with the initial condition

$$\begin{pmatrix} e(i_0) \\ \hat{x}(i_0) \end{pmatrix} = \begin{pmatrix} x(t_0) - \bar{x}_0 \\ \bar{x}_0 \end{pmatrix}. \qquad \textbf{6-475}$$

Defining the variance matrix of col $[e(i), \hat{x}(i)]$ as

$$E\left\{ \begin{pmatrix} e(i) - E\{e(i)\} \\ \hat{x}(i) - E\{\hat{x}(i)\} \end{pmatrix} (e^T(i) - E\{e^T(i)\}, \quad \hat{x}^T(i) - E\{\hat{x}^T(i)\}) \right\}$$
$$= \begin{pmatrix} Q_{11}(i) & Q_{12}(i) \\ Q_{12}^T(i) & Q_{22}(i) \end{pmatrix}, \quad i \geq i_0, \quad \textbf{6-476}$$

it can be found by application of Theorem 6.22 (Section 6.2.12) that the

6.6 Optimal Output Feedback Systems

matrices $Q_{jk}(i)$, $j, k = 1, 2$, satisfy difference equations, of which we give only that for Q_{22}:

$$\begin{aligned}Q_{22}(i+1) = {} & K(i)C(i)Q_{11}(i)C^T(i)K^T(i) \\ & + [A(i) - B(i)F(i)]Q_{12}^T(i)C^T(i)K^T(i) \\ & + K(i)C(i)Q_{12}(i)[A(i) - B(i)F(i)]^T \\ & + [A(i) - B(i)F(i)]Q_{22}(i)[A(i) - B(i)F(i)]^T \\ & + K(i)V_2(i)K^T(i), \qquad i \geq i_0, \end{aligned} \qquad \text{6-477}$$

with the initial condition

$$Q_{22}(i_0) = 0. \qquad \text{6-478}$$

Now obviously $Q_{11}(i) = Q(i)$, where $Q(i)$ is the variance matrix of the reconstruction error. Moreover, by setting up the difference equation for Q_{12}, it can be proved that $Q_{12}(i) = 0$, $i_0 \leq i \leq i_1 - 1$, which means that analogously with the continuous-time case the quantities $e(i)$ and $\hat{x}(i)$ are *uncorrelated* for $i_0 \leq i \leq i_1 - 1$. As a result, Q_{22} can be found from the difference equation

$$\begin{aligned} Q_{22}(i+1) = {} & K(i)[C(i)Q(i)C^T(i) + V_2(i)]K^T(i) \\ & + [A(i) - B(i)F(i)]Q_{22}(i)[A(i) - B(i)F(i)]^T, \end{aligned} \qquad \text{6-479}$$

$$Q_{22}(i_0) = 0.$$

When the variance matrix of col $[e(i), \hat{x}(i)]$ is known, all mean square and rms quantities of interest can be computed. In particular, we consider the criterion **6-470**. In terms of the variance matrix of col $(e, \hat{x})$ we write for the criterion:

$$\begin{aligned}\sigma = \bar{x}_0^T P(i_0)\bar{x}_0 + \text{tr} \bigg\{ & \sum_{i=i_0}^{i_1-1} \{R_1(i+1)[Q(i+1) + Q_{22}(i+1)] \\ & + F^T(i)R_2(i)F(i)Q_{22}(i)\} + P_1[Q(i_1) + Q_{22}(i_1)]\bigg\}, \end{aligned} \qquad \text{6-480}$$

where

$$R_1(i) = D^T(i)R_3(i)D(i), \qquad \text{6-481}$$

and $P(i)$ is defined in **6-248**. Let us separately consider the terms

$$\begin{aligned} & \text{tr} \bigg\{ \sum_{i=i_0}^{i_1-1} [R_1(i+1)Q_{22}(i+1) + F^T(i)R_2(i)F(i)Q_{22}(i)] + P_1 Q_{22}(i_1) \bigg\} \\ & = \text{tr} \bigg\{ \sum_{i=i_0+1}^{i_1-1} [R_1(i) + F^T(i)R_2(i)F(i)]Q_{22}(i) + [P_1 + R_1(i_1)]Q_{22}(i_1) \bigg\}, \end{aligned} \qquad \text{6-482}$$

where we have used the fact that $Q_{22}(i_0) = 0$. Now using the results of Problem 6.7, **6-482** can be rewritten as

$$\operatorname{tr}\left\{Q_{22}(i_0)\tilde{P}(i_0) + \sum_{j=i_0+1}^{i_1} \tilde{P}(j)K(j-1) \right. \\ \left. \cdot [C(j-1)Q(j-1)C^T(j-1) + V_2(j-1)]K^T(j-1)\right\}, \quad \textbf{6-483}$$

where $\tilde{P}$ satisfies the matrix difference equation

$$\tilde{P}(i-1) = [A(i-1) - B(i-1)F(i-1)]^T \tilde{P}(i) \\ \cdot [A(i-1) - B(i-1)F(i-1)] + R_1(i) + F^T(i)R_2(i)F(i), \quad \textbf{6-484}$$

$$\tilde{P}(i_1) = P_1 + R_1(i_1).$$

It is not difficult to recognize that $\tilde{P}(i) = P(i) + R_1(i)$, $i_0 + 1 \leq i \leq i_1$. By using this, substitution of **6-483** into **6-480** yields for the criterion

$$\sigma = \bar{x}_0^T P(i_0)\bar{x}_0 + \sum_{i=i_0}^{i_1-1} \operatorname{tr}\left\{R_1(i+1)Q(i+1) + [P(i+1) + R_1(i+1)]K(i) \right. \\ \left. \cdot [C(i)Q(i)C^T(i) + V_2(i)]K^T(i)\right\} + \operatorname{tr}[P_1 Q(i_1)]. \quad \textbf{6-485}$$

By suitable manipulations it can be found that the criterion can be expressed in the alternative form:

$$\sigma = \bar{x}_0^T P(i_0)\bar{x}_0 + \operatorname{tr}[P(i_0)Q_0] \\ + \sum_{i=i_0}^{i_1-1} \operatorname{tr}\left\{[R_1(i+1) + P(i+1)]V_1(i) + Q(i)F^T(i) \right. \\ \left. \cdot \{R_2(i) + B^T(i)[R_1(i+1) + P(i+1)]B(i)\}F(i)\right\}. \quad \textbf{6-486}$$

We can now state the following theorem.

Theorem 6.48. *Consider the stochastic output feedback regulator problem of Definition 6.20. Suppose that $V_2(i) > 0$ for all i. Then the following facts hold.*

(a) *The minimal value of the criterion* **6-470** *can be expressed in the alternative forms* **6-485** *and* **6-486**.

(b) *In the time-invariant case, in which the optimal observer and regulator problems have steady-state solutions as $i_0 \to -\infty$ and $i_1 \to \infty$, characterized by $\bar{Q}$ and $\bar{P}$, with corresponding steady-state gain matrices $\bar{K}$ and $\bar{F}$, the* *following* holds:

$$\lim_{\substack{i_0 \to -\infty \\ i_1 \to \infty}} \frac{1}{i_1 - i_0} E\left\{\sum_{i=i_0}^{i_1-1}[z^T(i+1)R_3(i+1)z(i+1) + u^T(i)R_2(i)u(i)]\right\}$$
$$= \lim_{i_0 \to -\infty} E\{z^T(i+1)R_3 z(i+1) + u^T(i)R_2 u(i)\}$$
$$= \operatorname{tr}[R_1\bar{Q} + (\bar{P} + R_1)\bar{K}(C\bar{Q}C^T + V_2)\bar{K}^T]$$
$$= \operatorname{tr}\{(R_1 + \bar{P})V_1 + Q\bar{F}^T[R_2 + B^T(R_1 + \bar{P})B]\bar{F}\}. \quad \textbf{6-487}$$

(c) *All mean square quantities of interest can be obtained from the variance matrix* diag $[Q(i), Q_{22}(i)]$ *of* col $[e(i), \hat{x}(i)]$. *Here* $e(i) = x(i) - \hat{x}(i)$, $Q(i)$ *is the variance matrix of* $e(i)$, *and* $Q_{22}(i)$ *can be obtained as the solution of the matrix difference equation*

$$Q_{22}(i+1) = [A(i) - B(i)F(i)]Q_{22}(i)[A(i) - B(i)F(i)]^T$$
$$+ K(i)[C(i)Q(i)C^T(i) + V_2(i)]K^T(i), \quad i \geq i_0, \quad \textbf{6-488}$$

$$Q_{22}(i_0) = 0.$$

The proof of part (b) of this theorem follows by application of part (a).

The general stochastic regulator problem can be specialized to tracking problems, regulation problems for systems with disturbances, and tracking problems for systems with disturbances, completely analogous to what we have discussed for the continuous-time case.

6.6.4 Nonzero Set Points and Constant Disturbances

The techniques developed in Section 5.5 for dealing with time-invariant regulators and tracking systems with nonzero set points and constant disturbances can also be applied to the discrete-time case. We first consider the case where the system has a nonzero set point z_0 for the controlled variable. The system state difference equation is

$$x(i+1) = Ax(i) + Bu(i) + w_1(i), \quad i \geq i_0, \quad \textbf{6-489}$$

the controlled variable is

$$z(i) = Dx(i), \quad i \geq i_0, \quad \textbf{6-490}$$

and the observed variable is

$$y(i) = Cx(i) + Eu(i) + w_2(i), \quad i \geq i_0. \quad \textbf{6-491}$$

The joint process col (w_1, w_2) is given as in Definition 6.20 (Section 6.6.3). From Section 6.4.6 it follows that the nonzero set point controller is specified by

$$u(i) = -\bar{F}\hat{x}(i) + H_c^{-1}(1)\hat{z}_0, \quad \textbf{6-492}$$

where $\bar{F}$ is a suitable feedback gain matrix, and

$$H_c(z) = D(zI - A + B\bar{F})^{-1}B \qquad \text{6-493}$$

is the (square) closed-loop transfer matrix (assuming that dim (z) = dim (u)). Furthermore, $\hat{x}(i)$ is the minimum mean square estimator of $x(i)$ and $\hat{z}_0$ that of z_0.

How $\hat{z}_0$ is obtained depends on how we model the set point. If we assume that the set point varies according to

$$z_0(i + 1) = z_0(i) + w_0(i), \qquad \text{6-494}$$

and that we observe

$$r(i) = z_0(i) + w_s(i), \qquad \text{6-495}$$

where col (w_0, w_s) constitutes a white noise sequence, the steady-state optimal observer for the set point is of the form

$$\hat{z}_0(i + 1) = \hat{z}_0(i) + \bar{K}_r[r(i) - \hat{z}_0(i)]. \qquad \text{6-496}$$

This observer in conjunction with the control law **6-492** yields a zero-steady-state-error response when the reference variable $r(i)$ is constant.

Constant disturbances can be dealt with as follows. Let the state difference equation be given by

$$x(i + 1) = Ax(i) + Bu(i) + v_0 + w_1(i), \qquad \text{6-497}$$

where v_0 is a constant disturbance. The controlled variable and observed variable are as given before. Then from Section 6.4.6, we obtain the zero-steady-state-error control law

$$u(i) = -\bar{F}\hat{x}(i) - H_c^{-1}(1)D(I - \bar{A})^{-1}\hat{v}_0, \qquad \text{6-498}$$

with all quantities defined as before, $\bar{A} = A - BF$, and $\hat{v}_0$ an estimate of v_0. In order to obtain $\hat{v}_0$, we model the constant disturbance as

$$v_0(i + 1) = v_0(i) + w_0(i), \qquad \text{6-499}$$

where w_0 constitutes a white noise sequence. The steady-state optimal observer for $x(i)$ and $z_0(i)$ will be of the form

$$\begin{aligned}\hat{x}(i + 1) &= A\hat{x}(i) + Bu(i) + \hat{v}_0(i) + \bar{K}_1[y(i) - C\hat{x}(i) - E(i)u(i)], \\ \hat{v}_0(i + 1) &= \hat{v}_0(i) + \bar{K}_2[y(i) - C\hat{x}(i) - E(i)u(i)].\end{aligned} \qquad \text{6-500}$$

This observer together with the control law **6-498** produces a zero-steady-state-error response to a constant disturbance. This is a form of integral control.

6.6 Optimal Output Feedback Systems

Example 6.25. *Integral control of the digital positioning system*

Consider the digital positioning system of previous examples. In Example 6.14 (Section 6.4.3), we obtained the state feedback control law

$$u(i) = -\bar{F}x(i) = -(110.4, \quad 12.66)x(i). \qquad \text{6-501}$$

Assuming that the servo motor is subject to constant disturbances in the form of constant torques on the shaft, we must include a term of the form

$$v_0 = \begin{pmatrix} 0.003396 \\ 0.06308 \end{pmatrix} \alpha \qquad \text{6-502}$$

in the state difference equation **6-26**, where α is a constant. It is easily seen that with the state feedback law **6-501** this leads to the zero-steady-state-error control law

$$\mu(i) = -\bar{F}\hat{x}(i) - \hat{\alpha}(i). \qquad \text{6-503}$$

The observer **6-500** is in this case of the form

$$\hat{x}(i+1) = \begin{pmatrix} 1 & 0.08015 \\ 0 & 0.6313 \end{pmatrix} \hat{x}(i) + \begin{pmatrix} 0.003396 \\ 0.06308 \end{pmatrix} [\mu(i) + \hat{\alpha}(i)]$$

$$+ \begin{pmatrix} k_1 \\ k_2 \end{pmatrix} [\eta(i) - (1, \quad 0)\hat{x}(i)],$$

$$\hat{\alpha}(i+1) = \hat{\alpha}(i) + k_3[\eta(i) - (1, \quad 0)\hat{x}(i)]. \qquad \text{6-504}$$

Here it has been assumed that

$$\eta(i) = (1, \quad 0)x(i) \qquad \text{6-505}$$

is the observed variable (i.e., the whole sampling interval is used for processing the data), and k_1, k_2, and k_3 are scalar gains to be selected. We choose these gains such that the observer is a deadbeat observer; this results in the following values:

$$k_1 = 2.6313, \qquad k_2 = 18.60, \qquad k_3 = 158.4. \qquad \text{6-506}$$

Figure 6.23 shows the response of the resulting zero-steady-state-error control system from zero initial conditions to a relatively large constant disturbance of 10 V (i.e., the disturbing torque is equivalent to a constant additive input voltage of 10 V). It is seen that the magnitude of the disturbance is identified after three sampling intervals, and that it takes the system another three to four sampling intervals to compensate fully for the disturbance.

546 Discrete-Time Systems

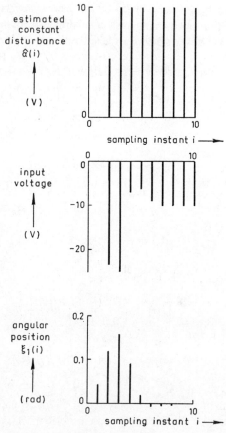

Fig. 6.23. Response of the digital positioning system with integral control from zero initial conditions to a constant disturbance.

6.7 CONCLUSIONS

In this chapter we have summarized the main results of linear optimal control theory for discrete-time systems. As we have seen, in many instances the continuous-time theory can be extended to the discrete-time case in a fairly straightforward manner. This chapter explicitly reviews most of the results needed in linear discrete-time control system design.

Although in many respects the discrete-time theory parallels the continuous-time theory, there are a few differences. One of the striking dissimilarities is that, in theory, continuous-time control systems can be made arbitrarily fast. This cannot be achieved with discrete-time systems, where the speed of action is restricted by the sampling interval. The fastest type of control that can be achieved with discrete-time systems is deadbeat control.

In this chapter we have usually considered linear discrete-time systems thought to be derived from continuous-time systems by sampling. We have not paid very much attention to what happens *between* the sampling interval instants, however, except by pointing out in one or two examples that the behavior at the sampling instants may be misleading for what happens in between. This is a reason for caution. As we have seen in the same examples, it is often possible to modify the discrete-time problem formulation to obtain a more acceptable design.

The most fruitful applications of linear discrete-time control theory lie in the area of computer process control, a rapidly advancing field.

6.8 PROBLEMS

6.1. *A modified discrete-time regulator problem*
Consider the linear discrete-time system

$$x(i+1) = A(i)x(i) + B(i)u(i), \qquad \text{6-507}$$

with the modified criterion

$$\sum_{i=i_0}^{i_1-1} [x^T(i)R_1(i)x(i) + 2x^T(i)R_{12}(i)u(i) + u^T(i)R_2(i)u(i)]$$
$$+ x^T(i_1)P_1x(i_1). \qquad \text{6-508}$$

Show that minimizing **6-508** for the system **6-507** is equivalent to a standard discrete-time regulator problem where the criterion

$$\sum_{i=i_0}^{i_1-1} [x^T(i+1)R_1'(i+1)x(i+1) + u'^T(i)R_2(i)u'(i)] + x^T(i_1)P_1x(i_1) \qquad \text{6-509}$$

is minimized for the system

$$x(i+1) = A'(i)x(i) + B(i)u'(i), \qquad \text{6-510}$$

with

$$R_1'(i) = \begin{cases} R_1(i) - R_{12}(i)R_2^{-1}(i)R_{12}^T(i), & i = i_0 + 1, i_0 + 2, \cdots, i_1 - 1, \\ 0, & i = i_1, \end{cases}$$

$$u'(i) = u(i) + R_2^{-1}(i)R_{12}^T(i)x(i), \qquad i = i_0, i_0 + 1, \cdots, i_1 - 1, \qquad \text{6-511}$$

$$A'(i) = A(i) - B(i)R_2^{-1}(i)R_{12}^T(i), \qquad i = i_0, i_0 + 1, \cdots, i_1 - 1.$$

6.2. *Stochastic state feedback regulator problems structured as regulator problems with disturbances*

Consider the linear discrete-time system

$$x(i+1) = A(i)x(i) + B(i)u(i) + v(i),$$
$$z(i) = D(i)x(i).$$
6-512

Here the disturbance variable v is modeled as

$$v(i) = D_d(i)x_d(i),$$
$$x_d(i+1) = A_d(i)x_d(i) + w_d(i),$$
6-513

where the $w_d(i)$, $i \geq i_0$, form a sequence of uncorrelated stochastic vectors with given variance matrices. Consider also the criterion

$$E\left\{\sum_{i=i_0}^{i_1-1}[z^T(i+1)R_3(i+1)z(i+1) + u^T(i)R_2(i)u(i)] + x^T(i_1)P_1x(i_1)\right\}.$$
6-514

(a) Show how the problem of controlling the system such that the criterion 6-514 is minimized can be converted into a standard stochastic regulator problem.

(b) Show that the optimal control law can be expressed as

$$u(i) = -F(i)x(i) - F_d(i)x_d(i), \quad i = i_0, i_0 + 1, \cdots, i_1 - 1,$$
6-515

where the feedback gain matrices $F(i)$, $i = i_0, \cdots, i_1 - 1$, are completely independent of the properties of the disturbance variable.

6.3. Stochastic state feedback regulator problems structured as tracking problems

Consider the linear discrete-time system

$$x(i+1) = A(i)x(i) + B(i)u(i),$$
$$z(i) = D(i)x(i).$$
6-516

Consider also a reference variable z_r, which is modeled through the equations

$$z_r(i) = D_r(i)x_r(i),$$
$$x_r(i+1) = A_r(i)x_r(i) + w_r(i),$$
6-517

where $w_r(i)$, $i \geq i_0$, forms a sequence of uncorrelated stochastic vectors with variance matrices $V_r(i)$. Consider as well the criterion

$$E\left\{\sum_{i=i_0}^{i_1-1}[z(i+1) - z_r(i+1)]^T R_3(i+1)[z(i+1) - z_r(i+1)] \right.$$
$$\left. + u^T(i)R_2(i)u(i)\right\}.$$
6-518

(a) Show how the problem of controlling the system such that the criterion **6-518** is minimized can be converted into a standard stochastic discrete-time optimal regulator problem.

(b) Show that the optimal control law can be expressed in the form

$$u(i) = -F(i)x(i) + F_r(i)x_r(i), \qquad i = i_0, i_0 + 1, \cdots, i_1 - 1, \quad \textbf{6-519}$$

where the feedback gain matrices $F(i)$, $i = i_0, \cdots, i_1 - 1$, are completely independent of the properties of the reference variable.

6.4. *The closed-loop regulator poles*

Prove the following generalization of Theorem 6.37 (Section 6.4.7). Consider the steady-state solution of the time-invariant linear discrete-time optimal regulator problem. Suppose that dim $(z) = $ dim (u) and let

$$H(z) = D(zI - A)^{-1}B,$$

$$\det [H(z)] = \frac{\psi(z)}{\phi(z)},$$

$$\phi(z) = z^{n-q} \prod_{i=1}^{q} (z - \pi_i), \qquad \text{with } \pi_i \neq 0, i = 1, 2, \cdots, q, \qquad \textbf{6-520}$$

$$\psi(z) = z^{s-p} \prod_{i=1}^{p} (z - \nu_i), \qquad \text{with } \nu_i \neq 0, i = 1, 2, \cdots, p,$$

and

$$R_2 = \rho N,$$

with $N > 0$ and ρ a positive scalar. Finally, set $r = \max(p, q)$. Then:

(a) Of the n closed-loop regulator poles, $n - r$ always stay at the origin.

(b) As $\rho \downarrow 0$, of the remaining r closed-loop poles, p approach the numbers $\hat{\nu}_i$, $i = 1, 2, \cdots, p$, which are defined as in **6-363**.

(c) As $\rho \downarrow 0$, the $r - p$ other closed-loop poles approach the origin.

(d) As $\rho \downarrow \infty$, of the r nonzero closed-loop poles, q approach the numbers $\hat{\pi}_i$, $i = 1, \cdots, q$, which are defined as in **6-364**.

(e) As $\rho \downarrow \infty$, the $r - p$ other nonzero closed-loop poles approach the origin.

6.5. *Mixed continuous-time discrete-time regulator problem*

Consider the discrete-time system that results from applying a piecewise constant input to the continuous-time system

$$\dot{x}(t) = A(t)x(t) + B(t)u(t). \qquad \textbf{6-521}$$

Use the procedure and notation of Section 6.2.3 in going from the continuous-time to the discrete-time version. Suppose now that one wishes to take into account the behavior of the system between the sampling instants and consider therefore the *integral* criterion (rather than a *sum* criterion)

$$\int_{t_{i_0}}^{t_{i_1}} [x^T(t)R_1(t)x(t) + u^T(t)R_2(t)u(t)] \, dt + x^T(t_{i_1})P_1 x(t_{i_1}). \quad \text{6-522}$$

Here t_{i_0} is the first sampling instant and t_{i_1} the last.

(a) Show that minimizing the criterion **6-522**, while the system **6-521** is commanded by stepwise constant inputs, is equivalent to minimizing an expression of the form

$$\sum_{i=i_0}^{i_1-1} [x^T(t_i)R'_1(i)x(t_i) + 2x^T(t_i)R'_{12}(i)u(t_i) + u^T(t_i)R'_2(i)u(t_i)]$$

$$+ x^T(t_{i_1})P_1 x(t_{i_1}) \quad \text{6-523}$$

for the discrete-time system

$$x(t_{i+1}) = \Phi(t_{i+1}, t_i)x(t_i) + \left[\int_{t_i}^{t_{i+1}} \Phi(t_{i+1}, \tau)B(\tau) \, d\tau\right] u(t_i), \quad \text{6-524}$$

where $\Phi(t, t_0)$ is the transition matrix of the system **6-521**. Derive expressions for $R'_1(i)$, $R'_{12}(i)$, and $R'_2(i)$.

(b) Suppose that A, B, R_1, and R_2 are constant matrices and also let the sampling interval $t_{i+1} - t_i = \Delta$ be constant. Show that if the sampling interval is small first approximations to R'_1, R'_{12}, and R'_2 are given by

$$R'_1 \simeq R_1 \Delta,$$

$$R'_{12} \simeq \tfrac{1}{2} R_1 B \Delta^2, \quad \text{6-525}$$

$$R'_2 \simeq (R_2 + \tfrac{1}{3} B^T R_1 B \Delta^2)\Delta.$$

6.6. *Alternative version of the discrete-time optimal observer problem*

Consider the system

$$x(i + 1) = A(i)x(i) + B(i)u(i) + w_1(i),$$
$$y(i) = C(i)x(i) + E(i)u(i) + w_2(i), \quad i \geq i_0, \quad \text{6-526}$$

where col $[w_1(i), w_2(i)]$, $i \geq i_0$, forms a sequence of zero-mean uncorrelated vector stochastic variables with variance matrices

$$\begin{pmatrix} V_1(i) & V_{12}(i) \\ V_{12}^T(i) & V_2(i) \end{pmatrix}, \quad i \geq i_0. \quad \text{6-527}$$

Furthermore, $x(i_0)$ is a vector stochastic variable, uncorrelated with w_1 and w_2, with mean $\bar{x}_0$ and variance matrix Q_0. Show that the best linear estimator of $x(i)$ operating on $y(j)$, $i_0 \leq j \leq i$ (not $i - 1$, as in the version of Section 6.5), can be described as follows:

$$\hat{x}(i + 1) = [I - K(i + 1)C(i + 1)][A(i)\hat{x}(i) + B(i)u(i)]$$
$$+ K(i + 1)[y(i + 1) - E(i + 1)u(i + 1)], \quad i \geq i_0. \quad \text{6-528}$$

Here the gain matrices K are obtained from the iterative relations

$$K(i+1) = [S(i+1)C^T(i+1) + V_{12}(i)][C(i+1)S(i+1)C^T(i+1)$$
$$+ C(i+1)V_{12}(i) + V_{12}^T(i)C^T(i+1) + V_2(i)]^{-1},$$

$$S(i+1) = A(i)Q(i)A^T(i) + V_1(i),$$

$$Q(i+1) = [I - K(i+1)C(i+1)]S(i+1) - K(i+1)V_{12}^T(i), \qquad \text{6-529}$$

all for $i \geq i_0$. Here $Q(i)$ is the variance matrix of the reconstruction error $x(i) - \hat{x}(i)$, and $S(i)$ is an auxiliary matrix. The initial condition for **6-528** is given by

$$\hat{x}(i_0) = [I - K(i_0)C(i_0)]\bar{x}_0 + K(i_0)[y(i_0) - E(i_0)u(i_0)], \qquad \text{6-530}$$

where

$$K(i_0) = Q_0 C^T(i_0)[C(i_0)Q_0 C^T(i_0) + V_2(i_0)]^{-1}. \qquad \text{6-531}$$

The initial variance matrix, which serves as initial condition for the iterative equations **6-529**, is given by

$$Q(i_0) = [I - K(i_0)C(i_0)]Q_0. \qquad \text{6-532}$$

Hint: To derive the observer equation, express $y(i+1)$ in terms of $x(i)$ and use the standard version of the observer problem given in the text.

6.7. *Property of a matrix difference equation*
Consider the matrix difference equation

$$Q(i+1) = A(i)Q(i)A^T(i) + R(i), \qquad i_0 \leq i \leq i_1 - 1, \qquad \text{6-533}$$

together with the linear expression

$$\operatorname{tr}\left[\sum_{j=i_0}^{i_1-1} Q(j)S(j) + P_1 Q(i_1)\right]. \qquad \text{6-534}$$

Prove that this expression can also be written as

$$\operatorname{tr}\left[Q_0 P(i_0) + \sum_{j=i_0+1}^{i_1} R(j-1)P(j)\right], \qquad \text{6-535}$$

where the sequence of matrices $P(j)$, $i_0 \leq j \leq i_1$, satisfies the matrix difference equation

$$P(i-1) = A^T(i-1)P(i)A(i-1) + S(i-1), \qquad i_0 + 1 \leq i \leq i_1, \qquad \text{6-536}$$

$$P(i_1) = P_1.$$

6.8. *Linear discrete-time optimal output feedback controllers of reduced dimensions*

Consider the linear time-invariant discrete-time system

$$x(i + 1) = Ax(i) + Bu(i) + w_1(i), \quad x(i_0) = x_0,$$
$$z(i) = D(i)x(i), \quad \text{6-537}$$
$$y(i) = C(i)x(i) + E(i)u(i) + w_2(i),$$

all for $i \geq i_0$, where col $[w_1(i), w_2(i)]$, $i \geq i_0$, forms a sequence of uncorrelated stochastic vectors uncorrelated with x_0. Consider for this system the time-invariant controller

$$q(i + 1) = Lq(i) + K_1 y(i),$$
$$u(i) = -Fq(i) - K_2 y(i). \quad \text{6-538}$$

Assume that the interconnection of controller and plant is asymptotically stable.

(a) Develop matrix relations that can be used to compute expressions of the form

$$\lim_{i_0 \to -\infty} E\{z^T(i) R_3 z(i)\} \quad \text{6-539}$$

and

$$\lim_{i_0 \to -\infty} E\{u^T(i) R_2 u(i)\}. \quad \text{6-540}$$

Presuming that computer programs can be developed that determine the controller matrices L, K_1, F, and K_2 such that **6-539** is minimized while **6-540** is constrained to a given value, outline a method for determining discrete-time optimal output feedback controllers of reduced dimensions. (Compare the continuous-time approach discussed in Section 5.7.)

(b) When gradient methods are used to solve numerically the optimization problem of (a), the following result is useful. Let M, N, and R be given matrices of compatible dimensions, each depending upon a parameter γ. Let $\bar{S}$ be the solution of the linear matrix equation

$$\bar{S} = M\bar{S}M^T + N, \quad \text{6-541}$$

and consider the scalar

$$\text{tr}\,(\bar{S}R) \quad \text{6-542}$$

as a function of γ. Then the gradient of **6-542** with respect to γ is given by

$$\frac{\partial}{\partial \gamma}[\text{tr}\,(\bar{S}R)] = \text{tr}\left(\frac{\partial R}{\partial \gamma}\bar{S} + \frac{\partial N}{\partial \gamma}\bar{U} + 2\bar{U}\frac{\partial M}{\partial \gamma}\bar{S}M^T\right), \quad \text{6-543}$$

where $\bar{U}$ is the solution of the adjoint matrix equation

$$\bar{U} = M^T \bar{U} M + R. \quad \text{6-544}$$

Prove this.

REFERENCES

B. D. O. Anderson (1966a), "The inverse problem of optimal control," Technical Report No. 6560-3, Stanford Electronics Laboratories, Stanford University, Stanford, Calif.

B. D. O. Anderson (1966b), "Solution of quadratic matrix equations," *Electron. Letters,* **2,** 10, pp. 371–372.

B. D. O. Anderson and D. G. Luenberger (1967), "Design of multivariable feedback systems," *Proc. IEE,* **114,** 3, pp. 395–399.

B. D. O. Anderson and J. B. Moore (1971), *Linear Optimal Control,* Prentice-Hall, Englewood Cliffs, N.J.

M. Aoki (1968), "Control of large-scale dynamic systems by aggregation," *IEEE Trans. Autom. Control,* **13,** 3, pp. 246–253.

K. J. Åström (1970), *Introduction to Stochastic Control Theory,* Academic Press, New York.

K. J. Åström, R. W. Koepcke, and F. Tung (1962), "On the control of linear discrete dynamic systems with quadratic loss," Research Report RJ-222, IBM, San Jose Research Laboratory, San Jose, Calif.

M. Athans and P. L. Falb (1966), *Optimal Control, An Introduction to the Theory and Its Applications,* McGraw-Hill, New York.

S. Barnett and C. Storey (1967), "Remarks on numerical solution of the Lyapunov matrix equation," *Electron. Letters,* **3,** p. 417.

S. Barnett and S. Storey (1970), *Matrix Methods in Stability Theory,* Nelson, London.

B. B. Barrow (1966), "IEEE takes a stand on units," *Spectrum,* **3,** 3, pp. 164–173.

R. W. Bass (1967), "Machine solution of high-order matrix Riccati equations," Douglas Paper No. 4538, Douglas Aircraft, Missile and Space Systems Division.

R. W. Bass and I. Gura (1965), "High order system design via state-space considerations," *Preprints, 1965 Joint Automatic Control Conference,* pp. 311–318, Rensselaer Polytechnic Institute, Troy, N.Y., June 22–25.

R. E. Bellman (1957), *Dynamic Programming,* Princeton Univ. Press, Princeton, N.J.

C. S. Berger (1971), "A numerical solution of the matrix equation $P = \phi P \phi^t + S$," *IEEE Trans. Autom. Control,* **16,** 4, pp. 381–382.

G. S. G. Beveridge and R. S. Schechter (1970), *Optimization: Theory and Practice,* McGraw-Hill, New York.

Th. A. Bickart (1968), "Matrix exponential: Approximation by truncated power series," *Proc. IEEE,* **56,** 5, pp. 872–873.

T. R. Blackburn (1968), "Solution of the algebraic Riccati equation via Newton–Raphson iteration," *Preprints, 1968 Joint Automatic Control Conference,* pp. 940–945, University of Michigan, Ann Arbor, Mich., June 26–28.

T. R. Blackburn and J. C. Bidwell (1968), "Some numerical aspects of control engineering computations," *Preprints, 1968 Joint Automatic Control Conference,* pp. 203–207, University of Michigan, Ann Arbor, Mich., June 26–28.

J. H. Blakelock (1965), *Automatic Control of Aircraft and Missiles,* Wiley, New York.

H. W. Bode (1945), *Network Analysis and Feedback Amplifier Design,* Van Nostrand, Princeton, N.J.

K. G. Brammer (1968), "Lower order optimal filtering of nonstationary random sequences," *IEEE Trans Autom. Control,* **13,** 2, pp. 198–199.

F. M. Brasch and J. B. Pearson (1970), "Pole placement using dynamic compensators," *IEEE Trans. Autom. Control*, **15**, 1, pp. 34–43.

R. W. Brockett (1970), *Finite Dimensional Linear Systems*, Wiley, New York.

A. E. Bryson and D. E. Johansen (1965), "Linear filtering for time-varying systems using measurements containing colored noise," *IEEE Trans. Autom. Control*, **10**, 1, pp. 4–10.

R. S. Bucy (1967a), "Global theory of the Riccati equation," *J. Comp. Systems Sci.*, **1**, p. 349–361.

R. S. Bucy (1967b), "Two-point boundary value problems of linear Hamiltonian systems," *SIAM J. Appl. Math.*, **15**, 6, pp. 1385–1389.

R. S. Bucy and J. Ackermann (1970), "Ueber die Anzahl der Parameter von Mehrgrössensystemen," *Regelungstechnik*, **18**, 10, pp. 451–452.

R. S. Bucy and P. D. Joseph (1968), *Filtering for Stochastic Processes with Applications to Guidance*, Interscience, New York.

S. Butman (1968), "A method for optimizing control-free costs in systems with linear controllers," *IEEE Trans. Autom. Control*, **13**, 5, pp. 554–556.

J. A. Cadzow (1968), "Nilpotency property of the discrete regulator," *IEEE Trans. Autom. Control*, **13**, 6, pp. 734–735.

P. E. Caines and D. Q. Mayne (1970), "On the discrete time matrix Riccati equation of optimal control," *Intern. J. Control*, **12**, 5, pp. 785–794.

P. E. Caines and D. Q. Mayne (1971), "On the discrete time matrix equation of optimal control—A correction," *Intern. J. Control*, **14**, pp. 205–207.

R. H. Cannon, Jr. (1967), *Dynamics of Physical Systems*, McGraw-Hill, New York.

S. S. L. Chang (1961), *Synthesis of Optimum Control Systems*, McGraw-Hill, New York.

C. T. Chen (1968a), "Stability of linear multivariable feedback systems," *Proc. IEEE*, **56**, 5, pp. 821–828.

C. T. Chen (1968b), "A note on pole assignment," *IEEE Trans. Autom. Control*, **13**, 5, p. 597–598.

C. F. Chen and L. S. Shieh (1968a), "A note on expanding $PA + A^T P = -Q$," *IEEE Trans. Autom. Control*, **13**, 1, pp. 122–123.

C. F. Chen and L. S. Shieh (1968b), "A novel approach to linear model simplification," *Preprints, 1968 Joint Automatic Control Conference*, pp. 454–461, University of Michigan, Ann Arbor, Mich., June 26–28.

M. R. Chidambara and R. B. Schainker (1971), "Lower order generalized aggregated model and suboptimal control," *IEEE Trans. Autom. Control*, **16**, 2, pp. 175–180.

J. B. Cruz and W. R. Perkins (1964), "A new approach to the sensitivity problem in multivariable feedback system design," *IEEE Trans. Autom. Control*, **9**, 3, pp. 216–222.

S. D. G. Cumming (1969), "Design of observers of reduced dynamics," *Electron. Letters*, **5**, 10, pp. 213–214.

W. B. Davenport and W. L. Root (1958), *An Introduction to the Theory of Random Signals and Noise*, McGraw-Hill, New York.

H. T. Davis (1962), *Introduction to Nonlinear Differential and Integral Equations*, Dover, New York.

E. J. Davison (1968a), "A new method for simplifying large linear dynamical systems," *IEEE Trans. Autom. Control*, **13**, 2, pp. 214–215.

E. J. Davison (1968b), "On pole assignment in multivariable linear systems," *IEEE Trans. Autom. Control*, **13**, 6, pp. 747–748.

E. J. Davison and F. T. Man (1968), "The numerical solution of $A'Q + QA = -C$," *IEEE Trans. Autom. Control*, **13**, 4, pp. 448–449.

E. J. Davison and H. W. Smith (1971), "Pole assignment in linear time-invariant multivariable systems with constant disturbances," *Automatica*, **7**, 4, pp. 489–498.

J. J. D'Azzo and C. H. Houpis (1966), *Feedback Control System Analysis and Synthesis*, 2nd ed., McGraw-Hill, New York.

C. A. Desoer (1970), *Notes for a Second Course on Linear Systems*, Van Nostrand Reinhold, New York.

J. J. Deyst, Jr. and C. F. Price (1968), "Conditions for asymptotic stability of the discrete minimum-variance linear estimator," *IEEE Trans. Autom. Control*, **13**, 6, pp. 702–705.

U. diCaprio and P. P. Wang (1969), "A study of the output regulator problem for linear systems with input vector," *Proc. Seventh Annual Allerton Conference on Circuit and System Theory*, pp. 186–188, Institute of Electrical and Electronics Engineers Catalog No. 69 C 48-CT.

J. L. Doob (1953), *Stochastic Processes*, Wiley, New York.

K. Eklund (1969), "Multivariable control of a boiler—An application of linear quadratic control theory," Report 6901, Lund Institute of Technology, Division of Automatic Control, Lund, Sweden.

O. I. Elgerd (1967), *Control Systems Theory*, McGraw-Hill, New York.

W. Everling (1967), "On the evaluation of e^{AT} by power series," *Proc. IEEE*, **55**, 3, p. 413.

J. B. Farison, F.-C. Fu (1970), "The matrix properties of minimum-time discrete linear regulator control," *IEEE Trans. Autom. Control*, **15**, 3, pp. 390–391.

A. F. Fath (1968), "Evaluation of a matrix polynomial," *IEEE Trans. Autom. Control*, **13**, 2, pp. 220–221.

A. F. Fath (1969), "Computational aspects of the linear optimal regulator problem," *IEEE Trans. Autom. Control*, **14**, 5, pp. 547–550.

W. H. Fleming (1969), "Controlled diffusions under polynomial growth conditions," in *Control Theory and the Calculus of Variations*, A. V. Balakrishnan, Ed., pp. 209–234, Academic Press, New York.

G. E. Forsythe and L. W. Strauss (1955), "The Souriau-Frame characteristic equation algorithm on a digital computer," *J. Math. Phys.*, **34**, pp. 152–156.

A. Fossard (1970), "On a method for simplifying linear dynamic systems," *IEEE Trans. Autom. Control*, **15**, 2, pp. 261–262.

J. S. Frame (1964), "Matrix functions and applications, Part IV," *Spectrum*, **1**, 6, pp. 123–131.

H. Freeman (1965), *Discrete-Time Systems*, Wiley, New York.

W. C. Freested, R. F. Webber, and R. W. Bass (1968), "The 'GASP' computer program—An integrated tool for optimal control and filter design," *Preprints, 1968 Joint Automatic Control Conference*, pp. 198–202, University of Michigan, Ann Arbor, Mich., June 26–28.

E. A. Gal'perin and N. N. Krasovskii (1963), "On the stabilization of stationary motions in nonlinear control systems," *J. Appl. Math. Mech.* (Translation of *Prikl. Mat. Mekh.*), **27**, pp. 1521–1546.

I. I. Gikhman and A. V. Skorokhod (1969), *Introduction to the Theory of Random Processes*, W. B. Saunders, Philadelphia.

T. L. Gunckel and G. F. Franklin (1963), "A general solution for linear, sampled-data control," *J. Basic Eng., Trans. ASME*, Ser. D, **85**, pp. 197–203.

P. Hagander (1972), "Numerical solution of $A^T S + SA + Q = 0$," *Information Sci.*, **4**, pp. 35–50.

P. H. Haley (1967), "Design of low-order feedback controllers for linear multivariable systems," Report CCS-10, Department of Engineering-Economic Systems, Stanford University, Stanford, Calif.

T. C. Hendricks and G. W. Haynes (1968), "The 'GASP' Computer Program," *Conference Record, Second Asilomar Conference on Circuits and Systems*, Pacific Grove, Calif.

M. Heymann (1968), "Comments on pole assignment in multi-input controllable linear systems," *IEEE Trans. Autom. Control*, **13**, 6, pp. 748–749.

T. Hida (1970), *Stationary Stochastic Processes*, Mathematical Notes, Princeton University Press, Princeton, N.J.

I. M. Horowitz (1963), *Synthesis of Feedback Systems*, Academic Press, New York.

C.-H. Hsu, C.-T. Chen (1968), "A proof of the stability of multivariable feedback systems," *Proc. IEEE*, **56**, 11, pp. 2061–2062.

IEEE Standards Committee (1971), "IEEE recommended practice: Rules for the use of units of the international system of units," Adopted December 3, 1970, reprinted in *Spectrum* **8**, 3, pp. 77–78.

International Organization for Standardization (various dates from 1958 to 1965), *Recommendations*, ISO/R31, Parts, I, II, III, IV, V, VII, and XI.

A. Jameson (1968), "Solution of the equation $AX + XB = C$ by inversion of an $m \times m$ or $n \times n$ matrix," *SIAM J. Appl. Math.*, **16**, 5, pp. 1020–1023.

A. H. Jazwinski (1970), *Stochastic Processes and Filtering Theory*, Academic Press, New York.

C. D. Johnson (1971a), "A unified canonical form for controllable and uncontrollable linear dynamical systems," *Intern. J. Control*, **13**, 3, pp. 497–518.

C. D. Johnson (1971b), "Accommodation of external disturbances in linear regulator and servomechanism problems," *IEEE Trans. Autom. Control*, **16**, 6, pp. 635–644.

C. D. Johnson and W. M. Wonham (1966), "Another note on the transformation to canonical (phase-variable) form," *IEEE Trans. Autom. Control*, **11**, 3, pp. 609–610.

G. W. Johnson (1969), "A deterministic theory of estimation and control," *IEEE Trans. Autom. Control*, **14**, 4, pp. 380–384.

T. L. Johnson and M. Athans (1970), "On the design of optimal constrained dynamic compensators for linear constant systems," *IEEE Trans. Autom. Control*, **15**, 6, pp. 658–660.

T. Kailath (1968), "An innovations approach to least-squares estimation—Part I: Linear filtering in additive white noise," *IEEE Trans. Autom. Control*, **13**, 6, pp. 646–654.

R. E. Kalman (1960), "Contributions to the theory of optimal control," *Bol. Soc. Mat. Mexicana*, **5**, pp. 102–119.

R. E. Kalman (1964), "When is a linear control system optimal?" *J. Basic Eng., Trans. ASME*, Ser. D, **86**, pp. 51–60.

R. E. Kalman (1966), "Toward a theory of difficulty of computation in optimal control," *Proc. Fourth IBM Scientific Computing Symposium*, pp. 25–43.

R. E. Kalman and J. E. Bertram (1960), "Control system analysis and design via the 'Second method of Lyapunov', I. Continuous-time systems." *J. Basic Eng., Trans. ASME, Ser. D*, **82**, 2, pp. 371–393.

References 557

R. E. Kalman and R. S. Bucy (1961), "New results in linear filtering and prediction theory," *J. Basic Eng.*, *Trans. ASME*, Ser. D, **83**, pp. 95–108.

R. E. Kalman and T. S. Englar (1966), "A user's manual for the automatic synthesis program," NASA Report CR-475.

R. E. Kalman, P. L. Falb, and M. Arbib (1969), *Topics in Mathematical System Theory*, McGraw-Hill, New York.

R. E. Kalman and R. W. Koepcke (1958), "Optimal synthesis of linear sampling control systems using generalized performance indexes," *Trans. ASME*, Ser. D, **80**, pp. 1820–1826.

D. L. Kleinman (1968), "On an iterative technique for Riccati equation computation," *IEEE Trans. Autom. Control*, **13**, 1, pp. 114–115.

D. L. Kleinman (1970a), "An iterative technique for Riccati equation computations," Technical Memorandum, Bolt, Beranek, and Newman, June 30.

D. L. Kleinman (1970b), "An easy way to stabilize a linear constant system," *IEEE Trans. Autom. Control*, **15**, 6. p. 692.

E. Kreindler (1968a), "On the definition and application of the sensitivity function," *J. Franklin Inst.* **285**, 1, pp. 26–36.

E. Kreindler (1968b), "Closed-loop sensitivity reduction of linear optimal control systems," *IEEE Trans. Autom. Control*, **13**, 3, pp. 245–262.

E. Kreindler (1969), "Sensitivity of time-varying linear optimal control systems," *J. Optimal Theory Appl.*, **3**, 2, pp. 98–106.

C. L. Krouse and E. D. Ward (1970), "Improved linear system simulation by matrix exponentiation with generalized order hold," *Preprints, 11th Joint Automatic Control Conference*, pp. 794–802, Georgia Institute of Technology, Atlanta, Georgia, June 22–26.

A. Kupperajulu and S. Elangovan (1970), "System analysis by simplified methods," *IEEE Trans. Autom. Control*, **15**, 2, pp. 234–237.

H. J. Kushner (1967), *Stochastic Stability and Control*, Academic Press, New York.

H. J. Kushner (1971), *Introduction to Stochastic Control*, Holt, Rinehart and Winston, New York.

H. Kwakernaak (1969), "Optimal low-sensitivity linear feedback systems," *Automatica*, **5**, 3, pp. 279–286.

H. Kwakernaak and R. Sivan (1971), "Linear stochastic optimal controllers of fixed dimension," *Proc. Fifth Annual Princeton Conference on Information Sciences and Systems*, Princeton, N.J., March 25–26.

H. Kwakernaak and R. Sivan (1972), "The maximally achievable accuracy of linear optimal regulators and linear optimal filters," *IEEE Trans. Autom. Control*, **17**, 1, pp. 79–86.

R. J. Leake (1965), "Return difference Bode diagram for optimal system design," *IEEE Trans. Autom. Control*, **10**, 3, pp. 342–344.

A. M. Letov (1960), "Analytical controller design I," *Autom. Remote Control*, **21**, pp. 303–306.

A. H. Levis (1969), "Some computational aspects of the matrix exponential," *IEEE Trans. Autom. Control*, **14**, 4, pp. 410–411.

S. Levy and R. Sivan (1966), "On the stability of a zero-output system," *IEEE Trans. Autom. Control*, **11**, 2, pp. 315–316.

M. L. Liou (1966a), "A novel method of evaluating transient response," *Proc. IEEE*, **54**, 1, pp. 20–23.

M. L. Liou (1966b), "Steady-state response of linear time-invariant systems," *Proc. IEEE*, **54**, 12, pp. 1952–1953.

M. L. Liou (1967), "Response of linear time-invariant systems due to periodic inputs," *Proc. IEEE*, **55**, 2, pp. 242–243.

M. L. Liou (1968), "Evaluation of state transition matrix and related topics," *Conference Record Second Asilomar Conference on Circuits and Systems*, Pacific Grove, Calif., Oct. 30–Nov. 1.

C. S. Lu (1971), "Solution of the matrix equation $AX + XB = C$," *Electron. Letters*, **7**, 8, pp. 185–186.

D. G. Luenberger (1964), "Observing the state of a linear system," *IEEE Trans. Mil. Electron.*, **8**, pp. 74–80.

D. G. Luenberger (1966), "Observers for multivariable systems," *IEEE Trans. Autom. Control*, **11**, 2, pp. 190–197.

D. G. Luenberger (1967), "Canonical forms for linear multivariable systems," *IEEE Trans. Autom. Control*, **12**, 3, pp. 290–293.

D. L. Lukes (1968), "Stabilizability and optimal control," *Funkcialaj Ekvacioj*, **11**, pp. 39–50.

N. H. McClamroch (1969), "Duality and bounds for the matrix Riccati equation," *J. Math. Anal. Appl.*, **25**, pp. 622–627.

A. G. J. MacFarlane (1963), "The calculation of functionals of the time and frequency response of a linear constant coefficient dynamical system," *Quart. J. Mech. Appl. Math.*, **15**, Pt. 2, pp. 259–271.

F. T. Man and H. W. Smith (1969), "Design of linear regulators optimal for time-multiplied performance indices," *IEEE Trans. Autom. Control*, **14**, 5, pp. 527–529.

K. Mårtensson (1971), "On the matrix Riccati equation," *Information Sci.* **3**, pp. 17–49.

E. J. Mastascusa and J. G. Simes (1970), "A method for digital calculation of linear system response," *Preprints, 11th Joint Automatic Control Conference*, pp. 788–793, Georgia Inst. of Technology, Atlanta, Georgia, June 22–26.

J. S. Meditch (1969), *Stochastic Optimal Linear Estimation and Control*, McGraw-Hill, New York.

R. K. Mehra (1969), "Digital simulation of multi-dimensional Gauss–Markov random processes." *IEEE Trans. Autom. Control*, **14**, 1, pp. 112–113.

J. L. Melsa (1970), *Computer Programs for Computational Assistance in the Study of Linear Control Theory*, McGraw-Hill, New York.

D. Mitra (1967), "The equivalence and reduction of linear dynamical systems," Ph.D. thesis, University of London.

J. B. Moore and B. D. O. Anderson (1968), "Extensions of quadratic minimization theory. I. Finite-time results," *Intern J. Control*, **7**, 5, pp. 465–472.

P. Chr. Müller (1970), "Solution of the matrix equation $AX + XB = -Q$ and $S^T X + XS = -Q$," *SIAM J. Appl. Math.*, **18**, 3, 682–687.

R. B. Newell and D. G. Fisher (1971), "Optimal, multivariable computer control of a pilot plant evaporator." *Preprints, Third International Conference on Digital Computer Applications to Process Control*, Helsinki, June 2–5.

G. C. Newton, L. A. Gould, and J. F. Kaiser (1957), *Analytical Designs of Linear Feedback Controls*, Wiley, New York.

B. Noble (1969), *Applied Linear Algebra*, Prentice-Hall, Englewood Cliffs, N.J.

J. J. O'Donnell (1966), "Asymptotic solution of the matrix Riccati equation of optimal control," *Proc. Fourth Allerton Conference on Circuit and Systems Theory*, pp. 577–586, University of Illinois, Urbana, Ill., Oct. 5–7.

J. B. Pearson (1965), "A note on the stability of a class of optimum sampled-data systems," *IEEE Trans. Autom. Control*, **10**, 1, pp. 117–118.

D. A. Pierre (1969), *Optimization Theory with Applications*, Wiley, New York.

J. B. Plant (1969), "On the computation of transition matrices for time-invariant systems," *Proc. IEEE*, **57**, 8, pp. 1397–1398.

M. Plotkin (1964), "Matrix theorem with applications related to multi-variable control systems," *IEEE Trans. Autom. Control*, **9**, 1, pp. 120–121.

E. Polak and E. Wong (1970), *Notes for a First Course on Linear Systems*, Van Nostrand Reinhold, New York.

B. Porter (1971), "Optimal control of multivariable linear systems incorporating integral feedback," *Electron. Letters*, **7**, 8, pp. 170–172.

J. E. Potter (1964), "Matrix quadratic solutions," *SIAM J. Appl. Math.*, **14**, 3, pp. 496–501.

J. E. Potter and W. E. VanderVelde (1969), "On the existence of stabilizing compensation," *IEEE Trans. Autom. Control*, **14**, 1, pp. 97–98.

H. M. Power (1969), "A note on the matrix equation $A'LA - L = -K$," *IEEE Trans. Autom. Control*, **14**, 4, pp. 411–412.

H. M. Power and B. Porter (1970), "Necessary conditions for controllability of multivariable systems incorporating integral feedback," *Electron. Letters*, **6**, 25, pp. 815–816.

B. Ramaswami and K. Ramar (1968), "Transformation to the phase-variable canonical form," *IEEE Trans. Autom. Control*, **13**, 6, pp. 746–747.

D. S. Rane (1966), "A simplified transformation to (phase-variable) canonical form," *IEEE Trans. Autom. Control*, **11**, 3, p. 608.

D. Rappaport and L. M. Silverman (1971), "Structure and stability of discrete-time optimal systems," *IEEE Trans. Autom. Control*, **16**, 3, pp. 227–233.

R. A. Rohrer (1970), *Circuit Theory: An Introduction to the State Variable Approach*, McGraw-Hill, New York.

H. J. Rome (1969), "A direct solution to the linear variance equation of a time-invariant system," *IEEE Trans. Autom. Control*, **14**, 5, pp. 592–593.

M. Roseau (1966), *Vibrations non linéaires et théorie de la stabilité*, Springer-Verlag, Berlin.

G. Rosenau (1968), "Höhere Wurzelortskurven bei Mehrgrossensystemen," *Preprints, IFAC Symposium on Multivariable Systems*, Düsseldorf, Oct. 7–8.

D. Rothschild and A. Jameson (1970), "Comparison of four numerical algorithms for solving the Liapunov matrix equation," *Intern. J. Control*, **11**, 2, pp. 181–198.

A. P. Sage and B. R. Eisenberg (1966), "Closed loop optimization of fixed configuration systems," *Intern. J. Control*, **3**, 2, pp. 183–194.

M. K. Sain (1966), "On the control applications of a determinant equality related to eigenvalue computation," *IEEE Trans. Autom. Control*, **11**, 1, pp. 109–111.

P. Sannuti and P. V. Kokotović (1969), "Near-optimum design of linear systems by a singular perturbation method," *IEEE Trans. Autom. Control*, **14**, 1, pp. 15–22.

R. Saucedo and E. E. Schiring (1968), *Introduction to Continuous and Digital Control Systems*, Macmillan, New York.

D. G. Schultz and J. L. Melsa (1967), *State Functions and Linear Control Systems*, McGraw-Hill, New York.

A. Schumitzky (1968), "On the equivalence between matrix Riccati equations and Fredholm resolvents," *J. Comp. Systems Sci.*, **2**, pp. 76–87.

R. J. Schwarz and B. Friedland (1965), *Linear Systems*, McGraw-Hill, New York.

W. W. Seifert and C. W. Steeg, Ed. (1960), *Control Systems Engineering*, McGraw-Hill, New York.

Y. -P. Shih (1970), "Integral action in the optimal control of linear systems with quadratic performance index," *Ind. Eng. Chem. Fundamentals*, **9**, 1, pp. 35–37.

C. S. Sims and J. L. Melsa (1970), "A fixed configuration approach to the stochastic linear regulator problem," *Preprints, 11th Joint Automatic Control Conference*, Atlanta, Georgia, pp. 706–712.

R. Sivan (1965), "On zeroing the output and maintaining it zero," *IEEE Trans. Autom. Control*, **10**, 2, pp. 193–194.

P. G. Smith (1971), "Numerical solution of the matrix equation $AX + XA^T + B = O$." *IEEE Trans. Autom. Control*, **16**, 3, pp. 278–279.

R. A. Smith (1968), "Matrix equation $XA + BX = C$," *SIAM J. Appl. Math.*, **16**, 1, pp. 198–201.

H. W. Sorenson (1970), "Least-squares estimation: From Gauss to Kalman," *Spectrum*, **7**, 7, pp. 63–68.

J. S. Tou (1964), *Modern Control Theory*, McGraw-Hill, New York.

E. Tse and M. Athans (1970), "Optimal minimal-order observer-estimators for discrete linear time-varying systems," *IEEE Trans. Autom. Control*, **15**, 4, pp. 416–426.

W. G. Tuel, Jr. (1966), "On the transformation to (phase-variable) canonical form," *IEEE Trans. Autom. Control*, **11**, 3, p. 607.

J. E. Van Ness (1969), "Inverse iteration method for finding eigenvectors," *IEEE Trans. Autom. Control*, **14**, 1, pp. 63–66.

D. R. Vaughan (1969), "A negative exponential solution for the matrix Riccati equation," *IEEE Trans. Autom. Control*, **14**, 1, pp. 72–75.

D. R. Vaughan (1970), "A nonrecursive algebraic solution for the discrete Riccati equation." *IEEE Trans. Autom. Control*, **15**, 5, pp. 597–599.

Y. Wallach (1969), "On the numerical solution of state equations," *IEEE Trans. Autom. Control*, **14**, 4, pp. 408–409.

O. H. D. Walter (1970), "Eigenvector scaling in a solution of the matrix Riccati equation," *IEEE Trans. Autom. Control*, **15**, 4, pp. 486–487.

L. Weinberg (1962), *Network Analysis and Synthesis*, McGraw-Hill, New York.

J. H. Westcott (1952), "The development of relationships concerning the frequency bandwidth and the mean square error of servo systems from properties of gain-frequency characteristics," in *Automatic and Manual Control*, A. Tustin, Ed., Butterworths, London.

D. E. Whitney (1966a), "Propagated error bounds for numerical solution of transient response," *Proc. IEEE*, **54**, 8, pp. 1084–1085.

D. E. Whitney (1966b), "Forced response evaluation by matrix exponential," *Proc. IEEE*, **54**, 8, pp. 1089–1090.

D. E. Whitney (1966c), "Propagation and control of roundoff error in the matrix exponential method," *Proc. IEEE*, **54**, 10, pp. 1483–1484.

W. A. Wolovich (1968), "On the stabilization of controllable systems," *IEEE Trans. Autom. Control*, **13**, 5, pp. 569–572.

W. A. Wolovich and P. L. Falb (1969), "On the structure of multivariable systems," *SIAM J. Control*, **7**, 3, pp. 437–451.

W. M. Wonham (1963), "Stochastic problems in optimal control," 1963 *IEEE Convention Record*, Part 2, pp. 114–124.

W. M. Wonham (1967a), "On pole assignment in multi-input controllable linear systems," *IEEE Trans. Autom. Control*, **12**, pp. 660–665.

W. M. Wonham (1967b), "On matrix quadratic equations and matrix Riccati equations," Report, Center for Dynamical Studies, Brown University, Providence, R.I.

W. M. Wonham (1968a), "On a matrix Riccati equation of stochastic control," *SIAM J. Control*, **6**, 4, pp. 681–697.

W. M. Wonham (1968b), "On the separation theorem of stochastic control," *SIAM J. Control*, **6**, 2, pp. 312–326.

W. M. Wonham (1970a), "Dynamic observers—Geometric theory," *IEEE Trans. Autom. Control*, **15**, 2, pp. 258–259.

W. M. Wonham (1970b), "Random differential equations in control theory," in *Probabilistic Methods in Applied Mathematics*, A. T. Barucha-Reid, Ed., pp. 131–212, Academic Press, New York.

W. M. Wonham and W. F. Cashman (1968), "A computational approach to optimal control of stochastic stationary systems," *Preprints, Ninth Joint Automatic Control Conference*, pp. 13–33, University of Michigan, Ann Arbor, Mich., June 26–28.

L. A. Zadeh and C. A. Desoer (1963), *Linear System Theory: The State Space Approach*, McGraw-Hill, New York.

AUTHOR INDEX

Boldface numbers indicate the page where the full reference is given.

Ackermann, J., 433, **554**
Anderson, B. D. O., 83, 85, 219, 280, 314, 322, **553, 558**
Aoki, M., 427, **553**
Arbib, M., 64, 65, 66, 79, 460, **557**
Åström, K. J., 100, 260, 502, 539, **553**
Athans, M., 219, 428, 531, **553, 556, 560**

Barnett, S., 104, **553**
Barrow, B. B., ix, **553**
Bass, R. W., 33, 34, 251, 324, **553, 555**
Bellman, R. E., 492, **553**
Berger, C. S., 472, **553**
Bertram, J. E., 24, **556**
Beveridge, G. S. G., 432, **553**
Bickart, Th. A., 14, **553**
Bidwell, J. C., 251, **553**
Blackburn, T. R., 251, 253, **553**
Blakelock, J. H., 292, **553**
Bode, H. W., 440, **553**
Brammer, K. G., 531, **553**
Brash, F. M., 431, **554**
Brockett, R. W., 26, **554**
Bryson, A. E., 352, 357, **554**
Bucy, R. S., 79, 219, 248, 341, 344, 364, 365, 433, **554, 557**
Butman, S., 372, **554**

Cadzow, J. A., 489, **554**
Caines, P. E., 497, **554**
Cannon, R. H., 4, **554**
Cashman, W. F., 253, **561**
Chang, S. S. L., 283, **554**
Chen, C. F., 104, 427, **554**
Chen, C. T., 46, 198, **554, 556**

Chidambara, M. R., 427, **554**
Cruz, J. B., 187, **554**
Cumming, S. D. G., 335, **554**

Davenport, W. B., 91, **554**
Davis, H. T., 217, **554**
Davison, E. J., 104, 198, 279, 427, **554, 555**
D'Azzo, J. J., 38, 53, 114, **555**
Desoer, C. A., 11, 12, 13, 19, 21, 24, 33, 34, **555, 561**
Deyst, J. J., 497, **555**
di Caprio, U., 270, **555**
Doob, J. L., 100, **555**

Eisenberg, B. R., 428, **559**
Eklund, K., 416, 417, **555**
Elangovan, S., 427, **557**
Elgerd, O. I., 4, **555**
Englar, T. S., 13, 219, 249, **557**
Everling, W., 14, **555**

Falb, P. L., 64, 65, 66, 79, 85, 219, 460, **553, 557, 561**
Farison, J. B., 489, **555**
Fath, A. F., 14, 251, **555**
Fisher, D. G., 277, **558**
Fleming, W. H., 390, **555**
Forsythe, G. E., 34, **555**
Fossard, A., 427, **555**
Frame, J. S., 103, **555**
Franklin, G. F., 539, **556**
Freeman, H., 442, **555**
Freested, W. C., 251, **555**
Friedland, B., 29, **560**
Fu, F.-C., 489, **555**

Gal'perin, E. A., 62, 555
Gikhman, I. I., 100, 555
Gould, L. A., 96, 150, 428, 559
Gunckel, T. L., 539, 556
Gura, I., 33, 34, 553

Hagander, P., 104, 556
Haley, P. H., 39, 556
Haynes, G. W., 251, 556
Hendricks, T. C., 251, 556
Heymann, M., 198, 556
Hida, T., 100, 556
Horowitz, I. M., 181, 556
Houpis, C. H., 38, 53, 114, 555
Hsu, C.-H., 46, 556

IEEE Standards Committee, ix, 556
International Organization for Standardization, ix, 556

Jameson, A., 104, 556, 559
Jazwinski, A. H., 344, 531, 556
Johansen, D. E., 352, 357, 554
Johnson, C. D., 84, 85, 280, 556
Johnson, G. W., 388, 556
Johnson, T. L., 428, 556
Joseph, P. D., 79, 219, 248, 554

Kailath, T., 361, 556
Kaiser, J. F., 96, 150, 428, 559
Kalman, R. E., 13, 24, 54, 64, 65, 66, 79, 217, 219, 231, 232, 233, 249, 284, 322, 341, 344, 364, 365, 460, 492, 497, 556, 557
Kleinman, D. L., 104, 252, 253, 557
Koepcke, R. W., 492, 502, 553, 557
Kokotovic, P. V., 428, 559
Krasovski, N. N., 62, 555
Kreindler, E., 187, 314, 315, 557
Krouse, C. L., 14, 557
Kuppurajulu, A., 427, 557
Kushner, H. J., 100, 260, 263, 390, 502, 503, 539, 557
Kwakernaak, H., 306, 423, 424, 428, 557

Leake, R. J., 327, 557
Letov, A. M., 247, 557
Levis, A. H., 14, 557
Levy, S., 309, 557
Liou, M. L., 14, 558
Lu, C. S., 104, 558

Luenberger, D. G., 83, 85, 335, 553, 558
Lukes, D. L., 237, 558

McClamroch, N. H., 252, 558
MacFarlane, A. G. J., 104, 558
Man, F. T., 104, 253, 555, 558
Mårtensson, K., 237, 558
Mastacusa, E. J., 14, 558
Mayne, D. Q., 497, 554
Meditch, J. S., 463, 531, 558
Mehra, R. K., 102, 558
Melsa, J. L., 14, 34, 38, 51, 194, 327, 428, 430, 558, 560
Mitra, D., 427, 558
Moore, J. B., 219, 280, 314, 322, 553, 558
Müller, P. Chr., 104, 558

Newell, R. B., 277, 558
Newton, G. C., 96, 150, 428, 559
Noble, B., 15, 19, 20, 376, 559

O'Donnel, J. J., 246, 247, 322, 326, 559

Pearson, J. B., 431, 515, 554, 559
Perkins, W. R., 187, 554
Pierre, D. A., 432, 559
Plant, J. B., 14, 559
Plotkin, M., 40, 559
Polak, E., 13, 33, 559
Porter, B., 277, 279, 559
Potter, J. E., 322, 388, 559
Power, H. M., 279, 472, 559
Price, C. F., 497, 555

Ramar, K., 84, 559
Ramaswami, B., 84, 559
Rane, D. S., 84, 559
Rappaport, D., 515, 559
Rohrer, R. A., 14, 559
Rome, H. J., 104, 559
Root, W. L., 91, 554
Roseau, M., 3, 32, 559
Rosenau, G., 287, 559
Rothschild, D., 104, 559

Sage, A. P., 428, 559
Sain, M. K., 40, 559
Sannuti, P., 428, 559
Saucedo, R., 444, 456, 560
Schainker, R. B., 427, 554

Schechter, R. S., 432, 553
Schiring, E. E., 444, 456, 560
Schultz, D. G., 327, 560
Schumitzky, A., 219, 560
Schwarz, R. J., 29, 560
Seifert, W. W., 96, 150, 560
Shieh, L. S., 104, 427, 554
Shih, Y.-P., 277, 560
Silverman, L. M., 515, 559
Simes, J. G., 14, 558
Sims, C. S., 428, 430, 560
Sivan, R., 306, 308, 309, 428, 557, 560
Skorokhod, A. V., 100, 555
Smith, H. W., 253, 279, 555, 558
Smith, P. G., 104, 560
Smith, R. A., 104, 560
Sorenson, H. W., 341, 560
Steeg, C. W., 96, 150, 560
Storey, C., 104, 553
Strauss, L. W., 34, 555

Tou, J. S., 502, 560
Tse, E., 531, 560

Tuel Jr., W. G., 84, 560
Tung, F., 502, 553

VanderVelde, W. E., 388, 559
Van Ness, J. E., 251, 560
Vaughan, D. R., 249, 325, 500, 560

Wallach, Y., 14, 560
Walter, O. H. D., 326, 560
Wang, P. P., 270, 555
Ward, E. D., 14, 557
Webber, R. F., 251, 555
Weinberg, L., 285, 299, 560
Westcott, J. H., 440, 560
Whitney, D. E., 14, 560, 561
Wolovich, W. A., 85, 198, 561
Wong, E., 13, 33, 559
Wonham, W. M., 62, 77, 84, 198, 200, 218, 219, 237, 253, 336, 351, 390, 556, 561

Zadeh, L. A., 11, 12, 13, 19, 21, 24, 33, 34, 561

SUBJECT INDEX

Boldface numbers indicate the page where the item is defined or introduced.

Accuracy, maximally achievable, of regulators and tracking systems, 306–310
Adjoint matrix differential equation, 440
Airplane, asymptotic regulation of the longitudinal motions of an, 310–312
   nonzero set point pitch control of an, 302–303
   pitch control of an, 291–293
   regulation of the longitudinal motions of an, 293–297
Aliasing, 458
Amplidyne, description of an, 114–115
   nonzero set point regulator for an, 321
   proportional feedback control of an, 116
   regulation of an, 320
Angular velocity control system, an observer for the, 374–375
   digital version of the, 521–522
   integral control of the, 439–440
   as an output feedback regulator problem, 438–439
   as an output feedback stochastic tracking problem, 439
   proportional feedback of the, 189–190
   as a regulator problem, 205–206
   solution of, the regulator problem for the, 212–216
   the Riccati equation for the, 220
   the stochastic tracking problem for the, 266–269
   steady-state solution of the regulator problem for the, 222–223
   as a stochastic tracking problem, 258–259
   reconsidered, 321

Autonomous system, **24**

Bandwidth, normalized, of a discrete-time control system, **481**
   of a control system, **145**
   normalized, of a discrete-time stochastic process, **481**
   of a stochastic process, **147**
Bode, **181**
Bode plot, 38
Break frequency, of a control system, **146**
   of a stochastic process, **147**
Brownian motion, **90**, 100
Butterworth, pole configuration, **285**
   polynomial, **299**
   transfer function, **299**

Cayley-Hamilton theorem, 84
Characteristic polynomial, closed-loop, 46, 274
   open-loop, 274
Complexity of output feedback control systems, 437
Computer control, 442
Computer program package for linear optimal control, 437–438
Constant disturbances, effect of, in control systems, 171–172, 191–192
   elimination of, in discrete-time regulators, 506–507
   in discrete-time output feedback control systems, 544
   in regulators, 277–280
   in output feedback control systems, 414–417
Continuous-to-discrete-time converter, **444**

567

Subject Index

Control instant, **522**
Control variable, *see* Input variable
Controlled variable, **122**, 128, 476
  shifted, **270**
Control law, asymptotically stable, **194**
  asymptotic properties of, discrete-time optimal, 509–516
  optimal, 281–312
  discrete-time nonzero set point optimal, **505**, 543
  discrete-time steady-state optimal, **497**
  discrete-time zero-steady-state-error optimal, **507**, 544
  interconnected with observer, 378–382
  discrete-time case, 536–537
  linear, **194**
  nonzero set point optimal, **273**, 411
  stability of, discrete-time steady-state optimal, 497
    steady-state optimal, 221, 233
    time-invariant steady-state optimal, 238
  steady-state optimal, **221**, 232
  time-invariant steady-state optimal, 238, 243
  zero-steady-state-error optimal, **416**
Controllability, 53–65
  complete, **54**
  of discrete-time linear systems, 459–462
  of the pair (A, B), **56**
  of time-invariant linear systems, 55–57
  of time-varying linear systems, 64–65
  uniform complete, **65**
  of discrete-time systems, 460–461
Controllability, canonical form, **60**
  of discrete-time systems, 461–462
  matrix, **55**
  of discrete-time systems, 460
Controller, **119**, 122
  closed-loop, **123**, 128–132, 476
  open-loop, **122**, 128, 476
Control problems, multivariable, 124
  terminal, 127
Control system, **123**, 131
Control systems, analysis of linear, 119–192
  discrete-time, 475–488
  stability of, 136–137, 184, 188, 477
  stabilization of, 137
  by open loop controllers, 184, 188
Covariance matrix, *see* Stochastic processes
Cutoff frequency, normalized, of a discrete-time control system, 481

  of a discrete-time stochastic process, **481**
  of a control system, **145**
  of a stochastic process, **147**
  of the i-th link, **156**
Deadbeat observers, **527**
Deadbeat response, output, 489, **513**, 515
  state, **489**, 537
Decoupling, 157
  static, 157
Degrees of freedom of a control system, 272, 279, 308, 415–416
Design objective, basic, 131, 471
Detectability, 76–78, 77
  of discrete-time systems, 465
  of the pair (A, C), 77
Difference function, **157**, 163, 164
Differential system, 2
Digital computation aspects of linear optimal control, 436–438
Digital positioning system, a deadbeat observer for the, 527–528
  description of the, 447–449
  frequency response of the, 458–459
  integral control of the, 545–546
  a nonzero set point regulator for the, 507–509
  observed variable of the, 524
  optimal state feedback of the, 494–495
  output deadbeat control of the, 516–517
  output feedback state deadbeat control of the, 537–538
  with proportional feedback, 483–487
  stability of the, 455
    with proportional feedback, 477–478
  state feedback deadbeat control of the, 490
Dimension of a linear system, 2
Direct link, **2**
  in discrete-time systems, 447, 524
Direct sum, **19**
Discrete-time equivalent of a continuous-time system, 445
Discrete-time systems, inherently, 442
  linear optimal control theory for, 442–552
Discrete-to-continuous-time converter, **444**
Disturbances, constant, effect of, in control systems, 167–174, 184–186, 188
  in discrete-time control systems, 487–488
  equivalent, at the controlled variable, **170**
  see Constant disturbances

Subject Index

Disturbance variable, **121**, 128, 476
Duality, of discrete-time systems, 465–466
   of discrete-time optimal observer and regulator problems, 533–534
   of optimal regulator and observer problems, 364–365
   of systems, 79–81
Equilibrium state, 24
Estimation of a constant, 346–347
Estimator, minimum mean square linear, 344, 531
Euclidean norm, 24
Exponentially correlated noise, break frequency of, 148
   cutoff frequency of, 148
   definition of, 88
   discrete-time, 469
   modeled by a first-order system, 105–106
   power spectral density, function of, 91
      of discrete-time, 472–473
   time constant of, 88
   variance of, 88
Exponential of a matrix, **13**, 20
   computation of the, by diagonalization, 15–17, 21
   by Laplace transformation, 33
   numerical computation of the, 13–14

Faddeeva's method, 34
Feedback, benefits of, 120
Feedback configuration, 43
Feedback link of a controller, 262–263
Feedforward link of a controller, 263
Frequency, normalized angular, 457
Frequency band, normalized, of a discrete-time stochastic process, **480**
   of a control system, **145**
   normalized, of the i-th link of a discrete-time control system, **481**
   of the i-th link, **156**
   of a stochastic process, **146**
Frequency response, 37–38
   of discrete-time systems, 457–458
Frequency response matrix, 38

Gain matrix, of an observer, **331**
   of a regulator, **194**

Heating system, 119, 120

Impulse response matrix, **13**
   of a time-invariant system, 14, 35
Input, computation of mean square, 132, 150
   dynamic range of, 149
   integrated square, **203**
   magnitude of, in regulator problems, 201
   mean square, **131**
   rms, **131**
   steady-stak mean square, 140–144, 264
Input variable, **2**, 121, 476
   shifted, **270**
Innovation process, **361**, 361–363, 401
   discrete-time, 533
Integral control, in discrete-time output feedback regulators, 544
   in discrete-time state feedback regulators, 507
   in output feedback regulators, 414–417
   in state feedback regulators, 277–280
Integrating action, **171**, 181, 191, 277, 417
Integral state, **277**
   for discrete-time regulators, 507
Interaction, 157
Interconnections, of discrete-time and continuous-time systems, 443–447
   of linear systems, 43–48
Interface system, 443
Inverted pendulum, an observer for the, 373–374
   controllability of the, 57
   control of the, with proportional feedback, 48–49
   description of the, 4–7
   detectability of the, 78
   modes of the, 17–18
      without friction, 22–23
   reconstructibility canonical form of the, 75–76
   reconstructibility of the, 67, 69–70
   stability of the, 26
   stabilization of the, by output feedback, 383–388
   by state feedback, 139–140, 195–196
   stable and unstable subspaces of the 30–31
   without friction, 31

Jacobian matrix, 3
Jordan normal form, 19–22, **20**

Kalman-Bucy filter, 341, 344; see also
    Optimal observer

Laplace transformation, 33
Leverrier's algorithm, **34**, 251, 456
Linearization, 2–3, 31–32
Loop gain matrix, **45**
Low-pass, stochastic process, 147
    transmission, 146
Lyapunov equation, **104**, 111, 251
    numerical solution of the, 104

Markov process, **117**
Matrix difference equation, property of a, 551
Measurement noise, 339; see also
    Observation noise
Mode, **16**, 22
Modes, hidden, in optimal regulators, 309

Nilpotency of a matrix, **489**
Nominal, input, 2
    plant transfer function, 179
    solution, 24
    trajectory, 2
Nonnegative-definiteness of a matrix, 87, 91
Null space, **19**
Numerator polynomial, 41
    numerical computation of the, 41–42

Observability, complete, **66**; see also
    Reconstructibility
Observation instant, **523**
Observation noise, **122**, 339, 476
    effect of, in control systems, 174–176
    in discrete-time control systems, 487–488
    in open-loop control systems, 186, 188
Observed variable, **122**, 128, 328, 476
Observer, asymptotic properties of the
    time-invariant optimal, 368–372
    determination of *a priori* data of the
        singular optimal, 376
    full-order, **330**
    full-order discrete-time, **525**
    interconnected with a control law, 378–382
    discrete-time case, 536–537
    optimal, 339–363

optimal discrete-time, 528–531
pole assignment in an, 332, 334
pole assignment in, a discrete-time, 526
a reduced-order, 336
poles, 332
reduced-order, **330**, 335–337
reduced-order discrete-time, 527
stability of the, 331–332
    discrete-time, 526
stabilization of a discrete-time, 526
stabilization of an, 334–335
steady-state optimal, **345**, 366–368
steady-state properties of, the discrete-time optimal, 535
the optimal, 345, 365–368
Observer problem, alternative version of the
    discrete-time optimal, 550–551
the colored noise optimal, 356–357
the discrete-time optimal, 528–531
the nonsingular, with correlated noises, 351–352
with uncorrelated noises, 341–346
the optimal, 340
the singular time-invariant, 352–356
Observer Riccati equation, 343–345
    asymptotic behavior of the solution of
        the, 370–372
    solution of the, 375–376
    steady-state solution of the, 345, 365–368
    the algebraic, 345, 367
Output equation, 2
    of discrete-time systems, 443
Output feedback control systems, constant
    disturbances in, 414–417
    discrete-time, 544
    nonzero set point, 409–413
    discrete-time, 543–544
    numerical determination of optimal
        reduced-order, 432–434
    optimal, 389–419
    optimal discrete-time, 536–546
    optimal reduced-order, 427–434
    discrete-time, 552
    pole assignment in, 388–389
    discrete-time, 537
    sensitivity of, 419–424
    stabilization of, 388–389
    a discrete-time, 537
    steady-state optimal, 396

structure of, 378–382
Output feedback regulator, evaluation of the performance of, the optimal, 391–397
  the optimal discrete-time, 540–543
  the optimal, 390–391
  the optimal discrete-time, 539–540
Output feedback regulator problem, the stochastic linear discrete-time optimal, **539**, 539–540
  the stochastic linear optimal, **389**, 389–402
Output feedback tracking systems, 402–405
Output variable, **2**

Parameter variations, effect of, in control systems, 178–181, 187–188, 488
Phase-variable canonical form, **82**, 82–85
  of discrete-time systems, 466–467
  dual, **84**
  of discrete-time systems, 467
Plant, **119**, 128
  dynamic range of a, 149
Poles, assignment of, in discrete-time regulators, 488–489
  in discrete-time observers, 526
  in discrete-time output feedback systems, 537
  in observers, 332, 334–335, 336
  in output feedback systems, 388–389
  in regulators, 194, 198–199
  asymptotic behavior of the closed-loop regulator, 281–289
  discrete-time case, 511, 513–515, 549
  asymptotic behavior of the observer, 368–370
  closed-loop, 51
  controllable, **61**
  of discrete-time systems, **462**
  distance to the origin of, closed-loop regulator, 285, 288–289
  closed-loop from Bode plot, 327
  faraway, 284, 287
  nearby, 289
  observer, **332**, 382, 537
  open-loop, 51
  patterns of closed-loop regulator, 281–289
  discrete-time case, 509–515, 549
  reconstructible, **75**
  of discrete-time systems, **465**
  regulator, **382**, 537
  stable, **29**
  of discrete-time systems, **462**
  of a system, **17**, 35
  of a transfer matrix, **35**
  uncontrollable, **61**
  of discrete-time systems, **462**
  unreconstructible, **75**
  of discrete-time systems, **465**
  unstable, **29**
Position servo, controllers for the, 133–136
  description of the, 124
  effect of, disturbances on the, 172–174
    observation noise on the, 176–178
    parameter variations on the, 181–183
  with position and velocity feedback, 134–135
  with position feedback only, 135
  with proportional feedback, 133–134
  settling time of the tracking error of the, 166–167
  stability of the proportional feedback scheme for the, 137–138
  tracking properties of the, 150–155
Positioning system, a colored noise observer for the, 357–360
  asymptotic properties of the optimal observer for the, 372–373
  integral control of the, 280–281
  integral output feedback control of the, 417–419
  nonzero set point control of the, 275
  nonzero set point output feedback control of the, 413–414
  an observer for the, 332–334
  an optimal observer for the, 347–351
  as an output feedback control problem, 382–383
  as an output feedback tracking problem, 405–409
  pole configuration of the optimal regulator for the, 290
  a reduced-order controller for the, 434–436
  a reduced-order observer for the, 337–338
  as a regulator problem, 206–207
  with a frictionless dc motor, 319
  sensitivity of, the optimal output feedback control system for the, 424–427

the optimal regulator for the, 317–318
stabilization of regulators for the, 319
steady-state solution of the regulator
  problem for the, 223–227
as a stochastic output feedback regulator
  problem, 397–400
as a stochastic regulator problem, 320–
  321
terminal control of the, 127
Power spectral density matrix, *see* Stochastic
  processes
Prefilter, 412
Process control, 547
Processing delay, 476, **523**
Pulse response matrix, **453**

Quadratic, expression for stochastic
  processes, 94–96
integral expressions, 108–111
sums for discrete-time stochastic
  processes, 471–472

Reconstructibility, 65–79
complete, **66**
of discrete-time systems, 462–465
of the pair (A, C), **69**
of time-invariant linear systems, 67–69
of time-varying linear systems, 78–79
uniform complete, **79**
uniform complete, of discrete-time
  systems, 463
Reconstructibility canonical form, **74**
of discrete-time systems, 464–465
Reconstructibility matrix, **67**
of discrete-time systems, 463
Reconstruction error, **331**, 340
mean square, **340**
Reduced-order output feedback
  controllers, 427–436
discrete-time, 552
Reel-winding mechanism, 234–237
Reference variable, **122**, 128, 476
constant part of the, **141**
variable part of the, **141**
Regulating error, integrated square, **203**
Regulators, asymptotic properties of
  nonzero set point, 297–302
discrete-time case, 512–513
asymptotic properties of optimal, 281–
  312

discrete-time case, 509–516
with incomplete and noisy measurements,
  389–402
discrete-time case, 539–543
with incomplete measurements, 378–389
discrete-time case, 536–537
nonzero set point, 270–275
discrete-time case, 504–506
pole assignment in, 194, 198–199
discrete-time, 488–489
poles of time-invariant optimal, 247, 282–
  283, 286
discrete-time case, 500, 509–510, 513
sensitivity of optimal, 312–317
discrete-time case, 520–521
steady-state properties of optimal, 230–243
Regulator problem, 123
choice of the weighting matrices in the
  optimal, 204
the deterministic linear optimal, 201–
  220, **203**
discrete-time case, 490–494, **491**
with disturbances, 253–255, 261–263
discrete-time case, 547–548
existence of, the solution of the optimal, 219
the optimal, 321–322
the optimal, discrete-time case, 547
frequency domain solution of the
  optimal, 326
the mixed continuous-time discrete-time
  optimal, 549–550
properties of the steady-state solution of
  the stochastic optimal, 263–265
solution of the optimal, 207–212
by diagonalization, 243–248
steady-state solution of the linear optimal,
  220–248
discrete-time case, 495–500
the stochastic linear optimal, 253–255,
  255, 259–265, 310
discrete-time case, **502**, 502–503
the stochastic linear optimal output
  feedback, 389, 389–402
discrete-time case, **539**, 539–540
the time-invariant deterministic linear
  optimal, 203
variational equations of the optimal, 209
Resolvent, **33**, 34
Return difference matrix, **45**, 186
asymptotic, 423

Riccati equation, **217**
  algebraic, **221**, 238, 243, 322–325
  derivation of the, 216–219
  discrete-time equivalent of the, 494
  existence of the solution of the, 219
  negative exponential solution of the, 325
  numerical solution of the, 248–253
    by diagonalization, 250–251
    by direct integration, 248–249
    by the Kalman-Englar method, 249
    by the Newton-Raphson method, 251–253
  observer, 343–345, 365–367, 375–376
  solutions of the algebraic, 322–325
  steady-state solution of the, 221, 231–232
  time-invariant, 237–238
Root loci, 51–53
  of optimal observer poles, 368–370
  of optimal regulator poles, 281–289
    discrete-time case, 511, 514–515, 549
Root-square locus, 283
Routh-Hurwitz criterion, 28

Sampler, **444**
Sampling, instant, **447**
  period, **447**
  rate, **447**
Satellite, revolving, 113–114
Savings bank account, 443
Sensitivity, of control systems to, disturbances, 167–172, 184–186, 188
  parameter variations, 178–181, 187–188
  of optimal output feedback control systems, 419–424
  of optimal state feedback control systems, 312–317
    discrete-time case, 520–521
Sensitivity function, **169**, 181
  of a discrete-time control system, 487
  a property of the, 440–441
Sensitivity matrix, **185**
  asymptotic, 423
  of a discrete-time control system, 487
Sensor, 119
Separation principle, 361, **390**
  proof of the, 400–402
Series connection, 43
Set point, **123**, 141, 270

nonzero, in output feedback control systems, 409–413, 417
  discrete-time case, 543–544
  in state feedback regulators, 270–275
    discrete-time case, 504–506
Settling time, 141, **165**
  a bound for the, 166
  discrete-time case, 483
Simulation of linear systems, 13–14
Smoothing problem, optimal, 361
Souriau's method, 34
Stability, 24–32
  asymptotic, **25, 26**, 28
    of discrete-time systems, 454
    in the large, **25, 26**, 28
      of discrete-time systems, 454
  of discrete-time systems, 454–455
  exponential, **26**, 28
    of discrete-time systems, 454–455
  of interconnections of systems, 46
  of linear systems, 25–26
  of a matrix, **28**
  of nonlinear systems, investigation of the, 31–32
  in the sense of Lyapunov, **24, 26**, 28
    of discrete-time systems, 454
  of solutions, 24–25
  of time-invariant linear systems, 27–29
Stabilizability, **62**, 62–64
  of discrete-time systems, 462
  of the pair (A, B), **63**
State augmentation technique, 43–44
State difference equation, **443**
  solution of the, 452–453
State differential equation, **2**
  linearized, **3**
  solution of the, for linear systems, 11–23
    by Laplace transformation, 33–35
State excitation noise, **339**
State feedback, 193–327
  of discrete-time systems, 488–522
  optimal, see Regulator problem
  stability improvement by, 193–201
    of discrete-time systems, 488–489
State reconstruction, 328–376
  for discrete-time systems, 522–536
  optimal, see Observer problem
  problem formulation for discrete-time, 522–524

574 Subject Index

State transformation, 10–11, 115, 116, 117
State variable, **2**
  augmented, 43
  shifted, **270**
Step response matrix, **13**
Steady-state analysis of control systems, *see* Tracking properties
Steady-state equivalent control scheme, open-loop, **183**, 183–188
Steady-state period, 141
Steady-state response, to a constant input, 38
  of discrete-time systems, 458
  to a harmonic input, 37
  of discrete-time systems, 457
Steady-state solution of regulator problems, *see* Regulator problem
Stirred tank, a decoupled control system for the, 190–191
  analysis of the steady-state tracking properties of the controlled, 158–165
  computation of, a quadratic integral criterion for the, 111–113
    the mean square concentration variation in the, 96
  controllability of the, 54–55, 61–62
  damping effect of the, 117
  description of the, 7–10
  discrete-time version of the, an optimal observer for the, 531–533
    as a regulator problem, 500–501
    as a stochastic regulator problem, 503–504
    description of the, 449
    with disturbances, 473–475
  frequency response matrix of the, 38–39
  impulse response matrix of the, 14
  modeling of the stochastic disturbances of the, 107–108
  nonzero set point regulation of the, 275–276
  pole configuration of the optimal regulator for the, 290–291
  proportional feedback control of the, 49–50
  a regulator system for the, 124–127
  solution of the stochastic regulator problem for the, 265–266
  stability improvement of the, 196
  stability of the, 27, 29
  stabilizability of the, 64
  steady-state solution of the regulator problem for the, 227–230
  step response matrix of the, 15
  with stochastic disturbances, 93–94
  as a stochastic regulator problem, 256–257
  transfer matrix of the, 36–37
  zeroes of the, 42
Stirred tank with time delay, deadbeat control of the, 517–520
  description of the, 449–452
Stochastic processes, 85–96
  covariance matrix of, **86**
    discrete-time, 468
  discrete-time, 467–469
  Gaussian, 87–88
  Gaussian discrete-time, 467
  independence of, **87**
  mean of, **86**
    discrete-time, 468
  modeling of, 106
  power spectral density function of, 90
  power spectral density matrix of, **90**, 90–91
    discrete-time, 468–469
  realizations of, 85
  response of linear systems to, 91–93
    discrete-time case, 469
  second-order joint moment matrix of, **86**
    discrete-time, 468
  second-order moment matrix of, **86**
    discrete-time, 468
  stationarity of, **85**
    discrete-time, 467
  uncorrelatedness of, **87**
  variance matrix of, **86**
    discrete-time, 468
  wide-sense stationarity of, **87**
    discrete-time, 468
  with uncorrelated increments, 88–90, 99–100
Subspace, controllable, **57**, 57–61, 116
  of discrete-time systems, 461
  invariance of a, **58**
  reconstructible, **75**
  stable, **30**
  of discrete-time systems, 455
  uncontrollable, **61**
  unreconstructible, **70**, 70–75

of discrete-time systems, 464
  unstable, 30
    of discrete-time systems, 455
Suspended pendulum, stability of the, 27
System equations, 2

Terminal control, 127
Terminal error, weighted square, **203**
Trace of a matrix, **95**
Tracking antenna, 119, 120, 124
Tracking error, **131**
  computation of the mean square, 131–132, 150
  mean square, **131**
  rms, **131**
  steady-state mean square, 140–144, **142**
Tracking problem, 121–123
  stochastic optimal, 257–258, 263
    discrete-time case, 548–549
  optimal linear, with incomplete and noisy measurements, 402–405, **403**
Tracking properties, steady-state, of open-loop control systems, 184, 188
  steady-state analysis of, 140–165
    discrete-time case, 478–482
    multiinput multioutput case, 155–158
    single-input single-output case, 144–150
  transient analysis of, 165–166
    discrete-time case, 482–483
  transient, of open-loop control systems, 184, 188
Transfer function, **35**
Transfer matrix, **35**
  closed-loop, of a time-invariant regulator, **272**
Transform analysis, 33–53
  of discrete-time systems, 455–458
Transient analysis of control systems, *see* Tracking properties
Transient period, 141

Transition matrix, **11**
  discrete-time case, **452**
  of a time-invariant system, 13–14
Transmission, **142**
  of a discrete-time control system, 479
  first-order, 146
  second-order, 146

White noise, **97**, 97–100
  discrete-time, **470**
  Gaussian, **100**
  Gaussian discrete-time, **470**
  integration rules for, 98
  intensity of, 97
  linear differential system driven by, 97–113
  linear discrete-time systems driven by, 470–472
Wiener, 361
Wiener process, **90**
Wiener-Lévy process, **90**

Zero, a system with a right-half plane, 304–306, 312
Zeroes, cancellation of open-loop, 289, 309
  effect on sensitivity of, right-half plane, 316–317
    output feedback control systems of right-half plane, 424
  open-loop, 51
  outside the unit circle, 515
  right-half plane, 300, 308–309
  of systems, 39–42, **41**
    discrete-time case, 457
  of transfer functions, 39
  of transfer matrices, 39–42
Zero-order hold, **444**
z-transform, 455–456
z-transfer matrix, **456**